PRIME ENG. & ARCH., INC.
470 OLDE WORTHINGTON ROAD
SUITE 325
COLUMBUS, OH 43082

MALCOLM PIRNIE, INC.
LIBRARY COPY
COLUMBUS, OHIO

INDUSTRIAL WASTE

Proceedings of the Thirteenth Mid-Atlantic Conference

C. P. Huang, Editor

ANN ARBOR SCIENCE
PUBLISHERS INC / THE BUTTERWORTH GROUP

INDUSTRIAL WASTE

Proceedings of the Thirteenth Mid-Atlantic Conference

June 29-30, 1981

The Mid-Atlantic Industrial Waste Conference is sponsored by
University of Delaware
Department of Civil Engineering
Division of Continuing Education
Newark, Delaware

In cooperation with
Bucknell University
Drexel University
The Johns Hopkins University
Villanova University
Delaware Department of Natural Resources
  and Environmental Control
Maryland Department of Natural Resources
Pennsylvania Department of Environmental Resources
U.S. Environmental Protection Agency, Region III

Copyright © 1981 by the University of Delaware
Newark, Delaware 19711

Published 1981 by Ann Arbor Science Publishers, Inc.
P.O. Box 1425, 230 Collingwood, Ann Arbor, Michigan 48106

Library of Congress Catalog Card Number 81-65971
ISBN 0-250-40473-7

Manufactured in the United States of America
All Rights Reserved

PREFACE

The mid-Atlantic region is an area not only of a high degree of industrialization but also of intensive activity and interest in industrial waste control. The Mid-Atlantic Industrial Waste Conference functions primarily as a forum for updating information on current developments of industrial waste management and control in the region. However, since industrial wastes are commonplace among all industrialized communities, information presented at the Mid-Atlantic Industrial Waste Conference should be applicable to other geographic localities. The Thirteenth Mid-Atlantic Industrial Waste Conference, held at the University of Delaware on June 29-30, 1981, therefore, was conceived with this thought. The focal points were, in addition to a dozen case studies, physical-chemical treatment, biological treatment, management and hazardous wastes.

Dr. Robert R. Harris, of Coement Associates, Inc., delivered the keynote address at the conference. He reviewed some aspects of public health problems in the chemical age. Mr. E. Neil Helmers, of DuPont Company, presented the luncheon address about chemical industries' assessment of pollution control regulations.

A total of 90 technical papers were presented at the conference. However, due to timing problems, only 66 of these were included in this volume. To provide a complete record of the conference, a supplementary volume of the proceedings was published by the Department of Civil Engineering, University of Delaware, to include the rest of the papers for distribution at the meeting.

The Organization Committee for the Conference was comprised of James Alleman, Joe Kavanaugh (University of Maryland), Charles R. O'Melia (The Johns Hopkins University), H. A. Elliott, A. F. Gaudy and C. P. Huang (University of Delaware), K. C. Goel (Goel Associates), Jack Fetch (Capital Controls Co.), Michael D. LaGrega (Bucknell University), Mirat D. Gurol (Drexel University), G. Lee Christensen (Villanova University), and Peter Slack (Pennsylvania Department of Environmental Resources).

Cooperating organizations in the conference included the Johns Hopkins University, the University of Maryland, Drexel University, Villanova University, Bucknell University, the Delaware Department of Natural Resources and Environmental Control, the Pennsylvania Department of Environmental Resources, and the U.S. Environmental Protection Agency—Region III.

Special thanks are due to Mr. David A. Bartley, Associate Conference Specialist, and a veteran of the Mid-Atlantic Industrial Waste Conference in the Division of Continuing Educaton at the University of Delaware, who arranged for the conference details, and to Mrs. Lorena Hobart, Technical Secretary of the Department of Civil Engineering, who handled preparation of the manuscript of the proceedings and assisted with conference coordination.

C. P. Huang

C. P. Huang is an Associate Professor of Civil Engineering at the University of Delaware. He received a BS in Civil Engineering from the National Taiwan University of Taiwan; and his MS and PhD in Environmental Engineering from Harvard University. Dr. Huang, a registered Professional Engineer, has had more than ten years of research and practical experience in the general field of environmental engineering. His major research and professional interests are physical-chemical treatment of water and wastewater, applied colloid and interfacial chemistry, and aqueous heavy metal chemistry. Dr. Huang is author or co-author of more than 50 publications.

CONTENTS

Session I
Physical-Chemical Treatment
(Metals and Inorganics)

Operation of an Automated Heavy Metals Wastewater Pretreatment
Facility 2
   Vincent M. DiTaranto, Jr.

Nationwide Survey of Heavy Metals in Municipal Sludge 15
   Stephen W. Bailey and David C. Zimomra

Concentration of Wastewater with High Calcium Scaling
Tendencies by Evaporation 26
   Seth Levine and Roberto Manas

Statistical Evaluation of Heavy Metals and pH Control in
Treated Effluents for Selected Inorganic Chemicals Industries 35
   Michael G. Warner, Ben C. Edmondson and Milford Walker, Jr.

Heavy Metals Removal: Comparison of Alternative Precipitation
Processes 43
   Karl A. Brantner and Edward J. Cichon

Predicting the Performance of a Lime-Neutralization/Precipitation
Process for the Treatment of Some Heavy Metal–Laden
Industrial Wastewaters 51
   A. R. Bowers, G. Chin and C. P. Huang

Design of Soluble Sulfide Precipitation System for Heavy Metals Removal 63
    James S. Whang, Daniel Young and Maurice Pressman

The Removal of Heavy Metals from Sewage Influent Waters by Foam Flotation. 72
    Chaiyuth Chavalitnitikul and Richard L. Brunker

Treatability of Cadmium(II) Plating Wastewater by Aluminosilicate Adsorption. 87
    C. P. Huang and P. K. Wirth

Properties Affecting Retention of Heavy Metals from Wastes Applied to Northwestern U. S. Soils 95
    Herschel A. Elliott and Michael R. Liberati

New Ozone Application System to the Cyanides Destruction in Plating Wastewaters 105
    Wojciech Kunicki

Application of Programmable Controllers to Batch Treatment of Chrome and Cyanide Wastes 115
    Tarik Pekin and Andrew P. Gagnon

Session II
Biological Treatment
(Principles and Applications)

Biologically Active Sand Filters. Part I—Laboratory and Field Studies . . . 128
    Roque A. Román-Seda, Héctor R. Fuentes and
    W. Wesley Eckenfelder, Jr.

Biologically Active Sand Filters. Part II—A Mathematical Model. 139
    Roque A. Román-Seda and David G. Wilson

Re-evaluation of the Oxygenation Coefficient–Temperature Relationship for Waste Treatment and Stream Modeling 151
    Allen C. Chao, Efren Galarraga and Robert H. L. Howe

Measurement and Validity of Oxygen Uptake as an Activated Sludge Process Control Parameter. 160
    Gary L. Edwards and Joseph H. Sherrard

Examination of the Performance of Activated Sludge Plants Using
Oxygen to Treat Pulp and Paper Mill Wastewaters 177
    David B. Buckley and James J. McKeown

Least Cost Design and Design Sensitivity of an Activated Sludge
Treatment System 189
    Robert B. Paterson and Morton M. Denn

Treatment of Tannery Beamhouse Waste with a Bench-Scale
Anaerobic Reactor. 197
    M. H. Tunick, A. A. Friedman and D. G. Bailey

Comparative Analysis of Cellulosic Substrate Utilization in a
Methane Fermentation Process. 208
    Jan Allen and Katherine Ruetenik

PACT Treatment of Synfuel Retort Waters. I. Influence of Process
Water Components on Oxygen Uptake Rates 217
    David H. Foster, Richard D. Noble and Oluafemmi T. Abudu

Assessment of Activated Sludge Process in Treating Solvent-Extracted
Coal Gasification Wastewaters 228
    Yung-Tse Hung, Guilford O. Fossum, Leland E. Paulson and
    Warrack G. Willson

Aerobic Treatment Process with By-product Utilization for
Wastewater Containing High Organic Solids 239
    P. Y. Yang and J. K. Lin

Performance of Three Types of Activated Sludge Processes under
Variable Organic and Pesticide Loadings 250
    John W. Smith, Hraj Khararjian and George Harvell

Composting of Industrial Residues 264
    Willibald A. Lutz

Session III
Case Studies

Integrated Hazardous Waste Management at a Specialty Organic
Chemicals Plant: A Case Study. 273
    Joseph J. Kulowiec

The Industrial Waste Cooperative Concept—Potential for the
Metropolitan Sewerage District, Boston, MA. 288
    Michael K. Prescott, Noel D. Baratta and Wayne T. Grandin

Procedures for Processing Atypical Industrial Wastes at a Publicly
Owned Treatment Works 295
    Robert T. Mohr and Paul H. Keenan

Developing Parks, Golf Courses and Other Recreation Areas on
Completed Landfills. 302
    Edward F. Gilman, Franklin B. Flower and Ida A. Leone

Development of a Chrome Residual Containment Project at Hawkins
Point, Baltimore City, Maryland. 316
    Clinton R. Albrecht

Leachate Detection in a Sole Source Aquifer: A Case Study 323
    William F. Graner and Thomas Hroncich

Company Solves Latex Waste Problem. 331
    James E. Kerr and Craig H. Lockhart

Session IV
Physical-Chemical Treatment
(Solid Residues and Organics)

Selection of Belt Presses for Sludge Dewatering. 341
    K. C. Goel, Harish K. Mital and Stephen Hsu

Sludge Management in the Poultry Processing Industry. 350
    William F. Ritter

The Influence of Pretreatment on Improving the Quality of Sludge. . . . 362
    Elizabeth F. Gormley and John V. DiLoreto

Performance of Final Clarifier as a Function of Wastewater
Characteristics 370
    Yeun C. Wu and Zahid Mahmud

Application of Hydroperm® Crossflow Filtration System in
Removing Suspended Solids 378
    Han-Lieh Liu

Upgrading the Dissolved Air Flotation Process 386
   J. Glynn Henry and Ronald L. Gehr

Effects of Accelerated Dynamic Response on Settleability of
Biological Solids in Activated Sludge Systems 396
   Mark J. Krupka and Jeffrey M. Thomas

Some Aspects of Iron and Lime versus Polyelectrolyte Sludge
Conditioning 404
   G. Lee Christensen and Stephen G. Wavro

Treatment of Industrial Wastewaters with Formaldehyde 416
   Franciszek S. Tuznik

UV Photodecomposition of Color in Dyeing Wastewater 427
   Calvin P. C. Poon and Bruno M. Vittimberga

Removal of Selected Pollutants with Potassium Ferrate 434
   Sergio Deluca, Allen C. Chao and Charles Smallwood, Jr.

Will Activated Carbon Work? 444
   William Brian Arbuckle

Session V
Hazardous Wastes
(Management and Control)

Industry Role in Hazardous Waste Facility Development 455
   Janis Stelluto

The Impact of Subtitle C of RCRA upon Industrial Wastewater
Treatment Facilities 460
   Felon R. Wilson

Development of the Hazardous Waste Management Program for
Pennsylvania Electric (PENELEC/GPU) Company 469
   Daniel F. Buss, Ronald Long-Sing Wen and James L. Greco

Remediation at an Inactive Municipal Disposal Site Containing PCBs. . . . 476
   Warren V. Blasland, Jr., G. David Knowles, Edward R. Lynch
   and Robert K. Goldman

Policy Toward Siting of Hazardous Waste Management Facility:
Effectiveness of State Preemptive Siting Authority 484
    Edward Yang and Amy Horne

Incineration of Hazardous Wastes at Sea 493
    Donald A. Oberacker, Lissa A. Martinez and Robert J. Johnson

Wet Oxidation of Toxic and Hazardous Compounds. 501
    Tipton L. Randall

In-house Treatment of Hazardous Wastes at Southern Illinois
University—Carbondale. 509
    John F. Meister

Public Participation: The Missing Ingredient for Success in
Hazardous Waste Siting. 517
    Eleanor W. Winsor

The Storage and Handling of Toxic and Hazardous Materials:
A Regulatory Approach 525
    J. H. Pim, W.C. Roberts and H. W. Davids

Session VI
Management
(Social, Legal and Economic Aspects;
Reuse and Recovery; Fate and Transport)

Citizen Response to Industrial Waste 533
    Paul D. Bush

Development of Effluent Guidelines for the Textile Industry 537
    Debra J. Moore

Environmental Impairment Liability Insurance 549
    C. Fred Gurnham

Dairy Industry Waste—Options and Economics 554
    ManMohan Varma and Ramesh C. Chawla

Prioritization of Production Related Wastewater Treatment Problems in the Textile Industry and Cost-Effective Recommendations for Their Solution 563
   Michael S. Bahorsky

Recovery of Organic Matter from Seafood Processing Wastewaters by Ultrafiltration as an Alternative Method 580
   Allen C. Chao and Gary Davis

Recycling Wastewater in Colorado—The Issues, The Problems and the Solutions 588
   Thomas G. Sanders

The Production of Insulating Fibers from Power Plant Wastes 596
   William H. Buttermore, David J. Akers and Edward J. Simcoe

The Relative Contribution of Point Sources of Ammonia to Surface Waters 608
   William J. Rue, Jr. and James A. Fava

Inactivation of Anaerobic Release of Phosphorus by Water Treatment Sludges. 618
   Yousef A. Yousef, Martin P. Wanielista, Harvey H. Harper and David B. Jellerson

Volatilization Problem: Land Disposal of Industrial Wastes 630
   Thomas T. Shen

Assessment of Remotely Sensed Data for the Detection of Impounded Chemical Wastes 639
   Mohan M. Trivedi, John M. Hill and Peter A. Allain

Index 647

SESSION I

PHYSICAL-CHEMICAL TREATMENT
(METALS AND INORGANICS)

OPERATION OF AN AUTOMATED HEAVY METALS WASTEWATER PRETREATMENT FACILITY

<u>Vincent M. DiTaranto Jr.</u>   American Hoechst Corporation
Leominster, Massachusetts.

INTRODUCTION

   The following paper will describe a unique wastewater treatment situation ocurring on a site which has effluent from a small sunglass frame plating facility and effluent from a styrene polymerization facility. The streams are very dissimilar and, in addition, the plating stream, being much smaller than the polymerization stream, requires its own unique system. The original installation attempted to treat both streams simultaneously using the same equipment. It became obvious over a period of time that this facility could not work effectively to remove the heavy metals, therefore, an investigation was launched into a pretreatment facility for the plating stream alone, whose treated effluent would then be combined with the raw polymerization waste stream for secondary clarification.
   The following was prepared as a daily operational log for the recently constructed and automated heavy metal pre-treatment facility. Included is a description of the automated processes including pH control, level control, chemical feed and sludge dewatering. A daily analysis of the plating influent and effluent was conducted on five days showing the removal efficiencies of various parameters through the heavy metal system, as well as the overall removal efficiencies through both systems. Major emphasis was placed on heavy metals and total suspended solids removal.
   The maximum daily removal efficiencies obtained through the heavy metal pretreatment system were: 96% for TSS, 97% for Cu, 97% for Ni and 98% for Zn. The overall removal efficiencies through both systems were: 96% for Cu, 99% for Ni, and 99% for Zn. The TSS was not evaluated due to the influx of these solids from the polymerization waste stream.
   The prime factors effecting control of these values were: pH, coagulant concentration, optimization of the sludge dewatering system, second stage floc formation, and

neutralization tank residence and mixing time required for production of the initial hydroxide floc.

BACKGROUND

An existing 200 gal/min pretreatment facility was designed to treat wastewater from a plastics manufacturing facility polymerizing styrene monomer twenty-four hours a day. (Note: The terms plastics and polymerization waste treatment system will be used interchangeably throughout the paper). The advent of a recent one-shift electroplating line from a sunglass production process, discharging heavy metals and miscellaneous wastewater to the existing facility, made it necessary to upgrade and automate the entire system. A heavy metal pretreatment system was added to the influent of the raw electroplating waste line. The treated effluent was then discharged to the influent surge tank of the polymerization waste treatment plant for additional treatment (Figure 1).

The heavy metal pretreatment system consists of a plate separator, a neutralization tank, a centrifuge and a central control panel. The new equipment was installed in order to segregate the plating waste for proper pH control, coagulation and sludge dewatering. The system will operate entirely independent of the polymerization wastewater system when raw plating wastes are discharged to the neutralization tank. Sludge is stored for thickening prior to dewatering through an automated centrifuge.

The polymerization wastewater pretreatment system's original level control equipment was modified to include automatic start-up and shut-down. A pH controller was added to the central control panel in order to provide additional alkalinity in the main surge tank when needed. The final pH and flow of the combined pretreated wastes discharged to the city sewers is continuously monitored and sampled.

PROCESS

The pretreatment of the heavy metals incorporates the basic fundamentals of alkaline neutralization, coagulation, flocculation, sedimentation, sludge dewatering, and effluent polishing through secondary clarification in the polymerization waste treatment system. (It should be noted that pH in the polymerization waste treatment system is regulated between 7.0-8.5 S.U. in order to maintain discharge limitations). Ferric sulfate ($Fe_2(SO_4)_3$) is added to the raw heavy metal wastes as a coagulant aid. Hydrated lime ($Ca(OH)_2$) is then added to the neutralization tank to increase pH and initiate coagulation by providing the insoluble metal hydroxide floc. A residence time of 40-60 minutes with continuous agitation produces a small floc. Anionic polyelectrolyte is then added in a rapid mix tank, which then overflows to a flocculation tank with slower mixing, to increase floc size. A plate separator with a maximum design flow rate of

60 gal/min is then used for sedimentation.
HEAVY METAL SYSTEM
Level Control

The level in the 5,000 gal. neutralization coagulation tank is controlled by a differential pressure cell, a two-mode controller, an electropneumatic valve positioner and a pressure control valve with positioner arm. The average flow

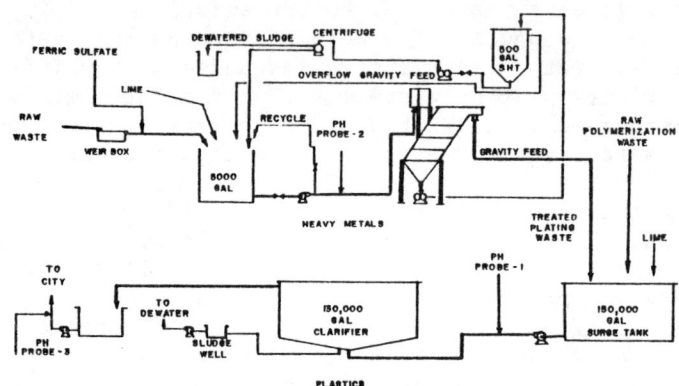

FIGURE I
HEAVY METAL AND PLASTICS PRETREATMENT SYSTEM

rate is approximately 40 gal/min and the operating level is maintained at 2,000 gallons. This level provides an average fifty minutes retention time with constant agitation for neutralization with hydrated lime.

The main feed pumps and the addition of coagulant aids utilizing rapid and slow mixing with SCR drives are controlled by the neutralization tank level control equipment. All systems begin operation at the 3.5 ft. level and shut down at 3.0 ft.

pH Control and Chemical Feed

Hydrated lime is added directly to the neutralization tank equipped with continuous slow agitation. A recycle line off the main feed pump back to the neutralization tank provides fast mixing capabilities. The addition of hydrated lime accomplished a dual function: 1) increases the pH of the acid plating wastes from 2.0 to 9.5; and 2) initiates the coagulation process by providing the hydroxide floc.

The pH is maintained utilizing measuring and reference probes with ultrasonic cleaner and temperature compensator, and a Luft multi-level controller. The lime feed is accomplished by an air-operated, double disphragm pump. Clogging has virtually been eliminated with this system, although daily flushing is usually practiced as a safeguard.

Sludge Dewatering System

Sludge dewatering is accomplished by a completely automated centrifuge. The dewatered filtrate is recycled back to the neutralization tank in order to maintain a closed system and prevent discharge of heavy metals to the polymerization waste treatment plant. Sludge draw-off from the plate separator is accomplished by an air-operated double-diaphragm pump, controlled by a timer. Sludge is pumped to a 500 gallon storage and thickening tank prior to dewatering. Overflow of the supernatant is recycled back to the neutralization tank. The dewatered sludge at 23% solids is presently disposed of through a state-approved disposal and recycling company.

OPERATION

To thoroughly evaluate the efficiency of the heavy metal pretreatment facility after the system had been in operation for sixteen weeks, a five-day monitoring program was conducted on the raw and treated plating waste. Emphasis was placed on measuring concentrations of heavy metals and total suspended solids discharged through this system (Figure 2). Since the treated plating effluent is then discharged to the raw influent of the polymerization waste treatment facility, the final combined treated effluent was monitored for residual heavy metal concentrations. Figure 3 represents the daily heavy metal concentrations in the raw plating wastes in addition to the residual heavy metal concentrations after secondary clarification through the polymerization waste treatment system.

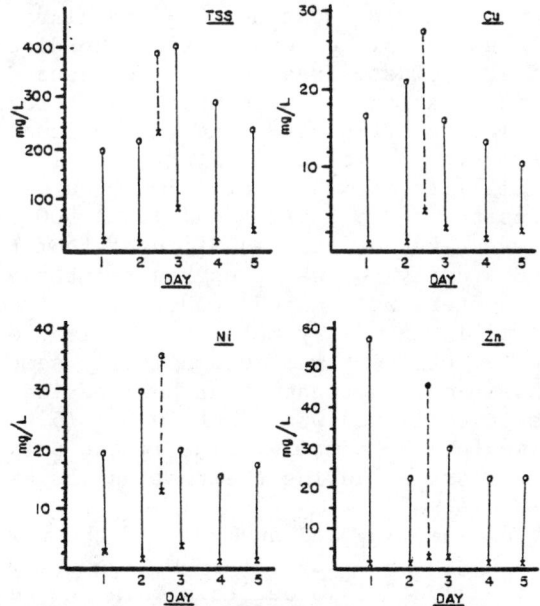

FIGURE 2. Daily concentrations of heavy metals and total suspended solids (TSS) in the raw plating waste (O) and in the treated effluent discharged to the polymerization pretreatment system (X). (Note excessive raw waste loading day 2, p.m. [dotted line] and day 3.)

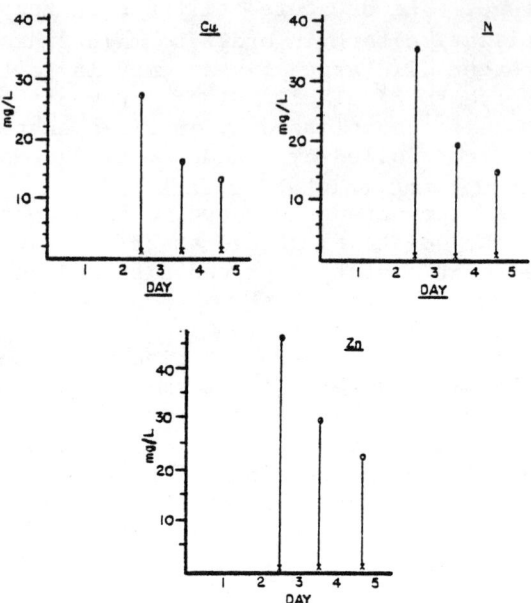

FIGURE 3. Daily concentrations of heavy metals in the raw plating waste (O) and in the combined effluent after pretreatment through both systems (X).

Sampling of the plating waste was conducted at three points: 1. the raw plating waste at the heavy metal system's flow monitoring station; 2. the clarified plating effluent through the plate separator before discharge to the polymerization waste treatment system; and 3. the final flow monitoring station of the combined and treated polymerization and plating effluents. The raw influent was sampled using an automatic sampler calibrated to collect approximately 120 milliliters (mls) every five minutes, producing a daily composite of 9.5 l (2.5 gals). A 500 ml hourly grab sample was taken of treated effluent from the plate separator and a daily log was prepared for the sample. The following parameters were noted: pH, color, incoming flow rate, and level in the heavy metal neutralization tank. The combined treated effluent was automatically sampled on a continuous basis over a twenty-four hour period. The total daily composite collected was about 19.0 l (5.0 gal). The residence time for the combined flow in the polymerization waste treatment system during the testing was approximately twenty-four hours.

The actual sampling and analysis of the raw and treated plating waste began on Friday, May 30, 1980. The raw waste was initially analyzed in order to obtain the general characteristics of the wastes and determine what pollutants should be monitored further (Table 1). It was determined that total suspended solids, as well as concentrations of copper, nickel and zinc, should be evaluated for removal efficiency through the heavy metal pretreatment system.

Table I. Day One
Raw Plating Waste Characteristics Through Heavy Metal Pretreatment System

Parameter	Units	Influent*	Effluent*
pH	S.U.	2.4	9.3
Alkalinity	mg/l as $CaCO_3$	0	43.9
Acidity	mg/l as $CaCO_3$	385	0
Total Cyanide	mg/l	N/A	0.06
Total Dissolved Solids	mg/l	832	1013
Volatile Dissolved Solids	mg/l	194	190
Total Suspended Solids	mg/l	202	21
Volatile Suspended Solids	mg/l	100	17
Turbidity	NTU	46	3.2
BOD	mg/l	8	10
COD	mg/l	205	66
Chloride	mg/l	102	83.3
Copper (Cu)	mg/l (Total)	17.2	0.9
Nickel (Ni)	mg/l (Total)	19.2	2.4
Zinc (Zn)	mg/l (Total)	56.4	1.3
Zinc (Zn)	mg/l (Soluble)	–	0.4
Silver (Ag)	mg/l (Total)	<0.05	<0.05
Lead (Pb)	mg/l (Total)	0.11	<0.10
Total Chromium (Cr)	mg/l (Total)	<0.10	<0.10

*Daily composite sample.

DAY ONE - Friday, May 30, 1980

The raw plating waste discharged to the pretreatment facility on day one was representative of a normal operating day (Table I). The total incoming flow was about 24,000 gals. The sludge holding tank was full of concentrated sludge, resulting in a dewatered filtrate very high in solids and metals, which was continuously recirculated through the pretreatment system.

The total dissolved solids (TDS) in the effluent increased by 22% (181 mg/l) due to the recycle of heavy filtrate from the centrifuge and also from the addition of hydrated lime and ferric sulfate coagulants. The total suspended solids (TSS) decreased 90% to 21 mg/l in the treated effluent, 17 mg/l of which was volatile suspended solids (VSS). Visible observation suggests that the majority of the VSS could be polyelectrolyte (this was not substantiated by further testing). The BOD was virtually non-existant, while COD showed a reduction of 68% or 66 mg/l in the treated effluent. The concentration of total cyanide (CN) in the treated effluent was 0.06 mg/l. A cyanide destruction unit was operated in the plating area utilizing alkaline chlorination to carbon dioxide, nitrogen gas, sodium chloride and water. The total metals analyzed were copper (Cu), nickel (Ni), zinc (Zn), silver (Ag), lead (Pb), and total chrome (Cr). Removal efficiencies were 97% Cu, 88% Ni, and 98% Zn (Table II). The concentrations of the influent and effluent for Ag, Pb, and Cr were <0.05, <0.10, and <0.10 respectively, during all five days of testing, therefore, no removal efficiencies were calculated.

The removal efficiency for nickel was 88% with a final concentration of 2.4 mg/l discharged to the polymerization waste treatment facility (Figure 2).

The pH, monitored hourly, showed that 90% of the operating day the pH remained at 9.0. This is low for maximum insolubility of nickel hydroxide, which should be maintained around 10 SU (1).

Table II.
Efficiency Of The Heavy Metal Pretreatment System % Removal*

Parameter	Day #1	Day #2 Am/Pm	Day #3	Day #4	Day #5
TSS	90	92/38	80	96	87
Cu	97	95/82	84	89	79
Ni	88	94/64	85	95	97
Zn	98	95/98	93	97	98
VSS	83	88/49	65	92	83
Turb	93	93/34	93	94	94

* % Removal = 100 (1-treated effluent concentration ÷ raw influent concentration)

A settling test of the floc tank prior to clarification through the plate separator yielded a 10% liquid sludge production. The sludge was withdrawn continuously to a holding and thickening tank prior to continuous dewatering.

Sludge production at 10% was double the normal amount due to a heavy dewatering filtrate discharged through the system. Although the filtrate on day one was not analyzed, a similar situation during the week was analyzed for TSS, VSS, Cu, Ni, and Zn (Table III). This showed a TSS of 37,500 mg/l and a metals concentration for Cu, Ni, and Zn of 436, 1042, and 1912 mg/l respectively. The total metals and solids concentration added to the system through filtrate recycle with a sludge dewatering feed rate to the centrifuge of 13 gal/min* would produce 876 Kg (1927 lbs) of TSS per eight-hour day. The daily metals concentration would be equivalent to: 10.3 Kg (22.7 lbs) of Cu, 24.7 Kg (54.26 lbs) of nickel and 45.3 Kg (99.6 lbs) of zinc, all per eight hour day.

The daily metals and TSS loading on the heavy metals system based on concentrations in the raw plating influent alone would produce 18.1 Kg (39.8 lbs) of TSS; 1.5 Kg (3.3 lbs) of Cu; 1.7 Kg (3.8 lbs) of Ni, and 5.1 Kg (11.1 lbs) of Zn, with an average daily flow rate of 44 gal/min.

*Filtrate recycle = sludge dewatering feed rate.

Table III.
Dewatered Filtrate Analysis - Day Five

	Morning	Mid-Morning	Afternoon
TSS	37,500	32,000	3150
VSS	14,400	12,000	1160
Cu	436	400	26.4
Ni	1,042	1,030	111.6
Zn	1,912	1,806	238
Ag	<0.1	<0.1	<0.1
Pb	1.0	0.8	0.2
Cr	0.4	0.2	0.2

(All results in mg/l)

DAY TWO - Monday, June 2, 1980

As a result of the analysis on day one, the emphasis was placed on removal of TSS, Cu, Ni, and Zn. A slight increase in the lime feed pump (to increase alkalinity) was made in order to improve the removal of Ni from the plating waste. Despite periodic high surges of 50 gal/min plus a flow of 10 gal/min for dewatered sludge filtrate return, the total of which is equivalent to the maximum capacity of the system 60 gal/min, the heavy metals system removal efficiencies during the morning for TSS, Cu, Ni, and Zn were 92%, 95%, 94% and 95% respectively (Table II).

The average flow rate between 6:00 a.m. through 12:00 noon was about 40 gal/min. Sludge production was 5% and surge tank level averaged about 4.0 ft. or 0.5 ft. above the controlled operating level.

At noon, a plating production emergency necessitated the discharge of miscellaneous tanks totalling approximately 625 gals in thirty minutes. These tanks included: 1) a 10% sulfuric acid pickle tank; 2) a deoxidizer for removal of solids from the zinc die cast frames; 3) an alkaline electrocleaning soap bath; 4) a floor drain sump; and 5) a nickel drag-out tank.

This action produced a very heavy solids concentration of about 400 mg/l (diluted composite), thereby shocking the system (Figure 2 - dotted line). Since sludge was being dewatered continuously, the actual solids loading on the plate separator was approximately 40,000 mg/l. Sludge production as a percentage of flow ranged from 40-50% for three hours, producing a heavy carry-over of solids to the polymerization waste treatment system. The neutralization tank level increased 1 ft. above operating level. The sludge withdrawal pump from the plate separator was operated at maximum between 15-20 gal/min.

Although the solids and metals discharge to the polymerization waste treatment system was considerable (Figure 2) the final discharge to the city was not affected by the upset due to the secondary clarification (Figure 3). Since the raw plating waste was a composite, only a diluted daily sample for TSS and metals could be analyzed.

The upset caused removal efficiency in the heavy metal pretreatment system to decrease substantially (Table II).

DAY THREE - Tuesday, June 3, 1980

Due to the upset to the heavy metal pretreatment system on day two, a very high solids concentration remained in the system (neutralization tank and sludge holding tank). Consequently, efficiency for day three was reduced due to the heavy solids loading remaining throughout the heavy metal system.

Jar tests were performed in order to determine if increasing the concentration of coagulant aids would improve effluent quality. It was found that doubling the feed rate for ferric sulfate and the anionic polyelectrolyte based on a raw influent flow rate of 40 gal/min would produce working concentrations of 200 mg/l and 1.0 mg/l respectively. This action improved efficiency significantly on the succeeding days four and five (Table II).

The average flow rate in the morning was 43 gal/min, sludge production was 20% of incoming flow. The sludge withdrawal rate was set at 10 gal/min in order to optimize removal efficiency, i.e., to remove a heavy dewatered sludge at 23% solids in the shortest possible time and reduce the overall solids return in the dewatered filtrate. It should be noted that the dewatering cycle is heavily dependent on sludge thickening storage time. These parameters are indirectly proportional to each other. Optimization of this technique was improved on day five, with a laboratory analysis of the filtrate, (Table III).

An overflow line from the sludge holding tank to the neutralization tank returns clear supernatant to the system when the dewatering cycle is optimized. This flow is regulated by the sludge withdrawal rate from the plate separator to the holding tank and the sludge dewatering feed rate to the centrifuge. Optimization of this system is a critical parameter in producing an efficient pretreatment operation.

At noon, about 150 gals of alkaline cleaning bath solution was dumped into the heavy metal pretreatment system.

This action, as in day two, was not a normal process day for the plating operation and was part of a two-day production problem. It should be noted here, that during a normal weekly operation, the emergency plating dumps do not occur and the tanks are bled off over a five to eight hour day, providing a consistent concentration of pollutants to the system.

The emergency dumping shocked the system and produced heavy solids carryover and reduced efficiency. The coagulant concentration was increased. Sludge production rose to 30% during the upset with excessive amounts of foam (soap bubbles). The centrifuge was shut down and all efforts were made to remove and store as much sludge as possible while the incoming raw waste continued. Sludge production was reduced back to 20% within 15 minutes of shutdown of the dewatering system.

The sludge withdrawal rate was set at 12 gal/min and polyelectrolyte was added to the sludge holding tank to enhance settling and thickening.

The total raw plating waste loading was not as concentrated as the prior afternoon; therefore, the overall impact to the heavy metal system was not as severe as in day two (Table II, Figure 2). The removal efficiencies through the heavy metal pretreatment system were obviously reduced for TSS (80%), Cu (84%), Ni (85%), and Zn (93%); however, overall removal efficiencies through the plastics system was not affected (Figure 3).

DAY FOUR - Wednesday, June 4, 1980

The average daily flow rate on day four was 53 gal/min. Sludge dewatering did not begin until 11:00 a.m.; however, sludge withdrawal from the plate separator to the holding tank continued at a rate of 10 gal/min. The heavy metal pretreatment system was operating at a maximum inflow of 63 gal/min. The TSS loadings was back to normal at 280 mg/l, and sludge production was 4% of the incoming flow. The total daily efficiency was high for all parameters: TSS (96%); Cu (89%); Ni (95%); and Zn (97%).

The removal efficiency through the heavy metals system for copper on day four was only 89%. This figure equates to a copper concentration of 1.6 mg/l in the treated effluent discharged to the polymerization waste treatment system. The overall removal efficiency for copper through both systems was 92% or 1.2 mg/l Cu in the combined treated plastics and heavy metals effluent discharged to the city sewer. The reduced efficiency for copper was probably the result of a pH of 10 maintained throughout the day, resolubilizing the copper hydroxide floc. The nickel removal efficiency through the heavy metal system was 95% or 0.4 mg/l Ni discharged to the plastics system. The overall removal efficiency for nickel through both systems was 99% or 0.2 mg/l Ni discharged to the city. The pH plays a significant role in precipitating Cu and Ni as an insoluble metal hydroxide complex.

Sludge dewatering began at 11:00 a.m., increasing the solids concentration to 10%; however, effluent clarity was excellent with very little carryover of visible solids. The TSS removal efficiency for the heavy metal system was highest on this day at 96% or 10 mg/l discharged to the polymerization waste treatment system.

The key pretreatment adjustments made to the system affecting high removal efficiency on day four were: 1) increased chemical coagulant dosage; 2) increased floc tank mixing; 3) use of a recycle mixing line from the main feed pump back to the neutralization tank; and 4) optimum use of the sludge holding tank, sludge thickening, dewatering cycle time, and feed rates.

It should be noted at this time that item no. 4 is critical in not only controlling solids and metals loading to the plate separator, but also the pH through the entire system. This is primarily due to the addition of residual alkalinity from the dewatered sludge filtrate and the supernatant overflow line from the sludge holding tank back to the neutralization tank.

DAY FIVE - Thursday, June 5, 1980

The main concern on day four was the concentration of the solids in the sludge holding tank and dewatering at a continuous efficient rate, thereby reducing the solids loading on the system and improving the efficiency. The fifth and last day of the program, the system ran with very little adjustments. No upsets occurred and the daily average flow rate was 40 gal/min. The total daily flow was about 25,500 gals. Sludge production was 12% from 7:00 a.m. to 12:00 noon.

The sludge dewatering cycle was reduced in successive stages causing a reduction in overall sludge production (Table III). The sludge production was reduced to 6% during the afternoon treatment, and influent solids were consistent throughout the day with no change in production processes (Figure 2). The only variable in the pretreatment system was filtrate concentration which affected a change in sludge production.

The continuous dewatering increased the pH to 10, and occasionally to 11, due to the addition of alkalinity from the filtrate sludge. The copper concentration in the treated effluent discharged to the plastics system was 2.2 mg/l, the highest during the normal operating days one, two (in the a.m.), four and five. The final concentration of copper in the combined effluent after secondary treatment was 1.1 mg/l. The reduced efficiency again as in day four was probably due to the higher pH causing the copper hydroxide floc to resolubilize through the heavy metal system. The pH range in the polymerization waste treatment system during the twenty-four hour composite sampling ranged from 7.0 to 8.5 S.U. It is also possible that complexing of the soluble copper ion might be occurring or an additional source of copper may be responsible, from the raw polymerization wastes. This can only be substantiated through further evaluation of both raw waste streams and production processes. It should also be noted that final concentrations of all metals were well within present discharge regulations to a publicly owned treatment works (POTW) during all five days of testing.

CONCLUSION

It was concluded that by segregating and optimizing the pretreatment process for the electroplating wastewater, maximum and consistent removal of heavy metals could be accomplished. Removal efficiencies that can be expected through the automated heavy metal pretreatment system during some of the more adverse conditions are: 96% for TSS; 97% for Cu; 97% for Ni; and 98% for Zn. It should be noted that due to the difference in pH for maximum insolubility of copper and nickel hydroxide (pH 9 and 10, respectively), that high removal efficiencies of 97% for both metals through the heavy metal pretreatment system is difficult to attain during the same day. Since the effluent from the heavy metal system is clarified and diluted in the polymerization pretreatment system, maximum overall removal efficiencies obtained during the testing period for Cu, Ni, and Zn were: 96%, 99.5%, and 99%, respectively. The efficiency of the heavy metal pretreatment system is contingent upon the rate and characteristics of the miscellaneous raw wastes discharged to the facility. A large surge tank would alleviate much of the upset conditions experienced during the program, producing a more diluted and consistent raw waste flow.

The pretreatment processes most critical in obtaining maximum efficiency of the heavy metal pretreatment system are: 1) pH; 2) coagulant concentration; 3) optimum operation of the sludge dewatering system when filtrate is returned to the system; 4) second stage floc production and mixing rate; and 5) residence time and premixing in the neutralization tank for initial hydroxide floc production. It should also be noted that due to the interrelationship of the above parameters for removal of the heavy metals, it was not possible to isolate one variable during the actual operating conditions of the heavy metal pretreatment system. Controlled laboratory conditions are necessary to evaluate each parameter in detail.

Some of the additional benefits of segregating the electroplating waste stream for pretreatment were: A 50% reduction of lime usage, flexibility in electroplating production process and reduced man hours of supervision for both systems.

ACKNOWLEDGEMENTS

Credits. The following engineers employed by the American Hoechst Corporation Plastics Division, Leominster, Massachusetts assisted in the design of the automated system. W. Mason, Adivsor and Environmental Controls Manager (retired); W. Williams, Instrumentation Engineer; D. Leger and T. Rouleau, Drafting; L. Grenier and Z. Richards, Grade III Wastewater Treatment Operators; H. Schrob, Manager of A.H.C. Engineering, providing the time and funding for the project.

AUTHOR. V.M. DiTaranto Jr., A. B., M. S., Environmental Engineer, American Hoechst Corporation, Leominster, Massachusetts.

REFERENCES

1. Metzner, A. V., "Removing Soluble Metals from Wastewater". Water & Sewage Works, (April, 1977).

2. "Development Document for Existing Source Pretreatment Standards for the Electroplating Point Source Category." Effluent Guidelines Division, Office of Water and Hazardous Materials, U. S. Environmental Protection Agency, Washington, D. C. 20460 (August, 1979).

NATIONWIDE SURVEY OF HEAVY METALS IN MUNICIPAL SLUDGE

<u>Stephen W. Bailey</u>. Booz, Allen & Hamilton, Inc., Bethesda, Maryland

<u>David C. Zimomra</u>. Booz, Allen & Hamilton, Inc., Bethesda, Maryland

INTRODUCTION

The heavy metal composition of sewage sludge has received considerable attention and study over the past decade in light of the efforts to control the release of toxic and hazardous materials to the environment. There is continued concern over the potential adverse effects on humans, animals, and crops from the toxic heavy metal contaminants found in sludge, especially for sludges which are land applied. As evidence of this concern, much effort has been directed toward understanding the hazards associated with sludge-borne metals. Several reviews of heavy metal composition of various sludges have been conducted, and numerous research articles have described the interactions of heavy metals in particular sludges with certain soils and select crops [1,2]. Although these studies have contributed significantly to the development of guidance and recommendations for low risk sludge management techniques, they are inherently regional- or site-specific. A nationwide perspective of sludge metal composition and corresponding disposal practices is necessary, particularly for those charged with the development of a responsible regulatory policy for municipal sewage sludge.

As part of the U.S. Environmental Protection Agency's comprehensive sludge regulatory program being developed under the authority of the Clean Water Act, the Resource Conservation and Recovery Act, and other pieces of environmental legislation, an effort was initiated to compile information on heavy metal composition of sewage sludges throughout the nation. In addition, EPA sought to determine how sludge-borne metals are currently being disposed of so as to better understand the extent of the hazard posed by toxic metals in sludge. This paper summarizes the results of EPA's nationwide survey of heavy metals in municipal sludges and the corresponding methods by which these sludges are being disposed.

DATA COLLECTION

Data suppporting this discussion was compiled for 511 publicly owned treatment works (POTW's) located in 39 states and the District of Columbia. Three basic approaches were used to compile the POTW sludge data. First, metal quality data which had previously been collected by EPA's Office of Solid Waste from various U.S. EPA Regional offices and state regulatory agencies since 1978 was assembled and screened for inclusion. A second source of data is based on a 1978 telephone survey of POTW sludge quality, quantity, and disposal practices conducted by EPA's Office of Solid Waste in support of their development of regulations governing application of sewage sludge on agricultural cropland [3]. Finally, the results of another telephone survey of 1008 randomly selected POTW's geographically distributed throughout the U.S., which was conducted in the last half of 1980 and early 1981, under the authority of the Office of Solid Waste, comprised the remaining source of data. This last set of data is being used by EPA in their ongoing efforts to develop a comprehensive sewage sludge management regulation governing all methods of sludge disposal.

A wide variety of POTW and sludge characteristics are represented in the data set. POTW's are included from all the geographic regions of the U.S. Plant sizes range from small rural plants treating 30,000 gallons per day (gpd) of wastewater to much larger urban facilities treating 1200 million gallons per day (mgd). There are facilities treating predominantly residential wastes, as well as treatment plants receiving primarily industrial wastes. Also, various types of wastewater and sludge treatment processes are represented, including sludges generated from primary treatment, aerobic biological treatment, and anaerobic digestion processes. Of particular significance is the distribution of sludge disposal practices represented by the data set. Table I indicates the range of disposal methods utilized by POTW's in the data set.

Table I. Distribution of Sludge Generated in the Data Set

Disposal Method	Percent of Sludge Generated
Food Chain Landspreading	13
Non-Food Chain Landspreading	6
Distribution and Marketing	20
Landfilling	27
Thermal Processing	15
Ocean Disposal	14
Other	6
	100%

The quantity of sludge represented by the data set represents approximately 20 percent of the estimated total quantity of sludge currently being generated in the U.S.

HEAVY METAL COMPOSITION OF SLUDGES

This review focuses on the presence of Cd, Pb, Cu, Ni, and Zn in sewage sludges, as these are considered to be the metals of primary concern when considering the adverse effects of land application of sludge [4]. Cadmium poses the greatest concern from the standpoint of human health because of its accumulation in various food chain crops. For lead, the greatest threat to animal or human health is from direct ingestion of soils amended with sludge or sludge adhering to forages. The metals copper, nickel and zinc are phytotoxic at excessive soil levels, and thus can cause reductions in crop yields.

Before discussing the metal composition data, it is important to make a point about the nature of the information as it affects analysis and interpretation. Each POTW is assigned a single value for the concentration of a particular metal in its sludge. As described above, these values were obtained either directly from the POTW or through U.S. EPA and state agencies. No attempt was made to ascertain whether these values were based on single samples taken randomly or were averages of multiple samples taken over a specified period of time. As a result, precise comparisons between sludges from different sources may not be valid. Instead, comparisons between general groups of POTW's, such as various size categories of facilities, or POTW's practicing different methods of sludge disposal, are probably more meaningful.

Table II shows the range, mean and median levels for the 511 POTW's in the data set, and compares these with previously reported values from numerous surveys of various sludges compiled by Chaney [1].

Table II. Concentration of Heavy Metals in Sewage Sludge (mg/kg)

Metal	511 POTW Data Set			Previously Reported Typical Medians [1]
	Range	Mean	Median	
Cd	0-1,320	46	13	15
Pb	4-10,800	541	374	500
Cu	8-23,124	1034	578	800
Ni	3-9,450	230	53	80
Zn	5-49,000	2183	1372	1700

The disparity between the medians and means for each constituent within the 511 POTW data set is the result of abnormally high values (as shown in the range) skewing the means. Consequently, the median may be a more adequate representation of central tendency [5]. For all metals, the medians of the 511 POTW data set are less than the previously reported levels in various sludges.

To gain some insight as to the impacts that industrial wastewater discharges to POTWs have on sludge quality metal concentration, statistics were also calculated for those facilities (51 POTW's) with no known industrial inputs. Table III indicates that, although metal concentrations are significantly lower for these non-industrial POTW sludges, the concentrations of metals still vary considerably. These constituents are most likely contributed by water treatment chemical additions, plumbing systems, urban runoff, or domestic sources such as household cleaners.

Table III.   Concentrations of Heavy Metals in Sewage
              Sludge Without Industrial Contributions
              (mg/kg)

Metal	Range	Median	Mean
Cd	0-30	6	9
Pb	9-1200	225	319
Cu	0-2600	407	586
Ni	3-443	33	63
Zn	59-49,000	850	1900

To further analyze the variability of metal constituents in sewage sludge, frequency distribution diagrams of sludge metals were prepared. The histograms presented in Figure 1, are all positively skewed (with the "tail" to the right) non-normal distributions. The skewness is quite similar for all five metal distributions, with lead concentrations being the least skewed and nickel concentrations exhibiting the largest skewness. In addition to skewness, the kurtosis, or measure of flatness, of the distribution was calculated. The histograms are all generally peaked, that is most POTW sludges are grouped within a narrow range of metal values. In relative terms, the cadmium concentrations have the least kurtosis (or exhibit the least flatness), and nickel has the highest kurtosis (the flatest of the distributions).

Another way to view the significance of the shape of the frequency distribution is to consider that most of the observations are closely grouped around the median values for each metal. Approximately 70 percent of the observations for Cd, Pb, Cu, Ni, and Zn are less than 25, 550, 950, 105, and 2000 mg/kg, respectively. These distribution characteristics have a significant bearing on EPA regulatory policy for municipal sewage sludge.

DISPOSAL OF SLUDGE-BORNE METALS

The sludges generated by the POTW's in the data set are disposed by a variety of methods. For the purposes of this analysis, these methods were grouped into seven categories:

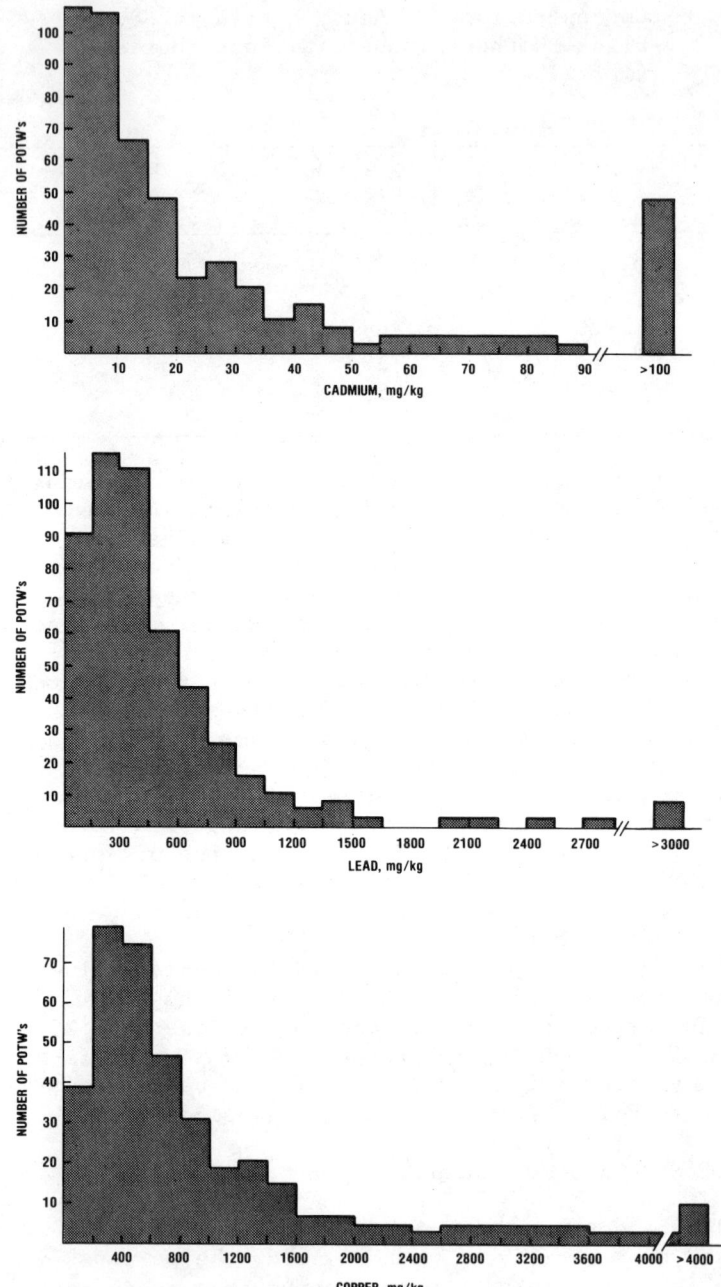

Figure 1: Frequency Distributions of Five Metals in POTW Sludge.

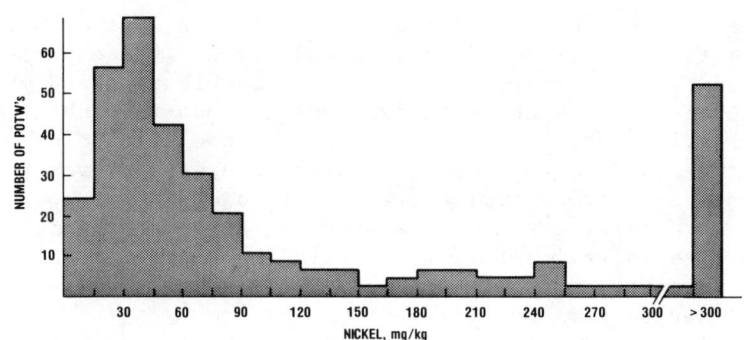

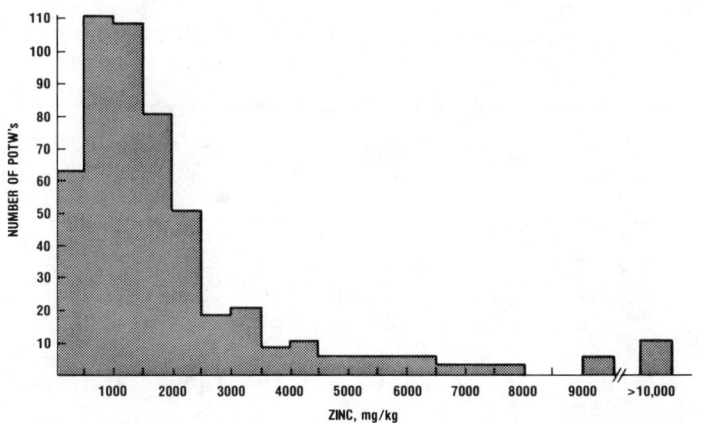

Figure 1 (Continued): Frequency Distributions of Five Metals in POTW Sludge.

- Distribution and marketing of sludge or sludge products (DM)
- Food chain landspreading (FC)
- Non-food chain landspreading (NFC)
- Landfilling (LF)
- Thermal processing (TP)
- Ocean disposal (OD)
- Other methods, such as long term lagooning or stockpiling (OTHER).

The methods where sludge-borne metals pose potentially significant risks to human and animal health and to crop productivity are primarily sludge distribution and marketing and food chain landspreading. Non-food chain landspreading offers potential risk from the standpoint of future land conversion to food chain uses, however the risks are not as imminent as for food chain landspreading. Metals in sludge disposed by the remaining methods pose relatively little risk, with the possible exception of potential groundwater contamination due to metals in leachates from landfills.

The median concentrations of heavy metals in sludges disposed by these seven methods are presented in Table IV.

Table IV. Median Concentrations of Heavy Metals in Sludges Disposed by Various Methods (mg/kg)

Metal	Disposal Method						
	DM	FC	NFC	LF	TP	OD	OTHER
Cd	12	10	17	16	28	45	10
Pb	334	377	348	430	517	821	320
Cu	414	638	270	677	974	1241	477
Ni	55	55	81	86	118	230	74
Zn	1252	1270	1502	1584	1684	1737	1200

This table indicates that sludges with generally lower concentration of metals are being disposed of by the "high risk methods" of distribution and marketing and food chain landspreading. The potentially more hazardous sludges with higher metal concentrations are currently disposed of by methods which are at lower risk to humans, animals, and crops.

A further examination of sludge-borne metals disposal can be made by calculating the mass loadings of metals disposed of by each method. Mass loadings are calculated from the concentrations of metals in sludges and the quantities of sludges containing those metals. Figure 2 presents the results of these calculations for the data set for each of the five metals in terms of kg of metal disposed annually by disposal methods. Metal loadings are relatively larger for the lower risk sludge disposal methods, such as landfilling, thermal processing, and ocean disposal. For the high risk disposal methods, the metal loadings from distribution and marketing of sludge predominate. The larger loadings from distribution and marketing are due to the fact that more sludges are being disposed by this method than by food chain landspreading, since the metal concentrations in these sludges are relatively equal as indicated above in Table IV.

In addition to total metal loadings (kg/yr) for the data set, median metal loadings were also determined for each disposal option, and are presented in Table V.

Table V. Median Loadings of Heavy Metals to Various Sludge Disposal Methods (kg/yr)

| Metal | DM | FC | Disposal Method | | | | |
			NFC	LF	TP	OD	OTHER
Cd	3	3	5	7	86	255	5
Pb	61	69	102	212	788	5152	535
Cu	70	279	277	494	2193	21541	758
Ni	16	26	136	95	849	4898	110
Zn	250	376	450	774	3151	13164	1186

As can be seen, sludges with the lowest median loadings are currently being disposed by methods which pose the highest risk in terms of potential hazards to human and animal health and crop productivity. Conversely, methods which pose relatively fewer risks are being used to dispose of sludges resulting in the highest median metal loadings. For example, the median cadmium loading for both distribution and marketing and food chain landspreading is 3 kg/yr, whereas for thermal processing or ocean disposal, the loadings increase by orders of magnitude to 86 and 255 kg/yr, respectively.

Although current practice shows that the highest risk disposal methods, that is those involving land application, receive sludges with the least relative amounts of metals, the outlook for the future may not be as reassuring. With the impending (December 1981) phase out of ocean disposal of sludge, the difficulties associated with the siting of new landfills, and the increased energy costs of sludge incineration, the

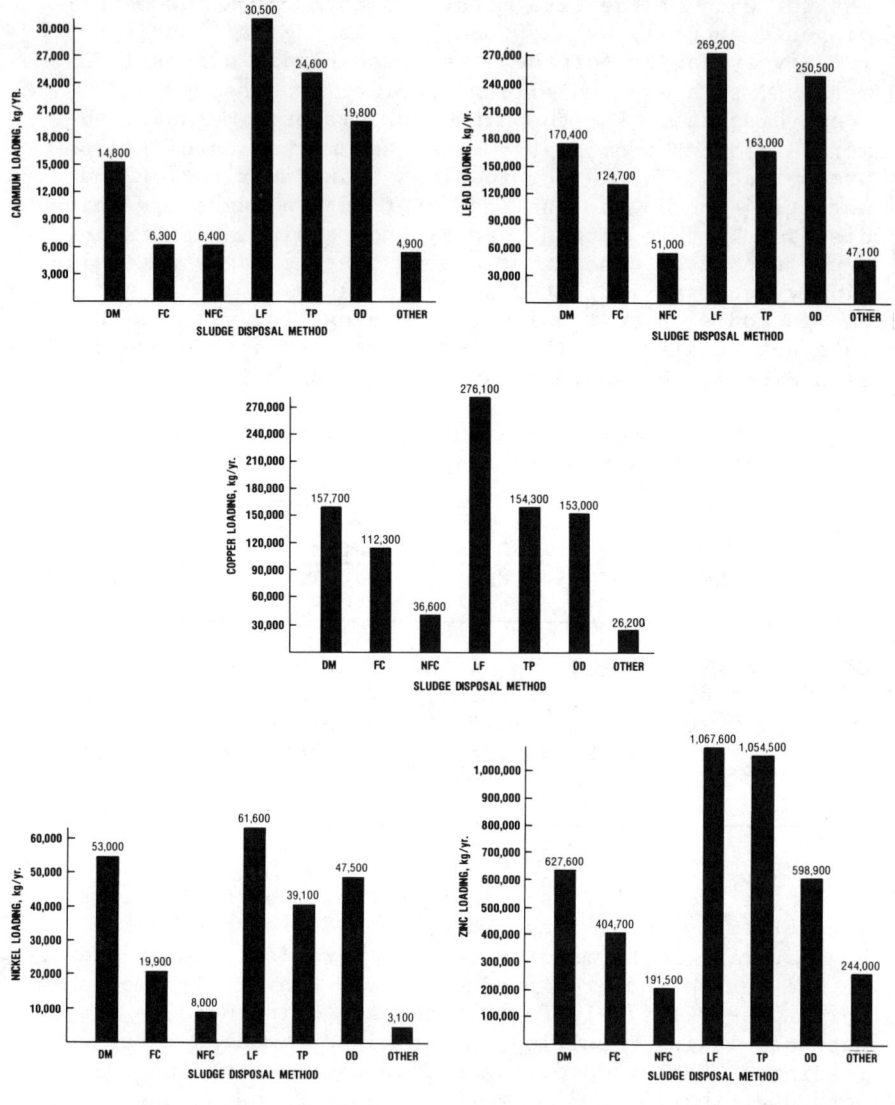

Figure 2: Total Loadings of Sludge-Borne Metals Disposed By Various Methods.

popularity of sludge disposal by land application can be expected to increase during the coming decade. Consequently, as the distribution of sludge disposal methods shifts toward land application options, metal loadings to these high risk methods may increase at much greater rates unless these elements can be removed from sludges.

Metals removal from POTW sludge is by no means a straightforward task. Since metals which accumulate in POTW sludges are contributed by non-industrial as well as industrial sources, in some cases adequate removal cannot be expected from industrial wastewater pretreatment programs alone. More far-reaching programs which achieve reduction of metals from other sources, such as water supply systems and urban runoff, may also be necessary.

REFERENCES

1. Chaney, R.L., "Health Risks Associated With Toxic Metals in Municipal Sludge," in Sludge: Health Risks of Land Application, (Eds. Gabriel Bitton, et.al), Ann Arbor Science Publishers, pp. 59-83, 1980.

2. Sommers, L.E., "Toxic Metals in Agricultural Crops", in Sludge: Health Risks of Land Application, (Eds. Gabriel Bitton, et. al.), Ann Arbor Science Publishers, pp 105-140, 1980.

3. "Environmental Impact Statement: Criteria for Classification of Solid Waste Disposal Facilities and Practices," U.S. Environmental Protection Agency, Office of Solid Waste, SW-821 (1979).

4. "Application of Sewage Sludge to Cropland: Appraisal of Potential Hazards of the Heavy Metals to Plants and Animals," Council for Agricultural Science and Technology Report 64, Ames, IA (1976).

5. Sommers, L.E., "Chemical Composition of Sewage Sludges and Analysis of Their Potential Use as Fertilizers," J. Environ. Qual. 6:225-232 (1977).

CONCENTRATION OF WASTEWATER WITH HIGH CALCIUM SCALING TENDENCIES BY EVAPORATION

Seth Levine, Unitech Division of Ecodyne, Union, New Jersey

Roberto Manas, Unitech Division of Ecodyne, Union, New Jersey

Water recovery and waste volume reduction have become increasingly important in process plant design in view of governmental discharge requirements and scarcity of water in some areas. The removal of inorganics from process streams, or desalting, will play an ever increasing role in the years ahead. The applications in the U.S. will involve responses to water shortages, return waterflow cleanup, reuse water cleanup, and drinking water cleanup. As a result of the Federal Water Pollution Control Acts, many process plants are being pushed towards zero discharge of pollutants. For example, many steam electric plants are being built with cooling water recycle rather than once through condenser cooling. As a result, cooling water loops require reuse water treatment. In most cases recycling is limited by scaling upon concentration. [1] In many cases the water within a plant can be reused several times. However, in the above case where zero discharge is anticipated, ultimate disposal of the cooling tower blowdown will be required. If ponds are not satisfactory, then further wastewater treatment will be required. Evaporation to accomplish waste volume reduction is one such approach. A cooling tower blowdown evaporator could reduce waste volumes considerably while recovering 90% or better in most cases of high purity water for recycle. This type of evaporator will be discussed later.

CHEMISTRY

In discussing desalting evaporators, let us first look at the chemistry of the major water treatment problems. Chemical compositions of typical cooling tower blowdowns, as well as brackish well waters, will include calcium, magnesium, sodium, chloride, sulfate, carbonate and silica amongst others. In many situations the process stream is saturated or near saturation in some of these constituents. Therefore, as evaporation occurs the scaling thresholds of some of these constituents may be reached. In some cases the ions will tend to form compounds which have solubilities that will decrease with temperature. When precipitation occurs, compounds may form hard scales. In heat exchangers, this will often decrease heat transfer to a point where chemical and mechanical cleaning is required. In the evaporator several water treatment steps are incorporated to protect the heat transfer surface from excessive scaling. Calcium carbonate and magnesium hydroxide are two such scales which are strongly affected by pH as well as temperature. At high enough temperatures bicarbonate will disintegrate to carbonate and form calcium carbonate scale when favorable conditions exist. Further disintegration of carbonate will form hydroxyl ions which can form magnesium hydroxide scales at favorable conditions of temperature and pH. These may be controlled though by acid treatment or chemical additives as will be discussed later. Calcium sulfate and its hydrates on the other hand will form relatively independent of pH. Due to its inverse solubility, at high enough temperatures precipitation will occur as concentration proceeds. Conventional treatment involves keeping solutions below the scaling thresholds of calcium sulfate. [2] As will be discussed later, the scaling threshold can be determined from a solubility product curve for the brackish water of concern. However, it becomes difficult to develop an accurate curve since varying amounts of other ions will affect the solubility. Another treatment is by a coprecipitation seeding technique which will be discussed later. Using this method will enable operation of the evaporator in the precipitation regime without scaling heat transfer surfaces. Lastly, silica will also present scaling problems at high enough concentrations. There has been success in handling silica at limited concentrations using the coprecipitation seeding technique.

DESCRIPTION OF THE MVR

With ever increasing energy costs, the MVR is finding wider use in evaporators for water recovery and desalination. The mechanical vapor recompression evaporator, or MVR, works on the principle of reusing the vapors driven off during evaporation. These vapors contain nearly as much energy as boiler steam. The difference between vapor and steam is, therefore, not the energy content but the quality of energy.

In order to transfer heat, a temperature difference is required. In two-phase systems, (liquid-vapor), this temperature difference must be accomplished by a pressure difference, thus the need to increase the pressure of the vapors. In MVR's this pressure increase is achieved with electrical or steam driven centrifugal compressors. The vapors given off in the evaporator, which contain approximately 1000 BTU/lb, are piped to the suction of a compressor where the vapor is compressed at a net expenditure of about 40 BTU/lb, depending on the compression ratio. The compressor then discharges these vapors to the steam chest of the evaporator to drive off more vapors to the compressor suction.

Centrifugal compressors are limited to low compression ratios. While some applications find compression ratios of 2.0, the vast majority of installations operate between 1.3 and 1.5

Because the heating surface in MVR's also functions as a condenser, little or no cooling water is needed. This can be of great advantage in many applications.

Process Description of the Crystallizing MVR

In water recovery applications, the crystallizing MVR is used as a means to control calcium scaling. Figure 1 shows the typical configuration for this type of evaporator. As a case study a brackish water evaporator has been selected for this discussion. Figure 4 is a photograph of the partially complete installation. A brief process description of this installation follows.

As the feed water enters the system, the pH is adjusted with sulfuric acid, and pumped to a tank for decarbonation and storage. As it is pumped to the preheater, the feed is dosed with an antiscalant. The purpose of the antiscalant is to protect the preheater from scaling, not the heating surface in the main evaporator.

In the preheater the temperature of the feed is raised to near its boiling point by transfer of the sensible heat contained in the hot distillate. The preheated feed then enters a counterflow deaerator, where carbon dioxide gas is removed and vented to atmosphere. Feed acidity is almost totally neutralized by removal of the carbon dioxide. Air is also removed to a level which precludes corrosion. Caustic soda may be injected if further neutralization is required.

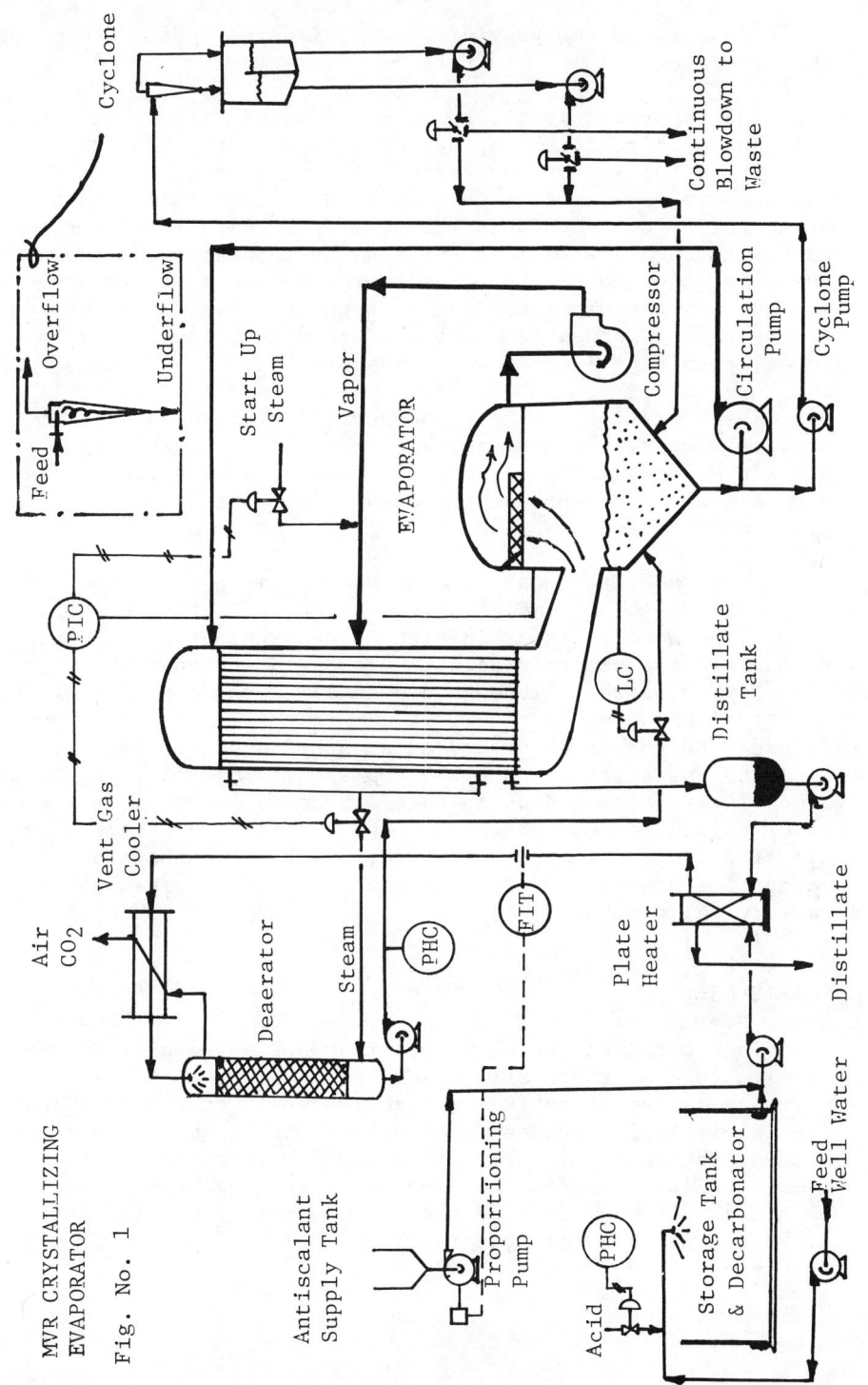

MVR CRYSTALLIZING EVAPORATOR
Fig. No. 1

The evaporator vents provide the required stripping steam for the deaerator.

The deaerated feed then mixes with the brine which is continually circulated through the evaporator. The brine enters the top liquor box of the heating element and is distributed to the inside of each tube as a thin film. As the brine falls down the tubes, water evaporates. The vapor passes through a wire mesh entrainment separator and enters the compressor where its temperature and pressure are upgraded. As the vapor condenses on the shell side of the tubes, its heat of condensation is transferred to brine inside the tubes, resulting in more water evaporating.

The brine inside the evaporator consists of a calcium sulfate slurry. Calcium sulfate crystals in the slurry provide an ideal site for crystal growth, preventing scaling in the evaporator through preferential crystallization on the existing crystals.

The evaporator must be continually blown down in order to purge the soluble salts which accumulate during evaporation. The need to maintain high concentrations of calcium sulfate in the slurry precludes blowing down the evaporator directly to waste. Although crystallization does take place, direct blowdown would result in almost total loss of crystals. In order to retain these crystals, a hydrocyclone is used to separate the slurry into two streams. The overflow is nearly free of crystals, and it is this stream which provides the major portion of the blowdown. The underflow contains nearly all of the crystals which are recycled back to the evaporator, except for a small portion which is purged to compensate for new crystal growth.

Case Study

To understand the chemical treatment approaches further, we will look at the system described earlier which is an existing system operating on brackish well water. A typical feed analysis for this system in shown in Table 1. As can be seen, carbonate scale prevention would be required. Acid treatment in this case was used as is shown in Figure 1. Sulfuric acid was added to the feedwater to lower the pH to 4.5-5.0. At this point, carbonates could be removed from the system as carbon dioxide by degasification.

Table 1. Typical Feedwater Analysis

pH	7.2
Chloride (as Cl)	5,250 ppm
"P" Alkalinity (as $CaCO_3$)	0
"M" Alkalinity (as $CaCO_3$)	83
Total Hardness (as $CaCO_3$)	3,340 ppm
Calcium (as Ca)	786 ppm
Magnesium (as Mg)	334 ppm
Sulfate (as SO_4)	3,200 ppm
Silica (as SIO_2)	39 ppm
Suspended Solids (ppm)	5 ppm
Total Dissolved Solids (%)	1.3
Organic and Volatile	---

As previously mentioned, scale formation in the heat transfer surface of the evaporator is prevented by controlled crystallization. Prior to starting the evaporator, a calcium sulfate slurry is prepared by mixing food grade calcium sulfate with water in an agitated tank, which is then transferred to the evaporator. In order for crystallization to take place, some degree of supersaturation is required. Fortunately, in order for spontaneous crystallization to occur, a higher degree of supersaturation level than for growth on existing crystal sites is required. This is the principle on which preferential crystallization works. By ensuring sufficient crystals, and thus sufficient surface area for rapid growth, supersaturation is kept under safe limits. The upper range to crystal concentration is limited by the equipment. The concentration of calcium sulfate crystals in the slurry is maintained within its specified range by adjusting the blowdown rate from the cyclone underflow.

At a concentration ratio of 9 to 1 under which this evaporator operates, silica scaling would become a problem for standard evaporators. However, because silica coprecipitates up to certain limits with calcium sulfate, the crystallizing MVR can extend the operating limits normally imposed by formation of silica scale.

Calcium sulfate though presented further problems prior to the evaporation loop. During preheating, hydrates of calcium sulfate began precipitating in the plate preheater. As can be seen from the solubility product curve of Figure 2, initial feedwater concentrations of calcium and sulfate ions were at near saturation. Therefore, as the temperature increased, solubility limits were exceeded. The solution to the problem involved the addition of a chemical to the feed

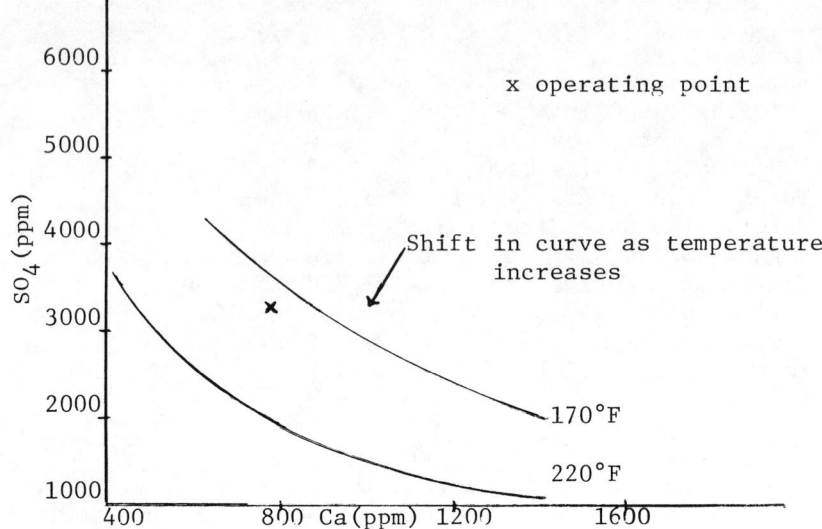

Figure 2. Solubility Product Curves for Calcium Sulfate in Brackish Water (Data from OSW R&D Progress Report No. 529)

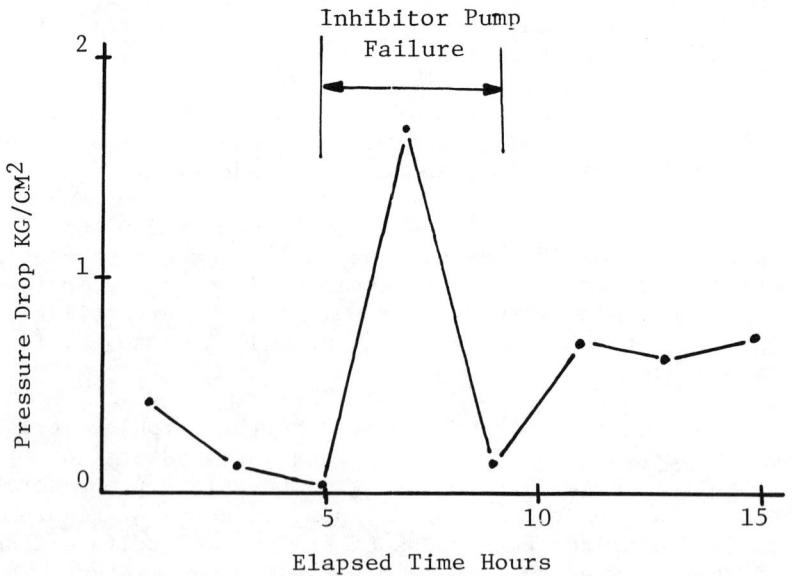

Figure 3. Preheater Pressure Drop

upstream of the preheater. Additives as mentioned earlier have been used in prevention of calcium carbonate scale such as polyphosphates rather than by acid pretreatment. However, in the control of calcium sulfate precipitation several problems existed. The additive had to remain stable at temperatures above 212°F. The additive had to be acceptable for potable water systems. Lastly, the additive had to be strong enough to prevent calcium sulfate scaling in the preheater but yet weak enough so as not to affect the coprecipitation seeding technique in a deleterious manner. One might ask why not use such an additive instead of coprecipitation? The answer is that none are available which are satisfactory at the concentrations maintained in the evaporator loop. The additive chosen was EL2438, a chemical manufactured by the Calgon Corp. The chemical contained phosphonate which acts as a threshold agent. Atoms of the phosphonate structure interject into the calcite lattice and prevent crystal growth. When the feed was dosed with additive, there was no further scaling in the preheater. Operation of the evaporator proceeded normally at capacity. The success of the pretreatment is further exemplified by Figure 3. The graph of pressure drop versus time shows that during a failure of the additive injection system, scaling occurred in the preheater as shown by increased pressure drop across the preheater. Once injection continued, the pressure drop returned to normal. This indicated that existing scale was flushed out and no further scaling occurred.

SUMMARY

Evaporator systems present themselves as an excellent means for concentrating wastewaters from a variety of industrial applications. Specifically, crystallizing mechanical vapor recompression systems have attained very high water recovery rates from wastewaters and brackish waters containing calcium scale forming compounds. Alkaline scales have been controlled well by conventional means such as acid treatment or chemical additives. In water systems where wastewaters are concentrated beyond calcium sulfate solubility limits, a seeding coprecipitation technique controls calcium sulfate scale. In systems where the feed enters saturated in calcium sulfate, chemical additives have been used to protect evaporator preheat sections without affecting the seeding coprecipitation technique. Two such systems which utilize these treatment schemes to produce potable water have been built for the Saudi National Electric Company.

Figure 4. Mechanical Vapor Recompression Installation
Unitech Division of Ecodyne

Bibliography

[1] Fluor Engineers and Constructors, "Desalting Plans and Progress", Office of Water Research and Technology 14-34-0001-7707, (1978)

[2] Howe, E.D., <u>Water Desalination</u>, Marcel Dekker, Inc., pp.35-37, 1974.

STATISTICAL EVALUATION OF HEAVY METALS AND pH CONTROL
IN TREATED EFFLUENTS FOR SELECTED INORGANIC CHEMICALS
INDUSTRIES

Michael G. Warner. Jacobs Engineering Group Inc.,
Pasadena, California

Ben C. Edmondson. Department of Management Science,
California State University, Fullerton, California

Milford Walker, Jr. Jacobs Engineering Group Inc.,
Pasadena, California

INTRODUCTION

Many inorganic chemicals manufacturers face similar problems in treating their process waste waters. Raw process wastes contain a variety of pollutants. "Conventional" pollutants include pH, biochemical oxygen demand (BOD), total suspended solids (TSS), oil and grease, and fecal coliform. These, along with certain nonconventional pollutants such as ammonia, fluoride, aluminum and iron, have been regulated by "best practical technology" (BPT) effluent limitations established by the EPA. For specific industries these regulations also included limitations on the dominant toxic pollutants associated with the manufacturing process. After the NRDC consent decree (1), the EPA had to evaluate the need for regulating any or all of 129 specific toxic chemicals on the basis of "best available technology economically achievable" (BAT).

Performance expected of BPT/BAT systems has increased considerably since most existing systems were designed and installed. In particular, the number of toxic metals subject to effluent limitations has increased and systems originally designed for optimum removal of one particular metal are now expected to perform respectably for several toxic metals (2,3).

For the inorganic chemicals industry, the significant difference between proposed regulations (45 FR 49450, July 24,

1980) and promulgated regulations is that fewer toxic metals were given specific effluent limits. The final selection of metals to be regulated was based on an evaluation of the treatment chemistry involved (4). In effluents treated using either hydroxide or sulfide precipitation, certain groups of metals exhibit roughly similar behavior as a function of the pH conditions. Figure 1 shows the theoretical solubility curves for metal oxides and hydroxides. Controlling the treatment pH between 8.5 and 9.5 controls four metals (chromium, copper, lead and zinc) near their minimum solubilities as hydroxides or oxides. Thus, setting limitations on one of the four ensures that the other three would also be controlled.

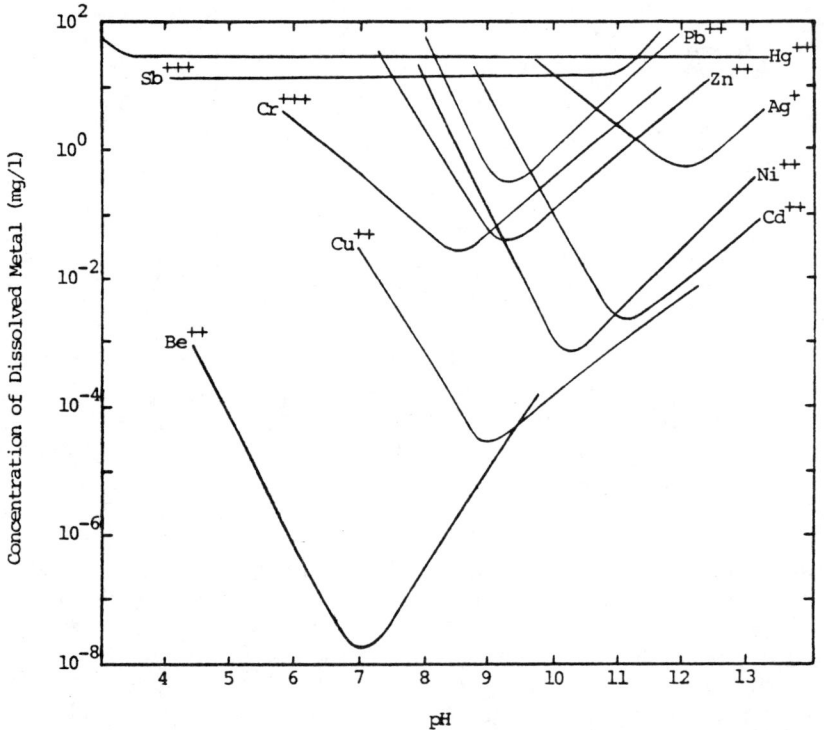

Figure 1. Theoretical solubilities of toxic metal hydroxides/oxides as a function of pH (5)

USE OF POLLUTION MONITORING DATA

Plants that have been granted NPDES permits are required to monitor designated parameters in these effluent streams, along with ancillary variables such as flow and temperature. Frequency of required monitoring at different plants varies to a large degree. Certain factors with regard to screening of data should be addressed. These include (a) nonrepresentative values, (b) reporting of values below detectable levels and (c) values of parameters recorded as zero.

Atypically large data values, while duly recorded for an accurate record of a plant's actual experience, need to be excluded from statistical calculations. Process or treatment system upsets, operator error, and the like do cause extreme concentrations of toxic pollutants in waste streams. Recorded measurements should be screened for such spurious or nonrepresentative values. Under the assumption of lognormal distribution of daily values, a statistical test for outliers, the student's t-test, classifies extreme values to be rejected as nonrepresentative (6).

Next, many measurements are recorded as "less than" a detection limit, e.g., $<5\mu g/l$. Such values should be included in statistical estimates with these observations taken as equal to the detection limit or threshold level. Converting the "less thans" to "equals" will raise a calculated average and lower a corresponding standard deviation.

Finally, sound engineering judgment dictates excluding values recorded as zero from statistical projections. (In reality, a zero value usually means "unable to determine," "no data available," or "lost sample" results.) Not only would their incorporation lower the estimated average, but in many cases would greatly lower the statistical estimates of variability.

In analysis of effluent monitoring data, measurements of pollutant concentrations are assumed to follow the lognormal distribution (6). This distribution, in which the concentration logarithms are distributed on a normal or "bell-shaped" curve, is widely used in pollution monitoring analysis (7). Day-to-day measurements were assumed to be statistically independent.

Effluent limitations expressed as the daily maximum and the maximum 30-day average do not depend solely upon the long-term average. Performance standards derived from statistical analysis are equivalent to estimates of upper percentiles of the distribution of concentrations. In the case of daily measurements following the lognormal distribution, the percentiles may be expressed as multiples of the long-term average. For this work, the ratio of the 99th percentile to the estimated long-term average is called the "variability factor", VF, which is related to the long-term average as: $VF = P/A$; and to system variability as: $\ln(VF) = 2.33\ S - S^2/2$; where

P = 99th percentile of concentration levels; A = Long-term average pollutant concentration; S = Standard deviation of the logarithms of pollutant concentrations; 2.33 = 99th percentile of standard normal distribution.

Although daily measurements are assumed lognormally distributed, that assumption does not directly extend to the statistical behavior of averages (8). When averages are taken over a reasonably large number of days, say eight or ten days/month, then the central limit theorem assures that concentration averages can be analyzed by the normal probability model.

If monitoring data is taken for a time series of sufficiently long-term so that several 30-day averages are available, then it's possible to develop variability factors directly from those averages. Under these conditions, the 30-day average values for several months behave approximately as random data from a normal distribution with mean A and standard deviation S''. Therefore, the probability is 95 percent that a monthly average will not exceed the performance standard P, where $P = A + 1.64 S''$. The variability factor is $VF = P/A = 1.0 + 1.64 (S''/A)$.

In the case where historical data do not encompass several months, then performance guidelines may be developed from daily data on the basis of estimating the probability that a (future) 30-day average will not exceed any specified concentration limitation. For the parameters involved in the treatability study, probability performance curves were developed (3). These curves indicate the probability that a 30-day average would be less than any given value for a range of concentration values.

INDUSTRIAL TREATMENT SYSTEM PERFORMANCE

Each industry designs for maximum removal and/or recovery of the major process-related waste substances and utilizes an appropriate treatment technology which is both reliable and cost effective. Optimum treatment conditions for the removal of a particular pollutant can rarely be achieved consistently, and any given set of conditions will be less than optimum for some of the treatable constituents.

The major sources of data on treated effluent quality include treatability data published in the various technical journals, monitoring data from NPDES reports submitted by individual plants in various industries, priority pollutant screening and verification results from three-day sampling episodes conducted by the EPA at representative plants in each industry, short-term bench-scale treatability studies, conducted by the EPA to demonstrate the feasibility of proposed BAT treatment options (9).

From a statistical point of view, both the long-term NPDES monitoring data and the short-term, EPA treatability results (e.g., sets of 12 to 15 daily data values) offer the advantage of providing estimated means and standard deviations from which variability factors can be calculated as the basis for effluent limitations. A comparison of statistical parameters derived from these two sources of data is presented in Table I for toxic metal concentrations in BPT/BAT treated effluents.

Table I. Toxic Metal Concentrations in BPT/BAT Treated Effluents (Historical Plant Data/EPA Treatability Data)

Industry	Mean Concentration (mg/l)	Variability Factors 30-Day Averages	Daily Averages
Chromium (III)			
TiO_2 (Chloride)	0.004/0.18	2.46/1.26	--/4.28
TiO_2 (Sulfate)	0.072/--	2.04/--	3.81/--
Chrome Pigments	0.44 /0.43	1.66/1.15	--/2.67
Chlorine (Diaphragm Cell)	--/0.071	--/1.11	--/2.19
Sodium Dichromate	--/0.26	--/1.24	--/3.97
Metal Finishing	0.15 /--	1.43/--	4.94/--
Nickel			
Nickel Sulfate	1.8/1.4	1.54/1.51	5.84/5.03
Copper Sulfate	--/0.25	--/1.19	--/3.28
Lead			
Chlorine (Diaphragm Cell)	0.032/0.46	1.21/1.80	3.62/10.2
TiO_2 (Chloride)	0.068/--	2.10/--	3.20/--
Chrome Pigments	0.42/--	2.12/--	--/--

Investigation of results for various parameters is not complete enough to estimate the degree of confidence in the outcome of a 12- to 15-day treatability study, but there is evidence that such studies will provide working estimates of achievable long-term averages and variability factors. The comparison does point out the need to consider individual plant and industry characteristics very carefully when attempting to make generalizations about BPT/BAT treatment system performance.

The paucity of toxic pollutant monitoring data has made it more difficult to estimate the achievable long-term average concentrations than to select reasonable variability factors. Some examples of the wide disparities that can exist from industry to industry are shown in Table II. Each value shown on the table represents the average performance of several plants in a particular industry derived from monitoring data, priority pollutant sampling data, or treatability studies.

Table II. Summary of Effluent Concentrations Averages for Five Toxic Metals Treated by Alkaline Precipitation and Clarification (4)

Chromium III (mg/l)	Source	Copper (mg/l)	Source	Lead (mg/l)	Source	Nickel (mg/l)	Source	Zinc (mg/l)	Source
0.040	IS	0.030	IS	0.017	HF	0.020	IS	0.020	OMD
0.050	NFM	0.030	CAD	0.10	MF	0.050	CAD	0.040	OMD
0.050	CAD	0.038	OMD	0.15	OMD	0.10	NFM	0.040	FI
0.070	TDS	0.060	OMD	0.19	NFM	0.17	HF	0.10	CAD
0.071	CAD	0.070	NFM	0.20	NFM	0.20	CS	0.11	TM
0.072	TDC	0.070	SEP			0.20	SDC	0.15	TDS
0.080	TM	0.080	OMD			0.25	CS	0.20	NFM
0.15	MF	0.090	OMD			0.26	CAD	0.24	IS
0.18	TDC	0.10	OMD			0.31	MF	0.25	IS
0.26	SDC	0.14	MF			0.33	IS	0.35	OMD
0.35	CAD	0.54	CAD			0.50	NFM	0.39	MF
0.36	IS	0.70	NFM			1.4	NS	0.54	CAD
0.43	CP	1.1	CS					0.55	HF
0.81	SDC	1.5	CS					0.60	NFM
1.8	IS							8.2	PM

Source Codes:

CAD	Chlor-Alkali, Diaphragm Cells	NFM	Nonferrous Metals
CAM	Chlor-Alkali, Mercury Cells	NS	Nickel Sulfate
CS	Copper Sulfate	OMD	Ore Mining and Dressing
CP	Chrome Pigments	PM	Paint Manufacturing
FI	Foundry Industry	SDC	Sodium Dichromate
HF	Hydrofluoric Acid	SEP	Steam Electric Power Generating
IS	Iron and Steel	TDC	Titanium Dioxide - Chloride Process
MF	Metal Finishing (including electroplating)	TDS	Titanium Dioxide - Sulfate Process
		TM	Textile Mills

EFFLUENT pH CONTROL

Just as pH control is an essential step within the treatment process, it is also required for final conditioning of the treated effluent prior to discharge. A data base for evaluation of final pH control performance was established by studying effluent pH records at eight plants with long-term continuous pH monitoring data (2,10). These plants were reviewed for type of waste water treatment employed, especially with regard to pH control stages for treatment and final discharge, and detailed records were made of effluent pH excursions outside the 6 to 9 range.

Excursion data gathered at each plant, normally covered a period of six months to a year, and included the following information: date, time, duration, peak pH value, reason (in the form of a code), and additional remarks. The pertinent values for these excursions, the peak pH value and duration, were extracted in such a manner as to maximize the effect of the excursion. The causes of excursions included the following: (a) treatment system upset/shutdown; (b) process upset; (c) spills or leaks; (d) storm water runoff; (e) emergency operation; (f) other, and (g) unknown - the most common cause being treatment system upsets.

Table III gives a plant-by-plant breakdown of the percentages of total time in excursion by pH range.

Table III. Excursion Times for Different Plants by pH Range

pH Range	Percent of Excursion Time							
	A	B	C	D	E	F	G	H
0-0.9			0.050		0.008		0.18	
1-1.9					0.010		0.03	
2-2.9					0.010	0.040	0.07	
3-3.9					0.040	0.180	0.05	0.12
4-4.9	0.001		0.001		0.006	0.006	0.03	0.07
5-5.9	0.003	0.080	0.004	0.18	0.050	0.030	0.12	0.51
9-9.9		0.005	0.004		0.010	0.240	0.09	
10-10.9			0.003					
11-11.9					0.060	0.050	0.07	1.07
12-12.9			0.003					
13-14.0			0.020					
Total % of Time in Excursion	0.004	0.085	0.085	0.18	0.194	0.60	0.63	2.06

A. Hydrofluoric acid and aluminum fluoride
B. Sodium silicate
C. Hydrofluoric acid and sulfuric acid
D. Sodium metabisulfite and sulfur dioxide
E. Sulfuric acid
F. Hydrochloric acid and chlorine/caustic
G. Hydrogen cyanide
H. Titanium dioxide (chloride process)

SUMMARY

Effective treatment of industrial wastewaters relies heavily on the pH control steps; however, our understanding of the chemical and physical pollutant removal mechanisms involved is far from adequate for accurately predicting the treatability of complex industrial waste streams.

REFERENCES

1. Natural Resources Defense Council, Inc. v. Train, 8 ERC 2120 (D.D.C. 1976) modified March 9, 1979; 12 ERC 1833.

2. U.S. Environmental Protection Agency, Proposed Development Document for Effluent Limitations Guidelines, New Source Performance Standards, and Pretreatment Standards for the Inorganic Chemicals Manufacturing Point Source Category, EPA 440/1-80/007b, July 1980.

3. Martin, E.E., Stigall, G.E., and Warner, M.G., Effluent Limitations Guidelines for the Inorganic Chemicals Industry, Proceedings of the 1980 Purdue Industrial Waste Conference, West Lafayette, IN.

4. U.S. Environmental Protection Agency, *Final Development Document for Effluent Limitations Guidelines, New Source Performance Standards, and Pretreatment Standards for the Inorganic Chemicals Manufacturing Point Source Category*, Effluent Guidelines Division, Washington, D.C., 1981. (In press.)

5. Pourbaix, M. et al., *Atlas of Electrochemical Equilibria in Aqueous Solutions*, Pergamison Press, Oxford, 1966.

6. Keeping, E.S., *Introduction to Statistical Inference*, D. Van Nostrand Company, Princeton, NJ, 1962.

7. Larsen, Ralph I., *A Mathematical Model for Relating Air Quality Measurements to Air Quality Standards*, National Environmental Research Center, Research Triangle Park, N.C.

8. *Development Document for Existing Source Pretreatment Standards for the Electroplating Point Source Category*, EPA 440/1-79/003.

9. U.S. Environmental Protection Agency, *Treatability Studies for the Inorganic Chemicals Manufacturing Point Source Category*, EPA 440/1-80/103.

10. JRB Associates, Inc., *An Assessment of pH Control of Process Waters in Selected Plants*, (Draft Report), March 1979.

HEAVY METALS REMOVAL: COMPARISON OF ALTERNATIVE
PRECIPITATION PROCESSES

<u>Karl A. Brantner</u>. Industrial Division, Metcalf & Eddy, Inc., Boston, Massachusetts.

<u>Edward J. Cichon, PhD</u>. Technical Specialist, Metcalf & Eddy, Inc., Boston, Massachusetts.

INTRODUCTION

Under proposed Best Available Technology (BAT) effluent limitations for industries with wastewaters known to contain heavy metals, a higher degree of treatment than clarification is likely to be necessary prior to discharge. Pilot plant studies were conducted to evaluate alternative treatment processes for effectiveness in achieving heavy metals removal in order to meet BAT standards. The wastewater treated was blowdown from electric arc furnace steelmaking. Three different process trains were piloted: hydroxide precipitation, carbonate precipitation, and sulfide precipitation.

Precipitation of heavy metals as hydroxides through the addition of lime is a well-established technology. Theoretically, hydroxide precipitation can result in very low residual metals concentrations for several metals by controlling pH. Although it is a relatively simple process, hydroxide precipitation has several limitations. Some metal hydroxides are amphoteric over a relatively narrow range of pH. The optimum pH values for minimum solubility differ for different metals, which may limit the degree of metals removed when several different metals are of concern. Another factor that affects the solubility includes the presence of other constituents in the wastewater matrix, particularly complexing agents which inhibit metal hydroxide precipitation. In addition, hydroxide precipitation generates large volumes of relatively low density sludge, which can present dewatering and disposal problems.

Precipitation of heavy metals through the addition of soda ash is a treatment process that has received very little attention. Carbonate precipitation has been addressed from both a theoretical standpoint and through laboratory testing (1). The most probable precipitates that are formed are a combination of metal hydroxides, metal carbonates and metal hydroxy-carbonates. Advantages that carbonate precipitation theoretically have over hydroxide precipitation include lower metal solubility, lower operating pH and lower volumes of significantly denser sludge.

Sulfide precipitation has been investigated in recent research and has been demonstrated for metal finishing wastewaters (2,3,4), although it has not yet been widely applied. Theoretically, lower effluent metals concentrations are achievable with sulfide precipitation than with hydroxide precipitation. In addition, one-step hexavalent chromium (Cr(IV)) removal can be achieved by chemical reduction to trivalent chromium (Cr(III)) and subsequent precipitation as chromium hydrous oxides. Sulfide may be added as either a soluble salt such as sodium sulfide or as a sparingly soluble salt such as ferrous sulfide. Both methods have their advantages and disadvantages, the most notable of which is probably that with a sparingly soluble sulfide salt, the potential of evolving hydrogen sulfide gas (H_2S) is minimized due to very low free sulfide concentrations.

MATERIALS AND METHODS

A trailer-mounted, physical/chemical treatment pilot plant was employed for demonstration of the alternative precipitation processes. The pilot plant consisted of facilities for chemical feed, flocculation, clarification, and filtration with various flow mode and bypass capabilities. The pilot plant was operated at 6 gallons per minute (gpm) through the chemical treatment and clarification system and 5 gpm through the dual media filter for polishing. At these flow rates the detention times were 10 minutes in the rapid mix tank, 8.3 minutes in the flocculator, and two hours in the clarifier. The clarifier overflow rate was 350 gallons per day per square feet (gpd/ft^2) or 0.25 gpm/ft^2, and the filter loading rate was 5 gpm/ft^2.

The mobile pilot plant was operated to treat electric arc furnace blowdown at a steel mill which operates three electric furnaces. The wastewater source was a sidestream of the blowdown from the gas cleaning recycle system which consists of venturi scrubbers and quenchers. Prior to blowdown, the recycled water had undergone clarification for suspended solids removal.

Hydroxide Precipitation. A slurry of hydrated lime (Ca(OH)$_2$) was applied to maintain pH 10 in the rapid mix tank. Lime addition was controlled by a pH controller with its input probe located in the rapid mix tank. The sludge produced was collected in the clarifier cone and wasted at the rate of production.

Carbonate Precipitation. Soda ash dosages were applied to the rapid-mix tank by a chemical feed pump at 650 mg/l

Na_2CO_3. The operating pH was that resulting for soda ash addition, which normally ranged from 8.3 to 8.7. No other pH adjustment was made. The sludge produced was collected in the clarifier cone, but none was wasted because the total sludge production did not reach the sludge storage capacity of the clarifier cone.

Sulfide Precipitation. Ferrous sulfide (FeS) was applied by a chemical feed pump which was manually set to maintain a dosage of 130 mg/l FeS. The pH was maintained at 8.0 - 8.5 by addition of lime. The FeS dosage of 130 mg/l was selected to be 1.5 times the calculated stoichiometric requirement. The excess FeS was incorporated in the sludge blanket, from which sludge was recycled to the flocculator for contact with the incoming wastewater. Therefore, the sludge blanket provided a reservoir of FeS should underdosing have occurred. Fresh FeS was routinely made up by reacting ferrous sulfate with sodium hydrosulfide under alkaline conditions.

Sample Collection and Analysis. During the operation of each precipitation process train, composite samples were collected for influent, clarifier effluent and filter effluent over eight-hour periods, resulting in three sets of samples per day. All samples were put on ice and shipped to the laboratory where they were analyzed for suspended solids, pH, zinc, antimony, arsenic, cadmium, chromium, copper, lead, nickel, and silver. All metals were analyzed by atomic absorption.

RESULTS AND DISCUSSION

Among the nine metals monitored, six were found to be in detectable concentrations in the pilot plant influent. The other three metals - antimony, arsenic and silver - either were not observed in detectable quantities or when they could be detected, were found in concentrations very near the detection limit. These metals as well as nickel, which was always in very low concentrations, did not provide a sufficient basis for evaluation. Therefore, the evaluation was based on comparisons of the removal of zinc, cadmium, chromium, copper, and lead. Analytical results for all three processes are summarized in Tables 1-3, presenting means and ranges. Selected test runs where the soluble metals fraction was determined are presented in Tables 4-6.

Zinc Removal. Effective removal of zinc was obtained by all three of the chemical precipitation processes tested. Hydroxide precipitation gave the most consistent performance while the sulfide and carbonate precipitation processes both resulted in a wide range of filtered effluent values.

The precipitation and clarification steps were responsible for removing most of the zinc during the hydroxide precipitation testing. Soluble metals data indicate that the zinc not removed by clarification or filtration was mostly in the particulate form. This demonstrated the fine particle size

TABLE 1. HYDROXIDE PRECIPITATION DATA SUMMARY

Parameter	Influent (ppm) Mean	Range	Clarifier effluent (ppm) Mean	Range	Filtered effluent (ppm) Mean	Range
Suspended Solids	42	20-63	22	9-33	9	4-14
pH	7.5	7.0-8.1	9.9	9.7-10.2	9.8	9.5-10.4
Total Cadmium	1.66	0.13-4.30	0.05	0.03-0.10	0.04	0.02-0.06
Total Chromium	1.11	0.07-2.90	1.04	0.07-2.80	0.97	0.06-2.90
Total Copper	0.29	0.12-1.50	0.03	0.02-0.03	0.03	0.02-0.03
Total Lead	1.7	0.8-2.6	0.2	0.1-0.3	0.2	0.1-0.3
Total Zinc	31	6-91	0.40	0.23-0.75	0.28	0.10-0.66

TABLE 2. CARBONATE PRECIPITATION DATA SUMMARY

Parameter	Influent (ppm) Mean	Range	Clarifier effluent (ppm) Mean	Range	Filtered effluent (ppm) Mean	Range
Suspended Solids	43	16-75	27	14-52	6	2-10
pH	7.1	6.7-7.7	8.3	8.1-8.4	8.1	7.8-8.5
Total Cadmium	1.37	0.26-2.90	0.14	0.02-0.27	0.04	0.02-0.06
Total Chromium	0.67	0.23-1.80	0.62	0.17-1.8	0.60	0.14-2.00
Total Copper	0.18	0.06-0.27	0.04	0.03-0.06	<0.03	<0.02-0.04
Total Lead	1.4	0.7-2.1	0.2	0.2-0.4	<0.1	<0.1-0.2
Total Zinc	26	1-67	3.2	0.37-5.0	1.18	0.19-5.00

TABLE 3. SULFIDE PRECIPITATION DATA SUMMARY

Parameter	Influent (ppm) Mean	Range	Clarifier effluent (ppm) Mean	Range	Filtered effluent (ppm) Mean	Range
Suspended Solids	90	30-210	37	10-63	5	2-8
pH	6.8	6.4-7.7	8.2	8.0-8.4	8.1	7.8-8.5
Total Cadmium	3.3	0.65-5.4	0.18	0.09-0.29	0.06	0.02-0.12
Total Chromium	0.52	0.02-1.9	0.12	0.08-0.20	<0.05	<0.05-0.05
Total Copper	0.35	0.19-0.56	<0.04	<0.04-0.05	<0.03	<0.02-0.03
Total Lead	4.5	2.3-8.1	0.4	0.2-0.5	<0.1	<0.1-0.2
Total Zinc	93	5.8-220	3.1	2.0-6.2	0.68	0.11-1.8

formed by the zinc hydroxide precipitate and the importance of the role of good flocculation. From the low soluble zinc levels in the clarifier overflow, it appears that zinc removal was not limited by the solubility, but by the effectiveness of liquid-solids separation.

In contrast to the hydroxide precipitation test runs where a relatively constant zinc level was obtained in the filtered effluent, the carbonate precipitation treatment train resulted in a broad range of zinc concentrations in the filtered effluent. Effluent zinc levels tended to be higher when

TABLE 4. TOTAL AND SOLUBLE METALS: HYDROXIDE PRECIPITATION

	Run 30						Run 33					
	Influent (ppm)		Clarifier Effluent (ppm)		Filtered Effluent (ppm)		Influent (ppm)		Clarifier Effluent (ppm)		Filtered Effluent (ppm)	
Parameter	Total	Soluble	Total	Soluble	Total	Soluble	Total	Soluble	Total	Soluble	Total	Soluble
Cadmium	0.16	0.12	0.03	<0.02	<0.02	<0.02	0.27	0.23	<0.02	<0.02	<0.02	<0.02
Chromium	2.9	2.8	2.8	2.8	2.9	2.9	1.4	1.3	1.1	1.1	1.1	1.1
Copper	1.5	0.04	0.03	<0.01	0.03	0.02	0.19	0.05	0.03	0.01	0.03	0.02
Lead	1.6	0.3	0.5	0.2	<0.1	<0.1	1.2	<0.1	<0.1	<0.1	<0.1	<0.1
Zinc	9.0	0.30	0.34	0.02	0.24	<0.02	8.0	0.57	0.26	0.03	0.19	0.03

TABLE 5. TOTAL AND SOLUBLE METALS: CARBONATE PRECIPITATION

	Run 39						Run 40					
	Influent (ppm)		Clarifier Effluent (ppm)		Filtered Effluent (ppm)		Influent (ppm)		Clarifier Effluent (ppm)		Filtered Effluent (ppm)	
Parameter	Total	Soluble	Total	Soluble	Total	Soluble	Total	Soluble	Total	Soluble	Total	Soluble
Cadmium	0.91	0.79	0.11	0.10	0.03	0.06	1.2	1.2	0.12	0.04	0.04	0.03
Chromium	0.52	0.31	1.0	0.89	0.71	0.42	0.33	0.22	0.29	0.27	0.28	0.24
Copper	0.22	0.02	0.06	0.03	0.03	0.02	0.10	0.01	0.05	<0.01	0.03	0.02
Lead	2.1	0.1	0.4	0.1	0.1	0.1	1.0	0.1	0.2	<0.1	<0.1	<0.1
Zinc	18	3.2	4.0	0.57	0.60	0.55	18	4.5	2.0	0.21	0.49	0.34

TABLE 6. TOTAL AND SOLUBLE METALS: SULFIDE PRECIPITATION

	Run 52						Run 54					
	Influent (ppm)		Clarifier Effluent (ppm)		Filter Effluent (ppm)		Influent (ppm)		Clarifier Effluent (ppm)		Filtered Effluent (ppm)	
Parameter	Total	Soluble	Total	Soluble	Total	Soluble	Total	Soluble	Total	Soluble	Total	Soluble
Cadmium	0.83	0.76	0.20	<0.02	<0.02	<0.02	1.5	1.4	0.14	<0.02	0.03	<0.02
Chromium	1.9	1.8	0.2	<0.05	<0.05	<0.05	0.31	0.23	0.08	<0.05	<0.05	<0.05
Copper	0.20	0.04	0.04	<0.02	<0.02	<0.02	0.30	0.04	0.04	<0.02	0.02	<0.02
Lead	2.6	0.6	0.5	<0.1	<0.1	<0.1	2.3	0.9	0.4	0.1	0.1	0.1
Zinc	5.8	1.5	2.3	0.25	0.17	0.13	23	14	2.3	0.10	0.38	0.09

influent levels were higher, indicating that the effluent level may be limited by the influent level when operating parameters are held constant. The inability of carbonate precipitation to consistently achieve low levels of zinc in the effluent may be attributed to the kinetics of zinc carbonate formation. Because the rate of zinc carbonate formation may be too slow to have been completed within the detention time of the pilot plant, the solubility of zinc hydroxide probably governs the removal of zinc (1). The less than optimum pH for zinc hydroxide precipitation during operation may account for the ineffectiveness of the zinc removal. Comparison of total and soluble zinc concentrations in the clarifier and filtered effluents indicated that a large proportion of zinc in the clarifier overflow was particulate and was subsequently removed by the filter.

Effluent zinc levels were higher than expected for sulfide precipitation. Influent zinc levels were also higher than during other testing phases, but influent and effluent zinc levels could not be correlated. Inspection of the soluble metals data shows that most of the zinc passing through the clarifier was due to solids overflow, while the zinc passing through the dual media filter was a combination of both particulate and soluble forms. The higher level of zinc in the clarified effluent may be attributed to the smaller particle size and poorer settleability of metal sulfide precipitates in the absence of polymers to promote coagulation and settling (4).

Chromium Removal. Effective removal of chromium was achieved only with the sulfide precipitation process. Both the hydroxide and carbonate precipitation processes demonstrated a minimal amount of chromium removal. The ability of the sulfide precipitation process to remove the chromium to low effluent levels is based on the reduction of Cr(VI) to Cr(III) in the presence of reducing agents such as sulfide ions or ferrous ions, with the resulting Cr(III) precipitated as $Cr_2O_3 \cdot nH_2O$ at nearly optimum pH. The lack of effective chromium removal by the hydroxide and carbonate precipitation processes stems from the absence of suitable reducing agents introduced to the wastewater during these process operations.

In addition to the inability of the hydroxide precipitation process to reduce Cr(VI) to Cr(III), the operating pH of 10 was not optimum for the removal of Cr(III). The fraction of trivalent chromium present would be limited by the solubility at this pH, which was approximately 0.2 to 0.3 parts per million (ppm). However, during the carbonate precipitation testing phase, where the pH approached the theoretical optimum for precipitation of the hydrous oxide, there still was not a significant removal of chromium from the wastewater. Therefore, it may be concluded that the chromium present was primarily in the hexavalent state.

Cadmium Removal. The removal of cadmium was comparable for each of the three precipitation processes tested. The mean filtered effluent level of cadmium was less than 0.10 ppm for each process. The precipitation and clarification steps were responsible for the greatest reduction. Nearly all of the residual cadmium in both clarifier and filter effluents was soluble, indicating that the residual level was limited by the solubility rather than the effectiveness of liquid-solid separation. While precipitation of cadmium was very effectively achieved by all three processes, the hydroxide precipitation treatment train was least dependent on the filter. This was probably due to the larger size of the cadmium hydroxide floc particles.

Copper Removal. The removal of copper was also comparable for each of the precipitation processes tested. Although effluent concentrations were consistently less than 0.05 ppm for all three processes, it should be noted that influent concentrations were low and mostly in the particulate form, preventing effective evaluation of precipitation. These fine particulates were effectively entrained in the floc formed during the precipitation process and removed by subsequent settling. The degree of entrainment appeared to be nonselective of the type of floc formed. Essentially all the copper was removed in the clarifier.

Lead Removal. The removal of lead was effectively accomplished by the carbonate and sulfide precipitation processes, with mean filtered effluent levels of <0.1 ppm. Hydroxide precipitation resulted in a mean effluent level of 0.2 ppm. The slightly better performance of the sulfide precipitation process may be due to the lower solubility of lead sulfide compared to lead hydroxide, while the improved performance of the carbonate precipitation process may be due to the formation of complex species such as $Pb_3(CO_3)_2(OH)_2$. As with copper, influent lead concentrations were mostly in the particulate form, preventing effective evaluation of precipitation. Effective entrainment of these particulates in the floc resulted in a high degree of removal which occurred mostly in the clarifier. However, some additional removal did occur across the filter.

Sludge Production. The average quantities of sludge produced over the operational periods of each precipitation process are presented in Table 7. Sludge quantities are shown on a per thousand gallons of wastewater treated basis.

TABLE 7. SLUDGE PRODUCTION

Precipitation process	Total solids (percent)	Dry lb sludge / 1,000 gal	Wet lb sludge / 1,000 gal
Hydroxide	2.6	14.1	540
Carbonate	9.1	3.2	35
Sulfide	2.0	6.8	350

Of the three precipitation processes, the lowest volume of sludge was produced by carbonate precipitation. It was a dense sludge that thickened well in the bottom of the clarifier. Hydroxide precipitation produced the greatest volume of sludge, closely followed by the sulfide precipitation sludge volume which was about two-thirds as much. Neither of the hydroxide and sulfide sludges thickened well in the clarifier, probably due both to their nature and disturbance from sludge pumping. Therefore, they would require additional thickening prior to dewatering and disposal.

Of the three sludge types generated, the hydroxide sludge would be expected to be the least stable when subjected to acidification. Sulfide sludge would probably be the most stable upon acidification, although it could be the most hazardous due to generation of H_2S. Sulfide sludges are also susceptible to oxidation to sulfate when exposed to air, resulting in resolubilization of the metals (5).

Operational Considerations. Of the three precipitation processes, the simplest process to operate was the carbonate precipitation process. Once the soda ash dosage rate was established, the chemical feed pump is set according to the flow rate. Make-up of soda ash solution was straightforward and without complications. However, process control cannot be assured to meet varying influent concentrations and conditions.

The most reliable process to operate was the hydroxide precipitation process. Automatic lime addition by pH control insured sufficient conditions for precipitation, providing proper monitoring of reactor pH. Only daily cleaning of the pH electrode was required to remove precipitated solids that coated the outside of the probe, reducing sensitivity and pH controller response. Make-up of lime slurry from hydrated lime presented little problem under rapidly mixed conditions.

Sulfide precipitation was the most complex operation, principally due to ferrous sulfide make-up, which required special attention because of its instability and the potential H_2S hazard. Considerable care was required in the handling of sodium hydrosulfide to minimize the generation of H_2S and prevent operator exposure. Because of the susceptibility of the ferrous sulfide slurry to oxidation when exposed to air, special handling and frequent batching was required to maintain the reactive integrity of the slurry for precipitation of heavy metals.

ACKNOWLEDGMENT

These studies were conducted for the U.S. EPA under Contract No. 68-02-3195 for demonstration of these processes as applied to iron and steel industry wastewaters.

REFERENCES

1. Patterson, J.W., Allen, H.E., Scala, J.J., "Carbonate Precipitation for Heavy Metals Removal," Jour. Water Poll. Control Fed., 49(12), 2397 (1977).
2. Robinson, A.K. and Sum, J.C., "Sulfide Precipitation of Heavy Metals," EPA-600/2-80-139 (1980).
3. Bhattacharyya, D., Jumawan, A.B., and Sun, G., "Precipitation of Heavy Metals with Sodium Sulfide: Bench Scale and Full-Scale Experimental Results," presented at the AIChE 73rd Annual Mtg., Chicago, Ill. (1980).
4. Schlauch, R.M., and Epstein, A.C., "Treatment of Metal Finishing Wastes by Sulfide Precipitation," EPA-600/2-77-04 (1977).
5. Lanouette, K., "Heavy Metals Removal," Chem. Engrg., 84(22) 73 (1977).

PREDICTING THE PERFORMANCE OF A LIME-NEUTRALIZATION/
PRECIPITATION PROCESS FOR THE TREATMENT OF SOME HEAVY
METAL-LADEN INDUSTRIAL WASTEWATERS

A. R. Bowers, G. Chin, and C. P. Huang, Environmental
Engineering Program, Department of Civil Engineering,
University of Delaware, Newark, Delaware 19711

INTRODUCTION

The presence of various heavy metals in industrial waste streams may cause serious damage to the environment when these metals are not removed from the wastewaters, but discharged directly into the environment. Often, treatment of these wastes is considered too costly and where permitted, they are simply discharged into the sanitary sewers. While discharge into the municipal sewers may result in satisfactory dilution and/or partial removal, the subsequent incorporation of these metals into the sludge during biological treatment operations may render the waste sludge unfit for land application, thereby greatly complicating disposal practices. Therefore, some degree of pretreatment is deemed necessary prior to the discharge of heavy metal laden wastewaters into the municipal sewerage system.

The most widely used method for heavy metals removal from wastewaters is simply lime addition-hydroxide precipitation, which has been shown effective in experimental and actual systems for a number of years (1,2,4,8). In simple systems, the resulting metal concentration-pH behavior may be predicted using available thermodynamic information for metal hydroxide precipitation (4,7). However, in very heterogeneous systems, co-precipitation and complexation of more than one species must be considered (7).

This work was conducted to examine the performance of a simple lime addition-precipitation process with actual multi-component wastewaters, and to show how theoretical calculations might be of value in predicting this performance.

Wastewater Characteristics

Four wastewater streams were obtained from separate industrial metal finishing and reclamation plants for use in this study. These streams contain various strong acids, high concentrations of iron and a variety of heavy metals. A summary of these constituents are given in Table 1. Although the components vary from time to time, these characteristics are believed to be highly typical of the normal wastewater streams.

Table I. Wastewater Characteristics

Waste Stream	A	B	C	D
Color	grey (opaque)	aqua green	dark green	rusty, green
Acids present	H_2SO_4	80% HNO_3 20% HF	H_2SO_4	HCl
Initial pH	0.26	2.36	0.81	2.44
Acidity[a]	22.2%	12.9%	19.2%	2.2%
TOC[b]	84	21	52	140
Metals present (ppm)				
Arsenic[c]	11.1	--	--	--
Chromium[d]	--	1078	38.3	28.7
Copper	19.2	70.6	--	22.8
Iron	610	15763	78810	13390
Lead	2.9	--	--	--
Nickel	--	6053	--	168
Zinc	--	--	3950	--

a: % as $CaCO_3$ by titration to pH 8.3 by NaOH according to Standard Methods (6).
b: Total organic carbon (mg/ℓ).
c: Present as AsO_4^{-3}.
d: Present as Cr(III).

Effluent Requirements

The wastewater streams were to be treated and discharged into a sanitary sewer system. The requirements were to be met prior to discharge as established by the sanitary district, these are listed in Table II.

Table II. Effluent Requirements

Metal	Concentration Limit (ppm)
Arsenic	0.2
Chromium	5.0
Copper	0.2
Iron	3.0
Lead	0.2
Nickel	2.0
Zinc	1.0

Materials and Methods

Actual wastewaters were used in this study, as previously stated and characterized.

A high calcium hydrated lime obtained from the bulk commercial lot was applied to the wastewaters as a 10% slurry, to mimic the actual treatment operation.

The pH of all samples was measured with a Fisher model 425 pH meter equipped with a combination electrode.

All metals analyses were performed on a Perkin-Elmer model 560 Atomic Absorption Spectrophotometer which was also equipped with a graphite furnace.

Samples were filtered with glass fiber filters and acidified prior to residual metals analysis.

Experimental Procedure

Wastewater samples of 50 ml were collected from storage and treated in batch by adding an appropriate amount of lime slurry. The samples were then slowly stirred for 30 minutes, then filtered. The pH of the filtrate was measured prior to acidification and A-A analysis.

Alkalimetric Titration

A 50 ml sample of each of the four wastewaters was titrated with the 10% lime slurry in order to evaluate the effectiveness of the $Ca(OH)_2$ compared to the NaOH used for the acidity determinations. Also, since large amounts of SO_4^{-2} and F^- ions were present, insoluble $CaSO_4$ and CaF_2 were expected to form. Therefore, these simple titration curves offer necessary data for predicting the availability of Ca^{+2}, SO_4^{-2}, and F^- ions, during the treatment operation. The titration curves are shown in Figure 1.

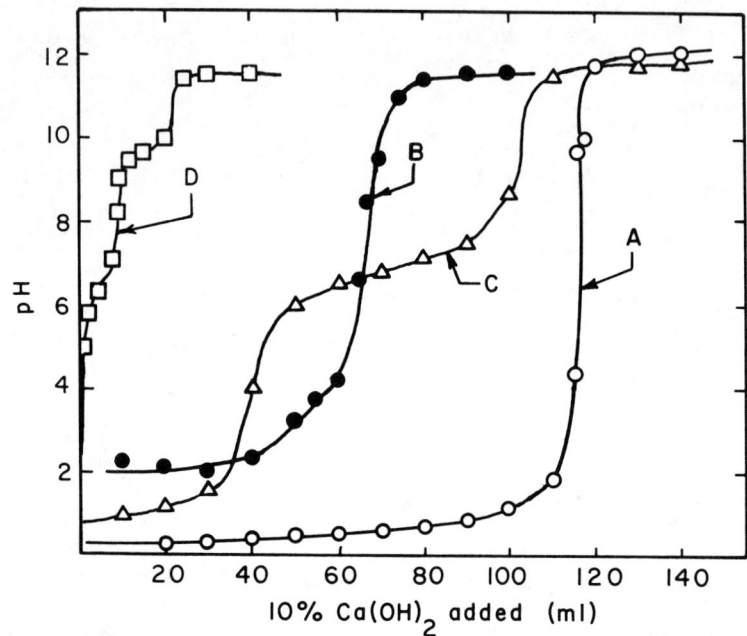

Figure 1. Lime titration curves for the wastewater streams used in this study.

Predicting Lime Addition Performance

The four wastewaters studied contained seven metals between them, as well as sulfate, nitrate, fluoride, and chloride ions which originated from the strong acids added during manufacturing operations. The addition of lime and the subsequent presence of calcium ions led to a variety of insoluble species, such as calcium sulfate, calcium fluoride and calcium arsenate. These were in addition to the various metal hydroxide precipitates formed as the pH was raised due to lime addition. It may be stated however, that lime addition is advantageous to NaOH addition even though more solids and resulting sludge were formed. The precipitation of SO_4^{-2} and F^- greatly decreased their availability to form metal complexes and thus facilitated the precipitation of metal hydroxides. Also, removal of the arsenate ion as $Ca_3(AsO_4)_2$ is an obvious advantage.

The pertinent chemical reactions and solubility products for these systems are as follows:*

(i) *sulfates*:

$$CaSO_4(s) \rightleftharpoons Ca^{+2} + SO_4^{-2} \; ; \; \log K_{so} = -2.9$$

or $\quad [Ca^{+2}][SO_4^{-2}] = 1.3 \times 10^{-3}$ (1)

$$PbSO_4(s) \rightleftharpoons Pb^{+2} + SO_4^{-2} \; ; \; \log K_{so} = -6.2$$

or $\quad [Pb^{+2}][SO_4^{-2}] = 6.3 \times 10^{-7}$ (2)

Sulfuric acid is a diprotic acid having one strong and one relatively weak dissociation constant ($pK_2 = 1.1$) and the fraction of SO_4^{-2} ions ($\alpha\, SO_4^{-2}$) is given by the expression:

$$\alpha\, SO_4^{-2} = \frac{K_2}{[H^+] + K_2} \quad (3)$$

(ii) *fluorides*:

$$CaF_2(s) \rightleftharpoons Ca^{+2} + 2F^- \; ; \; \log K_{so} = -10.4$$

or $\quad [Ca^{+2}][F^-]^2 = 4.0 \times 10^{-11}$ (4)

Hydroflouric acid is a weak acid ($pK_1 \approx 1.0$) and the fraction of F^- ions ($\alpha\, F^-$) is expressed as:

$$\alpha\, F^- = \frac{K_1}{[H^+] + K_1} \quad (5)$$

(iii) *arsenate*:

$$Ca_3(AsO_4)_2(s) \rightleftharpoons 3Ca^{+2} + 2AsO_4^{-3} \; ; \; \log K_{so} = -16.0$$

or $\quad [Ca^{+2}]^3 [AsO_4]^2 = 1.0 \times 10^{-15}$ (6)

---

*All of the solubility products and complex stability constants used were reported at 1M ionic strength and 25°C. In general activity coefficients and therefore these constants change very little between about 0.5 to 2.0 M and the use of 1M values should remain applicable for each wastewater studied. The activity of the H_3O^+ ion where needed, was assumed to be 0.8 (3).

Since arsenate is a weak triprotic acid ($pK_1 = 2.3$, $pK_2 = 6.4$, $pK_3 = 9.2$), the fraction of AsO_4^{3-} ($\alpha_{AsO_4^{3-}}$) is:

$$\alpha_{AsO_4^{-3}} = \frac{K_1 K_2 K_3}{[H^+]^3 + K_1[H^+]^2 + K_1 K_2 [H^+] + K_1 K_2 K_3} \quad (7)$$

(iv) *hydroxides*

All of the metal pollutants present may be precipitated out of solution as oxides or hydroxides, with the exception of arsenic. Pertinent reactions and solubility constants are given in Table III for all of the metals studied.

Table III. Solubility Constants for Metal Oxides and Hydroxides (5,7)

Metal	Reaction	Constant (Log values)
Chromium	$Cr(OH)_3(s) + 3H^+ \rightleftharpoons Cr^{+3} + 3H_2O$ $Cr(OH)_3(s) + 2H^+ \rightleftharpoons CrOH^{+2} + 2H_2O$ $Cr(OH)_3(s) + H^+ \rightleftharpoons Cr(OH)_2^+ + H_2O$ $Cr(OH)_3(s) + H_2O \rightleftharpoons Cr(OH)_4^- + H^+$	$*K_{so} = 12.2$ $*K_{s1} = 5.4$ $*K_{s2} = 2.1$ $*K_{s4} = -15.2$
Copper	$CuO(s) + 2H^+ \rightleftharpoons Cu^{+2} + 2H_2O$ $CuO(s) + H^+ \rightleftharpoons CuOH^+$ $CuO(s) + 2H_2O \rightleftharpoons Cu(OH)_3^- + H^+$	$*K_{so} = 7.7$ $*K_{s1} = -0.3$ $*K_{s3} = -19.1$
Iron	$Fe(OH)_3(s) + 3H^+ \rightleftharpoons Fe^{+3} + 3H_2O$ $Fe(OH)_3(s) + 2H^+ \rightleftharpoons FeOH^{+2} + 2H_2O$ $Fe(OH)_3(s) + H^+ \rightleftharpoons Fe(OH)_2^+ + H_2O$ $Fe(OH)_3(s) + H_2O \rightleftharpoons Fe(OH)_4^- + H^+$	$*K_{so} = 3.3$ $*K_{s1} = 0.5$ $*K_{s2} = -2.6$ $*K_{s4} = -18.5$
Lead	$PbO(s) + 2H^+ \rightleftharpoons Pb^{+2} + H_2O$ $PbO(s) + 2H_2O \rightleftharpoons Pb(OH)_3^- + H^+$	$*K_{so} = 12.7$ $*K_{s3} = -15.4$
Nickel	$Ni(OH)_2(s) + 2H^+ \rightleftharpoons Ni^{+2} + 2H_2O$ $Ni(OH)_2(s) + H^+ \rightleftharpoons NiOH^+ + H_2O$ $Ni(OH)_2(s) + H_2O \rightleftharpoons NI(OH)_3^- + H^+$	$*K_{so} = 13.5$ $*K_{s1} = 2.5$ $*K_{s3} = -18.0$
Zinc	$ZnO(s) + 2H^+ \rightleftharpoons Zn^{+2} + H_2O$ $ZnO(s) + H^+ \rightleftharpoons ZnOH^+$ $ZnO(s) + 2H_2O \rightleftharpoons Zn(OH)_3^- + H^+$	$*K_{so} = 11.2$ $*K_{s1} = 2.2$ $*K_{s3} = -16.9$

When the chemical equations are stated in the form as shown in Table III, an equation for any metal hydroxide species can be written in a linear form as:

$$-\log[Me(OH)_n^{m-n}] = (m-n)pH + p^*K_{sn} \qquad (8)$$

where: m = charge on base metal ion
n = number of hydroxides associated per metal ion (n = 0, 1, 2, 3)

(v) *Other Complex Species*

Any prediction of how well these metals will be removed by lime addition must also consider the solubilization which might occur due to complex formation with the anionic species present due to the various acids which are available.

The stability constants for metal-ligand complex formation are listed in Table IV.

Table IV. Stability Constants for Metal-Ligand Complexes Which May Occur in the Wastewater Streams[a] (3,5)

Metals Anions	Fe^{+3}	Ni^{+2}	Pb^{+2}	Zn^{+2}	Cu^{+2}
SO_4^{-2}	2.0 2.1				
HSO_4^-	0.8				
Cl^-	0.6 0.8	-0.6	0.9	-0.2	-0.1
F^-	5.2 4.0	0.5	1.4	0.8	0.9

a: Constants are for 1M ionic strength, in log values, and listed vertically as ML_1, ML_2.

A general equation for concentration of each metal-ligand species may be written as:

$$[MeL_n^{m+x}] = [Me^m][L^x] \cdot K_n \qquad (9)$$

where: m = charge on the base metal ion,
x = charge on the base ligand ion, and
n = number of ligands attached.

RESULTS AND DISCUSSION

The theoretical concentrations of these metals were calculated from pertinent solubility and complexation reactions for each wastewater stream. Calculations were done with simple iterations using a microprocessor in basic computer language and based on data collected from the alkametric titration curves presented in Figure 1.

(i) *Stream* A

The metals in this wastewater stream were considered removable as oxides or hydroxides, i.e. PbO(s), CuO(s) and Fe(OH)$_3$(s), with the exception of the anionic arsenic, which was removed as calcium arsenate (Ca$_3$(AsO$_4$)$_2$(s)). Lead sulfate was also considered along with calcium sulfate as other insoluble species. The total sulfuric acid concentration was estimated from the initial pH and Equation 3.

The calculated soluble species are all included in Figure 2 and some batch data points are also shown. From Figure 2, the results indicate that theoretically a pH of 9.0 ± 0.2 will

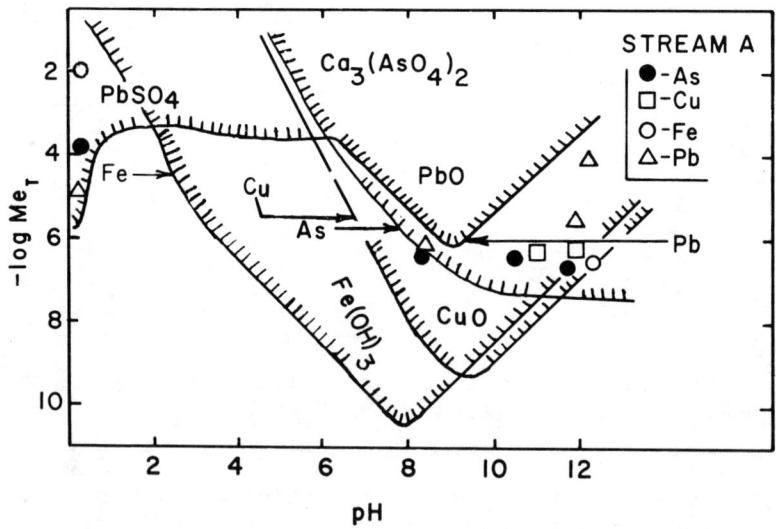

Figure 2. Predicted and experimental data for lime-addition treatment of wastewater A (Arrows indicate effluent requirements, Table 2).

be required to meet all of the effluent guidelines shown by the horizontal arrows. Although the data points do not necessarily correspond with the theoretical predictions, the predicted pH 9.0 for treatment operation appears adequate to meet the effluent regulations. It is interesting to note that theory predicts lead sulfate to be more soluble in basic than in acidic solution. In fact, the data showed more lead at the higher pH's than was present in the initial sample which was very acidic.

(ii) *Stream B*

The theoretical and experimental results for wastewater B are shown in Figure 3. The hydrofluoric acid concentration was estimated assuming the metals were solubilized from oxides, Equation 5, and knowing the acid used was 20% HF. Calcium fluoride was extremely insoluble so sufficient fluorides for complex formation occured only briefly for the first small additions of lime, i.e. pH < 3. The predicted requirement for treatment is 9.0 < pH < 11 to meet the indicated guidelines. The data do indeed agree well with the prediction. It is to be noted that the data for copper are not included in Figure 3 since the residual concentrations were less than detectable, as would be expected by the theoretical curve.

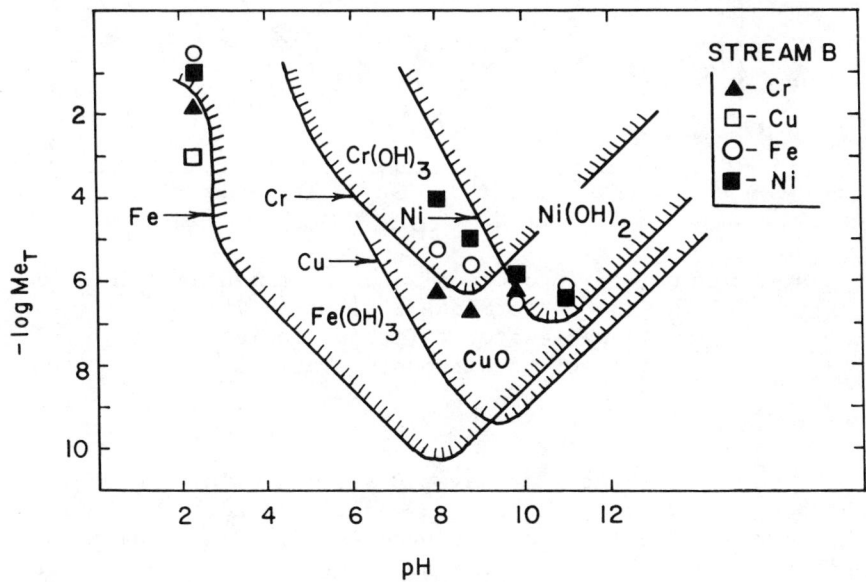

Figure 3. Theoretical and experimental values for the performance of lime-addition with wastewater B (Arrows indicate effluent requirements, Table 2).

(iii) *Stream C*

The results for wastewater C are shown in Figure 4. The sulfuric acid concentration was estimated from the metals initially present and Equation 3. Calcium sulfate precipitation was considered along with $Cr(OH)_3(s)$, $ZnO(s)$ and $Fe(OH)_3(s)$. Satisfactory performance is predicted at 7.8 < pH < 11.0. In fact, however, no detectable chromium or zinc were found in the experimental samples, possibly due to adsorption of these species onto the abundant iron hydroxide particles (7.9 g/ℓ).

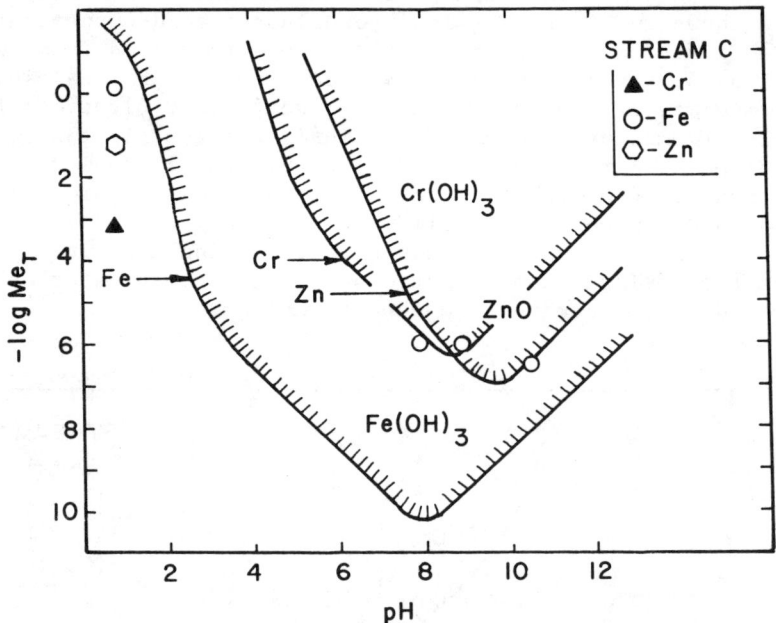

Figure 4. Theoretical and experimental values for the performance of lime-addition with wastewater C (Arrows indicate effluent requirements, Table 2).

(iv) *Stream D*

The results for wastewater D are shown in Figure 5. The hydrochloric acid concentration was estimated from the metals which were solubilized during processing. The predicted treatment pH is 9 < pH < 11. Once again, the experimental data falls within the effluent guidelines within this pH range.

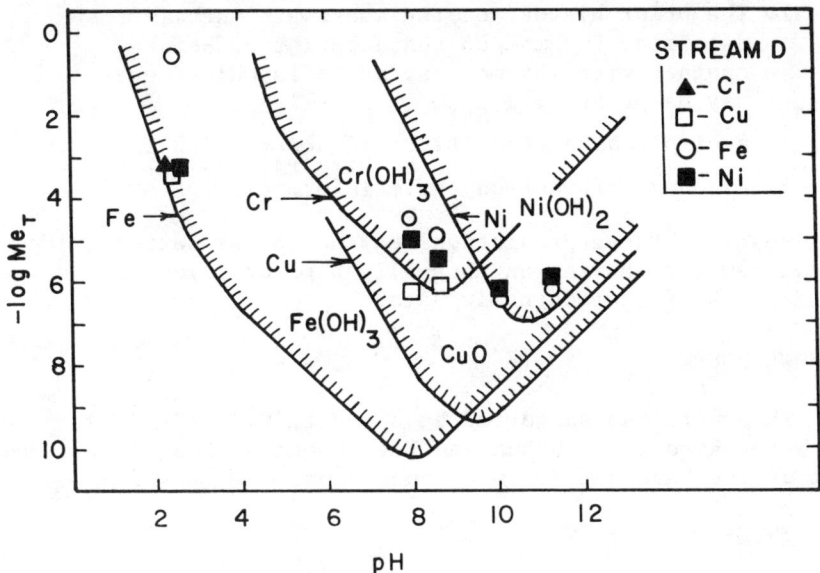

Figure 5. Theoretical and experimental values for the performance of lime-addition with wastewater D (Arrows indicate effluent requirements, Table 2).

EXPERIMENTAL DISCREPANCIES

Although effluent requirements may be met by operating at theoretically predicted pH's, many discrepancies between theoretical and experimental conditions do exist. Some of the soluble concentrations of heavy metals were found to be larger than theoretically predicted. This can be attributed to the presence of organics (reported as TOC) in the wastewaters that form complexes with the metals, thereby increasing the metal solubility. The fact that increased copper concentrations occurred in the waste streams with the highest TOC values, in wastewaters A and D, tends to support this hypothesis. Moreover, some of the iron may have existed initially as Fe(II), and although the presence of O_2 along with OH^- ions result in oxidation of Fe(II), this reaction may not have been completed due to poor aeration over the course of these experiments.

CONCLUSIONS

In general, theoretical predictions can be used to evaluate the treatability of heavy metal laden wastewaters by lime addition. Very little advance quantitative information is needed for successful predictions, but only the following are needed:

1. The prior history of the wastewater stream to determine the makeup characteristics before contact with the metals, and also which metals may be present;

2. a simple alkametric titration curve with lime;

3. a source of thermodynamic information.

However, these predictions may not be as satisfactory for wastewaters having an unusually high organic content, due to the lack of thermodynamic data.

ACKNOWLEDGMENT

This work was supported by a grant, ENG. 77-27379, from the Water Resources, Urban and Environmental Engineering Program of the National Science Foundation.

REFERENCES

1. J. G. Dean, F. L. Bosque, and K. H. Lanouette, "Removing Heavy Metals from Wastewater," Envir. Sci. and Tech., 6, 518 (1972).

2. Y. Inoue and M. Munemori, "Co-precipitation of Mercury(II) with Iron (III) Hydroxide," Envir. Sci. and Tech., 13, 443 (1979).

3. A. E. Martell and R. M. Smith, "Critical Stability Constants," Plenum Press, New York, NY (1974).

4. J. W. Patterson, H. E. Allen, and J. J. Scala, "Carbonate Precipitation for Heavy Metals Pollutants," J. Water Pol. Control Fed., 49, 2397 (1977).

5. L. G. Sillen and A. E. Martell, "Stability Constants," Special Publication No. 17, The Chemical Society, London (1964).

6. "Standard Methods for the Examination of Water and Wastewater," (14th Ed.) Amer. Public Health Assn., Wash., D.C. (1975).

7. W. Stumm and J. J. Morgan, "Aquatic Chemistry," Wiley-Interscience, New York, NY (1970).

8. M. J. Thomas and T. L. Theis, "Effects of Selected Ions on the Removal of Chrome (III) Hydroxide," J. Water Pol. Control Fed., 48, 2032 (1976).

DESIGN OF SOLUBLE SULFIDE PRECIPITATION SYSTEM FOR
HEAVY METALS REMOVAL

James S. Whang. AEPCO, Inc., Rockville, Maryland

Daniel Young. JRB Associates, McLean, Virginia

Maurice Pressman. Army MORADCOM, Fort Belvoir, Virginia

INTRODUCTION

As part of the Nation's effort to protect the environment from pollution, the U.S. EPA has begun to control industrial dischargers of wastewater to publicly owned treatment works (POTWs). Through the National Industrial Wastewater Pretreatment Regulations and Program under the Clean Water Act of 1977, EPA established a list of 34 categorical industries, for which national industrial wastewater pretreatment standards were to be promulgated. Among the 34 categorical industries, the electroplating industry is the first one, for which national pretreatment standards have been promulgated.

In order to comply with applicable wastewater pretreatment standards, electroplaters which discharge to POTWs must pretreat wastewater. There are several treatment technologies available for treating electroplating wastewater, each with its own advantages and disadvantages. This paper examines the design aspects of a pilot treatment plant which uses the soluble sulfide precipitation process to remove heavy metals from an electroplating wastewater.

NATURE OF THE PROJECT

Tobyhanna Army Depot in Tobyhanna, Pennsylvania ownes and operates an electroplating facility. Wastewater from the plating shop has been discharged in combination with other wastewaters from the Army Depot to a trickling filter plant. Characteristically high concentrations of metals in the wastewater have frequently caused upset of the trickling filter, and have contamined the plant's sludge. This has caused a need for special sludge disposal arrangements. Under the funding by the Army Mobility Equipment Research & Development Command (MORADCOM), this industrial wastewater was characterized and a treatability study was performed in order to develop and evaluate alternative treatment methods. These study efforts concluded that the soluble sulfide precipitation process would be most cost-effective and acceptable based on the following reasons.

- Low solubility of metal sulfides improves metal removal efficiency over the conventional hydroxide process.

- Metal sulfide precipitates are very fine. However, this problem can be effectively overcome by adding anionic flocculant to facilitate liquid-solids separation

- Metal sulfide sludge exhibits better thickening properties and dewaterability than metal hydroxide sludge from the hydroxide precipitation process.

- Metal sulfide sludge is three times less subject to leaching at pH=5 as compared to hydroxide sludge and therefore final disposal is easier and safer.

The Tobyhanna plating facility generates basically three wastewater streams. It is necessary to segregate these wastewater streams in order to effectively utilize the soluble sulfide process. These three wastewaters are (1) cyanide-bearing rinse water, (2) chromium-bearing rinse water, and (3) other alkaline and acid rinse waters. Substances present in each of the three wastewaters are as follows:

- Cyanide-bearing wastewater: cadmium, copper and cyanide.

- Chromium-bearing wastewater: sodium, chromate, and additives

- Other alkaline and acid wastewaters: sodium, nickel,

aluminum, tin, iron, lead, zinc, chloride, sulfate, nitrate, phosphate, and other additivies.

In addition to these three streams of continuous-flow wastewaters, the plating facility dumps spent solutions of various compositions periodically. The frequency of dumping ranges from once every month to once a year. These spent solutions have to be bled into the proposed pretreatment system at a slow, controlled, rate.

SOLUBLE SULFIDE TREATMENT PROCESS AND DESIGN CRITERIA

In order to effectively use the soluble sulfide precipitation process to remove metals, three prerequisite treatment steps are needed.

(1) Cyanide in the cyanide-bearing wastewater must be oxidized by chlorination to carbon dioxide and nitrogen gases, as shown by the following equations. This is necessary to prevent complexing of cyanide with metal ions in the subsequent treatment processes.

$$2NaCN + 2NaOCl \rightarrow 2NaCNO + 2NaCl \quad \text{(1st Stage)} \tag{1}$$

$$2NaCNO + 3NaOCl + H_2O \rightarrow 2CO_2\uparrow + N_2\uparrow + 2NaOH + 3NaCl \quad \text{(2nd Stage)} \tag{2}$$

(2) Chromium (VI) in the chromium-bearing wastewater must be reduced by sodium metabisulfite to chromium (III) as shown by the following equations to facilitate subsequent chromium precipitation by sulfide:

$$Na_2S_2O_5 + H_2O \rightarrow 2NaHSO_3 \tag{3}$$

$$2H_2CrO_4 + 3NaHSO_3 + 3H_2SO_4 \rightarrow Cr_2(SO_4)_3 + 3NaHSO_4 + 5H_2O \tag{4}$$

(3) Neutralization of combined flows of all three wastewater streams.

The neutralized wastewater is then treated with sodium bisulfide to form metal sulfide precipitates. The reactions for 2 valence metals and 3 valence metals are shown, respectively, as follows:

$$M^{+2} + HS^- + OH^- \rightarrow \underline{MS} + H_2O \tag{5}$$

$$2M^{+3} + 3HS^- + 3OH^- \rightarrow \underline{M_2S_3} + 3H_2O \tag{6}$$

Table I Design Criteria

Treatment Unit Process	Design Criteria*
Cyanide Oxidation by Chlorination	
First Stage	pH>10.5; 350mv<ORP<400mv; and t> 5 min
Second Stage	pH> 8.0; 580mv<ORP<620mv; and t= 0.5-1 hr
Chromium Reduction by Sodium Metabisulfite	pH< 3.0; 250mv<ORP<300mv; and t= 60 min
Soluble Sulfide Precipitation of Metal Ions	pH= 8.0; excess free sulfide ion > 1.0mg/l; and t = few seconds.

*ORP = Oxidation Reduction Potential in Milivolts.
 t = React Time.

Based on the results of the treatability study, the design criteria were developed and summarized in Table I.

DESIGN INPUT CONDITIONS AND EFFLUENT QUALITY REQUIREMENTS

The plating facility operates on 8-hour shift a day, 5 days a week. The design conditions for the treatment system and effluent quality requirements are summarized in Table II.

Table II Design Conditions and Effluent Quality Requirements

Wastewater Source	Daily Flow @ 8 Hrs/Day
Chromium-Bearing Wastewater	2,000 gpd
Cyanide-Bearing Wastewater	4,000 gpd
Acid/Alkaline Wastewater	12,000 gpd
TOTAL =	18,000 gpd

Characteristics of Wastewater Composite and Effluent Quality

Pollutant	Influent (mg/l)	Effluent Standard (mg/l)
Cu	17.0	1.8
Ni	5.2	1.8
Cr	42.0	2.5
Zn	2.6	1.8
Pb	4.4	0.3
Cd	5.9	0.5
Sn	5.0	1.0
Al	10.0	0.5
Fe	10.0	---
Total Metal	102.1	5.0
Cyanide	100.0	0.23

TREATMENT FACILITY DESIGN

Based on the design criteria derived from the treatability study, a continous flow-through treatment system was designed. The treatment process flow diagram is presented in Figure 1. The liquid process stream consists of two separate flow equalization sumps for chromium and cyanide wastewaters respectively; a 4-compartment cyanide oxidation unit which uses chlorination; a 2-compartment chromium reduction unit which uses sodium metabisulfite; an equalization basin to combine all three wastewater streams; a neutralization unit; a 2-compartment sulfide addition tank with pH adjustment; a flash mix tank; a flocculation tank; a laminar clarifier; a polishing filter; and a clear well equipped with a hydrogen peroxide feed system to oxidize excess residual sulfide. Polishing filter backwash water is diverted back to the equalization

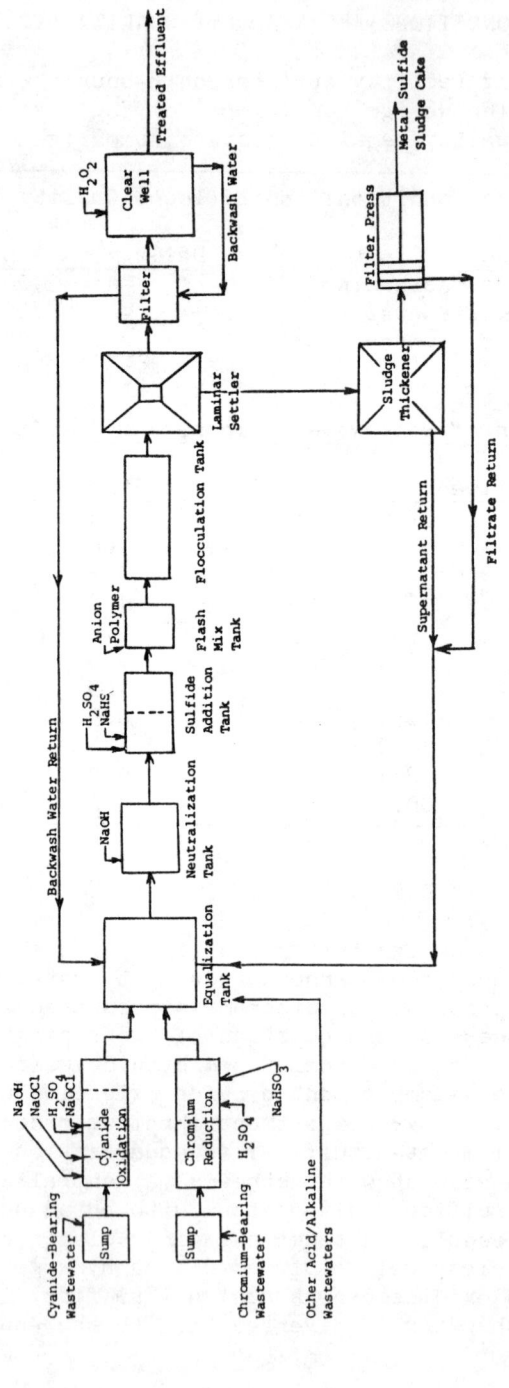

Figure 1 Treatment Process Flow Diagram (Soluble Sulfide Precipitation Process)

basin for further treatment.

The sludge processing stream consists of a gravity sludge thickener and a frame-type filter press. Both thickener supernatant and filter press filtrate are returned to the equalization basin for further treatment.

Table III summarizes major process design specifications of the system. Table IV summarizes the anticipated chemical consumption rates. The entire treatment system was designed to fit into a space 15 feet wide, 65 feet long, and 13 feet high.

SAFETY FEATURES OF THE DESIGN

There were a number of design features which were included based on personnel saftey considerations. These safety features included:

- A continuous ventilation system for the cyanide tank, the chromium tank, and the sulfide tank to exhaust toxic gas such as chlorine, hydrogen cyanide, and hydrogen sulfide which could occur due to abnormal operating conditions.

- Process control devices which are equipped with alarms to warn personnel of conditions conducive to hazards such as low pH in the cyanide tank, high and low pH in the neutralization tank, and low pH in the sulfide tank. There are also water level controllers and alarms.

- Oxygen masks, all purpose fire extinguishers, eye washers, emergency showers, automatic sprinkler system, and emergency exit markings.

CONCLUSIONS

Based on a comprehensive wastewater characterization program and a treatability study, the soluble sulfide precipitation process was selected for heavy metals removal from electroplating wastewater at the Tobyhanna Army Depot. This process is advantageous in terms of metals removal efficiency, solids and liquid separation, sludge thickening capability and dewaterability, and sludge stability for disposal by landfill. The treatment plant discussed in this paper is scheduled to be operational by September, 1981. Operating data will be presented, when it becomes available.

Table III Process Design Specifications

Unit Process	Liquid Volume	Detention Time	Other Pertinent Conditions
Cyanide Oxidation			
Stage I	125 gal	14 min @ 8.4 gpm	pH>11.0; 350mv<ORP<400mv
Stage II	505 gal	60 min @ 8.4 gpm	7.8<pH<8.2; 580mv<ORP<620mv
Chromium Reduction	420 gal	100 min @ 4.2 gpm	pH<2.5; 250mv<ORP<300mv
Neutralization	450 gal	12 min @37.5 gpm	7.8<pH<8.2
Sulfide Addition	450 gal	12 min @37.5 gpm	7.8<pH<8.2; free S$^=$ ion >1 mg/l
Flash Mix	50 gal	80 sec @37.5 gpm	Anion Flocculant Conc. = 10 to 15 mg/l
Flocculation	260 gal	6.9min @37.5 gpm	-------
Laminar Settling	2,400 gal	64 min @37.5 gpm	Surface Loading = 0.63 gpm/ft Tube Surface Loading Rate = 0.114 gpm/ft
Polishing Filter	N.A.	N.A.	Surface Area = 24 ft Loading Rate = 2.08 gpm/ft
Sludge Thickener	2,200 gal	4 days @ 540 gpd Sludge Flow	Continuous Decant
Filter Press	N.A.	N.A.	2 Cu.Ft. Capacity/Cycle

Table IV Chemical Consumption Rates

Unit Process	Chemical & Consumption Rate		
Cyanide Oxidation	NaOCl	100 gal/wk @ 15% Sol.	
	NaOH	25 gal/wk @ 10% Sol.	
	H_2SO_4	50 gal/wk @ 10% Sol.	
Chromium Reduction	$NaHSO_3$	55 gal/wk @ 3% Sol.	
	H_2SO_4	20 gal/wk @ 25% Sol.	
Neutralization	NaOH	60 gal/wk @ 40% Sol.	
Sulfide Addition	NaHS	100 gal/wk @ 20% Sol.	
pH Adjustment for Sulfide Process	H_2SO_4	70 gal/wk @ 25% Sol.	
Flocculation	Anionic Polymer	50 gal.wk @ 4% Sol.	
Oxidation of Residual Sulfide in Effluent	H_2O_2	Minimum	

THE REMOVAL OF HEAVY METALS FROM SEWAGE INFLUENT WATERS BY FOAM FLOTATION.

Chaiyuth Chavalitnitikul. Environmental Studies Institute, Drexel University, Philadelphia, Pa.

Richard L. Brunker, Department of Biological Sciences, Environmental Studies Institute, Drexel University, Philadelphia, PA.

INTRODUCTION

The presence of compounds or ions of heavy metals in municipal sewages has long been recognized. These metals occur in both sanitary wastewaters, storm sewage, and in combined sewer flows [1]. Heavy metals commonly found in sewages include: aluminum, arsenic, cadmium, chromium, copper, nickel mercury, lead, manganese, molybdenum, selenium, and zinc.

There are a number of natural and synthetic organic and inorganic complexing agents available to react with aqueous metal ions in addition to the solvent water. Natural water and wastewater systems frequently contain several important inorganic complexing ligands, such as Cl^-, NH_3, SO_4^{2-}, PO_4 and OH^- species. In much the same manner that reactions between metal ions and water yield various soluble and insoluble species, a similar suite of soluble ions and insoluble phases may result from reactions with inorganic anions, depending on the metal concentration, ligand concentration, and pH. A large variety of natural and synthetic organic matter is also present in natural waters and wastewaters. This includes the natural degradation products of plant and animal tissue, e.g. amino acids and humic acids, and organic species derived from chemicals applied either deliberately or inadvertantly by man. Examples of the latter case are detergents and cleaning agents, NTA, EDTA, pesticides, ionic and nonionic surfactants with various functional groups, and a large group of synthetic macrocyclic compounds. Metals are bonded to organic matter by way of (i) carbon groups yielding organometallic compounds, (ii) carboxylic group producing salts of organic acids, (iii) electron-donating atoms, O, N, S, P, etc., forming coordination complexes, or (iv) Π- electron-donating arrangements (olefinic bonds, aromatic ring, etc.) [70, 71].

Increasingly concentrated heavy metal loadings have been introduced into many municipal sewage treatment facilities as a direct result of extending service to the industrial community [2]. Although industry can be the major source of heavy metals in municipal sewage, it is not the only source. Several types of commercial establishments release heavy metals in significant amounts. In addition, it has been determined that wastewaters of purely domestic origin contained measurable and possibly significant concentrations of those heavy metals. Heavy metals in domestic wastewater are normally from water supply, human waste, food materials, nonfood household commodities, and corrosion and solution of pipes [1, 3].

The discharge of wastewater and disposal of sewage sludges which contain considerable amounts of toxic metals have been of great concern to both regulatory agencies and dischargers in recent years. Although some of the metals are essential dietary elements, others have been recognized as posing a major health and environmental problem because they can be harmful to plants, animals, and aquatic life [3, 19, 20]. Presently specific information concerning bioaccumulation phenomenon of heavy metals in plants is not fully understood. There is evidence indicating that bioaccumulation of toxic heavy metals differs considerably in crops such as lettuce, corn, potatoes, beets, spinach and soybeans. These common farm products have demonstrated a tendency to accumulate considerable amounts of elements, particularly cadmium, copper, lead, nickel, and zinc [21, 39]. Because of its extremely long retention time and considerable toxicity, cadmium is the greatest concern and has been intensely studied for its bioaccumulation in plants and its subsequent movement to human food chain [18, 31-39]. Some recent experiments have established that plants absorb substantial amounts of cadmium from Cd-amended or contaminated soils and that the levels of cadmium accumulation by plants depends upon plant species as well as substrate Cd levels and soil type [32-38].

There are considerable variations of metal removal efficiency of conventional secondary treatment processes. These range from zero to greater than 90 percent depending upon the metal, its concentration, the characteristics of the wastewater, the plant design, and how it is operated. High concentrations of metals often have adverse affects on biological treatment processes [2, 4-17].

The disposal of municipal wastewater sludges has become a source of public concern. Amounts of sludges produced at the U.S. municipal wastewater treatment facilities is increasing as the number of facilities converting to secondary treatment are incremented and expanded. An estimated 8 billion dry Kg. are expected to be produced annually by the early 1980s. The current national breakdown of municipal wastewater sludge disposal is estimated to be 35% by incineration, 15% by ocean disposal, 25% by landfill, and 25% by land application. Incineration is energy dependent and causes air pollution problems. Ocean dumping will be prohibited after 1981. Land application is being considered as a feasible alternative despite questions concerning its impact on the biota and the food chains [18].

Questions concerning the potential human health and environmental problems may cause regulatory agencies to adopt more stringent requirements for the effluent and sludge metal concentrations in the future [40]. If this occurs biological treatment processes will require supplemental, effective, and reliable procedures for the removal of these toxic metals.

Currently, numerous techniques exist to remove metal ions from aqueous solutions. These include chemical precipitation as insoluble salts, hydroxides, oxides, ion exchange, adsorption on activated carbon, distillation, reverse osmosis, electroosmosis, solvent extraction, and foam flotation [41]. Foam flotation appears to have some distinct advantages over the others when dealing with large volumes of waste containing heavy metals in the parts per billion or parts per million ranges. This method appears to be applicable in the removal of trace metals from the sewage influents providing for the protection of the aquatic environment, particularly potable water sources, from heavy metal contamination and preventing the accumulation of these metals in sludges. Residues, from this flotation process must also be studied to determine its characteristics and recommended method of disposal.

BACKGROUND

The use of foam flotation techniques for the removal of heavy metals from aqueous systems is well documented [42-47].

Karger et al [48], and Pinfold [49] have proposed nomenclature for the entire field of adsorptive bubble separation techniques. Foam separation relies upon a surfactant that causes a nonsurface-active material to become surface active forming a product that is then removed by bubbling a gas through the bulk solution to form a foam. The foam separations of interest here are ion flotation, precipitate flotation, and adsorbing colloid flotation.

Ion flotation is a process that involves the removal of a nonsurface-active ion (a colligend) by the addition of a surfactant (a collector) which yields an insoluble product. Ion flotation requires stoichiometric or greater amounts of the collector since the ion and the collector actually form a compound. Precipitate flotation is the process in which the nonsurface-active material forms a precipitate with something other than the collector, which in turn made surface active by the collector. Adsorbing colloid flotation is defined as the removal of dissolved material by adsorption on colloidal particles followed by the removal of colloid particle, dissolved material, and the collector by flotation. Precipitate and adsorbing colloid flotation do not require stoichiometric amounts of the collector since only electrical attraction exists between the collector and the precipitate but sufficient surfactant must be used to form stable foam [48-50].

Rubin and his colleagues compared precipitate and ion flotation of several metal ions [51-55]. Their investigations concerned the efficiencies of lead (II), copper (II), iron (III) and zinc (II) from aqueous solutions using sodium lauryl sulfate as primary collector and ethyl alcohol as the frothing agent. The results showed that the factor that determined whether the removal process is precipitate flotation or ion flotation depended on the nature of the metal-collector product, the sublate. Factors affecting the solubility of the sublate, and hence the mechanism of removal, included the coordination behavior of the metal towards the collector, the nature of the collector, and the pH value of the solution. Other factors that may have had a significant effect on the removal process included the rate of gas flow, the concentrations of the metal and the collector, and the ionic strength of the solution. Sebba et al [56-58] investigated the separation of strontium and aluminum from aqueous solutions by ion flotation in which α-sulphopalmitic acid and long-chain fatty amines (especially tetradecylamine) were used as collectors for strontium and aluminum removal respectively. They reported that approximately 99.9% of strontium and 90% of aluminum were removed. Grieves et al [59-62] devoted much effort to applying flotation techniques to waste treatment problems. They studied both dispersed-air and dissolved-air ion flotation of dichromate ion using a cationic surfactant, ethylhexadecyl-dimethylammonium (EHDA) bromide. A nonionic polymer was also used in dissolved-air ion flotation as a flocculant aid. They determined that on the basis of batch and continuous experiments the ion flotation of dichromate appeared to be a feasible process for the separation and concentration of hexavalent chromium from aqueous solution. The primary advantage of the dissolved-air ion flotation process was found to be the concentration of dichromate, surfactant, and polymer in a small liquid volume of collapsed foam which was less than 1% of the total feed throughput. Foam concentrations were of the order of 10,000 to 20,000 mg/liter of dichromate. Kobayashi [63-64] published the effects of anionic and cationic surfactants on the ion flotation of copper(II) and cadmium(II). He reported that sodium α-sulfolaurate was determined to be most effective for the flotation of both Cu^{2+} and Cd^{2+} ions achieving as high as a 95% and 97% removal respectively. About 97% of Cd^{2+} ions could be floated at a pH of 11.3 when cationic surfactant was used with bentonite, but little Cu^{2+} was removed. Kobayaski [65] also investigated the removal of Cd^{2+} ions from an aqueous solution by adsorbing particle flotation using bentonite and a cationic surfactant, hexa-decyltrimethylammonium chloride. It has been found that the most suitable method was to conduct flotation in the region of the coagulation flotation of bentonite.

Huang and Wilson [66] studied precipitate flotation and adsorbing colloid flotation of mercury(II) and cadmium(II) from aqueous systems. Sodium lauryl sulfate and hexadecyltrimethylammonium bromide (HTA) were used as collectors. Fe(OH)$_3$, Al(OH)$_3$, FeS, and CuS were used as adsorbing colloids. They found that floc foam flotation of both metals with CuS and HTA was very effective resulting in residual Hg(II) levels as low a 5 ppb and residual Cd(II) levels as low as 20 ppb. Floc foam flotation of Cd(II) with FeS and HTA yielded residual Cd(II) levels as low as 10 ppb. Ferguson et al [50] investigated the foam separation techniques such as ion flotation, precipitate flotation, and adsorbing colloid flotation to determine if these procedures would be feasible for removing lead(II) and cadmium(II) from highly contaminated waste water. They reported that increased ionic strength, calcium (II), and phosphate interference made ion flotation impractical. Precipitate flotation offers a great improvement over ion flotation, but this process would not reduce the lead(II) or cadmium(II) in solution to levels that are considered safe for discharge into streams and water ways. Adsorbing colloid flotation gave excellent results. Lead sulfide and cadmium sulfide were adsorbed to ferrous sulfide which was then removed by foaming with hexadecyltrimethylammonium bromide. Zeitlin et al [67-70] investigated a host of adsorbing colloid flotation separations including molybdate, uranyl, zinc, and copper with Fe(OH)$_3$ and sodium dedocyl sulfate, to name a few, from sea water. They found that the average recovery of uranium and molybdenum from sea water were 82% and 95.3% (S.D. + 2.6%) respectively after addition of ferric hydroxide and cationic surfactant, dodecylamine, and air, were 94.0% (S.D. ± 6.1%) and 95.0% (S.D. ± 6.1%) respectively. Thackston et al [41] also investigated the adsorbing colloid flotation of lead(II) using sodium lauryl sulfate as a collector, and ferric chloride (FeCl$_3$) to form a precipitate of ferric hydroxide (Fe(OH)$_3$. They found that with the chemical and hydraulic conditions optimum, the system was easily able to produce effluents with lead concentration of less than 0.5 mg/l (a simulated lead-bearing waste contained approximately 20 mg/l of lead from lead nitrate).

Thus far, no application of adsorptive bubble separation techniques has been made to remove heavy metals from municipal wastewater. Research efforts were initiated to generate information that would provide the initial information concerning whether the removal of heavy metals from municipal sewage influents is conceptually possible and feasible. The goal was to generate data that would provide insights into the possible structuring of a pilot model that will remove these heavy metals from municipal sewage influents.

MATERIALS AND METHODS

Grab samples of municipal wastewater influents were obtained from the Philadelphia North East Treatment Plant. Anionic surfactants used including: 0.01M and 0.1M sodium lauryl sulfate (NLS) ($CH_3(CH_2)_{10}CH_2OSO_3Na$) Lab Grade; and 0.1M sodium laurate (NL) or lauric acid, 99-100% pure, Sigma Grade.

The apparatus employed (Fig. 1) utilized house compressed air which was passed through an air filter (Whatman Gramma-12) to remove any water or other impurities. This air was then passed through a flotation cell which was an assembly of a Buckner funnel with a 600 ml capacity, equipped with a sintered glass plate with a No. 4 porosity.

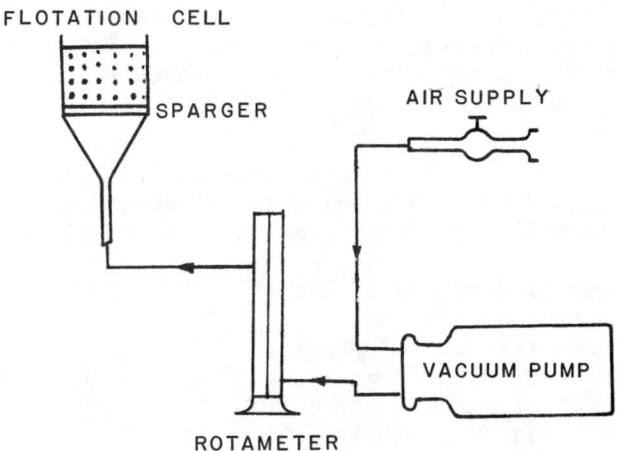

Figure 1. Apparatus employed in ion flotation studies. The air filter simply removed any water from the compressed air supply. The vacuum pump was not turned on but the air flow, pressure regulators, and pressure indicators were used. Other items were secured to a ring stand. Tygon tubing with a 1/4" I.D. was employed throughout the system.

RESULTS AND DISCUSSION

1) Flotation without adding surfactant (typical early trial procedure). Conditions used:
   - A grab sample of 500 ml.
   - Air pressure at 1 atm. (14.7 psi).
   - Air flow rate at 100 ml./min.
   - No surfactant was added.
   - One hr. bubble interval.

   Foam production from the sample was observed. However, heavy metal removal was not detected in this procedure. (Table 1).

Table I. Heavy Metals Data (mg/l)

Metal	Cd	Cr	Cu	Ni	Pb	Zn
Before	<0.01	0.19	0.19	0.08	0.06	0.74
After flotation	<0.01	0.20	0.19	0.08	0.04	0.72

The results indicated that there were some surfactants in the sewage but they were present in insufficient amounts to remove the metals or they had the wrong characteristics. Conditions such as the air flow rate, pressure or sparger pore may have been improper for such a removal. It was speculated from observing the amounts of foam provided that the amount of surfactants in the sewage probably varied widely with time throughout the day.

2) Flotation with 0.01M sodium lauryl sulfate (NLS). Three procedures were undertaken with different concentrations of 0.01M NLS to the grab samples to be bubbled.
   Conditions:
   - Grab samples of 500 ml each.
   - Air pressure at 1 atm.
   - Air flow rate at 100 ml/min.
   - Procedure I with 3 ml. 0.01 M NLS
               II with 15 ml. 0.01 M NLS
              III with 20 ml. 0.01 M NLS
   - Time: 1 hr. each

Foams were produced in all three procedures. There was no detectable reduction in heavy metal concentrations in the procedure that employed 3 ml of 0.01 M NLS (17.8 mg/l in a 500 ml sample, Table II). However, when the amount of 0.01 M NLS was increased to 15 ml (86.4 mg/l in a 500 ml sample) reductions of Cr, Cu, Pb, Zn, and BOD were achieved. A considerable reduction in the concentrations of all six metals and BOD levels was realized when the volume of added 0.01 M NLS was increased to 20 ml (115.2 mg/l in a 500 ml sample).

Table II. Heavy Metals and BOD Data (mg/l)

Metal	Cd	Cr	Cu	Ni	Pb	Zn	(BOD)
Before	0.02	0.52	0.52	0.37	0.22	1.68	380
After Flotation I (3 ml 0.1M NLS)	0.02	0.54	0.54	0.38	0.22	1.68	370
% Removal	(-)	(-)	(-)	(-)	(-)	(-)	(3)
After Flotation II (15 ml 0.1M NLS)	0.02	0.40	0.38	0.37	0.18	0.18	306
% Removal	(-)	(23)	(27)	(-)	(18)	(30)	(21.5)
Before	0.2	2.93	0.63	0.22	0.28	3.15	(328)
After Flotation III (20 ml 0.1M NLS)	<0.1	1.85	0.26	0.09	0.10	1.42	(192)
% Removal	(50)	(36)	(59)	(59)	(64)	(55)	(50)

3) Flotation with 0.1 M sodium lauryl sulfate (NLS) vs. 0.1 M sodium laurate (NL).
   Two experiments were run to compare the effectiveness, 2 ml of 0.1 M NLS, 0.1 M NLS (2 ml) and 0.1 M NL (8 ml), (115.2 mg/l and 356,5 mg/l in 500 ml sample respectively).
   Conditions:
   - Grab sample of 500 ml each
   - Air Pressure at 1 atm.
   - Air flow rate at 45 ml/min
   - Experiment I with 0.1 M NLS, 2 ml
                  II with 0.1 M NL, 8 ml
   - Time: 1 hr each

Both experiments similar results in removal of six metals, see Table III. The removal of the first experiment (with NLS) ranged from 60% of Cd to 75% of Cr and Pb, compared to the second experiment (with NL) ranged from 60% removal of Cd to 83% of Pb. Neither procedure demonstrated a notable reduction of BOD. To the contrary the second procedure gave a negative reduction of -164 percent.

It appeared that NLS has some advantages over NL in

terms of the effectiveness and costs. It should be noted here that these two experiments used the lower air flow rate (45 ml/min) compared to the preceding procedures.

Table III. Heavy Metals and BOD Data mg/l

Metal	Cd	Cr	Cu	Ni	Pb	Zn	(BOD)
Before	0.05	0.36	0.31	0.14	0.12	1.10	280
After flotation I (2 ml 0.1 NLS)	0.02	0.09	0.08	0.04	0.03	0.30	275
% Removal	(60)	(75)	(74)	(71)	(75)	(73)	(2)
After flotation II (8 ml 0.1M NLS)	0.02	0.08	0.08	0.05	0.02	0.30	740
% Removal	(60)	(78)	(74)	(64)	(83)	(73)	(-164)

4) Flotation with Lower Air Flow Rate (25 ml/min)
Two procedures were carried out with a low air flow rate at 25 ml/min but different flotation times: 1:15 hr. vs 18 min.
Conditions:
- Grab sample of 400 ml each
- Air pressure at 1 atm.
- Air flow rate at 25 ml/min.
- 0.1M NLS, 4 ml each (or 230.4 mg/l in sample)
- Time: Experiment I. 1:15 hr.
           II.   18 min.

The experiment I resulted in a better removal of four metals, i.e. Cr, Cu, Pb, and Zn, but less removal of Ni than the experiment II, see Table IV. Cd removal of both experiments could not be compared at this time. There were negative BOD removals in both experiments.

Table IV. Heavy Metals and BOD Data mg/l

Metal	Cd	Cr	Cu	Ni	Pb	Zn	(BOD)
Before	<0.01	0.30	0.38	0.06	0.07	0.95	266
After flotation I (Time: 1 hr)	<0.01	0.17	0.18	0.04	0.03	0.31	270
% Removal	(?)	(43)	(53)	(33)	(57)	(67)	(-1.5)
Before	<0.01	0.27	0.23	0.04	0.13	0.80	300
After flotation II (Time: 18 m.)	<0.01	0.19	0.17	0.02	0.10	0.44	420
% Removal	(?)	(30)	(26)	(50)	(23)	(45)	(-40)

Because of the conflicting results of Ni removal of both experiments, these procedures will be re-run to determine if the results can be repeated.

These experiments suggested that the air flow rate and flotation time are critical factors in this process.

CONCLUSION

The experiments which have been accomplished as of this writing removed Cd, Cr, Cu, Ni, Pb, and Zn at levels of 60, 75, 74, 71, 75 and 73 percents respectively, but have yielded inconclusive data concerning BOD removal. The experiments utilized different amounts of 0.01 M - 0.1 M sodium lauryl sulfate in 500 ml sewage sample. The air flow rate, at a constant pressure (14.7 psi), generally ranged from 25 to 100 ml/min. The pH and temperature of samples were approximately 7.0 and 11-15°C respectively. There were no pH and temperature adjustments made throughout the studies. A one hour flotation time was generally used. A liquid volume of collapsed foam has not been investigated.

These studies have demonstrated that foam flotation may be a method of choice to remove trace heavy metals from the sewage influents and resultant sludges. Further studies should be undertaken to determine the optimal conditions of each parameter or variable, namely dimensions of flotation column (may need to be re-designed), surfactants (anionic and/or cationic types), concentrations of surfactants, pH and temperature of sewage, air pressure, air flow rate, flotation times, and liquid volume of collapsed foam. BOD removal should also be seriously considered.

Investigations into the design of a pilot model concerned with the removal of toxic heavy metals from domestic sewage influents seems reasonable. The utilization of foam flotation procedures appears to be exploitable as a feasible answer to this vexing problem.

REFERENCES

1. Gurnham, C.F.; Rose, B.A.; Ritchie, H.R.; Fetherston, W.T.; and Smith, A.W., "Control of Heavy Metal Content of Municipal Wastewater Sludge," National Science Foundation (1979).
2. Brown, H.G.; Hensley, C.P.; McKinney, G.L.; and Robinson, J.L., "Efficiency of Heavy Metals Removal in Municipal Sewage Treatment Plants," Environ. Letters, 5(2), 103 (1973).
3. Maruyama, T.; Hannah, S.A.; and Cohen, J.M., "Metal Removal by Physical and Chemical Treatment Processes," JWPCF, 47(5), 962 (1975).
4. Cheng, M.H.; Patterson, J.W.; and Minear, R.A., "Heavy Metals Uptake by Activated Sludge," JWPCF, 47(2), 362 (1975).
5. Barth, E.F.; English, J.N.; Salotto, B.V.; Jackson, B.N. and Ettinger, M.B., "Field Survey of Four Municipal Wastewater Treatment Plants Receiving Metallic Wastes," JWPCF, 37, 1101 (1965).
6. Adams, C.E., Jr.; Eckenfelder, W.W., Jr.; and Goodman, B.L., "The Effects and Removal of Heavy Metals in Biological Treatment," Paper presented at Heavy Metals in the Aquatic Environment Conf., Vanderbilt Univ., Nashville, TN (Dec 407, 1973).
7. Ghosh, M.M., and Zugger, P.D., "Toxic Effects of Mercury on the Activated Sludge Process," JWPCF, 45, 424 (1973).
8. Barth, E.F.; Ettinger, M.B.; Salotto, B.V.; and McDermott, G.N., "Summary Report on the Effects of Heavy Metals on the Biological Treatment Processes," JWPCF, 37, 86 (1965).
9. Esmond, S.E.; and Petrasek, A.C., Jr., "Removal of Heavy Metals by Wastewater Treatment Plants," Paper presented at Wat. and Wastewater Equipment Manufacturers Assn. Ind. Wat. and Poll. Conf. and Exposition, Chicago, IL (March 14-16, 1973).
10. Reid, G.W., Nelson, R.Y.; Hall, C.; Bonilla, U.; and Reid, B., "Effects of Metallic Ions on Biological Waste Treatment Processes," Wat. & Sew. Works, 115(7), 320 (1968).

11. Stones, T., "The Fate of Chromium During the Treatment of Sewage," Jour. Inst. Sew. Purif., 4, 345 (1955).
12. Stones, T., "The Fate of Copper During the Treatment of Sewage, Jour. Inst. Sew. Purif., 1, 82 (1958).
13. Stones, T., "The Fate of Nickel During the Treatment of Sewage," Jour. Inst. Sew. Purif., 2, 252 (1959).
14. Stones, T., "The Fate of Zinc During the Treatment of Sewage, Jour. Inst. Sew. Purif., 2, 254 (1959).
15. Stones, T., "The Fate of Lead During the Treatment of Sewage," Jour. Inst. Sew. Purif., 2, 221 (1960).
16. Jackson, S., and Brown, V.M., "Effect of Toxic Wastes on Treatment Processes and Watercourses, Wat. Poll. Control Eng. (G.B.), 69, 292 (1970).
17. Bailey, D.A., and Robinson, K.S., "The Influence of Trivalent Chromium on the Biological Treatment of Domestic Sewage," Wat. Poll. Control Eng.(G.B.), 69, 100 (1970).
18. Pahren, H.R., Lucas, J.B.; Ryan, J.A.; and Dotson, G.K., "Health Risks Associated with Land Application of Municipal Sludge, JWPCF, 51(11), 2588 (1979).
19. Hyde, H.C.; Page, A.L.; Bingham, F.T.; and Mahler, R.J., "Effect of Heavy Metals in Sludge on Agricultural Crops," JWPCF, 51(10), 2475 (1979).
20. Chaney, R.L., "Crop and Food Chain Effects of Toxic Elements in Sludges and Effluents," in Recycling Municipal Sludges and Effluents on land," Natl. Assoc. of State Universities and Land-Grant Colleges, Wash., D.C. pp. 129-141 (1973).
21. Bowen, J.E., "Absorption of Copper, Zinc, and Manganese by Sugarcane Leaf Tissue," Plant Physiol., 44, 255 (1969)
22. Chaney, R.L., In: Factor Involved in Land Applications of Agricultural and Municipal Waste, USDA-ARS, Beltsville, MD, pp 97-120 (1974).
23. Chumbley, G.G., Permissible Levels of Toxic Metals in Sewage Used on Agricultural Land, ADAS Advisory Paper No. 10, Ministry of Agriculture, Fisheries and Food, Middlesex, England (1971).
24. Page, A.L., Fate and Effects of Trace Elements in Sewage Sludge When Applied to Agricultural Lands, US EPA Project No. EPA-670/2-74-005, Wash. (1974).
25. Andersson, A., and Nilsson, K.O., "Enrichment of Trace Elements From Sewage Sludge Fertilizer in Soils and Plants," AMBIO, 1, 176 (1972).
26. Webber, J., "Effects of Toxic Metals in Sewage on Crops," Wat. Poll. Contr., 71, 404 (1972).
27. Le Riche, H.H., "Metal Contamination of Soil in the Woburn Market-Garden Experiment Resulting from the Application of Sewage Sludge," J. Agr. Sci., 71, 205 (1968).
28. Dowdy, R.H., and Larson, W.E., "The Availability of Sludge-Borne Metals to Various Vegetable Crops," J. Env. Qual., 4(2), 278 (1975).

29. "Applications of Sludge to Cropland," in *Veterinary and Human Toxicology*, 19(1), 47 (1977).
30. Bradford, G.R.; Page, A.L.; Lund, L.J.; and Olmstead, W. "Trace Element Concentrations of Sewage Treatment Plant Effluents and Sludges: Their Interactions with Soils and Uptake by Plants," *J. Environ. Qual.*, 4(1), 123 (1975).
31. Flick, D.F.; Kraybill, H.F.; and Dimitroff, J.M., "Toxic Effects of Cadmium: A Review," *Environ. Res.*, 4, 71 (1971).
32. Page, A.L.; Bingham, F.T.; and Nelson, C., "Cadmium Absorption and Growth of Various Plant Species as Influenced by Solution Cadmium Concentration," *J. Environ. Qual.*, 1(3), 288 (1972).
33. Turner, M.A., "Effect of Cadmium Treatment on Cadmium and Zinc Uptake by Selected Vegetable Species, *J. Env. Qual.*, 2, 118 (1973).
34. John, M.K., "Cadmium Uptake by Eight Food Crops as Influenced by Various Soil Levels of Cadmium," *Environ. Pollut.*, 4(1), 7, (1973).
35. Haghiri, F., "Cadmium Uptake by Plants," *J. Environ.Qual.* 2, 93 (1973).
36. Bingham, F.T.; Page, A.L.; Mahler, R.J.; and Ganje, T.J., "Growth and Cadmium Accumulation of Plants Grown on Soil Treated with a Cadmium-Enriched Sewage Sludge,: *J. Env. Qual.*, 4(2), 207 (1975).
37. Bingham, F.T.; Page, A.L.; Mahler, R.J.; and Ganje, T.J. "Yield and Cadmium Accumulation of Forage Species in Relation to Cadmium Content of Sludge-Amended Soil," *J.Environ.Qual.*, 5(1), 57 (1976).
38. Baker, D.E.; Amacher, M.C.; and Doty, W.T., *Monitoring Sewage Sludges Soils and Crops for Zinc and Cadmium*, Land as a Waste Management Alernative, Ann Arbor Science, Ann Arbor, MI, pp. 261-281 (1977).
39. Melsted, S.W.; Hinesly, T.D.; Tyler, J.J.; and Ziegler, E.L., *Cadmium Transfer from Sewage Sludge-Amended Soil to Corn Grain to Pheasant Tissue*, Land as a Waste Management Alternative, Ann Arbor Sci., pp 199-208 (1977).
40. Metcalf and Eddy, Inc., *Wastewater Engineering: Treatment, Disposal, Reuse*, 2nd Ed., McGraw-Hill Book Co. p. 7 (1979).
41. Thackston, E.L.; Wilson, D.J.; Hanson, J.S.; and Miller, D.L., Jr., "Lead Removal with Adsorbing Colloid Flotation, *JWPCF*, 52(2), 317 (1980).
42. Sebba, F., *Ion Flotation*, Elsevier, Amsterdam, The Netherlands (1962).
43. Lemlich, R.,(Ed.), *Adsorptive Bubble Separation Techniques*, Academic Press, New York, NY (1972).
44. Somasundaran, P., Foam Separation Methods, *Separat. Purif. Methods*, 1, 117 (1972).

45. Somasundaran, P., "Separation Using Foaming Techniques," Separat. Sci., 10, 93 (1975).
46. Karger, B.L., and De Vivo, D.G., "General Survey of Adsorptive Bubble Separation Processes, Separat. Sci., 3, 393 (1968).
47. Grieves, R.B., "Foam Separations: A Review," Chem. Eng. Jour., 9, 93 (1975).
48. Karger, B.L.; Grieves, R.B.; and Lemlich, R., "Nomenclature Recommendations for Adsorptive Bubble Sepration Methods," Separat. Sci., 2, 401 (1967).
49. Pinfold, T.A., "Adsorptive Bubble Separation Methods," Separat. Sci., 5(4), 379 (1970).
50. Ferguson, B.B.; Hinkle, C.; and Wilson, D.J., "Foam Separation of Lead(II) and Cadmium(II) from Waste Water," Separat. Sci., 9, 125 (1974).
51. Rubin, A.J., and Johnson, J.D., "Effect of pH on Ion and Precipitate Flotation Systems," Anal. Chem., 39, 298 (1967).
52. Rubin, A.J., "Removal of Trace Metals by Foam Separation Processes," Jour. AWWA, 60, 832 (1968).
53. Rubin, A.J., and Lapp, W.L., "Foam Separation of Lead(II) with Sodium Lauryl Sulfate," Anal. Chem., 41, 1133 (1969).
54. Rubin, A.J.; Johnson, J.D.; and Lamb, J.C., "Comparison of Variables in Ion and Precipitate Flotation," Ind. Eng. Chem. Proc. Design Level, 5, 368 (1966).
55. Rubin, A.J., and Lapp, W.L., "Foam Fractionation and Precipitate Flotation of Zinc(II), Separat. Sci., 6, 357 (1971).
56. Davis, B.M., and Sebba, F., "Removal of Trace Amounts of Strontium from Aqueous Solutions by Ion Flotation-I. Batch Experiments," J. Appl. Chem., 16, 293 (1966).
57. Davis, B.M., and Sebba, F., "Removal of Trace Amounts of Strontium from Aqueous Solutions by Ion Flotation-II. Continuous Scale Operation," J. Appl. Chem., 16, 297 (1966).
58. Lusher, J.A., and Sebba, F., "The Separation of Aluminum from Beryllium by Ion Flotation of an Oxalato-Aluminate Complex," J. Appl. Chem., 15, 577 (1965).
59. Grieves, R.B., and Schwartz, S.M., "Continuous Ion Flotation of Dichromate," A.I.Ch.E. Jour., 12, 746 (1966).
60. Grieves, R.B., "Ion Flotation of Chromium(VI) Species: pH, Ionic Strength, Mixing Time, and Temperature," Sepat. Sci., 8, 501 (1973).
61. Grieves, R.B., and Schartz, S.M., "Batch and Continuous Ion Flotation of Hexavalent Chromium," J. Appl. Chem. 16, 14 (1966).
62. Grieves, R.B., and Ettelt, G.A., "Continuous, Dissolved-Air Ion Flotation of Hexavalent Chromium," A.I.Ch.E. Jour., 13(6), 1167 (1967).

63. Kobayashi, K., "Effects of Anionic and Cationic Surfactants on the Ion Flotation of Cu^{2+}," <u>Bull. Chem. Soc. JPN</u>, 48, 1180 (1975).
64. Kobayaski, K., "Effects of Anionic and Cationic Surfactants on the Ion Flotation of Cd^{2+}," <u>Bull. Chem. Soc. JPN</u>, 48, 1745 (1975).
65. Kobayaski, K., "Studies of the Removal of Cd^{2+} Ions by Adsorbing Particle Flotation," <u>Bull. Chem. Soc. JPN</u>, 48, 1750 (1975).
66. Huang, S., and Wilson, D.J., "Foam Separation of Mercury(II) and Cadmium(II) from Aqueous Systems," <u>Separat. Sci.</u>, 11, 215 (1976).
67. Kim, Y.S., and Zeitlin, H., "A Rapid Adsorbing Colloid Method for the Separation of Molybdenum from Sea Water," <u>Separat. Sci.</u>, 6, 505 (1971).
68. Kim, Y.S., and Zeitlin, H., "The Separation of Zinc and Copper from Sea Water by Adsorbing Colloid Flotation," <u>Separat. Sci.</u>, 7, 1 (1972).
69. Kim, Y.S., and Zeitlin, "Separation of Uranium from Sea Water by Adsorbing Colloid Flotation," <u>Anal. Chem.</u>, 43, 1390 (1971).
70. Sawyer, C.N., and McCarty, P.L., <u>Chemistry for Environmental Engineering</u>, McGraw-Hill Book Company, NY (1978).
71. Rubin, A.J., <u>Aqueous-Environmental Chemistry of Metals</u>, Ann Arbor Sci., Ann Arbor, MI (1974).

TREATABILITY OF CADMIUM(II) PLATING WASTEWATER BY ALUMINOSILICATE ADSORPTION

C. P. Huang, Environmental Engineering Program, Department of Civil Engineering, University of Delaware, Newark, Delaware 19711

P. K. Wirth, Environmental Engineering Program, Department of Civil Engineering, University of Delaware, Newark, Delaware 19711

INTRODUCTION

The presence of cadmium(II) in, particularly, the aquatic environment has been recognized as detrimental. This is evidenced by the numerous literature reports on the toxic effects of cadmium (II) to animals and aquatic life [1]. Although many sources can be attributed as cause of cadmium contamination, the plating industry is probably one of the largest contributors of cadmium-containing wastewater. In realizing the high toxic potential of cadmium(II), regulatory agencies are keen in restricting the disposal of cadmium-containing wastewater to aquatic systems including sewers. For example, in Philadelphia, nearly all of the cadmium plating firms were forced to discontinue the cadmium plating operation for not being able to meet the City's Wastewater Control Regulations which became effective on January 1, 1977 calling for a cadmium limit of 0.1 ppm in the wastewater stream to be achieved by July 1, 1977. Options to save the cadmium plating industries and the environment appear to be limited to the development of an economically feasible process for cadmium removal. The authors have conducted feasibility studies with batch reactors on the removability of cadmium (II) from synthetic cadmium plating wastewater by activated carbon adsorption processes [2,3]. While preliminary results appear promising on the use of activated carbon, separate studies of the adsorption capability of other adsorbents seems necessary for comparison. The major objective of this study was, therefore, to investigate the adsorption characteristics of cadmium (II) onto some commercially available aluminosilicate adsorbents. Important factors controlling the Cd(II) removability such as type of aluminosilicate and pH were thoroughly studied. An equilibrium model was also proposed to describe the adsorption characteristics of Cd(II) on aluminosilicate.

MATERIALS AND METHODS

Solids - The solids used for this study were obtained from the J. M. Huber Corporation. Each aluminosilicate was composed of a different (Si/Al) ratio. The important characteristics of the solids are given in Table.

Table 1. Properties of Alumino Silicates

Solid[1]	Si/Al	Surface Area[2] (m^2/gm)	Diameter (μm)	pH_{zpc}
Zeolite A (c)[3]	1.0	1.7 (2-4)	4	7.0
Blazer (a)	1.4	71 (80-120)	0.04	8.0
Mordenite (c)	5	324 (325-375)	3	1.3
Zeolex 23 (a)	6	41 (65-80)	0.04	1.0
Zeo 49 (a)	80[4]	177 (200-300)	0.02	1.3
Zeosyl 100 (a)	400[4]	81 (120-150)	0.02	2.0

[1] Information provided by manufacturer unless otherwise indicated.
[2] Calculated from BET gas adsorption measurements; values in parentheses were provided by manufacturer.
[3] a - amorphous, c - crystalline
[4] Zeo 49 and Zeosyl 100 are hydrated silicas, aluminum occurs only as an impurity.

Chemicals - All metal solutions were prepared from reagent grade chemicals. Three types of cadmium solutions were used throughout the study: (1) cadmium perchlorate ($Cd(ClO_4)_2$) solution, (2) synthetic cadmium fluroborate ($Cd(BF_4)_2$), and (3) cadmium cyanide ($Cd(CN)_2$) plating waste solutions. The composition of the solutions are given in Table 2.

Adsorption Experiments - The procedure for the batch adsorption experiments is as follows. First, appropriate quantitied of the ionic strength solution ($NaClO_4$ to give an I = 0.01 M), distilled-deionized water, and cadmium were added to 125 ml polypropylene bottles. The solids were added from stock suspensions of 10 or 25 gm/liter which had been prepared at least one week prior to use. Solid addition was made just prior to initial pH adjustment with 1.0 and 0.1 M $HClO_4$ or NaOH using a Corning Model 12 Research pH Meter

Table 2. Stock Solutions (Cd(II) 10^2 M)

Cadmium Perchlorate $Cd(ClO_4)_2$		Cadmium Fluoroborate $Cd(BF_4)_2$	
CdO	1.284 grams/ℓ	CdO	1.284 grams/ℓ
$HClO_4$	10. ml 1 M/ℓ	$HClO_4$	10. m. 1 M/ℓ
		HBF_4	3.608 grams (49%)/ℓ
Cadmium Cyanide $Cd(CN)_2$		$NH_4(BF_4)$	0.711 grams/ℓ
CdO	1.284 grams/ℓ	$H_3(BO_3)$	0.320 grams/ℓ
NaCN	4.901 grams/ℓ		

equipped with a Beckman combination pH Electrode. The pH range for the batch samples varied from pH 3 to pH 10. After pH adjustment, the samples were shaken in a reciprocating shaker for 12 hours at room temperature. After shaking, the final equilibrium pH was recorded. The solids were then removed by centrifugation at 15,000 rpm for 15 minutes at 25°C with a DuPont Sorvall temperature controlled centrifuge. Metal analysis of the supernatant was performed by means of flame atomic absorption spectroscopy using a Jarrell-Ash Model 810 Atomic Absorption Spectrophotometer. Calibration standards of similar ionic strength were prepared in order to avoid matrix interference effects.

RESULTS AND DISCUSSION

Effect of pH and adsorbent type - Fig. 1 shows the effect of pH on the Cd(II) adsorption capacity of various aluminosilicates. The results generally indicate that tht extent of Cd(II) removal increases with increasing pH. The effect of pH on Cd(II) adsorption is mainly due to the acid-base behavior of the aluminosilicate and the Cd(II) species. According to both thermodynamic calculation and experimental observation, $Cd(OH)_2$ (s) starts to precipitate at pH > 9. The results shown in Fig. 1 can be attributed to adsorption as the removal mechanism, at least at pH < 9. Another observation, probably the most striking one, is that among all aluminosilicates tested, Mordenite, a crystalline solid, appears to exhibit the greatest Cd(II) adsorption capacity at pH < 5 and performs as well as other aluminosilicates such as Aerogen 2000, Zeolex 23, and Blazer at pH > 5.

Further detailed experiments, emphasizing Mordenite, were undertaken to observe the Cd(II) removal capability from three Cd(II) containing solutions, namely, $Cd(ClO_4)_2$, $Cd(BF_4)_2$ and $Cd(CN)_2$.

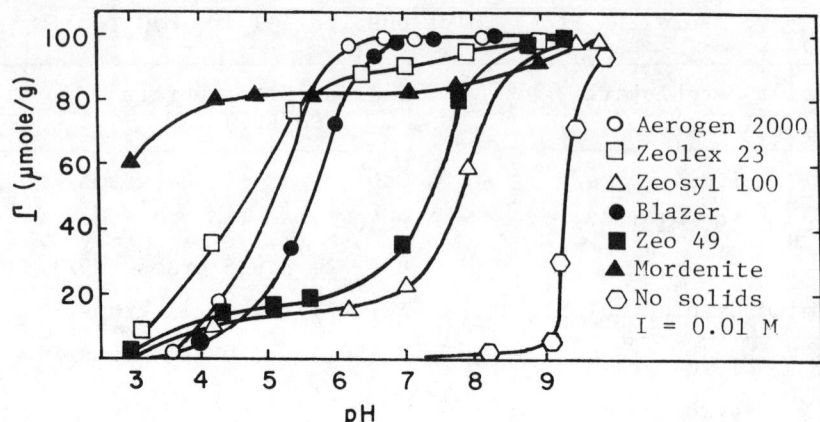

Figure 1. The Removal of Cd(II) from $Cd(ClO_4)_2$ Solution by Various Aluminosilicates. Note that the lowest bottom curve is a blank in the absence of solid.

Fig. 2 shows the results of Cd(II) removal capacity of Mordenite. It is clearly seen that the removal characteristics of $Cd(ClO_4)_2$ and $Cd(BF_4)_2$ system are essentially identical, indicating the chemical behavior of Cd(II) species in a $Cd(ClO_4)_2$ solution is similar to that in a $Cd(BF_4)_2$ plating solution, at least in its response to Mordenite adsorption. For both solutions, the percentage of Cd(II) removal is extremely high (near 100%) when the adsorbent dose is 3 to 5 g/ℓ.

The results also show a trend of decreasing Cd(II) adsorption at ph < 3, a feature promising for regeneration of Cd(II)-laden Mordenite with strong acid. It is noted that Mordenite is quite acid resistant. The response of Mordenite adsorption of Cd(II) in $Cd(CN)_2$ solution is quite different from that of the two above systems. The optimum pH range for Cd(II) removal is approximately between 4 and 6; the percentage of Cd(II) removal drops drastically, particularly on the alkaline side, out of this pH region. The pronounced decrease in Cd(II) adsorption in the alkaline region can be attributed to the unfavorable electrostatic effect between the negatively charged Mordenite surface and the formation of anionic Cd-Cn complexes such as $Cd(CN)_3^-$ and $Cd(CN)_4^{2-}$.

In another separate study, Elliott and Huang [4] reported that Mordenite was unable to adsorb anionic Cu(II) complexes such as $CuNTA^-$ (Cu(II)-Nitrilotriacetate) but extremely effective in removing cationic Cu(II) complexes such as $CuGly^+$ (Cu(II)-glycine). This substantial cation adsorption capacity of Mordenite in particular and aluminosilicates in general is attributed to their ionic exchange characteristics which have two components: (1) an <u>intrinsic</u> and invariant cation

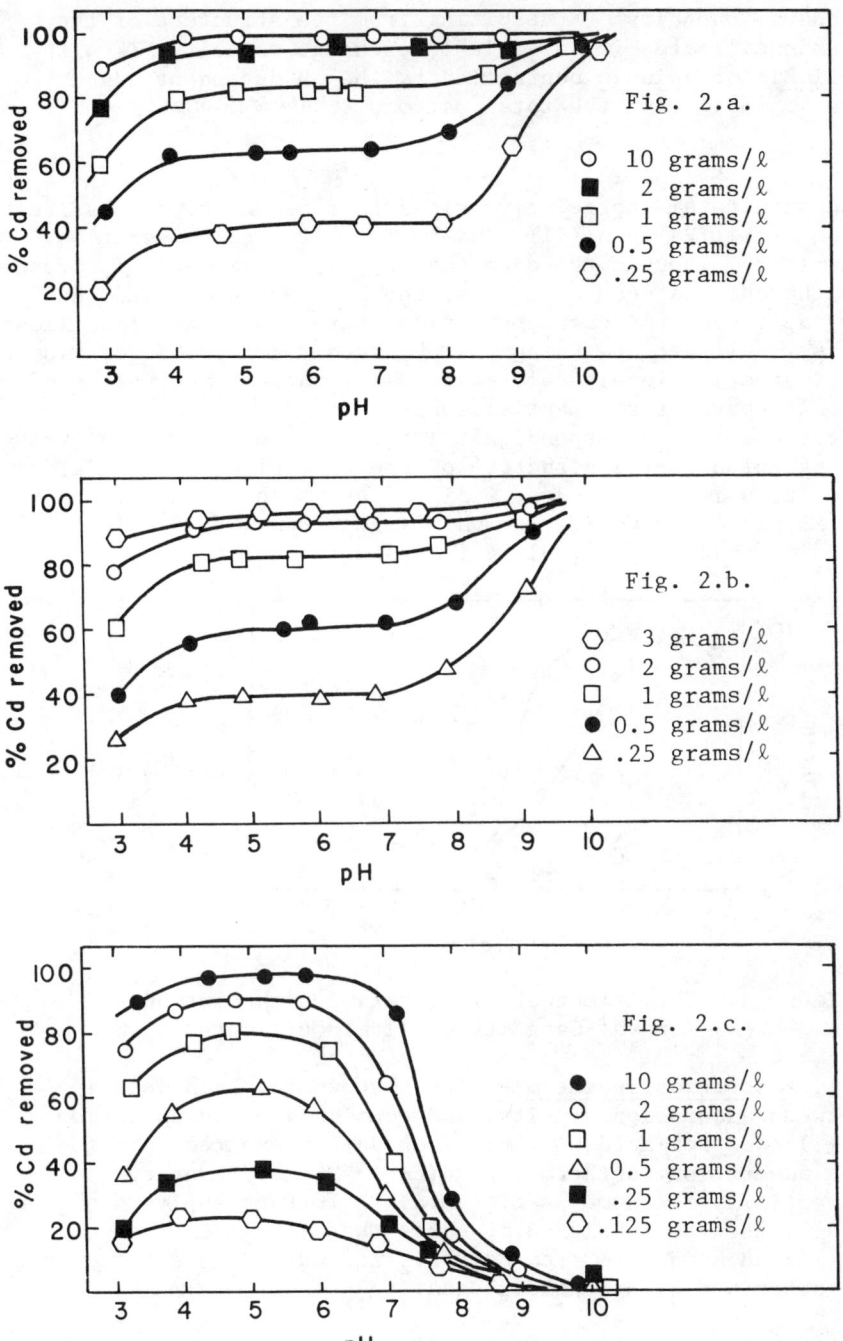

Figure 2. The Cd(II) Removal Capacity of Mordenite from: (a) $Cd(ClO_4)_2$ Solution, (b) $Cd(BF_4)$ Solution, and (c) $Cd(CN)_2$ Solution.

exchange capacity, σ_I, resulting from the structure of the aluminosilicate, (2) a variable exchange capacity, σ_+, either <u>cationic</u> or <u>anionic</u> controlled by the pH-dependent edge charge [5]. Thus the total cation exchange capacity (CEC) is

$$CEC = \sigma_I + \sigma_+$$

The structural charge, σ_I, arises from isomorphous substitution of Si(IV) by Al(III) and is not affected by solution chemistry; whereas the edge charge, σ_+, is directly influenced by the chemical composition of the bulk phase and indirectly by the structural characteristics of the solid aluminosilicate phase. Elliott and Huang [4] observed that the CEC of aluminosilicates is closely related to Si/Al ratio. Low values of Si/Al represent substantial substitution by Al(III) in the lattice and a correspondingly large σ_I; whereas, the σ^+ value is affected by the structure of the electrical double layer and often is pH-dependent. Fig. 3 shows that the optimum Si/Al ratio for Cd(II) removal ranges from 1 to 6 or corresponding to Zeolite A, Blazer, Mordenite and Zeolex 23.

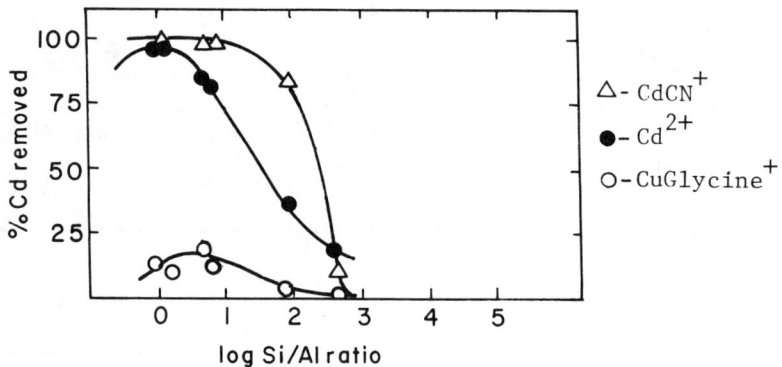

Figure 3. The Effect of Si/Al ratio on the Cationic Removal Capacity of Aluminosilicate

<u>Adsorption isotherm</u> - Fig. 4 shows the relationship between adsorption density, and equilibrium concentration of Cd(II), C_e for Mordenite based on data presented in Fig. 2.b. The adsorption isotherm for Cd(BF$_4$)$_2$/Mordenite system is essentially the same as Fig. 3.a. A further analysis of the data presented in Fig. 3.a. shows that the adsorption of Cd(II) on Mordenite from Cd(BF$_4$)$_2$ and Cd(ClO$_4$)$_2$ solution can be described by a Langmuir equation

$$\Gamma = \frac{\Gamma_m C_e}{K + C_e} \ ;$$

where Γ_m and K represent adsorption density at monolayer

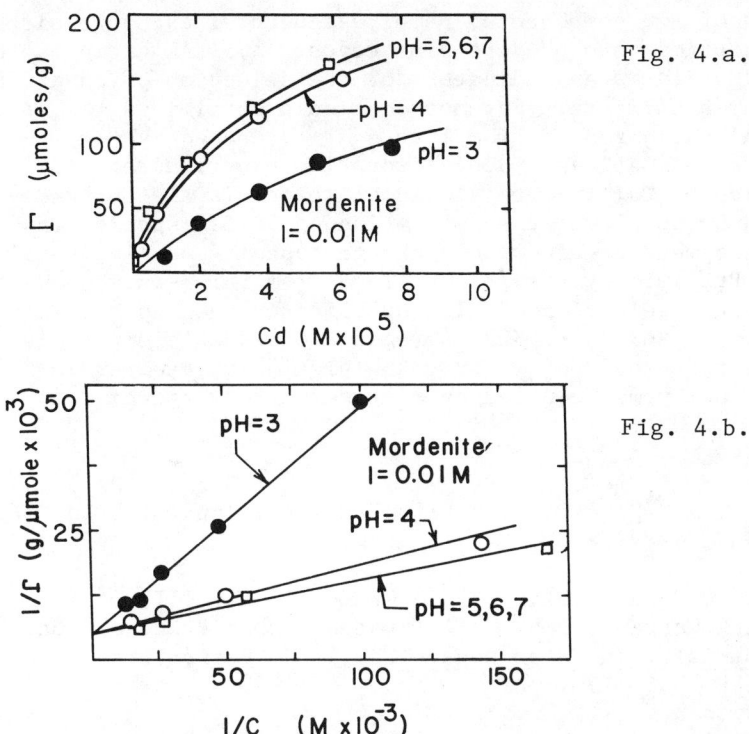

Figure 4. A Typical Adsorption Isotherm for Cd(II) and Mordenite

coverage and Langmuir constant, respectively. Furthermore, the Langmuir constant, K, is related to the pH value by the expression, $K = K_m (1 + [H^+]/K_H)$. The Langmuir isotherm is therefore modified to $\Gamma = \Gamma_m C_e / K_m(1 + [H^+]/K_H) + C_e$, typical of a competitive type of adsorption affected by pH. A linearized competitive Langmuir isotherm is shown by Fig. 4.b. The results show that the Langmuir parameter, K_m, K_H, and Γ_m, are 18 μM/gm, 250 μM/ℓ, and 200 μM/gm, respectively. Huang and Smith also found this same equation to be applicable to some Cd(II) plating wastewater using activated carbon as the adsorbent. The values of these parameters are in the same order of magnitude as those of activated carbon reported by Huang and Smith [2]. For example, the values of K_m, K_H, and Γ_m are 68 μM/gm, 162 μM/g and 35 μM/ℓ respectively for $Cd(BF_4)_2$-Nuchar SA system.

SUMMARY

The adsorption of Cd(II) ions onto aluminosilicates can be primarily interpreted as an ion exchange process which consists of two components, namely structural exchange sites and adsorption onto pH-dependent edges. A Si/Al ratio may be used primarily as an indicator for the selection of an effective aluminosilicate. For most cationic metal ions the optimum Si/Al ratio ranges from 1 to 6 as evidenced by the adsorption of Cd^{2+} and $CdCN^+$ adsorption experiments. The adsorption of Cd(II) onto aluminosilicates is also markedly affected by pH due to the speciation of Cd(II) species and the development of edgewise exchange capacity on aluminosilicates. Preliminary results obtained from this study show that with careful selection of aluminosilicates such as Mordenite (for pH < 5) and Zeolex 23, Aerogen 2000 and Blazer, it is possible to remove Cd(II) from the plating wastewater to level (< 0.1 ppm) required by most regulatory agencies.

REFERENCES

1. Tucker, A., "The Toxic Metals," Earth Island Lt., London, England, 1972.

2. Huang, C.P. and Smith, E.H., "Removal of Cd(II) from Plating Wastewater by an Activated Carbon Process," Ch. 7 in Chemistry in Water Reuse (Ed., W.J. Cooper), Ann Arbor Sci., p. 355, 1981.

3. Huang, C.P., Wirth, P.K., and Smith, E.H., "Removal of Heavy Metals by the Activated Carbon Process - The Case of Cadmium Plating Wastewater," National Conference on Hazardous Waste and Toxic Waste Management, New Jersey Institute of Technology, Newark, NJ, 1980.

4. Elliott, H.A. and Huang, C.P., "Adsorption of Cu(II) Complexes on Aluminosilicate Zeolites. Effect of Si/Al Ratio," Water Research (in press), 1981.

5. Parks, G.A., "Aqueous Surface Chemistry of Oxides and Complex Oxide Minerals," Advan. Chem. Ser. 67, p. 121, 1967.

ACKNOWLEDGMENT

This work was supported by a grant, ENG. 77-27379, from the Water Resources, Urban and Environmental Engineering Program of the National Science Foundation.

PROPERTIES AFFECTING RETENTION OF HEAVY METALS FROM WASTES APPLIED TO NORTHEASTERN U. S. SOILS

<u>Herschel A. Elliott</u>. Agricultural Engineering Department, University of Delaware, Newark, Delaware.

<u>Michael R. Liberati</u>. Agricultural Engineering Department, University of Delaware, Newark, Delaware.

INTRODUCTION

Application of industrial wastes and sludges to land is becoming increasingly attractive as economic factors and environmental regulations limit disposal by alternate methods. Besides accomplishing ultimate disposal, land application often allows utilization of waste constituents to improve soil fertility or favorably modify soil structure. Clearly however, spreading industrial wastes on land is not without potential hazards. Plants grown on the waste-amended soils can take up toxic materials and incorporate them in plant tissue. Such substances, therefore, may potentially enter the food chain and result in adverse physiological effects to animals and man. Additionally, the possibility exists that harmful chemicals applied to land will leach through the soil and contaminate groundwater supplies, on which nearly 50% of all Americans rely for drinking purposes.

The extent to which these hazards are manifested depends, to a large degree, on the ability of the soil to adsorb and retain such pollutants. The soil retention capacity is, in turn, a function of soil system parameters such as pH, oxidation-reduction potential, mineralogical composition, and surface change characteristics, to name a few.

Of the heavy metal pollutants, cadmium (Cd) is one whose presence in the environment is chiefly due to industrial operations [1, 2] such as electroplating, solder and battery operations and combustion of fossil fuels. Cadmium is of special concern for a number of reasons. It is a nonessential and nonbeneficial element in mammalian nutrition [3] and, in fact, low-level exposure to Cd appears to be linked to such conditions as emphysema, gastric and intestinal disorders, anemia and hypertensive heart disease [4]. Furthermore, unlike Zn, Cu and Ni which become phytotoxic before plant tissue reaches levels inimical to human health, Cd can exist in plants at sub-phytotoxic levels which are deleterious to animals and humans.

The purpose of the present study was to investigate the pH-dependence of Cd adsorption on surface and subsurface samples from eight northeastern U. S. soils. The relative importance to Cd retention of various soil components was studied by selective removal of organic matter and iron oxides. Values of the cation exchange capacity (CEC) were compared with the extent of adsorption to assess the validity of employing CEC as a criterion for determining maximum Cd loading rates for land application of wastes.

MATERIALS AND METHOD

Materials

Samples of surface and subsurface horizons of eight soils from the northeastern U. S. were used in this study. The soils were collected for use by the USDA Regional Research Project NE-96, "Soil Properties Affecting Sorption of Heavy Metals from Wastes." The soils were selected to provide various chemical, physical and mineralogical characteristics of those found in the northeast agricultural areas. Values in Table I for CEC, % organic matter, % clay fraction, and % Fe were provided by the National Soil Survey Laboratory in Lincoln, NE. Work on heavy metal adsorption by these soils has been reported elsewhere [5].

The cadmium solutions were prepared by dissolving cadmium metal in nitric acid, and diluting with distilled-deionized water. Reagent grade chemicals were obtained from Fisher Scientific Company.

Methods

Batch adsorption experiments were carried out by adding 80 ml of distilled water, 10 ml of 10^{-4} M $Cd(NO_3)_2$, and 10 ml of 0.25 M $NaNO_3$ to 125 ml polyethylene bottles. Metal solutions were prepared from nitrate salts, and $NaNO_3$ was used as an inert electrolyte because NO_3^-, unlike Cl^-, has a low tendency to complex cadmium in solution. One gram of soil was added to the solutions just prior to pH adjustment with 0.01 - 1.0 M HNO_3 and NaOH using an Orion model 901 pH meter. After pH adjustment, the samples were shaken at room temperature on an Eberbach shaking machine. A three hour shaking time was found adequate to reach equilibrium. Final pH values were then determined, and a portion of the sample was centrifuged to remove solids.

Table I. Cadmium Adsorption and Selected Properties of Soils

Soil Series	% Cadmium Adsorbed				CEC (meq/100 gm.)	% Organic Matter	% Clay Fraction	% Iron
	pH 4	pH 5	pH 6	pH 7				
Christiana A	17	54	76	88	4.0	0.98	9.9	1.4
Christiana B	6	40	64	82	5.3	0.05	28.4	4.7
Evesboro A	15	43	74	90	2.4	0.67	4.1	0.2
Evesboro B	10	22	37	80	1.3	0.12	3.2	0.3
Groveton A	47	86	94	96	23.2	4.25	6.6	1.3
Groveton B	17	70	90	95	7.6	0.62	3.7	0.7
Lima A	65	82	92	96	11.3	1.37	18.8	1.1
Lima B	65	77	87	91	9.3	0.63	22.6	1.6
Mardin A	41	70	85	93	11.7	1.54	18.9	1.4
Mardin B	24	46	65	85	7.2	0.24	20.8	1.2
Paxton A	46	83	92	95	16.4	3.81	10.6	1.0
Paxton B	19	57	85	93	8.9	1.51	7.6	1.1
Pocomoke A	24	70	90	82	7.2	2.05	7.7	*
Pocomoke B	11	38	56	72	2.7	0.15	2.0	*
Vergennes A	61	82	90	93	20.3	2.83	7.3	1.6
Vergennes B	18	71	86	92	24.0	0.57	17.2	2.1

* trace amounts

Metal analysis of the supernatant was performed by means of flame atomic absorption spectroscopy using a Perkin-Elmer Model 5000 Spectrophotometer. The difference between initial amount of metal in solution and the amount remaining after the reaction period was assumed to be adsorbed by the soil.

The electrophoretic mobility of the soils was determined with a Zeta Meter (Zeta Meter, Inc.), using the method detailed elsewhere [6].

Soil organic matter was removed by additions of H_2O_2 to a 1:1 soil to 1N NaOAc mixture. Organic matter was considered totally removed when the effervescence caused by the H_2O_2 additions ceased. Free iron removal was accomplished by the sodium dithionite citrate treatment method [7]. Soils were washed with distilled water after organic matter and free iron removal procedures.

RESULTS AND DISCUSSION

Effect of Solution pH

The adsorption of Cd(II) on surface (A horizon) and subsurface (B horizon) samples of the eight soils is presented in Table I for equilibrium system pH values of 4, 5, 6, and 7. For the soils tested, a similar pH-dependent trend was observed: the extent of Cd(II) uptake was least at low pH, increased rather abruptly to nearly maximum adsorption as the pH approached neutrality, and leveled off or decreased slightly at more alkaline conditions. The strong influence of pH can also be seen in Figure 1, which shows the adsorption of Cd(II) on untreated Christiana A, a sandy soil indigenous to Delaware and Maryland.

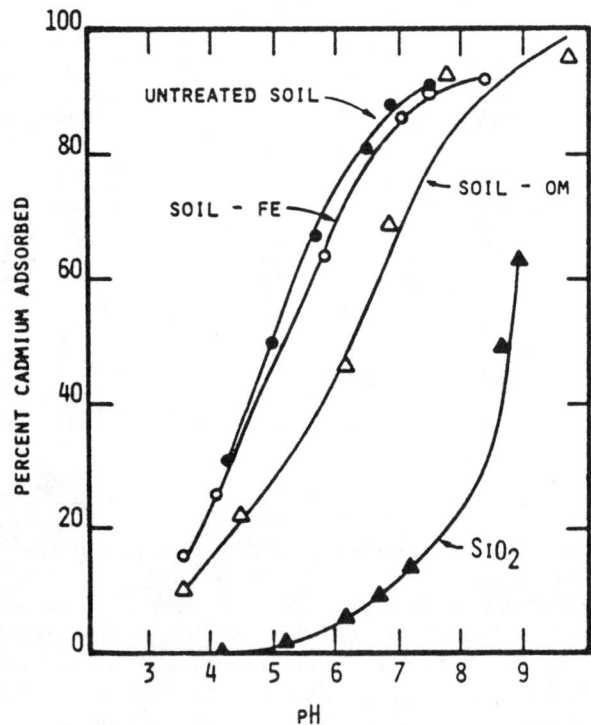

Figure 1. The pH-dependent adsorption of Cd(II) by pure silica (▲) and Christiana A as untreated soil (●), with iron removed (○), and with organic matter removed (△).

The crucial role of pH in affecting heavy metal uptake has been recognized [8], but it is difficult to precisely define a single mechanism responsible for this behavior. In certain systems, the abrupt increase in heavy metal retention with increasing pH appears to be related to the formation of soluble metal hydrolysis products [9], e.g., $CdOH^+$. Such a mechanism cannot be invoked in the current context since cadmium is present as Cd^{2+} (aq) over the entire pH range of study (Figure 2a). Additionally, it was experimentally confirmed that precipitation of cadmium hydroxide or carbonate was not removing metal from solution.

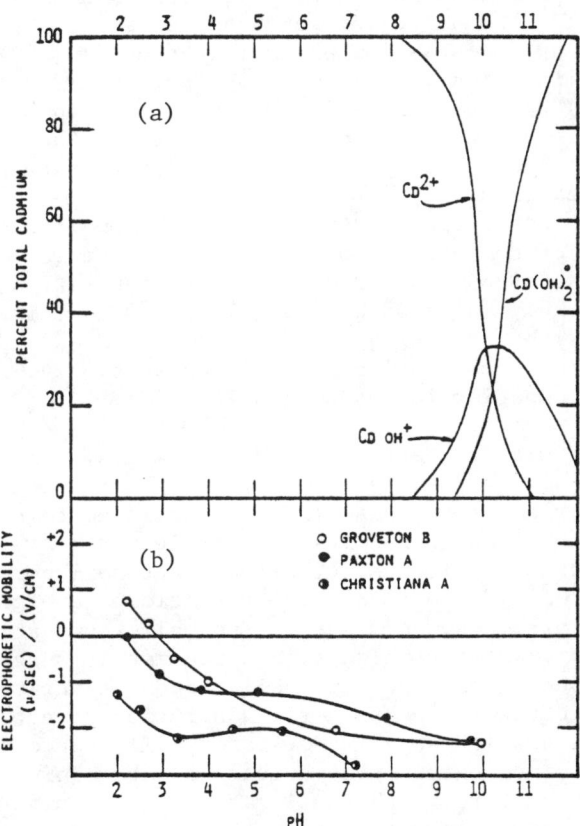

Figure 2. (a) The pH-dependent speciation of Cd(II) and (b) the electrokinetic behavior of selected soils.

One possible explanation for the pH-dependent adsorption behavior was the increasing negative soil surface change which characterizes these soils as the solution pH becomes more alkaline. Figure 2b shows the electrophoretic mobility of three of the soils as a function of pH. It can be seen that the electrophoretic mobility, and hence the soil surface change, is negative over the entire pH range of concern in natural systems. This behavior is common among soils and indicates a progressively larger electrostatic attraction between the negatively charged surface and the cationic cadmium, Cd^{2+}, (Figure 2a) as the pH increases.

Although the pH range encountered in this study had little effect on Cd speciation (Figure 2a), it will appreciably influence aluminum ions which occupy soil exchange sites. Aluminum (Al^{3+}) bound to soil exchange sites will hydrolyze or precipitate as the solution pH is increased [10]. Each time a bound aluminum ion is complexed by OH^-, one additional negative site is then available for accommodating another cationic species. Since aluminum hydrolysis occurs in the pH range studied, a concomitant increase in ability to retain Cd^{2+} should be realized.

The lower adsorption under more acidic conditions can also be a result of competition of protons (H^+) with Cd^{2+} for available exchange sites. At pH 4, the H^+ concentration is tenfold that of cadmium; $[H^+]=10^{-4}M$, $[Cd(II)]=10^{-5}M$. This effect diminishes with increasing pH since as the pH exceeds 5.0 the cadmium has a concentration, as well as a valency, advantage over H^+ in competing for exchange sites.

Influence of Organic Matter, Fe-oxides, Clays

As complex heterogeneous mixtures, soils contain many components, each of which has its own particular heavy metal adsorption characteristics. The important soil components purported to play a role in metal retention are organic matter, iron-oxide materials, and clays. The well-known ability of organic compounds to complex transition metals suggests that organic matter is an important soil component influencing heavy metal uptake. The specific adsorption of cations by sesquioxides has led some investigators [11] to conclude that iron and aluminum oxides play a role in determining the movement of cations through the soil matrix. Clay minerals, as a principal contributor to the cation exchange capacity of a soil, should be related to heavy metal retention.

The percentage composition of these three components for the sixteen soil samples studied is provided in Table I. The soils used embrace a wide range of compositions: organic matter (0.05-4.25%), iron oxides (trace - 4.7%) and clays (2.0-28.4%).

Elucidation of the influence of these soil components was attempted by selectively removing the organic matter or the iron oxides prior to performing an adsorption experiment. The results obtained for the Christiana A soil are shown in Figure 1. It can be noted that removal of Fe-oxides had only a slight effect on Cd uptake. Removal of soil organic matter demonstrated a greater ability to inhibit Cd retention, although in its absence the soil was still capable of substantial adsorption. The behavior exhibited by Christiana A was typical for the majority of the soils tested.

Figure 1 also shows the adsorption of Cd on pure silica as a function of pH. Silica was chosen since it is probably the principal primary mineral in most soils. It is clear from the results that, at typical field pH conditions, the quartz fraction of a soil contributes little to Cd adsorption. From this finding, along with the other data included in Figure 1, it was concluded that clay minerals were responsible for a majority of the Cd uptake. For example, at pH 6, the quartz fraction contributed approximately 5% of the total Cd removal, organic matter about 25%, Fe-oxides 5%, and clay minerals nearly 45% of the total adsorptive capacity. The high retention capacity of pure montmorillonite and kaolinite for Cd [12] lends credence to the important role of clay minerals in this context. In order of importance in influencing Cd uptake, the trend clay>organic matter>Fe-oxides was valid for the majority of the soils studied.

The finding that organic matter is not the sole determinant in cadmium retention is in agreement with the work of Kuo and Baker [13]. Organic matter does, however, seem to be important in explaining the consistently greater extent of adsorption for the surface soils over subsurface soils (Table I). For example, Figure 3 shows the pH-dependent Cd adsorption of the Paxton A and B soils, with and without organic matter removed. The adsorption curves tended to coincide when the organic matter was removed from both horizons. Among the other soils, this behavior seems to be followed particularly when the clay content of the A and B horizons was nearly equal.

Cadmium Adsorption and CEC

Since soils vary considerably in their ability to retain heavy metal pollutants such as cadmium, it is beneficial to have a knowledge of important adsorption-related parameters on which to base maximum permissible waste loading rates. The U. S. Environmental Protection Agency has established criteria which limit the amount of Cd which can be applied to soils where food-chain crops will be grown [14]. These criteria are based on two parameters: soil pH and cation exchange capacity (CEC).

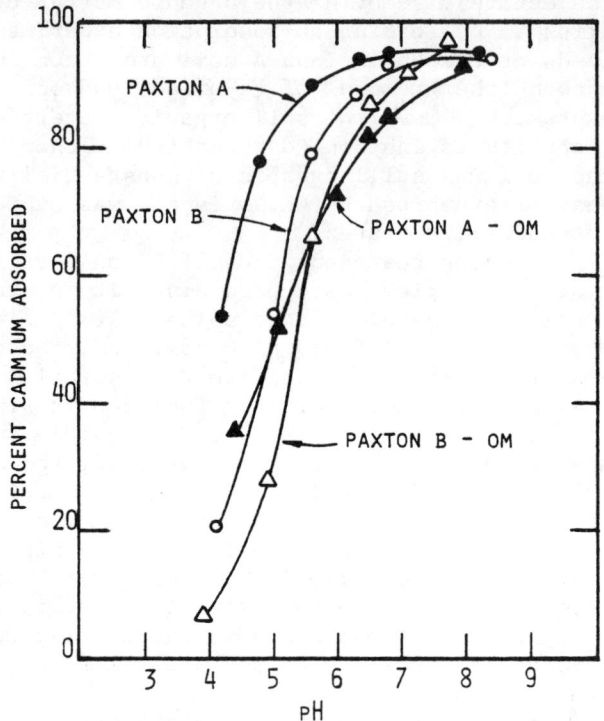

Figure 3. Effect of organic matter removal on the adsorption of Cd(II) by Paxton A and Paxton B soils.

In order to assess the functional relationship between CEC and Cd adsorption, the soil CEC's were compared with the percent cadmium adsorption at various pH values (Table I). It can be seen that there was no consistent correlation between CEC and extent of adsorption. The reported CEC values were determined by the conventional ammonium acetate method in which the soil is buffered to pH 7. However, as noted by Leeper [8], NH_4^+ cannot displace heavy metals which occupy exchange sites, and accordingly, CEC is not an accurate guide to predicting the soil's capacity to retain cadmium.

Since clay plays a major role in Cd retention, then perhaps their quantitative measurement would serve as a useful predictor of a soil's metal holding capacity. Unfortunately, the conventional clay fraction determination is unsatisfactory. Consider the adsorption of Cd at pH 7 by the Christiana A and B soils (Table I). The percent adsorption was 88% (Christiana A) and 82% (Christiana B) while the percent clay fraction was

9.9 and 28.4%, respectively. Values for percent clay fraction (Table I) are representative of the percentage of total soil, regardless of mineralogical composition, which passes through a 2 μm sieve. This, of course, includes fine quartz particles which, as discussed, do not materially affect a soil's ability to adsorb cadmium. A finely divided sandy soil could have a high percent clay fraction but yet contain little clay minerals. As a guide to heavy metal uptake, knowledge of the amounts and specific types of clays (montmorillonite, kaolinite, etc.) would be superior to the general percent clay fraction information.

SUMMARY

The adsorption characteristics of heavy metals, as exemplified by Cd(II), on surface and subsurface samples of eight northeastern U. S. soils has been investigated in the laboratory. For the pH range typically encountered in natural systems, all soils exhibited a common pH-dependent trend: the extent of adsorption was least at low pH values (4-5), increased to nearly 100% in the neutral pH range, and leveled off or diminished slightly under more alkaline conditions (pH 8-9). This behavior has been attributed to an increasing negative soil surface charge with increasing pH, hydrolysis/precipitation of aluminum occupying exchange sites, and competition between H^+ and Cd^{2+}.

By selective removal of organic matter and Fe-oxides from the soils, it was concluded that the importance of various soil components in influencing Cd retention followed the order: clays>organic matter>Fe-oxides. Primary minerals (e.g., quartz) seemingly contribute little to the Cd adsorptive capacity of these soils. The consistently greater Cd(II) uptake by surface over subsurface samples was apparently due to differences in organic matter content, inasmuch as organic matter removal from both resulted in similar adsorption characteristics.

There was no strong correlation between Cd(II) uptake and the soil cation exchange capacity (CEC) as routinely measured by the ammonium acetate method. The EPA criteria for maximum Cd application to soils for growing food-chain crops are based on soil pH and CEC. The data reported herein suggest that the importance of CEC, as conventionally determined, may be overstated. A knowledge of the clay mineralogy in conjunction with pH may be more reliable than CEC as a guide to predicting heavy metal retention by soils.

REFERENCES

1. Gurnham, C. F., Rose, B. A., Ritchie, H. R., Fetherston, W. T., and Smith, A. W., "Control of Heavy Metal Content

of Municipal Wastewater Sludge," Report to Natl. Sci. Found. NTIS-PB-295917, 1979.
2. Jones, R. L., Hinesly, T. D., and Ziegler, E. L. "Cadmium Content of Soybeans Grown in Sewage-Sludge Amended Soil," J. Enviro. Qual., 2,351 (1973).
3. Schroeder, H. A. and Balassa, J. J., "Cadmium: Uptake by Vegetables from Superphosphates in Soil," Science, 140,819 (1963).
4. Lagerwerff, J. V., "Lead, Mercury and Cadmium as Environmental Contaminants," in Micronutrients in Agriculture (Eds. Mortvedt, J. J., Giordano, P. M. and Lindsay, W. L.) Soil Sci. Soc. Amer., Inc. p. 594, 1972.
5. Harter, R. D., "Adsorption of Copper and Lead by Ap and B2 Horizons of Several Northeastern United States Soils," Soil Sci. Soc. Am. J., 43,679 (1979).
6. Elliott, H. A. and Sparks, D. L., Electrokinetic Behavior of a Paleudult Profile in Relation to Mineralogical Composition, Soil Sci. (accepted), (1981).
7. Mehra, O. P., and Jackson, M. L., "Iron Oxide Removal from Soils and Clays by a dithionite-citrate system buffered with sodium bicarbonate," Clays Clay Miner, 5, 317 (1960).
8. Leeper, G. W., Managing the Heavy Metals on the Land, Marcel Dekker, 1978.
9. Elliott, H. A. and Huang, C. P., "The Effect of Complex Formation on the Adsorption Characteristics of Heavy Metals," Environ. Int., 2, 145 (1979).
10. Cavallaro, N. and McBride, M. B., "Activities of Cu^{2+} and Cd^{2+} in Soil Solutions as Affected by pH," Soil Sci. Soc. Am. J., 44, 729 (1980).
11. Kinniburgh, D. G., Jackson, M. L., and Syers, J. K., "Adsorption of Alkaline Earth Transition, and Heavy Metal Cations by Hydrous Oxide Gels of Iron and Aluminum," Soil Sci. Soc. Am. Proc., 40, 776 (1976).
12. Liberati, M.,"Soil Properties Affecting Adsorption of Cadmium," Undergraduate Thesis, University of Delaware, Newark (1981).
13. Kuo, S., Baker, A. S., "Sorption of Copper, Zinc and Cadmium by Some Acid Soils," Soil Sci. Soc. Am. J., 44:969-974, (1980).
14. "Criteria for Classification of Solid Disposal Facilities and Practices: Final, Interim Final, and Proposed Regulations," Federal Register,44,53438 (1979).

NEW OZONE APPLICATION SYSTEM TO THE CYANIDES
DESTRUCTION IN PLATING WASTEWATERS

Wojciech Kunicki. Environmental Protection Dept.
Institute of Precision Mechanics, Warsaw, Poland

INTRODUCTION

The high oxidation potential of ozone and the wide range of organic and inorganic groupings suscetable to oxidation suggest application of this oxidant to eliminate the objectionable compounds from water and wastewater.

Ozone application for cyanide destruction has been practiced in Europe and the United States for thirty last years. In the US, first full-scale operations plant was installed in 1967 at the Boening Company in Wichita, Kansas.

In Poland, we have started with the first basic studies in 1973 and finally we designed and installed first industrial ozone treatment plant. It was started up in September, 1980.

CHARACTERISTIC OF ELECTROPLATING WASTEWATERS

The electroplating industry produces quantities of wastewater containing cyanide and heavy metals salts. Usually concentrations of cyanide occurences up to 80 mg/dm^3 and heavy metals concentration up to 50 mg/dm^3 /for each metal: copper, cadmium, zinc, silver/. The cyanide waste may also contain hard-to-decompose cyanide complexes such as iron complexes.

Range of pH cyanide wastewater varries from 7.5 to 12.5. Cyanide wastewater is traditionally treated by chemical means such as oxidation with gaseous chlorine or oxidation with hypochlorites. The ozone destruction of cyanide waste is an additional promissing method.

REVIEW OF OZONE DESTRUCTION OF CYANIDE WASTE

The first publication on the oxidation of cyanides with ozone was presented by Neurwith in 1933 /1/ while in 1951 Tyler /2/ published the results of the ozonation of spent cadmium plating solutions. The reaction taking place was presented as follows:

$$2\ KCN + 2O_3 \longrightarrow 2\ KCNO + 2O_2 \qquad /1/$$

$$2\ KCNO + H_2O + 3O_3 \longrightarrow 2\ KHCO_3 + N_2 + 3O_2 \qquad /2/$$

$$2\ KCN + H_2O + 5O_3 \longrightarrow 2\ KHCO_3 + N_2 + 5O_2 \qquad /3/$$

The above publication confirmed complete absorption of ozone by the solution up to the moment when cyanide traces remained. The stoichiometric reaction can be expressed as follows:

$$CN^- + O_3 \longrightarrow CNO^- + O_2 \qquad /1.85\ g\ O_3/g\ CN^-/ \qquad /4/$$

$$3\ CN^- + O_3 \longrightarrow 3\ CNO^- \qquad /0.62\ g\ O_3/g\ CN^-/ \qquad /5/$$

In reality the ozone requirement is higher and ranges from 2 to 3 g O_3/g CN^-.

Further work however showed that in general the order of the reaction changes in the different stages of the ozonation process. Figure 1 shows the dependence of total concentration of cyanides on the time of ozonation.

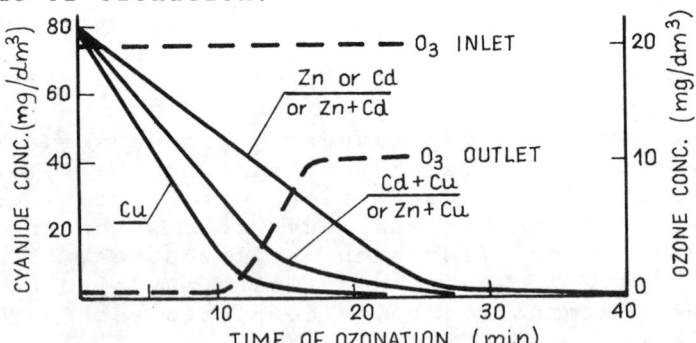

Figure 1. Changes of cyanides concentration during ozonation cyanide wastes containing heavy metals.

One can distinguish 2 or 3 stages of ozonation /3/. As it was showned, first and second stage of reaction are rapid, but the last stage indicates a reaction of zero order /concentration of cyanide below 2 mg/dm^3/ and rate of the reaction is close to zero.

The studies carried out since 1973 /4/ at the Institute of Precision Mechanics on the oxidation of cyanide plating effluents with the aid of ozone, have brought about the basis for a comprehesive solution of all problems related to this method /technology, equipment, measuring and control devices/. The basic studies defined the type and range of parameters that influence the process of ozonation in a cruicial way, i.e.:
- initial concentration of cyanides and ozone,
- qualitative composition of effluents,
- pH of effluents,
- type of phase contact.

The studies were performed on a laboratory, pilot - plant, and full-scale industrial plant. It was found that both the concentration of cyanides and ozone influence the kinetics of the reaction, however of decisive significance is the qualitative composition of effluents especially its heavy-metal content /5/.

Copper containing effluents is oxidized much faster than zinc and cadmium containing effluents /6/. Copper is a typical catalyst for the ozonation process. The pH of effluents should be varried from 7.5 to 10.5, which does not effect significantly the ozonation process.

EQUIPMENT FOR OZONATION PROCESS

Ozone generators

Ozone is generated on-site in the effluent treatment plant by ozone generators. Ozone is produced by passing dry air or oxygen through a high voltage corona discharges.

Usually ozonators are rated at 2% by weight ozone concentrations when air is used. New types of ozone generators operate at low voltage /10 - 12 kV/ and high frequency /600 Hz to 3 kHz/ /7/. This contributes to a lower power the consumptions about 12 to 16 Wh/g O_3, on air feed. Some of the characteristic parameters of ozonators developed at the Institute of Precision Mechanics /Warsaw - Poland/ are given in Tabele I.

Table I. Some technical data of IMPOZ ozone gener.

Specification	Unit	mini	labor	700	2000 HF
Ozone output	g/h	0.15	14.0	700	2000
Power consumpt.	kW	0.03	0.60	15	35
Cooling water	m^3/h	–	0.2	2.5	7.0
Weight	kg	5.5	60	2000	3500

Ozone contactors

The efficiency of ozone contactors will determine the lenght of the treatment time, as well as the losses on unconsumed ozone which will take place when poor ozone contactors are used.

Last time several companies have been developed high efficiency ozone contactors depending on the ozone technology adopted, e.g. bubble column, ejectors, Film Layer Purifing Chamber, static mixers, Frings turbine, etc. /8/

Beside highly efficient ozone generator, the second most important factor are the ozone contactors and system of ozonation process.

Control and Measuring devices

The number of control and measuring devices depends on the size of instalation and the level of the automatisation adopted. According to our experience, oxidation - reduction potential measurement /ORP/ is good indication of the progress of the ozone - cyanide reaction /see Fig. 2/.

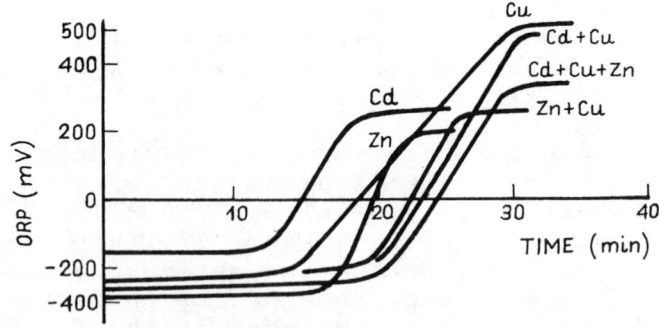

Figure 2. Changes of oxidation-reduction potential /ORP/ during ozonation cyanide wastewaters.

It is also possible to check any time the cyanide content in the treated effluents leaving the system by analitical methods. The initial concentration of ozone coming from generator is determined by the photometric method or periodically by the iodometric method.

TECHNOLOGY OF CYANIDE EFFLUENTS OZONATION

Up until recently the ozonation of cyanide waste waters presented many defects as compared with conventional technologies. On one hand, there was the obstacle of the high cost of equipment and problems of technological nature, on the other: low ozone concentration achieved by ozonators, complicated reactors for achieving adequate inter-phase contact, low rate of reaction and large losses of residual ozone. To date, the majoroty of research works have tended toward improving the efficiency and economy of ozone production. At the same time, efforts were made to choose parameters of absorption systems in such a way as to achieve appriopriate efficiency and utilization of ozone. Depending on the size of the treatment plant, the technical conditions and quality of effluents /rinse-water or spent concentrated solutions/, different ozonization technologies are used.

Apart from the specific systems for ozonation of hard-to-oxidize cyanide complexes /9/ is new technology: oxyphotolysis /ozonation with ultraviolet radiation/.

Now we have got several experiences with four basic ozonation technologies as follows:

Ozonation in a Semi-Periodic System

This method is used mainly on a pilot scale, less frequently in industry. It consists in ozonizing wastes in periodically filled columns, into which ozone is fed continuously. The efficiency of such a system depends to a large extend on phase contact, hence application of packing of the absorption column. Residual ozone which appears in the second kinetic stage of the reaction have to be absorbed in the tank of raw effluents, thereby becoming pre-oxidized. The limiting flow capacity of the system can be defined as 1.0 to 2.0 m^3/h cyanide wastes.

The Ozonation - Chlorination System

This system deserves attention chefly because of the simplicity of the installation. It consists in a two-stage cyanide oxidation: first - ozonation to the point of concentration of cyanides in the first kinetic stage /4-6 mg/dm^3/ signalled by the appearance of residual ozone, second - chlorination in the periodic or continuous /automatic/ system up to the total removal of cyanides /below 0.1 mg/dm^3/. The ozonation process can be carried-out directly in the effluents storage tank, diffusing the ozone with a bubbler similar to the pneumatic mixer in plating tanks /10/.

Ozonation in a Continuous System

This is the most commonly used system in the newly designed treatment plants. Both effluents and ozone are fed into the absorption column continuously, in a counter-current way. Treated effluents flow out of the column automatically after reaching a specified amount of excess ozone dissolved. Cyanide effluents flowning from the plating shop had been subjected to pH regulation /which turned out to be unnecessary/ and directed to the absorption column /capacity 2 m^3/, where ozonation took place.

Outflow effluents from the column occured on redox potential signal equivalent to about 0.2 mg/dm^3 of ozone excess dissolved in the effluent. In case of deficency of oxidizer the outflow of effluents was stopped to increase to ozone doze. Practice has shown that the measurement of the redox potential occures to be a sufficient signal of cyanides removal.

After mixing with acidic-alkaline effluents and final pH regulation effluents were directed to the settling tank and then discharge into the sewage system.

The principal adventage of this system is total automatisation of the ozonation process.

Re-circulatory ozonation System

The most interesting method of ozonation technology is the recirculatory ozonation system shown in Figure 3 which has been developed primarly for the existing large treatment plants operating in the conventional periodic system. /11/.

The system consists of a small absorption column /capacity 1-2m³/, ozone generator and an installation which allows connection of the circulation with pumps and storage tanks. The capacity of the system is about 4-6m³/h.

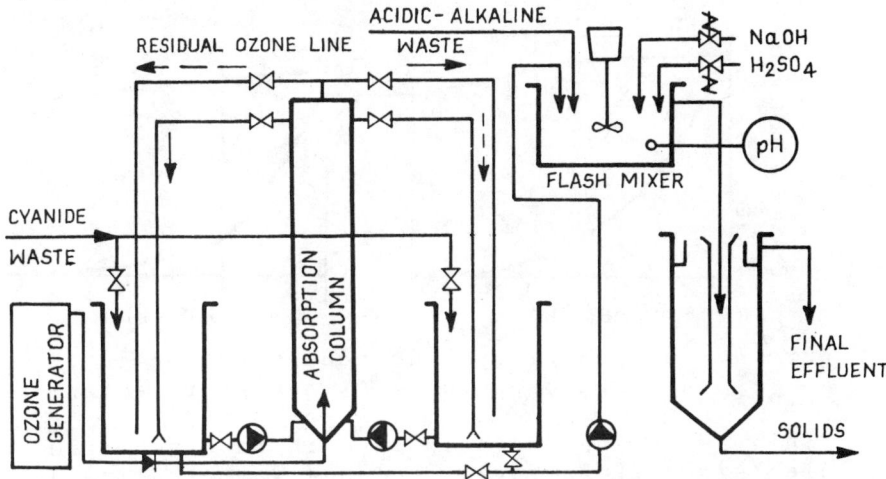

Figure 3. Flow-diagram plating waste treatment with recirculatory system of cyanide waste ozonation

Ozonation takes place in the absorption column, into which ozone is fed continuously while effluents are circulated; column-storage tank /conflow with ozone/.

The ozone outlet to the absorption column is connected with the storage tank of raw effluents, so that are aerated in the first phase and pre-ozonized with residual ozone in the second kinetic stage of reaction.

After the removal of cyanide, recirculation cycle turns up to the second storage tank.

Studies of the system led to the development of a modified version with a variable flow rate of effluents through the column, which combined with the preliminary ozonation of cyanides made possible a few-fold increase in the capacity of the system in relation to the semi-periodic system.

In optimal situation increased effluents flow should correspond with decreasing concentration of cyanides during ozonation process. This problem was too difficult for practical application in the industrial scale. Settlement by compromise was to employ two or three increase flows of effluent to be treated.

According to the graph figure 4, the final effect of this system was the increase of reaction time of about 20-25%, decrease of ozone consumption /2.2 to 2.8 gO_3/gCN^-/ and possibility of complete ozone utilization.

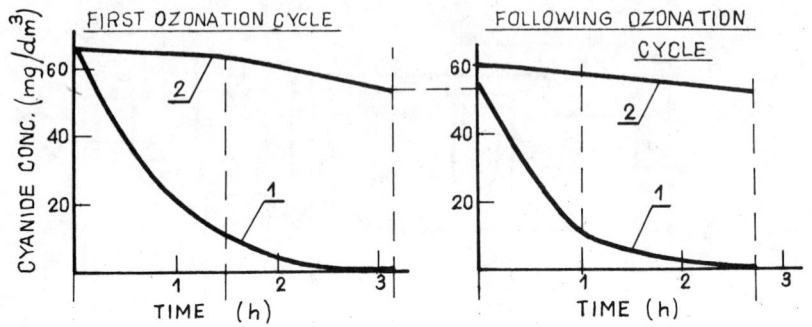

Figure 4. Changes of cyanide concentration in the absorption column /1/ and storage tank /2/

This system has been applied in the industrial scale /Aircraft Factory - Poland/ for alkaline cyanide waste waters containing up to 80 mg/dm^3 cyanide and heavy metals: copper, zinc and cadmium. Flow under normal condition 80 to 120 m^3 daily.

Ozone generators applied capacity was up to 20kg of ozone daily /with air feed/ . Air preparation unit provides the clean, oil free, particles free, dry air needed for the generation of ozone.

Absorption column has 0.6 m diameter and 5.5 m hight /1.2 m^3 of volume/. It co-operates with two storage tanks of 18 m^3 capacity each.

The total time of one cycle of ozonation lakes usually 2-3 hours to the final level of cyanide reaches 0.1 mg/dm^3.

After complete destruction of cyanide, effluents are pumped to the neutralization tank, mixed with acidic and alkaline waste waters and subjected to the final pH adjustment. The metal oxides and hydroxides settle out in settleing tank. The solid waste is removed periodically, as necessary and dewatering on the filterprasse. /two times a week/.

A failuse of any equipment, such as pump, control instruments and ozone generator is indicated by an audio-visual alarm system.

ECONOMIC ASPECT OF THE OZONATION CYANIDE WASTES

Investigations carried-out with cyanide effluents

treatment plant using ozone shows that the costs of ozone treatment is 20-30% lower than the cost of sodium hypochlorite treatment. However, the total of ozonation are higher than that of conventional methods used, this occures mainly because of high investment outlays in respect of ozone generators and installations which represent 50 to 60% of the cost of the whole tretment plant.

According to the economic analyses the total operation cost of cyanide oxidation with ozone for large plants are some how lower that for the conventional technologies /for treatment plant capacity above 600 g CN/h/ie. 1.5 kg O_3/h.

In addition, the arguments in favour of ozone are the advantages for the natural environment, consisting primarily in low salt content in effluents, savings of reagents, decreased COD and BOD and a better general quality of effluents /decomposition of phenols, detergents, organic compounds etc./.

One should also bear in mind the effects arising from elimination of transport and storage of the toxic reagents and losses resulting from their partial decomposition /e.g. in the case of NaOCl/.

According to our experience, the destruction of cyanide in plating waste water by ozone is simply, rapid and effective.

REFERENCES

1. Neuwirth, F.,Berg-u, Huttenman, Jabrb.Montan. Hohschule Loeben 81, 126-131, /1933/.
2. Tyler, R.G., Maske, W., Westing, M.J., and Mathews, W., Sewage and Industr. Wastes, 23,1150-1153 /1951/.
3. Sondak, N.E., Dodge, B.F., "The oxidation of cyanide-bearing plating wastes by ozone" Plating, 48, 280-284, /1961/.
4. Kunicki, W., Kołodko, K., Tuznik, F., "Oxidation of cyanide from plating wastes by ozone" Raport IMP/208.00.0001 /1975/.
5. Kołodko, K., Tuznik, F., "Ozone neutralization of the cyanide electroplating effluents", Powłoki Ochronne 3/19/, 33-38, /1976/.
6. Fabian, Ch., "Abwasser Reinigung nach dem Chemisier - Verfahren. Galvanotechnik, 2166, 100-107, /1975/.
7. Bollyky, L.J., "Ozone treatment of cyanide and Plating Waste", First Symposium IOI, Washington, DC /1973/.

8. Stopka, K., "New conceptin ozonation", Ozonair Corp. publ. /1977/.
9. Prengle, H.W., Jr, Mauk, C.E., Charles, E., "Ozone / UV Chemied oxidation wastewaters process for metal complexes, organic species and desinfection" 83rd National Meeting of AICh.E, Houston, Tx /1977/.
10. Bahenski, V., Zika, "Treating Cyanide Wastes by Oxidation with Ozone", Koroze Ochrana Mater., 10 /1/, 19-21 /1966/.
11. Kunicki, W., "Utilisation of galvanic cyanide effluents with ozone - technology and installations", Powłoki Ochronne, 4-5, /44-45/, 10-15, /1980/.

APPLICATION OF PROGRAMMABLE CONTROLLERS TO BATCH TREATMENT OF CHROME AND CYANIDE WASTES

Tarik Pekin, Senior Engineer

Andrew P. Gagnon, Senior Engineer

Camp Dresser & McKee Inc.
Boston, Massachusetts

INTRODUCTION

This paper discusses the application of a programmable logic type controller to batch treatment of chrome and cyanide wastes.

Toxic hexavalent chrome (Cr VI) and cyanide (CN) are found in wastewaters generated by many electroplating operations. The Environmental Protection Agency (EPA) pretreatment standards (1) for existing sources (PSES) limit the amount of these and other toxic wastes that can be discharged into publicly owned treatment works (POTW). Table 1 below summarizes the EPA pretreatment standards.

Table 1. Pretreatment Standards (mg/l)

Pollutant	1-day max.	4-day avg.
Less than 10,000 gal/day Discharge		
Cyanide amenable to chlorination	5.0	2.7
Lead	0.6	0.4
Cadmium	1.2	0.7
10,000 gal/day or More Discharge		
Total Cyanide	1.9	1.0
Copper	4.5	2.7
Nickel	4.1	2.6
Chromium	7.0	4.0

Zinc	4.2	2.6
Lead	0.6	0.4
Cadmium	1.2	0.7
Total Metals	10.5	6.8

The EPA regulations should be referred to for additional information including applicability of standards, compliance dates and mass-based optional limitations. Effluent limitations applicable to direct discharge to surface waters have not been promulgated yet, but it should be expected that they will only be more stringent than pretreatment standards. In any case, chrome and cyanide in most electroplating wastewaters occur at levels well above present and expected future regulatory limits.

TREATMENT METHODS

Hexavalent chrome or cyanide cannot be removed by conventional (alkaline) precipitation, so, wastestreams carrying these contaminants are segregated into chrome and cyanide wastestreams for appropriate pretreatment prior to precipitation.

The remaining electroplating wastestreams are commonly termed as acid/alkali wastes. Copper, nickel, zinc and cadmium in the acid/alkali wastes can all be precipitated as hydroxides at a pH of about 10.5. Lead will also precipitate at the same time as a lead carbonate if carbonates are present or added to the wastestream (2).

Segregated chrome and cyanide wastes generally constitute more concentrated wastestreams, as compared to the acid/alkali wastes, and occur at lower flow rates. Therefore, these concentrated wastes are more economically treated by batch type equipment (4) up to a flow of about 20 gpm. This paper addresses batch type equipment only. Detailed discussions of the chrome reduction and cyanide oxidation methods follow.

Chrome Reduction

Chrome in its trivalent state is virtually insoluble at a pH range of 8.0 to 9.9 (3). Therefore, reduction of hexavalent chrome to trivalent chrome followed by precipitation has been widely utilized and proven to be an economical treatment method.

The most widely used chrome reducing agents are gaseous sulfur dioxide (SO_2) and sodium metabisulfite ($NA_2S_2O_5$). Sodium metabisulfite (or its hydrolyzed form) is generally more suitable for small installations and batch treatment systems. As the oxidation-reduction reaction with $NA_2S_2O_5$ is highly pH-dependent, sulfuric acid is added to maintain the pH at a desired level of about 2.0. Sodium metabisulfite hydrolyzes to sodium bisulfite (3) and subsequently dissociates to generate sodium hydroxide (NaOH) which adds to the acid requirements. The overall reaction for reduction of hexavalent chrome to trivalent chrome with the addition of sodium metabisulfite and sulfuric acid is as follows:

$$4H_2CrO_4 + 3NA_2S_2O_5 + 3H_2SO_4 \rightarrow 3NA_2SO_4 + 2Cr_2(SO_4)_3 + 7H_2O$$

The reduced chrome is most easily precipitated as a hydroxide by addition of an alkali such as lime or NaOH.

Reduction of hexavalent chrome with sodium metabisulfite can be automated by monitoring two parameters: pH and oxidation-reduction potential (ORP). Figure 1 shows the typical relationship between ORP levels and sodium metabisulfite addition when the pH is maintained at 2.0. Before reagent addition, the ORP level would be about 600 mv. As the reagent is added, the ORP level will drop, and a level of 300 mv will indicate that the reaction is complete. The overall reaction may require about 15 minutes for completion.

Cyanide Oxidation

Cyanide can be oxidized to nitrogen gas and carbon dioxide by reacting with elemental chlorine or hypochlorite. Sodium hypochlorite (NaOCl) is generally more suitable for small systems. The first step of the reaction is oxidation of cyanide to toxic cyanogen chloride (CNCl), which is instantaneous at all pH levels (3):

$$NaCN + HOCl \rightarrow CNCl + NaOH$$

The second step is the oxidation of cyanogen chloride to cyanate (CNO); this step is slow at pH levels below 8.0, but it goes to completion in 30 minutes at levels above pH 8.5:

$$CNCl + 2NaOH \rightarrow NaCNO + 2H_2O + NaCl$$

The third step of the reaction is favored by alkaline conditions, and a pH range of 8.5 to 9.0 is generally provided:

$$2NaCNO + 3NaOCl + H_2O \rightarrow 3NaCl + N_2 + 2NaHCO_3$$

Cyanate is broken down into nitrogen gas and carbon dioxide (in carbonate form). In summary, all reactions can be carried out at a pH range of 8.5 to 9.0.

Oxidation of cyanide can be automated by monitoring the same parameters as for the chrome reduction: pH and ORP. Figure 2 shows the typical relationship between ORP values and chlorine (or hypochlorite) addition when the pH is held at 8.5 to 9.0. Before chlorine addition, the ORP level would be at about -300 mv. Following chlorine addition, an ORP level of above +300 mv will indicate that the oxidation has been com-

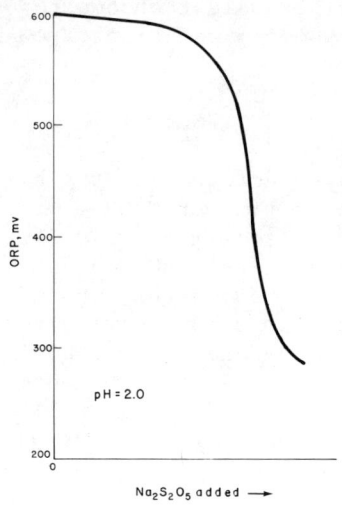

FIGURE 1
CHROME REDUCTION
ORP vs. $Na_2S_2O_5$ ADDED

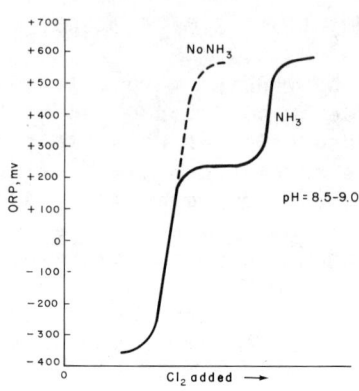

FIGURE 2
ALKALINE CHLORINATION OF CYANIDE
ORP vs. Cl_2 ADDED

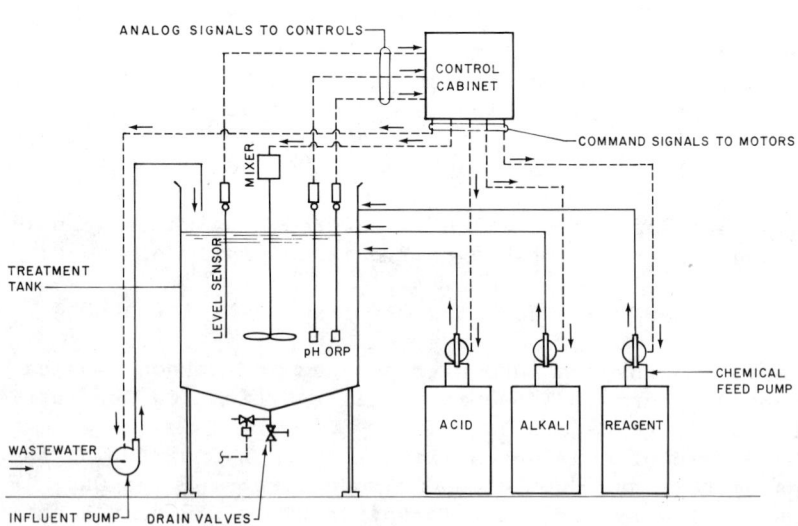

FIGURE 3 - TYPICAL BATCH TREATMENT SYSTEM

pleted. The actual amount of chlorine needed will be higher than the stoichimetric requirement for cyanide oxidation alone when other oxygen-demanding contaminants -- such as ammonia -- are present.

BATCH TREATMENT EQUIPMENT

Figure 3 shows a typical batch treatment system. The major equipment components are similar for chrome or cyanide treatment because of the similarities in treatment steps and control parameters.

The raw hexavalent chrome or cyanide wastes are pumped from a sump or storage tank into the batch treatment tank, which is equipped with a mechanical mixer. The level sensor provides information to the control system that causes acid (H_2SO_4) or alkali (lime) pumps to turn on or off to provide the desired pH levels. The ORP level is monitored in the same manner, and the reagent pump is turned on or off until the design ORP levels are reached. Precipitation in the batch treatment tank can be provided when required. Sludge or treated effluent is discharged by opening the drain valves. Automation of the treatment system, therefore, involves control of mechanical equipment (mixer, valves and pumps) operations as a function of the liquid levels and process parameters of pH and ORP. The following summarizes how a programmable logic controller was used in a specific project to accomplish the automatic control task. Some of the project details have been generalized for the purposes of this paper.

THE CONTROL SYSTEM

Overall Control Concept

The project involved chrome treatment and cyanide treatment systems of 1,500-gallon batch capacity each and flocculation/clarification facilities for 70,000 gallons per day of acid/alkali wastes. Although the wastewater pretreatment facility is relatively small, it was determined early in the design phase of the project that centralized control and monitoring would be the most effective and economical approach. Some consideration was given to a small computer system, but it was quickly recognized that this would introduce a higher level of sophistication than was actually required and would not be cost effective.

Because of the limited number of loops required, it was decided to use dedicated analog controllers for all the pH and ORP control loops. It was also determined that conventional analog indicators and strip chart recorders would adequately define the condition of the process and provide adequate historical data for plant record keeping.

The control alternatives for the batch processing system

then were narrowed down to a choice between conventional relay logic and a programmable logic controller (P.L.C.). There was no clear-cut cost advantage in using a P.L.C. for the batch sequence only, but extending the use of the P.L.C. to perform other facility control and monitoring functions, such as influent pump controls and process alarming, made it the most cost-effective choice.

Central Control Facilities

The instruments and controls for the entire waste treatment facility are contained in a central control "console" with two sloping mounting surfaces, as illustrated in Figure 4. Removable rear panels and front door provide access to internally mounted equipment and connections. The console is located in a centrally located office/control room enclosed with glass viewing panels, enabling the operator to oversee most of the facilities. It contains all necessary recorders, indicators and controllers as well as all automatic and manual controls. A graphic display of the entire waste treatment process is also included on the face of the console. The instrument and control systems in the plant are totally "electronic", and all analog signals are standard 4-20 mA signals.

Programmable Logic Controller

Beginning with the initial design phase, it was the intent to utilize the programmable logic controller to the maximum extent possible. With this basic design approach, it was determined that the P.L.C. could be effectively utilized for the following purposes:

A. As a substitute for all control relays, timers and time delay relays.
B. As a substitute for current trip, or alarm switches.
C. As a substitute for an alarm annunciator.

The P.L.C. and its peripheral equipment is illustrated in Figure 5. The system is available from the manufacturer with all interconnecting cables, and includes Central Control Unit, Power Supply, Analog Input Models, Discrete Input Modules, Discrete Output Modules, Read/Write Programmer, Timer Access Module, Loop Access Module, and Cassette Tape Loader. The Central Control Unit is a 16 bit microprocessor with 4,000 word memory. Since the actual memory required for the project is estimated at 1,500 to 2,000 words, there is adequate capacity for any future expansion requirements.

The actual input-output requirements for the system are relatively small; i.e., approximately 32 discrete inputs, 18 discrete outputs, 9 analog inputs, and no analog outputs. By comparison, the system provided is capable of expansion to

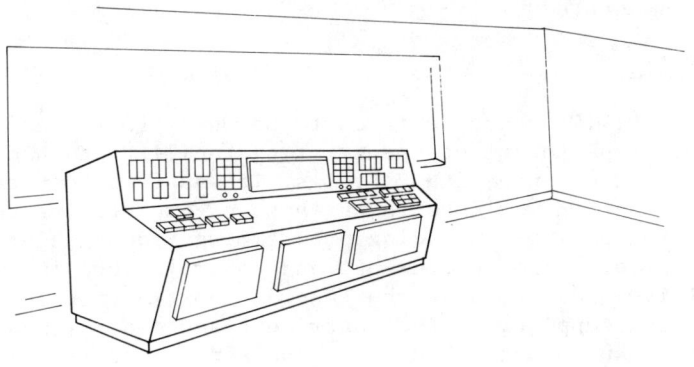

FIGURE 4
CENTRAL CONTROL CONSOLE

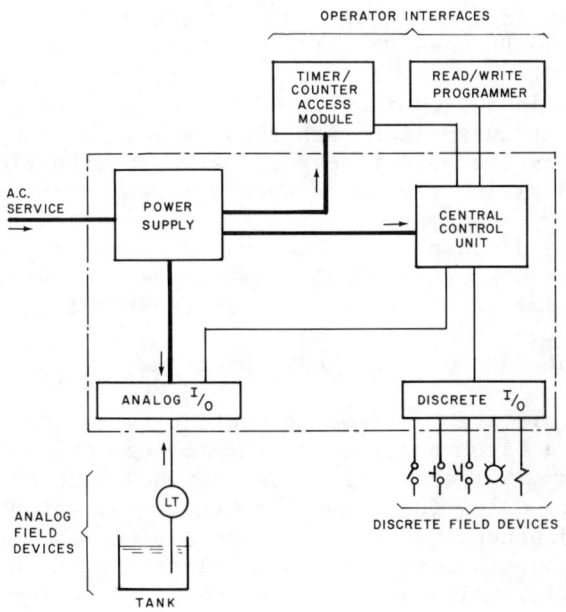

FIGURE 5
PROGRAMMABLE LOGIC CONTROLLER
SYSTEM BLOCK DIAGRAM

accept up to 64 analog inputs or outputs and up to 96 discrete inputs or outputs, and it can functionally replace up to 99 timers or counters.

Program Logic

Most P.L.C.'s are designed to be user programmed, but because plant personnel at the existing facility had had little practical experience with P.L.C.'s, it was felt that the best course of action would be to purchase a complete instrument and control package including programming from a major instrument supplier. The instruments and controls were included for competitive bidding under the general construction contract. Because the supplier of the instrument and control system would be unknown until bid time, and since most manufacturers use different programming techniques, it was decided to establish and write the logic for all treatment systems in a "neutral" format or language rather than a detailed ladder diagram format. The format chosen had to convey all the functional requirements of the systems and provide sufficient information so that an instrument supplier could easily translate this into the ladder diagram format required by the actual equipment. In this case, logic requirements were illustrated using symbology from ISA Standard 55.2 entitled "Binary Logic Diagrams for the Process Operations." This format was chosen because of the fact that it is a published standard by a nationally recognized organization with clearly defined symbology. For a user unfamiliar with this symbology, however, the logic can be just as effectively illustrated with flow charts using Yes/No decision blocks or even written functional descriptions.

A detailed discussion of the controls and programming logic is given below covering only the chrome and cyanide batch treatment systems of the wastewater pretreatment facility.

Batch Treatment Controls and Logic Diagrams

The controls are arranged to provide two basic modes of operation for a batch program: either manual or semi-automatic. Each pump, valve and mixer is provided with a "HAND-OFF-AUTO" switch to allow full manual control of a batch program completely independent of the programmable controller. For semi-automatic operation, all H-O-A selector switches are left in the automatic position. Semi-automatic operation also requires pushbutton initiation for filling, starting the batch sequence and draining the treated waste. (These pushbutton controls are the same for chrome and cyanide batch treatment and are illustrated in Figure 6.) All other control functions and interlocks for the batch sequence are then carried out automatically through the programmable controller.

The logic sequence diagram for batch treatment of chrome wastes is shown in Figure 7. The same diagram was provided in the project construction specifications. (The logic diagram for cyanide treatment is very similar in concept and therefore is not presented here.)

The logic diagram at first glance may appear to be quite complex. Once a few basic symbols are understood, however, it becomes apparent that the complex diagram is nothing more than a series of relatively simple logic statements. There are four basic logic symbols used in this example. There are a number of other symbols that may be utilized for more complex sequences, but these four symbols will adequately define the batch process we are concerned with here.

These symbols and their definitions are illustrated in Figure 8. The reader should become familiar with these symbols before review of the logic sequence diagram in Figure 7. Batch treatment of chrome wastes proceeds as follows according to the logic in Figure 7:

Initiation

The process is initiated by first transferring the waste from storage to the batch tank. This is accomplished by depressing the "Fill Tank" pushbutton (see Figure 6). This will start the transfer pump but only if the storage tank level is above a certain minimum level and the batch tank is empty. During this process, the "Tank Filling" light is illuminated. When the batch tank is full, the transfer pump is automatically stopped, and the "Tank Full - Start Batch" light is illuminated, signalling to the operators that the batch treatment can be started.

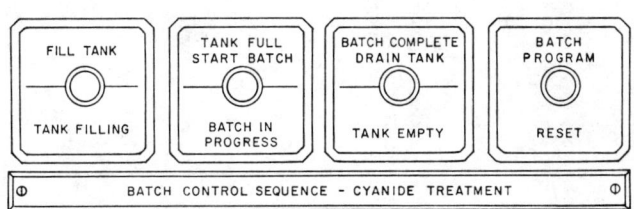

FIGURE 6
ILLUMINATED PUSHBUTTON CONTROLS FOR SEMI-AUTOMATIC BATCH SEQUENCE

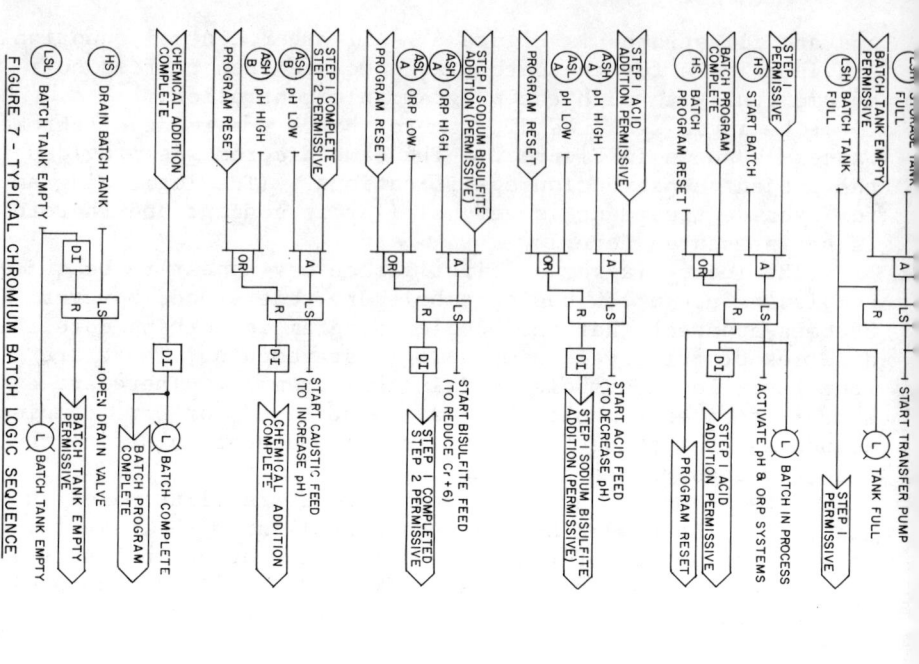

FIGURE 7 - TYPICAL CHROMIUM BATCH LOGIC SEQUENCE

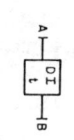

AND SYMBOL

Output D exists only if all inputs A, B, and C exist (equal to series circuit)

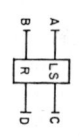

OR SYMBOL

Output D exists only if one or more inputs A, B, and C exist (equal to parallel circuit)

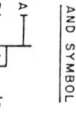

MEMORY SYMBOL

Output C exists as soon as input A exists (set memory) and will continue to exist until input B exists (reset memory). Output D exists only if C does not.

TIME ELEMENT SYMBOL

Existence of input A for time t causes output B to exist when time t expires. B terminates when A terminates.

FIGURE 8 - BASIC LOGIC SYMBOLS

124

Chrome Reduction (Step 1)

Depressing the "Start Batch" pushbutton illuminates the "Batch in Progress" light, starts the tank mixer and activates the tank pH and ORP systems. After a time delay period, the acid feed pump is started to lower the pH. The pump is allowed to run until pH reaches a predetermined minimum value (approx. 2.0). The pump will restart if the pH goes above a set high level (approx. 2.5). After the pH has remained within this range for a pre-set time period, the sodium bisulfite pump is started to reduce the hexavalent chrome to trivalent chrome. The pump is allowed to run until the ORP level reaches a certain minimum value (approx. 300 mv).

Clarification (Step 2)

After the ORP has remained at this low value for a pre-set time period (approx. 15 min.) the caustic feed pump is started to increase pH. The pump is allowed to run until the pH reaches a pre-set value of about 8.5, which enables precipitation of trivalent chrome.

After this pH value has been maintained for a pre-set time period (approx. 2 hours), the "Batch Complete - Drain Tank" light is illuminated, indicating to the operator that the treatment is complete and the chrome sludge can be removed.

The operator may hold the treated batch in the batch tank as long as desired.

After the sludge is pumped through a separate manually operated sludge drain valve, the tank can be emptied by depressing the "Drain Tank" pushbutton which will open the main tank drain valve. A low level switch will be activated when a pre-set low level is reached. After a pre-set time period (approx. 1 min.), the valve is closed, the "Batch Tank Empty" light is illuminated and all circuits are reset, allowing a new batch to be processed.

In addition, the entire program can be reset during any step by pushing the program reset button.

CONCLUSION

The choice of a P.L.C. for batch treatment application such as we have described is a natural one, because it provides the timing and logic capability necessary for the various control sequences required and also provides convenient means of process alarm sensing and annunciation.

The same programming logic can be used for continuous treatment of chrome or cyanide wastes. In fact, continuous treatment would involve fewer programming steps because separate tanks for each treatment step inherently provide the treatment sequence and timing.

One of the major advantages of a P.L.C. over hard wired relay logic is that the P.L.C. is "forgiving;" that is, it is much easier to correct programmed logic in software than it is to change or require hardware. Even so, changing software programs and debugging can be time consuming and costly. Therefore, it is still very important to plan carefully to ensure that the P.L.C. receives all necessary inputs and outputs and has sufficient memory capacity to accept any anticipated expansion. It is anticipated that the P.L.C. will provide trouble-free operation and ultimately help to reduce operating costs, by minimizing spare parts inventories and service and maintenance requirements.

REFERENCES

1. Environmental Protection Agency, "Effluent Guidelines and Standards; Electroplating Point Source Category; Pretreatment Standards for Existing Sources," Federal Register, 46 (18), 1-28-81, p. 9462.

2. Sittig, M., Electroplating and Related Metal Finishing, Pollutant and Toxic Materials Control, Noyes Data, p. 270, 1978.

3. Eckenfelder, Jr., W.W., Industrial Water Pollution Control, McGraw-Hill, P. 121, 119, 130, 1966

4. Environmental Protection Agency, "Economics of Wastewater Treatment Alternatives for the Electroplating Industry," EPA 625/5-79-016, p. 6 (1979).

SESSION II

BIOLOGICAL TREATMENT
(Principles and Applications)

BIOLOGICALLY ACTIVE SAND FILTERS
PART I
LABORATORY AND FIELD STUDIES

Laboratory Studies:
Roque A. Román Seda. Department of Civil Engineering, University of Puerto Rico, Mayaguez, Puerto Rico.

Héctor R. Fuentes. Civil and Environmental Engineering Department, Vanderbilt University, Nashville, Tennessee.

W. Wesley Eckenfelder, Jr. Civil and Environmental Department, Vanderbilt University, Nashville, Tennessee.

Field Studies:
Roque A. Román Seda and W. Wesley Echkenfelder, Jr.

Laboratory and field studies were performed with an upflow sand filter at hydraulic loads of 2, 4, 6 and 8 gallons per minute per square foot (gpm/sq. ft.) for polishing secondary effluents. Biological activity was observed through dissolved organics (TOC and BOD) and dissolved oxygen (DO) uptake as a function of filter depth and filtration time. Other filtration variables such as suspended solids (SS), headlosses (ΔH), and water temperature (Tw) were also observed.

LABORATORY STUDIES

Biological activity has been observed at German and French water treatment plants(1); yet no systematic data on biological activity on sand filters is known to the authors other than the one presented herein.

Sand filtration of secondary effluents involves the removal of suspended solids which are essentially composed of active biomass residues not removed by secondary settling processes. In well-operated activated sludge plants these residues are usually accompanied by dissolved organic concentrations of aproximately the same magnitude (which is to say 20-25 mg/ℓ or a food to microorganism ratio (F/M) of 1.0).

Therefore, if the remaining dissolved organic fraction were biodegradable enough and the DO in the wastewater were restored, one could expect aerobic biological activity to take place. However, as filtration proceeds, biological solids accumulate within the sand filter, decreasing the F/M ratio with time and consequently increasing the overall removal of biodegradable dissolved organics. It was the purpose of the first phase of the laboratory studies to investigate the extent to which biological activity might take place.

Laboratory Studies With Air Aeration

Procedure: Filtration studies were performed in the laboratory using a 3 inch I.D. translucent 8 feet high acrylic column filled with 4 feet of white silica sand(Figure 1.)of 0.88 mm. effective size (E.S.)and 1.16 uniformity coefficient, with an overall average porosity of 0.43. This depth was chosen based on recommendations from other investigators (2,3,4,5,6) in order to prevent premature bed lifting. A secondary effluent was simulated in the laboratory by utilizing a 75% by weight feed of dextrose, 25% of peptone, and a trace of diammonium hydrogen phosphate $(NH_4)_2 HPO_4$ as nutrients. This **feed** was diluted to approximately 25 mg/ℓ dissolved BOD in a 55 gallon(gal)aeration tank prior to filtration. A concentrated mixed liquor suspended solids(MLSS) suspension was also diluted to approximately 25 mg/ℓ suspended solids. This mixture was aerated in order to bring dissolved oxygen levels as close to saturation as possible. This mixture was then pumped into the upflow filter at constant flow.

Influent and effluent dissolved organics(TOC basis),and suspended solids were monitored with filtration time, as well as depth profiles of dissolved oxygen for hydraulic loads of 2, 4, 6, and 8 gpm/sq. ft.

Results: Results of all these runs are shown on Figure 2. Suspended solids removal efficiencies were very high, as expected. Water temperatures ranged from 22°- 25°C. All reported values represent time averages over the entire run.

Removal efficiencies of 26,40,38, and 21 percent(TOC basis) for dissolved organics were observed at 2, 4, 6, and 8 gpm/sq.ft, respectively.(See Figure 3.) When the observed influent and effluent values were converted to BOD basis (using a theoretical ratio for the synthetic waste of 0.375 grams of carbon per gram of BOD) the above removal efficiencies were seen to represent 85,91,88 and 56 percent at 2,4,6, and 8 gpm/sq.ft, respectively, which was judged to be very encouraging.

Dissolved Oxygen profiles revealed that the bulk of oxygen removal took place within the first foot of filter

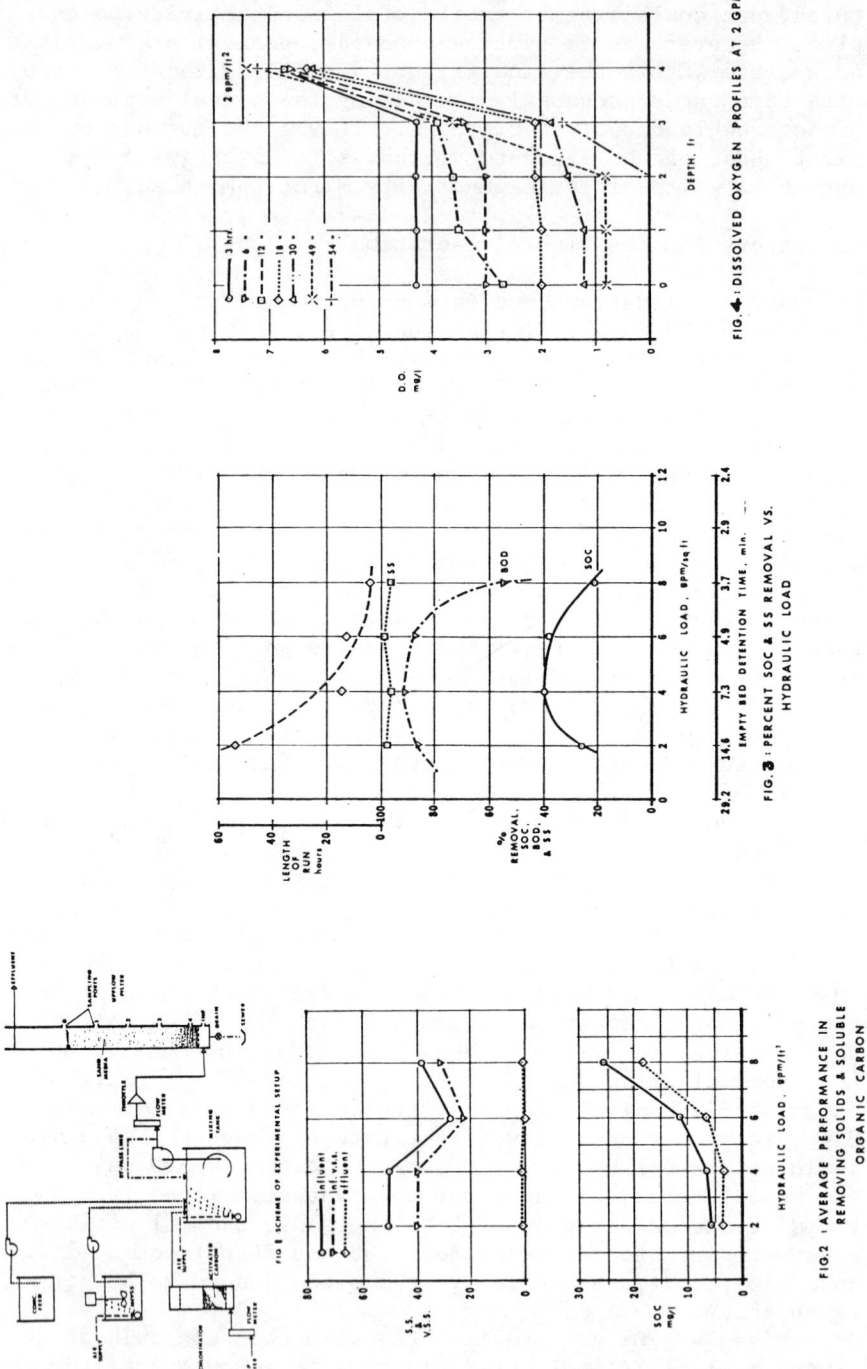

FIG. 4: DISSOLVED OXYGEN PROFILES AT 2 GPM/FT²

FIG. 3: PERCENT SOC & SS REMOVAL VS. HYDRAULIC LOAD

FIG. 1 SCHEME OF EXPERIMENTAL SETUP

FIG. 2: AVERAGE PERFORMANCE IN REMOVING SOLIDS & SOLUBLE ORGANIC CARBON

for the runs at 2 and 4 gpm/sq.ft. (See Fig. 4) This pattern of course meant that the biological solids were accumulating mainly within this first foot of filter. This result was somewhat surprising to us since we had expected the solids to be more evenly distributed throughout the filter bed because of the upflow mode. Removal of oxygen for the runs at 6 and 8 gpm/sq ft. followed similar patterns except that penetration was shown to be significant in the first 1.5 feet, as could be expected from the increased driving force.

At the end of these experiments it was decided to investigate whether higher oxygen tensions in the water (introduced through pure oxygen aeration) would help improve dissolved organic removal efficiency.

Laboratory Studies With Pure Oxygen Aeration

Procedure: A similar series of experiments was performed at 2, 4, 6, and 8 gpm/sq. ft. using pure oxygen as the aerating gas. Sand, filter, and waste characteristics were identical to the ones utilized for the original series of experiments, except that this time the aeration tank was enclosed and had a gas exhaust line connected to a ventilated laboratory safety hood.

This time profiles of dissolved organics and headlosses were made as a function of filtration time in addition to dissolved oxygen. Influent and effluent suspended solids were also observed, in addition to water temperature.

Results: Suspended solids removal efficiencies were again observed to be very high, as shown in Figure 5. Water temperatures ranged from 19° to 25°C.

Influent dissolved oxygen concentrations varied from 11-12 mg/ℓ at the end of a run. Average influent dissolved oxygen concentrations were 19,19,21 and 16 for 2,4,6, and 8 gpm/sq. ft. respectively.

Dissolved organic removal efficiencies are shown in Figure 6 compared to those observed using air aeration. As can be seen, removal efficiencies were essentially the same in both cases for all hydraulic loads tested. These results led us to conclude that higher oxygen tensions do not have any effect on dissolved organic uptake. This is really not so surprising when one recalls that at concentrations greater than 0.5 mg/ℓ oxygen is not a limiting substance.

Once again the dissolved oxygen profiles (See Figure 7) show that biological activity was concentrated within the first foot of the filter. This conclusion was supported by the headloss data shown in Figures 8 and 9.

Control Run

The experiments with both air and oxygen aeration

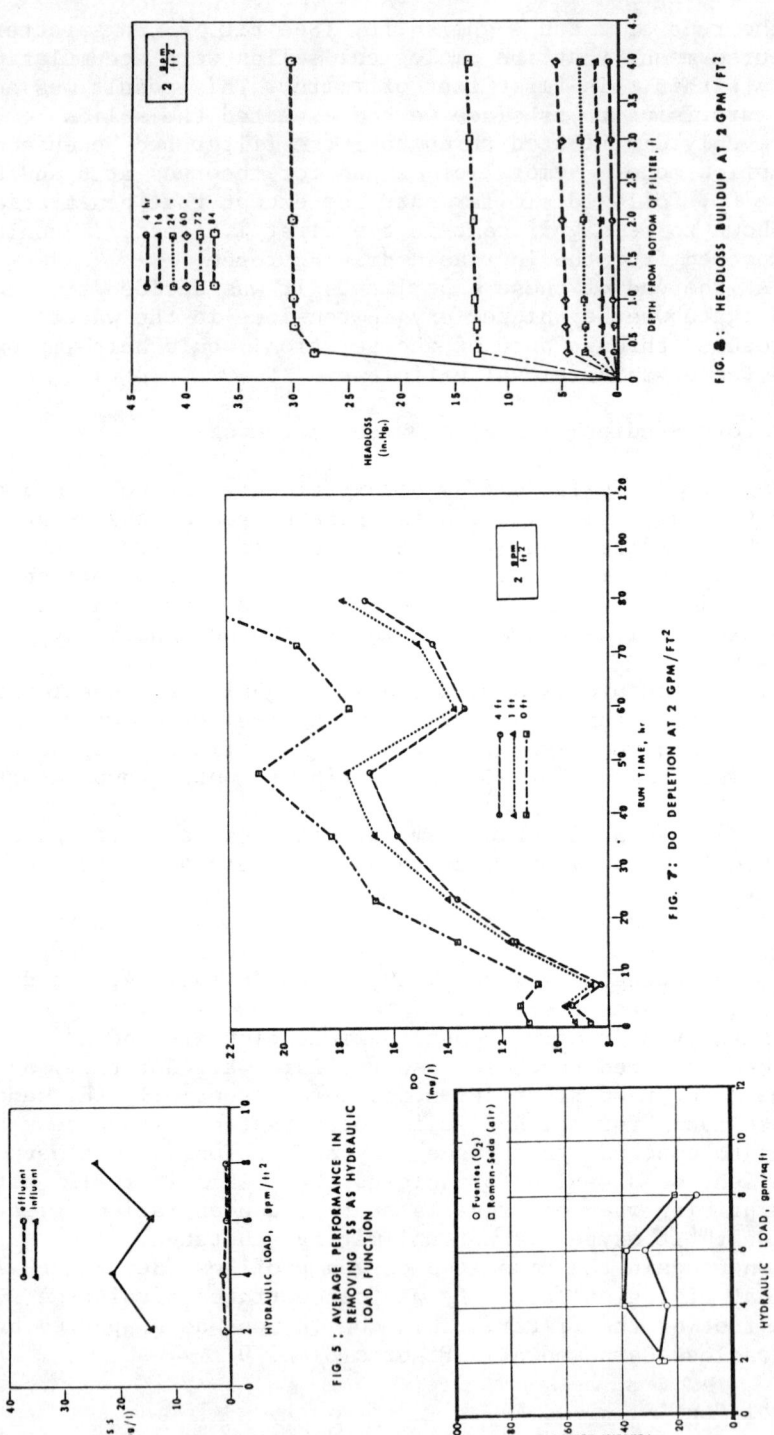

FIG. 5: AVERAGE PERFORMANCE IN REMOVING SS AS HYDRAULIC LOAD FUNCTION

FIG. 7: DO DEPLETION AT 2 GPM/FT²

FIG. 6: HEADLOSS BUILDUP AT 2 GPM/FT²

FIG. 8: REMOVAL EFFICIENCIES FOR DISSOLVED ORGANICS USING PURE OXYGEN AERATION

demonstrated that significant biological activity took place within the filter(as evidenced by the depletion of dissolved oxygen), and that simultaneous removal of dissolved organics also took place. But the obvious inference that the microorganisms were indeed responsible for the organic uptake is not totally valid unless one can demonstrate that other physical-chemical removal mechanisms are not responsible for the disappearance of dissolved organics from the water.

A control run was therefore executed in which no aeration was performed prior to filtration and in which all oxygen was purged from the simulated secondary effluent. This condition was meant to reflect the near oxygenless condition of secondary effluents upon emergence from the secondary clarifiers.

The run was executed at 5 gpm/sq. ft. since the data from both series of experiments led one to expect a maximum uptake of dissolved organics to take place at that hydraulic load.

The rationale behind this run was that if no significant uptake of dissolved organics was observed, then surely biological activity had to be the responsible mechanism for the dissolved organic removal observed in the other runs. The run was performed under identical sand, waste, and filter characteristics, except as has already been noted.

Observed influent and effluent values of dissolved organic carbon (SOC) are shown in Figure 10. Average removal (TOC basis) was observed to be 6.4 percent, which when compared to the observed removals of all other runs can be considered negligible. (Average water temperature was around 23°C) As can be seen, no removal was exhibited for the first 12 hours, while removal took place only in the last six hours of the run. We hypothesize that the small amount of removal shown at the end of the run was probably due to anaerobic activity. We therefore concluded that aerobic biological activity must indeed be responsible for removal of dissolved organics in the previous two series of filtrations runs.

We were now in the position to proceed with a detailed study of biological activity in sand filters under actual field conditions.

FIELD STUDIES

Experimental Procedure

An 8 ft. high, 11.5 inch I.D., clear acrylic filter column was designed and built on a 4 ft. x 6 ft. platform on wheels (see Figure 11). A plastic 400 gal holding tank was placed next to the filter column for aeration purposes. A secondary effluent (prior to chlorination) from the treatment plant at which the filter had been located was pumped

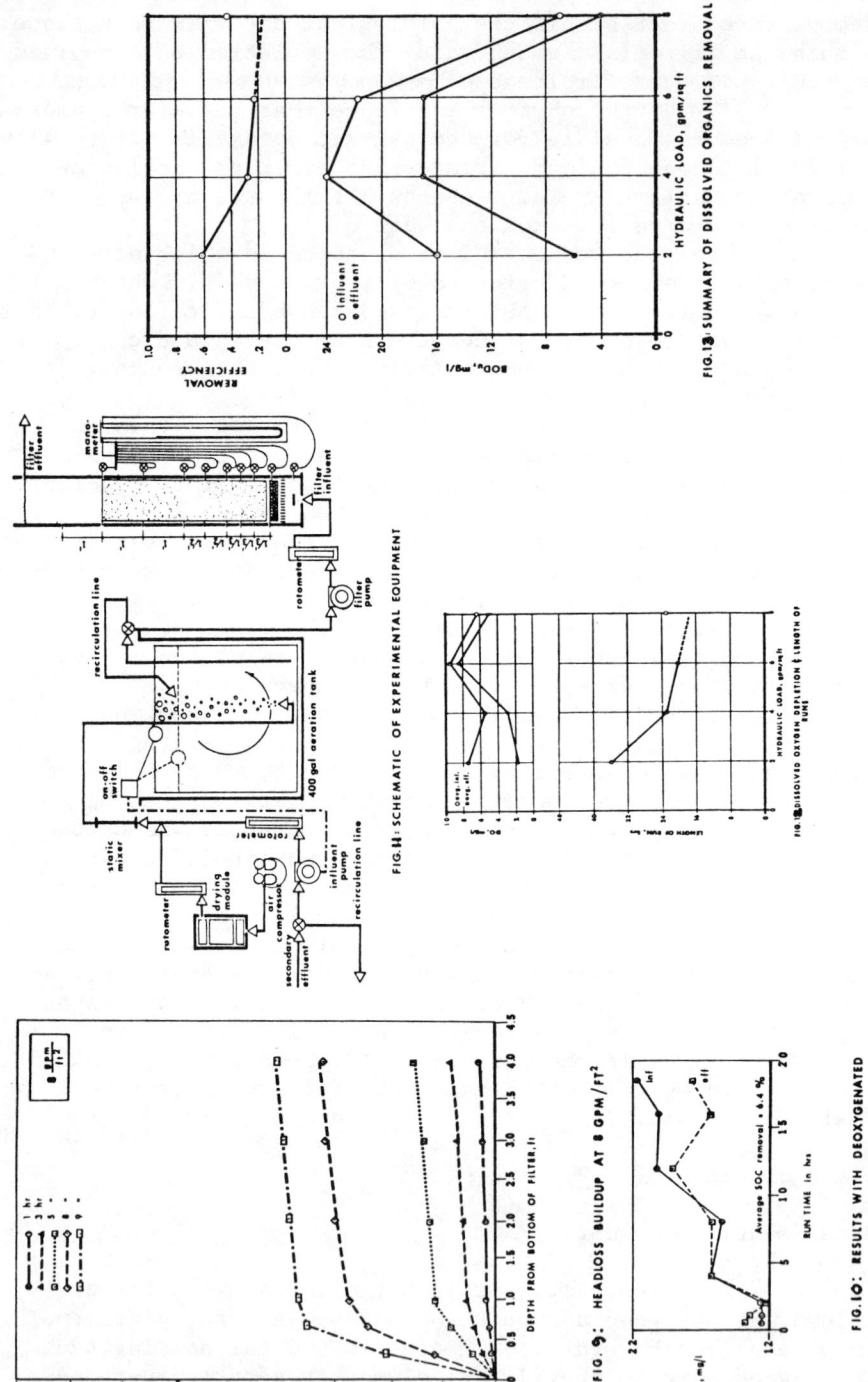

to the aeration tank. A compressor enjected air into the waste just ahead a static in-line mixer and the mixture was then discharged at the bottom of the aeration tank. A level switch controlled the influent pump so as to maintain water levels within the tank in a narrow runge. Both water and air flows were monitored through calibrated rotometers.

The aerated mixture was brougth to as near normal dissolved oxygen saturation as possible and was then pumped in the upflow mode into the filter. Runs were performed at constant flow.

Sampling ports were located at the influent, 1/3, 2/3, 1, 2, 3, and 4 ft.(effluent) levels in the filter column. Water was continuously run through the sampling ports in order to avoid disturbing the filter bed when taking samples. Pressure lines were brought from these same depths to a 30-inch differential pressure manometer filled with mercury.

The filter was taken to the Bellevue Sewage Treatment Plant which services a suburban community 30 minutes west of Metropolitan Nashville on Highway 70. The plant consisted of an activated sludge system without primary clarification which had been designed to treat 500,000 gal/day,(gpd)but which was treating approximately 250,000 gpd during the period of the field studies. The plant had a microstrainer to polish its secondary effluent but this was purposely put out of operation during the days in which the experiments were being performed. The plant discharged after chlorination into the Harpeth river, which runs south and west of Nashville.

The filter bed utilized white silica sand, but this time the sand had an effective size of 0.85 mm, a uniformity coefficient of 1.7 and a median size of 1.31 mm. The average sand porosity was 0.48 and because of the sand geometry, the non-sphericity correction factor Ψ was approximately equal to 0.7(7).

Runs were again performed at 2,4,6, and 8 gpm/sq ft. Dissolved organics (BOD basis), suspended solids, dissolved oxygen and headlosses were monitored as a function of depth and time. Water temperature was also observed.

Results of Field Studies

Dissolved organic removal for all runs is summarized in Figure 12. As can be seen, biological activity was present at all hydraulic load rates, but only at 2 gpm/sq ft. was the removal large enough to be of practical use. This conclusion is also supported by the summary of dissolved oxygen data in Figure 13. The data suggests that acceptable removals may be obtained at 3 gpm/sq. ft.

The observed data also show that the most significant accumulation of solids occurs within the first foot of filter, as evidenced by Figures 14,15, and 16. These figures

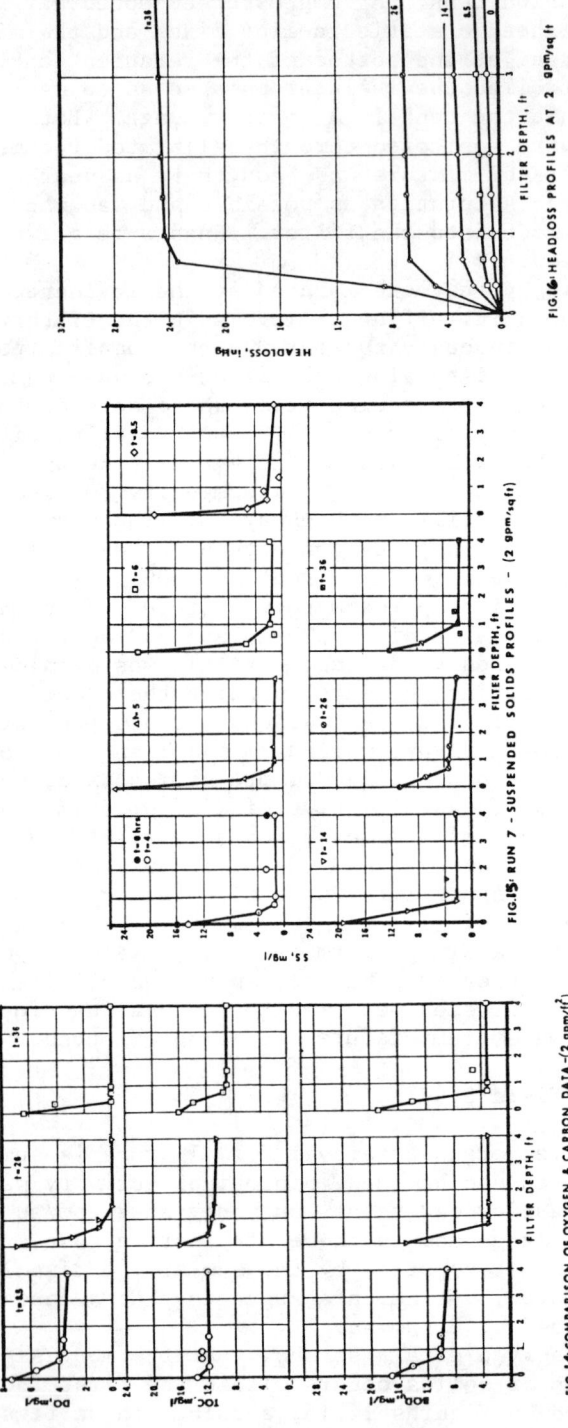

illustrate conditions for 2 gpm/sq ft., but the trends are fairly typical for all hydraulic loads, except that at 6 and 8 gpm/sq ft. significant penetration reaches 1.5 to 2 ft. of filter bed.

Control Run

A control run at 2 gpm/sq ft. was performed in which the secondary effluent was not aerated prior to filtration. As is shown by Table 1, significant removal only took place at the very end of the run, with a total average removal for dissolved organics of 10%. Again this reduction was probably due to anaerobic activity within the filter in the regions where the dissolved oxygen was zero or close to zero, which would explain why the removal did not take place earlier in the run, since the oxygen would inhibit those anaerobic organisms.

CONCLUSIONS

The results of the series of studies performed in the laboratory and in the field permit us to draw the following conclusions:

1- Biological activity will take place within a sand filter during a normal filtration run while polishing a secondary effluent if there are biodegradable dissolved organics present as well as dissolved oxygen in sufficient quantity to satisfy the oxygen demand exerted by the microorganisms.

2- Dissolved oxygen levels near saturation are sufficient to satisfy the oxygen demand exerted by the normal average residuals of 20-25 mg/ℓ dissolved BOD_u. If amounts much less than saturation are utilized the dissolved oxygen will quickly drop to zero and useful aerobic activity will stop.

3- The effectiveness of the biological activity in lowering residuals of dissolved organics will depend on the relative biodegradability of that fraction. The data from the field studies indicated that for domestic wastes useful removals with a potential for full scale applications occur in the narrow range of 2-3 gpm/ sq ft.

4- The data from both laboratory and field studies indicated that only the first foot of filter was active in the removal of suspended solids, dissolved organics, and dissolved oxygen. Therefore it seems that for domestic wastes with average suspended solids for 9-40 mg/ℓ and average BOD_u concentrations up to 30 mg/ℓ a substantially shallower upflow filter may be used, if corrective measures are taken to prevent premature lifting of the filter bed. Wastes the strength of which exhibit very large fluctuations from the mean may still warrant the use of deep upflow

TABLE 1
DISSOLVED OXYGEN AND BIOCHEMICAL OXYGEN DEMAND
IN CONTROL RUN

Rum Time (hrs.)	DO in (mg/1)	DO out (mg/1)	BOD_5 in (mg/1)	BOD_5 out (mg/1)
4	1.5	1.1	10	10
12	1.3	0.8	25	25
20	1.5	0.6	5	4
29	1.6	0.5	3	3
40	1.5	0.4	12	11
52	1.4	0	12	11
64	1.7	0	32	24

filters. Filtration studies in the downflow mode and treating industrial wastes of various kinds should be performed in order to determine the entire range of applicability of biologically active sand filters.

A mathematical model for the filtration processes (including biological activity) was proposed and tested with the data from the field studies. The results of this modelling effort is detailed in Part II of this series of papers.

List of References

1. Miller, G. W. and Rice, R.G. "European Water Treatment Practices-The Promise of Biological Activated Carbon", Civil Engineering-ASCE, 81 (Feb. 1978)
2. Ives, K.J. "Discussion of Russian Water Supply and Treatment Practices", Jour. ASCE,San.Eng.Div.,861:SA4:101-103(July 1960).
3. Naylor, A.E., Evans, S.C. and Duscombe, K.M. "Recent Developments on the Rapid Sand Filters at Luton, Water Pollution Control (British), 309-320(1967).
4. Calise, V.J. and Homer, W.A. "Russian Water Supply and Treatment Practices", Jour.ASCE, San.Eng.Div.,86:1(Mar 1960).
5. Smit, P. "Upflow Filters", Jour.AWWA,55:804(June 1963).
6. Hamman, C.L., and McKinney, R.E."Upflow Filtration Process", Jour. AWWA, 60:1023 (Sept. 1968).
7. Fair, D.M., Geyer, J.C.,and Okun, D.A."Water and Wastewater Engineering",Volume 2. Water Purification and Wastewater Treatment and Disposal", John Wiley & Sons,Inc.,27-12(1968).

BIOLOGICALLY ACTIVE SAND FILTERS
PART II
A MATHEMATICAL MODEL

Roque A. Román-Seda, Civil Engineering Department
University of Puerto Rico, Mayaguez, Puerto Rico.

David G. Wilson, Chemistry Department
Vanderbilt University, Nashville, Tennessee.

INTRODUCTION

A dynamic mathematical model of filtration is proposed herein in which removal of dissolved organics and depletion of dissolved oxygen is mathematically described, in addition to removal of suspended solids and the buildup of pressure losses.

DESCRIPTION OF THE MODEL

Elementary mass balances about a thin filter slab (Figure 1) and unit area of filter will yield the equations that describe the rate of change with time of suspended solids and dissolved organics in the bulk liquid.

Suspended Solids

$$\frac{\partial C_{ss}}{\partial t} = \frac{Q(t)}{\varepsilon(x,t)} \frac{\partial C_{ss}(x,t)}{\partial x} - \frac{\lambda(x,t)}{\varepsilon(x,t)} C_{ss}(x,t) \quad (1)$$

where $C_{ss}(x,t)$ = bulk suspended solids concentration M/L^3 as a function of filter depth x and filtration time t.
t = filtration time, T
$Q(t)$ = face velocity (or hydraulic load), L/T
$\varepsilon(x,t)$ = remaining filter porosity at x and t, dimensionless
$\lambda(x,t)$ = filter modulus at x and t, $1/T$

The first term on the right describes the accumulation due to differences between the material transported in and out of the filter slab by bulk liquid motion. The second term is the removal term, the original form of which was proposed by Iwasaki [1]. This term indicates that removal is a function of the concentration of suspended solids left in the bulk liquid. It attempts to lump in a single parameter all of the transport and attachment mechanisms which are responsible for actual removal at the microscopic level. Longitudinal dispersion has been left out of the equation since the partitioning of the filter bed into slabs which is required by the numerical evaluation method will adequately compensate for these effects [2].

The filter modulus was originally conceived of as being a constant, but it was very quickly realized that this assumption was a mistake. It was found [3] that λ not only reflects the initial conditions in the filter bed (grain diameter, $d_o(x)$, filter porosity, ε_o, and media surface characteristics) but initial conditions in the incoming solids suspension as well (filter face velocity, $Q(t)$, suspended solids diameter, d_{ss}, water temperature, T, and suspended solids surface characteristics). In addition, the filter modulus has to reflect all of those characteristics and conditions as the filter clogs, which means continuously changing filter characteristics (grain diameter, $d(x)$, porosity, $\varepsilon(x,t)$, and filter deposit surface characteristics). Therefore λ has to be variable.

The difficulty, therefore, in mathematically describing removal of suspended solids lies in the difficulty of finding an appropiate expression for λ which reflects all the changing conditions inherent in the clogging process In order to surmount these difficulties the following assumptions are made:

1. The filter bed consists of a packed bed of spheres of uniform diameter within the filter slab under consideration at the time [4]. See Figure 2.

2. The solids removed are deposited evenly over the surface of the spherical grains in a filter slab, such that the grain diameter grows with time.

3. The transport and attachment mechanisms (Figure 3) which eliminate the filtration process can be taken into account by making the filter modulus directly proportional to the media surface area.

4. There exists a scouring velocity v_{max} at which no further solids deposition is possible.

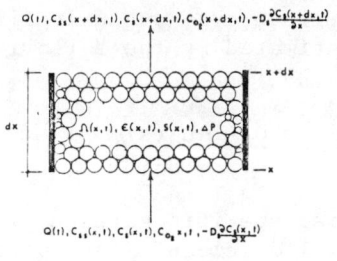

FIG.1 : SCHEMATIC ELEMENTARY FILTER SLAB

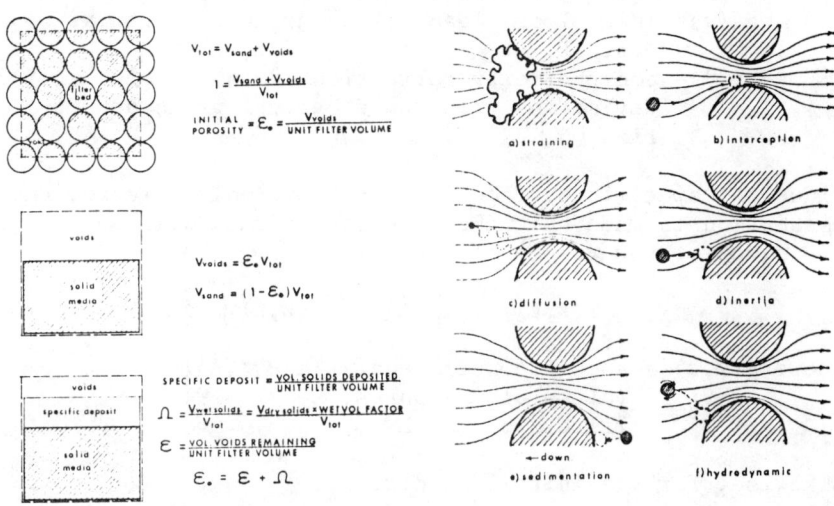

FIG. 2 : RELATIONSHIP BETWEEN POROSITY AND SPECIFIC DEPOSIT

FIG. 3 : TRANSPORTATION MECHANISMS OF FILTRATION

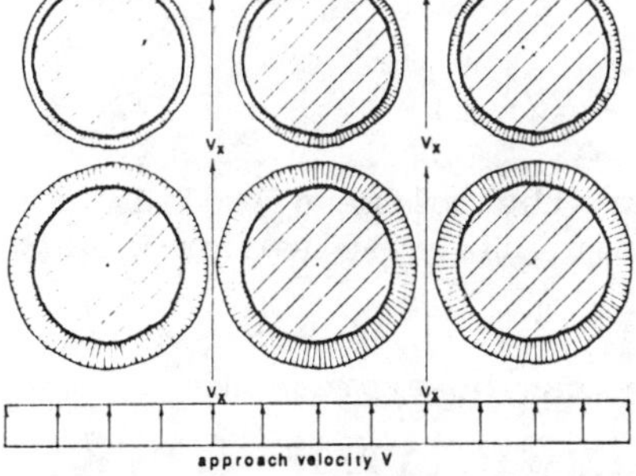

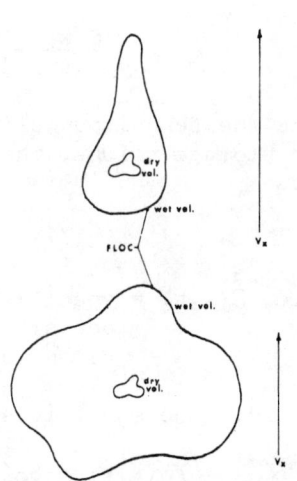

FIG. 4 : SCHEMATIC OF MECHANISM FOR LOCAL VELOCITY

FIG. 5 : EFFECT OF LOCAL VELOCITY ON WET VOLUME FACTOR

The initial surface area per unit filter volume (specific surface) can be estimated by the following relation by Ives [5]:

$$S_o = \frac{6\,(1-\varepsilon_o(x))}{\psi\,d_o(x)} \qquad (2)$$

where S_o = initial specific surface $1/L$
$\varepsilon_o(x)$ = initial or clean filter porosity, volume of pores per unit filter volume.
$d_o(x)$ = initial or clean filter grain diameter at depth x, L.
ψ = nonsphericity correction factor to account for deviations from the spherical grain assumption [6].

Since the specific surface is an area to volume ratio, the assumption is made that the specific surface will vary as the 2/3 power of the volume of the pores:

$$S(x,t) \;\alpha\; V_p^{2/3} = [\varepsilon(x,t)V]^{2/3} \qquad (3)$$

where $S(x,t)$ = specific surface at x and t, $1/L$
V_p = volume of the pores, L^3
V = filter volume, L^3

Likewise, for the initial conditions,

$$S_o(x)\,\alpha\; [\varepsilon_o(x)V]^{2/3} \qquad (4)$$

therefore

$$\frac{S(x,t)}{S_o(x)} = \left[\frac{\varepsilon(x,t)}{\varepsilon_o(x)}\right]^{2/3}$$

Since the deposited solids simply use up part of the available storage volume, the following relation of conservation holds:

$$\varepsilon_o(x) = \Omega(x,t) + \varepsilon(x,t) \qquad (5)$$

where $\Omega(x,t)$ = specific deposit, the volume of solids deposited per unit filter volume at x and t, dimensionless.

Therefore the specific surface Ω can be expressed

$$S(x,t) = S_o(x)\left[\frac{\varepsilon_o(x)-\Omega(x,t)}{\varepsilon_o(x)}\right]^{2/3} \qquad (6)$$

As the clogging process develops, deposits ($\Omega(x,t)$) build up within the filter and the available porosity ($\varepsilon(x,t)$) decreases. For constant flow filtration that means that the available open area for fluid flow is decreasing and that therefore there must follow a corresponding increase in the local fluid velocity. Thus as deposits build up the local velocity increases and the shear stresses and friction losses caused by the drag forces must therefore also increase. There will come a point in which the local velocity ($v(x,t)$) is of such magnitude that solids can no longer attach themselves to the available surfaces. Under those conditions effective removal migrates to a region of the filter where the local velocities are still below the scouring level (v_{max}). See Figure 4. The net effect of the local velocity is to decrease removal efficiency and therefore to decrease λ.

If the initial conditions were to define an initial value λ_0 for λ, the mathematical expression for λ could assume the following form:

$$\lambda(x,t) = \lambda_0 \, S(x,t) \left[1 - \frac{v(x,t)}{v_{max}}\right]^X \quad (6a)$$

where X = exponent to account for deviations from idealized geometry.

which takes into account all of the effects discussed above which affect the value of λ.

Due to the great variability that exists in some waste streams [7] it was felt that the standard assumption that $\partial C_{ss}/\partial t = 0$ is not a valid one for many cases of interest. Therefore, it was decided to retain the $\frac{\partial C_{ss}}{\partial t}$ term in order to make the equation truly dynamic and applicable to all relevant cases.

Dissolved Organics

$$\frac{\partial C_s}{\partial t} = \frac{Q(t)}{\varepsilon(x,t)} \frac{\partial C_s(x,t)}{\partial x} - K\Omega(x,t) C_s(x,t) \cdot \left[1 - \frac{v(x,t)}{v_{max}}\right]^{X_s} \quad (7)$$

where $C_s(x,t)$ = biodegradable dissolved organic (BOD) concentration at depth x and filtration time t, M/L^3.
K = biochemical reaction rate constant, $1/T$
v_{max} = local velocity at which capture of organic molecules by floc is impossible, L/T.

The above equation is also the result of a simple mass balance taken about thin filter slab. The first term on the right reflects the advective mass transport while the second term is the kinetic law which describes the disappearance of biodegradable dissolved organics from the bulk liquid due to biological activity within the filter. This disappearance is tied to the biomass present in the system such that removal will be zero for a clean filter bed.

Increases in local velocity as the filter clogs will also limit the extent to which organic molecules can be captured by the deposited solids. Therefore, a scouring velocity term $\left[1-\dfrac{v(x,t)}{v_{max}}\right] X_s$ is made an integral part of the removal term.

Accumulation of Solids

$$\frac{\partial \Omega}{\partial t} = \left[V_y K\Omega(x,t)\, C_S(x,t) + \frac{\lambda_o}{\rho_{ss}}\, S(x,t) C_{SS}(x,t) \right] W_{vf} \cdot \left[1 - \frac{v(x,t)}{v_{max}}\right] X \quad (8)$$

where V_y = volumetric yield coefficient, volume of biomass produced/mass of substrate removed, L^3/M.
ρ_{ss} = floc density, dry mass basis, M/L^3
W_{vf} = wet volume/dry volume conversion factor, dimensionless

In a biologically active filter there are two sources of deposits:

1. New biomass synthesized within the filter bed.
2. Suspended solids removed by the filter.

Both sources are reflected in Equation 8. A volumetric yield coefficient and a floc density parameter are utilized to nondimensionalize each of the deposit terms, since the specific deposit Ω itself is dimensionless. It was found in the simulations described further on, though, that these last two conversions did not fully account for all the entrained water in the floc deposits, as reflected by a failure of the headloss model to produce significant growth of headlosses. Once an entrained water correction factor, W_{vf}, was introduced into the equation, the headloss model began to adequately reproduce the observed rise in headlosses.

The entrained water correction factor was originally conceived as a constant. However, this led to a condition where the local velocities increased too fast and zones of zero removal appeared where the actual filter still exhibited removal. Therefore, a variable entrained water correction factor was formulated according to the equation:

$$W_{vf} = W_{vf_o} - c \left[\Delta H(x,t) - \Delta H(x)_o\right] \quad (8a)$$

where W_{vf_o} = initial value of entrained water factor, dimensionless.
c = linear decay constant, $1/L$
$\Delta H(x,t)$ = headloss at depth x and filtration time t, L
$\Delta H(x)_o$ = initial headloss at depth x, L

The physical reason for this variation in W_{vf} was that as local velocities increased the increase in shear forces would "squeeze" out some of the entrained water in the floc, thus causing a net decrease in the floc volume, as suggested by Figure 5.

The concept of the entrained water correction factor and the scouring velocity factor provided the means, within the model, to simulate the process by which the very growth of deposits within the filter causes changes in the local flow regime that limits any further growth of deposits.

Dissolved Oxygen

$$\frac{\partial C_{o2}}{\partial t} = \frac{Q(t)}{\varepsilon(x,t)} \frac{\partial C_{o2}(x,t)}{\partial x} - RK\Omega(x,t) C_S(x,t) \qquad (9)$$

where R = oxygen utilization coefficient, mass of oxygen consumed/mass of organics removed, dimensionless.

The above is the result of a straightforward mass balance where the removal term is simply the assumed kinetic law for removal of organics times an appropiate oxygen utilization coefficient.

Headlosses

$$\frac{\partial P}{\partial x} = k' \frac{\mu}{\rho} Q(t) \left[\frac{\sigma s}{d(x)}\right]^2 \frac{[1-\varepsilon(x,t)]^2}{\varepsilon(x,t)^3} \qquad (10)$$

where P = pressure, ML/T^2-L^2
 k' = Kozeny coefficient, $1/L$
 μ = absolute viscosity, $M/L-T$
 ρ = water density, M/L^3
 σ = shape factor, equals $6/\psi$; correction for deviation from spherical geometry, L.
 d(x) = mean diameter of media at depth x, L

The above relationship is, of course, the Kozeny equation in pressure units. But the difference of Equation 10 with the traditional Kozeny equation is that the porosity parameter is a variable in depth and time, whereas in the original equation it is a constant (ε_0).

The above system of equations was evaluated numerically utilizing a predictor-corrector method of the following general form:

Initial Prediction Function:

$$y_i(\Delta t) = y_i(0) + \Delta t \frac{dy_i(0)}{dt} \qquad (11)$$

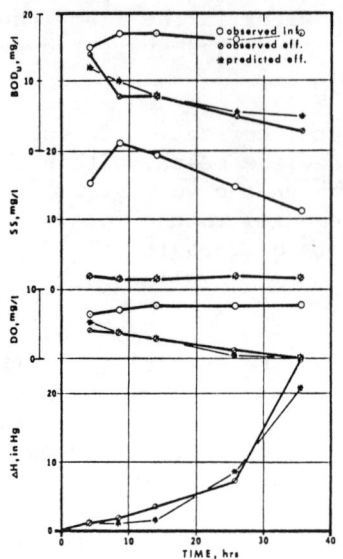

FIG. 6: COMPARISON OF OBSERVATIONS & MODEL PREDICTIONS
(2 gpm/sq ft)

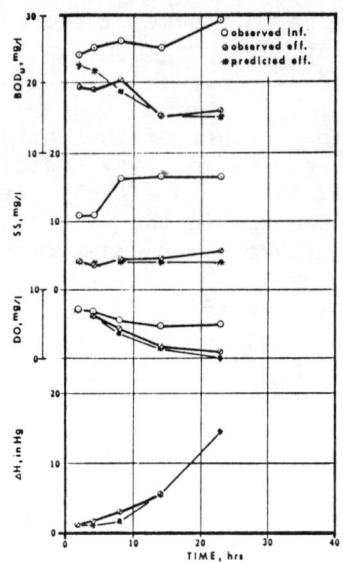

FIG. 7: COMPARISON OF OBSERVATIONS & MODEL PREDICTIONS
(4 gpm/sq ft)

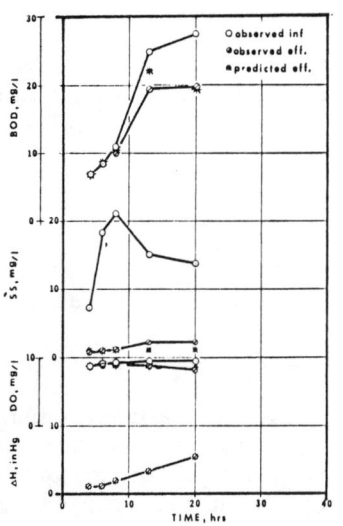

FIG. 8: COMPARISON OF OBSERVATIONS & MODEL PREDICTIONS
(6 gpm/sq ft)

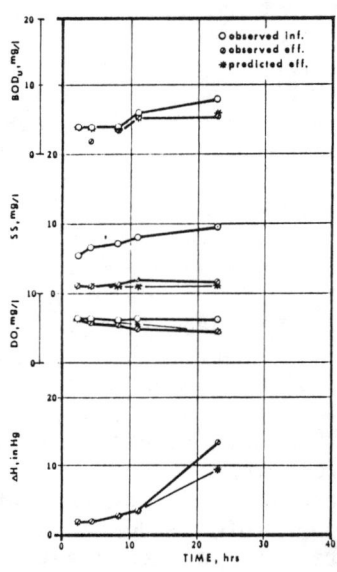

FIG. 9: COMPARISON OF OBSERVATIONS & MODEL PREDICTIONS
(8 gpm/sq ft)

General Predictor Function:

$$\overset{*}{y}_i(t+\Delta t) = y_i(t-\Delta t) + 2\Delta t \frac{dy_i(t)}{dt} \qquad (12)$$

General Corrector Function:

$$y_i(t+\Delta t) = y_i(t) + \frac{\Delta t}{2}\frac{dy_i(t)}{dt} + \frac{dy_i^*(t+\Delta t)}{2} \qquad (13)$$

Details of the numerical method are discussed elsewhere [8], as well as the methods for evaluation of model parameters.

Simulations were made at 2, 4, 6, and 8 gallons per minute per square foot (gpm/sq ft) utilizing identical filter characteristics as well as influent conditions to those observed in the field studies discussed under Part I [9] of these series of studies. Initial conditions assumed a clean filter bed filled with clean water such that Ω, C_{ss} and C_{O2} were all zero.

Comparison of Results of Simulations with Observations

Figure 6, 7, 8 and 9 each illustrate the comparison for biodegradable dissolved organics, dissolved oxygen, suspended solids and headlosses between simulations and observed results at hydraulic loads of 2, 4, 6, and 8 gpm/sq ft, respectively. Influent and effluent values shown are time-averaged values for both observed and simulated data. It can be seen from the effluent data that the model adequately reproduced the observed effluent levels and total headlosses.

Simulation With Independent Data

A simulation was performed of one of a series of independent experiments performed by Mr. Héctor Fuentes, who was directly responsible for the laboratory studies on biologically active filters dealing with pure oxygen aeration [9].

The simulation was performed at 4 gpm/sq ft at identical filter characteristics and influent conditions as those in the actual laboratory experiment. The results are shown in Figures 10, 11, 12, and 13. It can be seen that in all the simulated filtration times, the predicted effluent level followed the observed level very closely, including dissolved organics, suspended solids, dissolved oxygen and headlosses.

CONCLUSIONS

The modelling effort of the biologically active sand filters study series produced results that indicate the following:

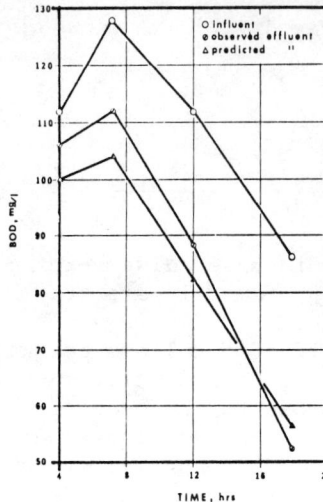

FIG. 10 · COMPARISON OF OBSERVED & PREDICTED BOD
INDEPENDENT DATA -(4 gpm/sq ft)

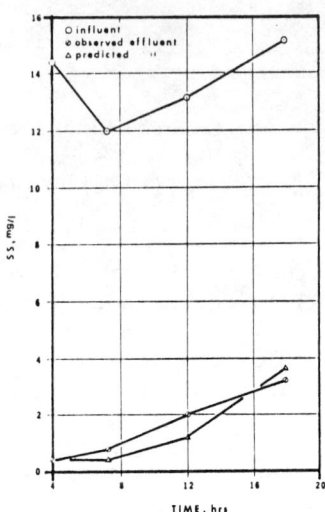

FIG. 11 · COMPARISON OF OBSERVED & PREDICTED SS
INDEPENDENT DATA -(4 gpm/sq ft)

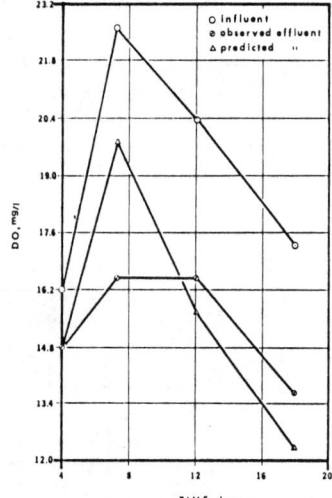

FIG. 12 · COMPARISON OF OBSERVED & PREDICTED DO
INDEPENDENT DATA -(4 gpm/sq ft)

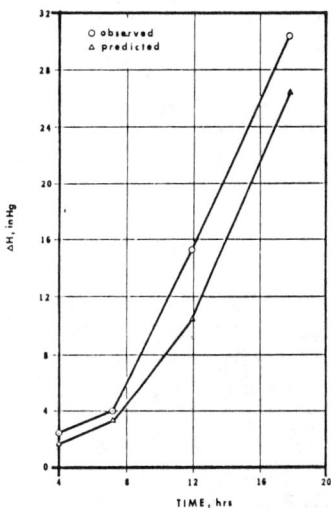

FIG. 13 · COMPARISON OF OBSERVED & PREDICTED
HEADLOSSES - INDEPENDENT DATA

1. The proposed mathematical model for a biologically active sand filter adequately simulated observed effluent levels for suspended solids, biodegradable dissolved organics, dissolved oxygen, and headlosses.

2. Simulations of effluent levels with an independent set of data produced from laboratory studies seemed to confirm the ability of the model to reproduce effluent levels.

3. Analysis of the simulated profiles (which was not presented in this paper due to space limitations) revealed that the model adequately reproduced the observed profiles up to 4 gpm/sq ft for all observed parameters. Further refinement of the model is necessary in order to reproduce adequate profiles at 6 and 8 gpm/sq ft. Such work is presently being carried on.

4. The model can be a very useful design tool if appropiate calibration data are obtained through the use of a 3-inch diameter filter at the hydraulic loads of interest.

5. A numerical algorithm for the evaluation of the model has been tested and utilized in a high speed computer at the Vanderbilt University Computer Center which yields information useful for design and other purposes.

Acknowledgements

A grateful acknowledgement is made to the Civil and Environmental Engineering Department of Vanderbilt University which provided the funds that made these studies possible.

List of References

1. Iwasaki, Y. "Some Notes of Filtration", Jour. Am. Wtr. Wks. Ass., 29, 1591 (1937).

2. Wilson, D.G. "Theory of Adsorption by Activated Carbon II. Continuous Flow Columns", Separation Science and Technology, 14, 415 (1979).

3. Ives, K.J. and Sholji, I. "Research on Variables Affecting Filtration", Jour. San. Eng. Div., Proc. ASCE, 91:SA4:1 (Aug 1965).

4. Weber, W.G., Jr. "Physicochemical Processes for Water Quality Control", Chapter IV, Wiley-Interscience, New York (1972).

5. Ives, K.J. "Theory of Filtration", Special Subject No. 7, Proc. Internal. Water Supply Congress Exhib., Vienna (1969).

6. Fair, G.M., Geyer, J.C., and Okun, D.A. "Water and Wastewater Engineering Volume 2. "Water Purification and Wastewater Treatment and Disposal", John Wiley & Sons, Inc., 27-12 (1968).

7. Eckenfelder, W.W. "Principles of Water Quality Treatment", CBI Publishing Co., Inc. (1980).

8. Kelly, L.G. "Handbook of Numerical Methods and Applications", Addison-Wesley Publishing Co. (Reading, Massachussetts) 186 (1967).

9. Román-Seda, R.A., Fuentes, H.R., and Eckenfelder, W.W. "Biologically Active Sand Filters. Part I. Laboratory and Field Studies", Proc. 13th. Mid-Atlantic Industrial Waste Conference, Ann Arbor Science Publishers, Inc. (to be published).

RE-EVALUATION OF THE OXYGENATION COEFFICIENT - TEMPERATURE RELATIONSHIP FOR WASTE TREATMENT AND STREAM MODELING

<u>Allen C. Chao</u>. Department of Civil Engineering, North Carolina State University, Raleigh, North Carolina

<u>Efren Galarraga</u>. Escuela Politecnica de Quito, Quito, Ecuador

<u>Robert H. L. Howe</u>. Eli Lilly Company, Lafayette, Indiana

INTRODUCTION

The oxygenation coefficient (K_2) is temperature dependent and an increasing K_2 value at higher temperature has been used for stream reaeration. Recently, Howe (1977) proposed that the K_2 value decreases as the liquid temperature increases. His modification is counter to the well recognized increasing oxygenation coefficient at higher temperatures. This situation is quite thought provoking, regarding the K_2 - temperature relationship. The writers have conducted studies to measure the oxygen transfer coefficient at various liquid temperatures using a laboratory diffused air system. Preliminary results have shown that the K_2 - temperature relationship is dependent on the magnitude of K_2. When the K_2 value is below 40 days^{-1}, it decreases at increasing liquid temperatures while other factors remain constant. Howe's modified equation seems to be valid for oxygenation conditions of relatively low air flow rates or power input for aeration.

EQUIPMENT AND METHOD

Equipment

A 220 liter (22" ϕ, 36" deep) insulated nalgene tank was used. A constant temperature water bath was employed to generate thermo flow which was continuously pumped through a copper coil submerged in the tank. The temperature of the thermo flow could be maintained from 15°C to 35°C, and the temperature of the aerated liquid controlled within $\pm$ 0.1°C of a preselected temperature. A mixing paddle (2-½" by 2-3/4") which rotated at 200 rpm and extended 12" below the water surface was used to mix the aerated liquid.

Compressed air was used for aerating the liquid. It was depressurized to 10 psi and its flow rate controlled with a needle valve. The air temperature was controlled at the same level as the liquid temperature by passing through a ¼" copper coil submerged in the aerated liquid. The air flow rate was controlled at 60 ml./min. and 120 ml./min.

The dissolved oxygen was monitored with a D.O. Meter (Yellow Spring Instrument Co., Model 54). The D.O. response curve was recorded on a 10" chart (1" = $1^{mg}/1$ D.O.). The D.O. probe was calibrated with water samples of known D.O. values which were determined by use of the Modified Winkler Method.

For each test series, all conditions were held constant except the liquid temperature which varied from $14^{\circ}C$ - $32^{\circ}C$. The aeration tests were also done with a smaller reactor of 2 liter volume which was submerged in the constant temperature water bath for control of the liquid temperature. The air flow rates were maintained at 3000 ml/min and 3800 ml/min.

The reactor was filled with tap water and the liquid temperature adjusted to a preselected level. The thermo flow was pumped continuously to maintain the temperature. Nitrogen gas was bubbled through the liquid until the D.O. was reduced to below 1 mg/l. Afterward, the nitrogen gas was stopped and air was turned on to aerate the content.

Methods of Analysis

The response curve was analyzed by use of a linear regression analysis of the log deficit vs. time [Redman, et. al. 1979; Salmit & Redman, 1975]. The saturation D.O. value was obtained from "Standard Methods" [AWWA, APHS & WPCF]. The K_2 value was calculated and the temperature coefficient (θ) was then determined from a set of K_2 values for various temperature levels by use of a SAS computer program.

RESULTS AND DISCUSSION

K_2 - Temperature Relationship

The results seem to indicate that the K_2 - temperature relationship is dependent of the ratio of air flow rate to liquid volume used in the aeration study, or the magnitude of the measured K_2 values. When 60 ml/min and 120 ml/min air was employed to aerate 200 liters of liquid, the measure K_2 values are less than 4 day^{-1}. As shown in Figure 1, they decrease for increasing liquid temperature, thus confirming the validity of Howe's modification of the K_2 - temperature relationship.

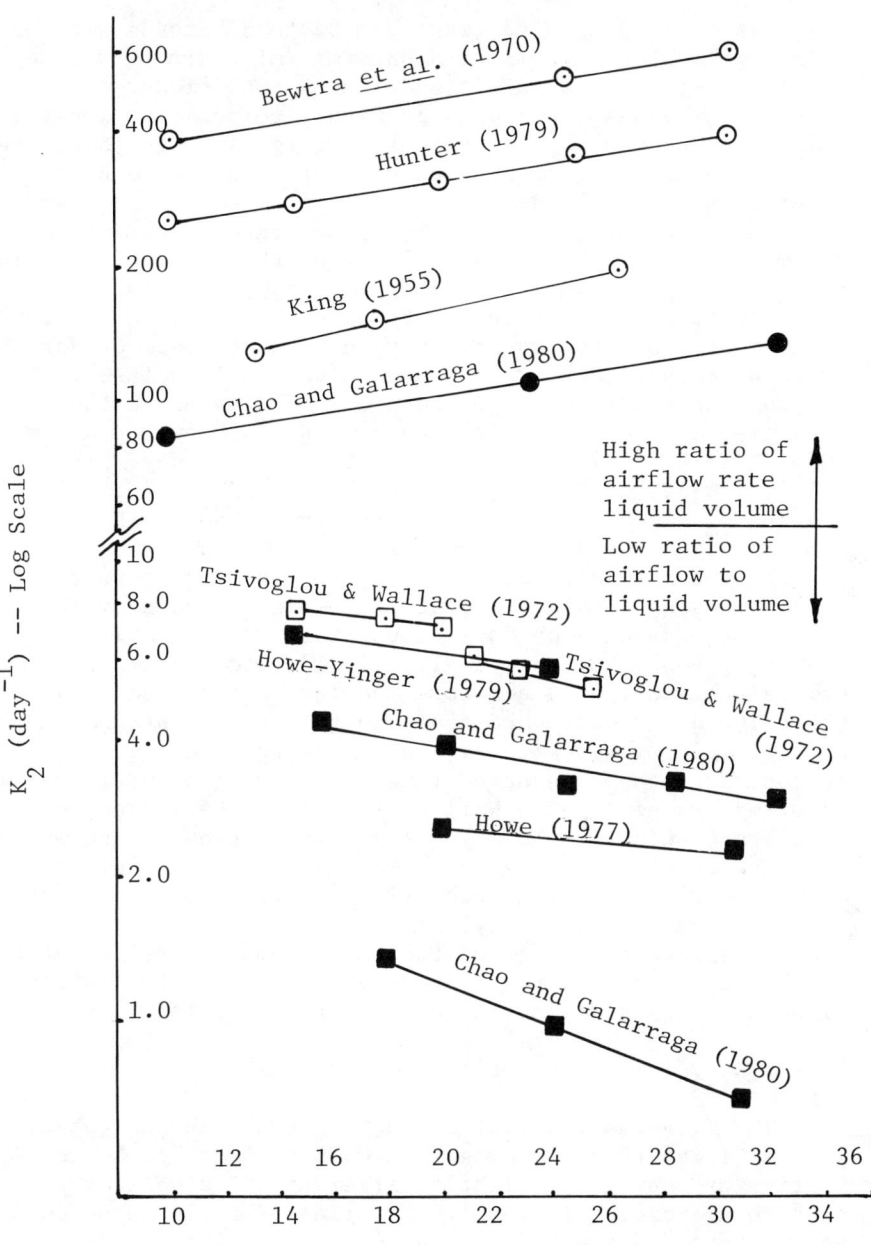

Figure 1. Influence of Temperature on K_2

⊙ Increasing K_2 vs. Temperature for Aerators-Literature data
● Increasing K_2 vs. Temperature - Author's data
☐ Decreasing K_2 vs. Temperature for Streams-Literature data
■ Decreasing K_2 vs. Temperature - Author's data

Higher K_2 values ranging from 88 to 328 day^{-1} were obtained with 3000 ml./min. and 3800 ml./min. air flow rate to aerate 2 liters of sample. Under such conditions, the measured K_2 values increases at higher temperature levels. This fits the old K_2 - relationship (Fig. 1). Further, the results indicate that the θ value increases toward higher or lower measured K_2 values.

The dependence of the K_2 - temperature relationship on the magnitude of the measured value or the ratio of air flow rate to liquid volume is also supported by K_2 data published in literature. Many researchers have conducted laboratory experiments to determine the oxygen transfer rate of various types of mechanical aerators and diffused air systems. The oxygen transfer rates reported are generally above the 100 day^{-1} level. Some examples are 304 day^{-1} [Conway & Kumke, 1966], 115-475 day^{-1} [Gilbert, 1978], 119-194 day^{-1} [King, 1955], 120-140 day^{-1} [Lakin & Salzman, 1979], 163 day^{-1} [Kalinske, et. al., 1968], 105-194 day^{-1} [Namie, 1977], and 313-832 day^{-1} [Namie & Nelson, 1979]. The magnitude of these oxygen transfer rates are in the high range; they increase for increasing liquid temperature. This is shown by the plot of the results of K_2 - temperature relationship reported by King [1955], Bewtra, et al., [1970], and Hunter [1979] on several commonly used aerators having reported K_2 values above 100 day^{-1} which are equivalent to those obtained in a diffused air system with high air flow to liquid volume ratio. In contrast, the reaeration rates of several natural streams reported by Tsivoglou & Wallace [1972] and Tsivoglou, et al. [1965] are below the 10 day^{-1} level, and have been shown to decrease at higher temperature levels (shown in Fig. 1). The K_2 values reported by Metzger [1968] which are in the intermediate range of 18-21 day^{-1} (Fig. 1) seem to be relatively less sensitive to change of the water temperature than those in the higher or lower K_2 regions. Unfortunately, no air flow rate/liquid volume was noted for evaluation!

Stream K_2 vs. Temperature

The K_2 value for most U. S. rivers have been reported by Thomas [1948] to be ranging from 0.06-0.96 day^{-1}. Other investigators may have slightly different K_2 values but generally they do not exceed 10 day^{-1}. Examples of field measured K_2 values reported are 0.6-0.7 day^{-1} [Bayer, 1975], 0.3-3.2 day^{-1} [Churchill, et al., 1962], 0.47-5.4 day^{-1} [Juliano, 1969], 0.28-0.36 day^{-1} [O'Conor, 1961], and 0.5-8 day^{-1} [Tsivoglou & Wallace, 1972; Tsivoglou et al., 1968]. Based on the aforementioned findings, the reaeration rate of all natural streams should follow Howe's proposed modification. Yet this seems to be refuted by the data in literature.

Early studies of the K_2 - temperature relationship were

conducted with small scale laboratory glassware [Adney, 1920; Becker, 1923], and the results cannot be transferred directly to field applications. Streeter [1926] and Streeter et al., [1936] pioneered the studies on stream reaction and formulated the widely recognized K_2 - temperature equation which has been extensively used in stream modeling and reaeration [Bayer, 1975; Pigry & Whinston, 1973]. Their data, however, are not conclusive to indicate an increasing stream K_2 value at higher water temperature.

In another report, Streeter [1926] determine the K_2 value of the Ohio River by measuring the D.O., B.O.D. and K_1 values of the upstream and the downstream ends of a selected stretch. Using the D.O. deficit equation, the K_2 value was calculated. When the stream D.O. in Summer months was checked by using the calculated K_2 value, the calculated D.O consistently showed much higher values than the actually measured D.O. values. The simplified conditions of Streeter's D.O. deficit equation may cause some random discrepancies between the measured and calculated stream D.O. But these consistently higher results indicated an inadequate use of a higher K_2 for summer months, which was converted with the old K_2 - temperature equation. This inconsistent D.O. profile of a stream, caused by use of the converted K_2 from a different temperature level, was also observed by Hull [1961].

K_2 Temperature Relationship for Waste Treatment

It is expected that the reported oxygen transfer rates for aerators in literature are much higher than those obtained with aerators in a full-scale aeration system. This is mainly because many researchers employed a relatively small aeration tank to accomodate the aerators or diffused air system to be tested. The relatively small aeration tank results in a relatively short test period to reach the saturation D.O. It was noted that most studies were completed within 8-20 minutes resulting in high reaeration rates. If the same aeration equipment is placed in a full-scale aeration basin, the time required to approach saturation D.O. is much longer, leading to lower oxygen transfer rates.

This lower oxygen transfer rate associated with full-scale treatment plants may change the K_2 - temperature relationships. However, since the aeration capacity of an aeration tank is oversized to provide a proper mixing for suspending the solids, the oxygen transfer rate is not a critical factor. Thus, whether the oxygen transfer rate in an aeration basin increases or decreases at higher temperature is of little importance to affect the operation of an aeration basin. The findings that the K_2 and the θ values increase for higher rate of air flow rate to liquid volume may be useful in scrubbing of gas because the results indicate that more gas can be dissolved in water at higher temperature. The significance of air flow rate/liquid volume is indeed obvious!

Comments on Literature Data

Many researchers have worked on developing a rationale basis for relating the K_2 value to the physical characteristics of a stream. In their work, these researchers simply assumed that the K_2 value increases for higher temperature and used a formula similar to that developed by Streeter for calculating the K_2 value at various temperature levels [O'Conor, 1961; Dobbins, 1964; Issacs & Gaudy, 1968; Krenkel, 1962; Krittel & Kochtitzky, 1947].

Several researchers have tried to use laboratory facilities to determine the K_2 - temperature relationship for streams. Their work may indicate a tendency of increasing K_2 for increasing temperature, but does not produce convincing evidence. Downing & Truesdale [1955] used a small flume of 92 cm long, 30 cm wide and 38 cm deep submerged in a constant water bath to measure the oxygen transfer rate. Elmore [1961] employed 1 liter per minute air flow rate to aerate 18 liters of liquid to simulate stream reaeration. The reaeration coefficient obtained was much higher than that may be expected from a natural river.

Direct evidence of an increasing stream reaction rate at higher temperature can be found in the raw data reported by Tsivoglou et al. [1972] and Tsivoglou [1968] for small sections of streams having similar flow conditions but significant temperature difference. The plot shown in Fig. 1 was excerpted by the authors.

The above discussions indicate the lack of convincing evidences in literature to support higher stream reaeration coefficients for increasing temperature and that use of the old K_2 - temperature relation is inadequate to predict the K_2 value at higher temperature. Although the findings of this study and some limited literature data support Howe's proposed K_2 - temperature relationship, more studies should be initiated to resolve the issue.

CONCLUSION

Our preliminary test results have shown that the K_2 - temperature relationship is dependent on the magnitude of K_2 observed. If the K_2 value is less than 40 day^{-1}, it has a tendency to decrease for increasing liquid temperature. Thus, Howe's modification of the K_2 - temperature relationship is valid. The K_2 value of most streams is below the 40 day^{-1} level, his suggestion should be used for calculating the K_2 value at 28°C for stream modeling and simulation. This will greatly reduce the oxygen transfer rate of streams at the critical temperature level. Since the issue is so controversial and the consequence has so much impact on stream quality, more research should be done to re-evaluate the K_2 - temperature relationship and establish a suitable formula for calculating stream K_2 values at higher temperature levels.

REFERENCES

Adeney, W. E. and Becker H. G. "The Determination of the Rate of Solution of Atmospheric Nitrogen and Oxygen by Water, Part II" Philosophical Magazine 39, 385 (1920).

Bayer, M. B. A Water Quality Optimization Model for Non-serial River Systems in Mathematical Models for Environmental Problems, Proceedings of the International Conferences held at the University of Southampton, England September 8-12, 1975. (Ed. Brebbia) p 253, John Wiley & Sons, New York (1975).

Becker, H. G., "A Simple Form of Apparatus for Observing the Rate of Reaeration between Gas and Liquids, and Its use in Determining the Rate of Solution of Oxygen by Water Under Different Conditions of Mixing," Philosophical Magazine, 45, 581 (1923).

Berthouex, P. M. and Hunter, W. G. "Problems Associated with Planning BOD Experiments". Jour Sanit. Engng. Div., Proc. ASCE, 97 SA3, 333 (1971).

Bewtra, J. K., Nicholas, W. R. and Polvowski, L. R., "Effect of Temperature on Oxygen Transfer in Water," Wat Res, 4, 115 (1970).

Boyle, W. C., Berthouex, P. M. and Rooney, T. C., "Pitfalls in Parameter Estimation for Oxygen Transfer Data," Jour of Environ Engng Div., Proc ASCE, 100, (EE2), 391 (1974).

Calderbank, B. H., "Physical Rate Processes in Industrial Fermentation Part II. Mass Transfer Coefficients in Gas-Liquid Contacting with and without Mechanical Agitation," Trans. Instn. Chem. Engrgs. 37, 173 (1959).

Chevereau. "Mathematical Model for Oxygen Balance in Rivers in Models for Environmental Pollution Control," (Ed. Deininger, R. A.), Ann Arbor Science Publishers, Ann Arbor, Michigan, (1973).

Conway, R. A. & Kumke, G. W., "Field Techniques for Evaluating Aerators," Jour Sanit Engng, Proc ASCE, 92 (SA2), 21 (1966).

Churchill, M. A., Elmore, H. L. and Buckingham, G. A., "The Prediction of Stream Reaeration Rates, Jour of Sanit Engng, Proc ASCE, 88 (SA4), Pt. 1, 1 (1962).

Dobbins, W. E. "BOD & Oxygen Relationships in Streams," Jour of Sanit Engng, Proc. ASCE, 90 (SA3), 53 (1964).

Downing, A. L. and Truestale, G. A., "Some Factors Affecting the Rate of Solution of Oxygen in Water," *Jour of Appl. Chem.*, 5, 570 (1955).

Elmore, H. L. and West, W. F., "Effects of Water Temperature on Stream Reaeration," *Jour of Sanit Engng Div, Proc ASCE* 87, (SA6) 59 (1961).

Gilbert, R. G., "Measurement of Alpha & Beta Factor in Proceedings Workshop toward an Oxygen Transfer Standard," epa-600/9-78-021, p. 147 (1979).

Howe, R. H. L., "K_2 - temperature Relation," *Jour of the Environ. Engng. Proc ASCE*, 103, (EE4), 729 (1977).

Hull, C. H. J., "Discussion on Oxygen Balance of an Esturasy, by O'Conor, D. J.," (1961), *Trans. ASCE*, 126, p 556 (1961).

Hunter, J. S. III, "Accounting for the Effects of Water Temperature in Aeration Test Procedures in Proceedings Workshop toward an Oxygen Transfer Standard," EPA-600/9-78-021, p 85 (1979).

Issac, W. P. and Gaudy, A. F. Jr., "Atmospheric Oxygenation in a Simulated Stream, *Jour of the Sanit Engng Div., Proc. ASCE*, 94 (SA2), 319 (1968).

Juliano, D. W., "Reaeration Measurements in an Estuary," *Jour of the Sanit Engng, Proc ASCE*, 95 (SA6), 1165 (1969).

Krenkel, P. A. and Orlob, G. T., "Turbulent Diffusion and the Reaeration Coefficient," *Jour of the Sanit Engng, Proc ASCE* 88 (SA2), 53 (1962).

King, H. R. "Mechanics of Oxygen Absorption in Spiral Flow Aeration Tanks II. Experimental Work," *Sewage & Ind. Wastes*, 27 Pt. 2, 1007 (1955).

Kittrell, F. W. and Kochtitzky, O. W. Jr., "Natural Purification Characteristics of a Shallow Turbulent Stream," *Sewage Works Jour* 19, 1032 (1947).

Kalinske, A. A., Shell, G. L. and Lash, L. D., "Hydraulics of Mechanical Surface Aerator, "*Water & Wastes Engineering* 5, (4), 65 (1968).

Metzger I. "Effect of Temperature on Stream Aeration," *Jour of the Sanit Engng Div., Proc ASCE* 94, (SA6), 1153 (1968).

Metzger, I. and Dobbins, W. E., "Role of Fluid Properties in Gas Transfer," *Environ. Sci. & Tech.* 1, 57 (1967).

Naimie H. and Burns, D., "Cobalt Interference in the Non-Steady State Clean Water Test for the Evaluation of Aeration Equipment for Various Natural Water Systems," *Wat Res*, 11, 667 (1977).

Namie, H. and Nelson, S., "Influence of pH and Iron and Manganese Concentrations on the Non-Steady State Clean Water Test for the Evaluation of Aeration Equipment," in Proc Workshop Toward an Oxygen Transfer Standard, EPA-600/9-78-021, p 91 (1979).

Neal, L. A., "Use of Tracers for Evaluation of Oxygen Transfer," in *Proc*. Workshop Toward an Oxygen Transfer Standard EPA-600/6-78-021, p 210 (1979).

O'Conor, D. J., "Oxygen Balance of an Esturary," Trans ASCE, 126 (Pt. 2), 556 (1961).

Pingry, D. E. and Whinston, A. B., "A Regional Planning Model for Water Quality Control in Models for Environmental Pollution Control" (Ed. R. A. Deininger), Ann Arbor Science Publishers, Inc., Ann Arbor, Michigan (1973).

Redman, D. T., Wren, J. P. and Mandt, M. G., "Oxygen Transfer Data Interpretation, Non-steady State Clean Water Tests," in Proceedings Workshop toward an Oxygen Transfer Standard EPA-600/9-78-021, p 128, (1977).

Salzman, R. N. and Lakin, M. B., "Influence of Mixing in Aeration," in Proceedings Toward an Oxygen Transfer Standard, EPA-600/9-78-021, p 59, (1979).

Schmit, F. L. and Redman, D. T., "Oxygen Transfer Efficiency in Deep Tanks," Jour Wat Poll Cont. Fed. 47, 2586 (1975).

Strepter, H. W., "The Rate of Atmospheric Relation of Sewage Polluted Streams," Transc. ASCE, 89, 1351 (1926).

Streeter, H. W., Wright, C. T. and Kehr, R. W., "Measurement of Natural Oxidation in Polluted Streams, III. An Experimental Study of Atmospheric Reaction under Stream-flow Condition," Sewage Works Journal, 8 282 (1936).

Stukenberg, J. R., Wahbeh, V. N. and Mckinney, R. E., "Experience in Evaluating and Specifying Aeration Equipment," Jour Water Poll Cont. Fed., 49, 66 (1977).

Thomas, H. A., "Pollution Load. Capacity of Streams," Water and Sewage Works, 95, 409 (1948).

Tsivoglou, E. C. and Wallace, J. R., "Characterization of Stream Capacity," EPA-R3-72-012, October 1972.

Tsivoglou, E. C., Cohen, J. B., Shearer, S. P., and Godsil, P. J., "Tracer Measurement of Stream Reaeration II. Field Studies," Jour Water Poll Cont Fed, 40, 285 (1968).

Standard Methods for the Examination of Water & Wastewater, AWWA, APHS & WPCF, 14th Ed., 1975.

MEASUREMENT AND VALIDITY OF OXYGEN UPTAKE AS AN ACTIVATED SLUDGE PROCESS CONTROL PARAMETER

Gary L. Edwards. Malcolm Pirnie, Inc., Newport News, VA.

Joseph H. Sherrard. Associate Professor of Civil Engineering, VPI&SU, Blacksburg, VA.

INTRODUCTION

Several methods have been proposed and utilized to control the activated sludge process. Three common control methods currently in use are (a) constant mixed liquor suspended solids level, (b) constant specific substrate utilization rate, and (c) constant sludge age [1]. To overcome some of the inadequacies and disadvantages of these methods, the utilization of oxygen uptake rate (r), or specific oxygen uptake rate (K_r), have been proposed as control parameters.

The primary goal of this investigation was to resolve the question of whether or not specific oxygen uptake rate is a meaningful control parameter. Five different twenty-four hour studies were conducted at the Blacksburg and Virginia Polytechnic Institute Treatment Plant to study the effects of variable influent organic and hydraulic loadings on the specific oxygen uptake rate in the aeration basins and the corresponding effects on effluent quality. Results from three of these periods are presented and discussed in this paper. From the data obtained, it was determined that specific oxygen uptake rate is not a valid activated sludge process control parameter.

LITERATURE REVIEW

The typical domestic waste treatment plant has been shown to exhibit a continual diurnal pattern in influent hydraulic, nitrogen and organic loadings [2,3]. Both dissolved oxygen (DO) concentrations and oxygen uptake rates in the aeration basin exhibit a direct response to this change in influent loadings [4,5]. This is to be expected as oxygen requirements increase with an increase in influent loadings. The change in oxygen uptake rate, r, and specific oxygen uptake rate, K_r, are due to an increase in activity of the aeration basin microorganisms. With a constant mixed liquor suspended solids, MLSS, concentration, oxygen uptake rates at the inlet of an aeration basin "should reflect the varying biochemical oxygen demand (BOD) in the incoming primary effluent" [5]. At a domestic waste treatment plant in East Lansing, Michigan, Schulze and Kooistra showed that r and K_r varied with influent BOD over time and that each exhibited a peak and valley for every daily period studied. Their study not only showed a daily variation in influent BOD, r and K_r, but also a seasonal variation. A dramatic increase in influent loadings and subsequent oxygen uptake rates occurred with the return of students to Michigan State University at the beginning of the fall term [5].

Oxygen uptake rates have been shown to be independent of DO concentrations and directly related to nitrification. A study by Kalinske [6] has shown that oxygen uptake rates are independent of DO concentrations in a well-mixed activated sludge system over a DO concentration range of 0.5 mg/l to 35 mg/l. Lawrence and Brown [7] found that the "rate of nitrification will not be affected by an oxygen concentration above a mixed liquor DO concentration of 1.5 to 2.0 mg/l." Degyansky and Sherrard [8] concluded that both oxygen requirements for nitrification and alkalinity destroyed during nitrification are not constant values but are dependent on mean cell residence time, Θ_c and the ratio of BOD_5 to ammonia-nitrogen (NH_3-N). In a theoretical study using casein as a substrate, Sherrard [9] showed that as the amount of nitrogen available for nitrification increases, the oxygen uptake rate increases.

Oxygen uptake rates have been found to be highest at the inlet and decreasingly lower toward the outlet of a completely-mixed activated sludge aeration basin [5]. The ratio between inlet and outlet oxygen uptake rates has been shown to be as high as 10:1. To explain this phenomenon, Duggan and Cleasby [4] concluded that there were two components to oxygen uptake, i.e., exogenous and endogenous respiration. For the exogenous phase, microorganisms utilize influent substrate whereas the endogenous phase consists of the degradation of internally stored substrate. Exogenous respiration fluctuates with the influent loading whereas the endogenous reaction, typical of the effluent end of the aeration basin, is much slower.

Effluent quality appears to have little relationship to oxygen uptake rates. In a study by Duggan and Cleasby [4] over a twenty-four hour time period, K_r was found to be highly variable with little change occurring in effluent quality.

In a further study by Wilkinson and Hamer [10], an immediate increase in respiration rate was noted for step-feed increases of a methanol, phenol, acetone and isopropanol substrate mixture with no change in effluent organic concentrations.

To overcome some of the disadvantages of current control parameters, oxygen uptake rate has been proposed as a control parameter for the activated sludge process [1,11,12,13]. Benefield et al. [1] proposed three distinct advantages to the use of oxygen uptake rate:
- a) The effect of fluctuating solids level in the secondary clarifier is minimized.
- b) Any change in the influent substrate concentration will be reflected immediately in the oxygen uptake rate.
- c) It is possible to compensate for loading fluctuations by varying the interval between control periods.

Another advantage as stated by Haas is the use of reagentless assays [11]. The disadvantages cited for oxygen uptake rate were the increased amount of lab study, higher degree of operator attention and more mathematical manipulations required over other control parameters [1].

Opposition to the use of oxygen uptake as an activated sludge control parameter has risen [9,14]. Khararjian [14] believes that use of the oxygen uptake rate would tend to confuse plant operators and engineers [14]. Sherrard, [9] citing the study of Duggan and Cleasby [4], has opposed use of oxygen uptake rate on the basis of highly variable and fluctuating oxygen uptake rates with little or no change in effluent quality. Sherrard [9] maintains that a better method of control would be to maintain a high mean cell residence time and a dissolved oxygen concentration above the growth limiting concentration of 1 mg/l to 2 mg/l at all times.

As demonstrated above, considerable controversy exists over the use of oxygen uptake as a control parameter. To add further light to this subject a study of K_r and its relationship to influent and effluent quality was conducted.

SITE LOCATION, TEST PERIODS, AND MATERIALS AND METHODS

To determine the validity of K_r as a control parameter, five 24 hr time periods of expected highly variable influent flow rate and composition were chosen to be studied. Specific waste characteristics were selected to monitor influent and effluent quality. Separate descriptions of the site location, sampling periods methods of sampling and data analysis are described below.

Site Location

This study was conducted at the Blacksburg and Virginia Polytechnic Institute Sanitation Authority Stroubles Creek Wastewater Treatment Plant. The Stroubles Creek Plant serves a population of 30,000 from the Town of Blacksburg, Virginia. Included in this population are approximately 20,000 students attending Virginia Tech. The Stroubles Creek Plant is a conventional flow activated sludge waste treatment plant that has two side-by-side aeration basins with a total hydraulic capacity of 6 mgd.

Test Periods

Test periods for this study consisted of five different 24 hr periods between September 3, 1980 and October 12, 1980. Results from only three of these 24 hr periods are reported herein. The first test reported was conducted on Wednesday and Thursday, September 15 and 16, 1980, before the University fall term began. The second test was conducted on Thursday and Friday, October 2 and 3, 1980. During this period, the University was in normal session. The third test was conducted on Saturday and Sunday, October 11 and 12, 1980. A Virginia Tech Homecoming football game was played on Saturday afternoon with over 40,000 people in attendance.

Materials and Methods

During each 24 hr test period, influent and effluent parameters, as well as oxygen uptake rates, were measured every 2 hr. Influent parameters were determined from grab samples taken from the primary clarifier effluent weir overflow and effluent parameters were determined by the use of grab samples from the secondary clarifier weir overflow. Influent and effluent parameters monitored were BOD_5, COD, alkalinity, NH_3-N and TKN. NO_3-N concentrations were also determined in effluent samples. Measurement of all parameters was made in accordance with "Standard Methods" [15].

Oxygen uptake rates were measured every two hours at three sampling stations in the west aeration basin. Sampling points were located adjacent to the three mechanical aerators in the aeration basin. The first sampling point was approximately twenty-five feet from the basin inlet, the second sampling point was in the middle of the basin, and the third sampling point was twenty-five feet from the effluent weir. The basin itself is 160 feet long in length. During any twenty-four hour period only the aerators nearest the influent and effluent were in operation. At each two-hour interval a grab sample was taken first at station 1 and analyzed for the oxygen uptake rate, then at station 2, and finally at station 3. Samples were taken at the surface and transferred to BOD bottles for immediate dissolved oxygen (DO) determinations, by use of a YSI 54 Dissolved Oxygen Meter. After the sample was placed in the BOD bottle, dissolved oxygen readings were taken every thirty seconds for a ten minute period. The samples were then stored for later MLSS determinations.

Plots of dissolved oxygen concentration versus time were made from the data obtained and the line of best fit was found for each. The oxygen uptake rate was determined as the slope of the line of best fit. Typical oxygen uptake rate plots are depicted in Figures 1 and 2. Specific oxygen uptake rates were calculated by dividing oxygen uptake rate by the corresponding MLSS concentration.

RESULTS AND DISCUSSION

In this section, experimental results are presented graphically for three of the five 24 hr time periods studied. Results for the two 24 hr periods omitted are similar to those reported in the three 24 hr periods selected for presentation and are only omitted due to space limitations.

Shown in Figures 3A through 3F are the data for September 15-16, 1980. In Figure 3A the influent flow is seen to follow a diurnal pattern with a peak flow at 1 p.m. and a lesser peak at 10 p.m. Influent BOD_5 and COD concentrations remain relatively low throughout the period as seen in Figure 3B. Effluent BOD_5 and COD concentrations remained relatively constant and very low throughout the period Nitrification is evident in Figure 3D as influent NH_3-N and TKN concentrations are high whereas effluent NH_3-N and TKN concentrations are low and NO_3-N concentrations are high. As seen in Figure 3E, the dissolved oxygen concentrations remained very high throughout the period. The inverse relationship of DO to influent BOD_5 and COD is evident as DO concentrations decreased between 12 a.m. and 4 p.m. as influent BOD_5 and COD concentrations increased during the period. A direct response of specific oxygen uptake rate to the decrease in DO is noted in Figure 3F as K_r increased between the hours of 12 a.m. and 4 p.m. K_r then decreased for the rest of the study period as DO concentrations increased and influent BOD_5 and COD decreased.

The data for the period of October 2-3, 1980 are presented in Figures 4A through 4F. Influent flows for October 2 and 3, 1980 were unavailable due to equipment malfunction, but as can be seen in Figure 4A, influent flow to the plant increased sharply with the return of students to Virginia Tech for the fall session. Influent BOD_5 and COD concentrations also increased with the return of students as seen in Figure 4B. A diurnal pattern for influent COD is evident whereas influent BOD_5 concentrations steadily increased until 12 p.m. then decreased through the night. Effluent BOD_5 and COD concentrations were again very low and relatively constant throughout the period. Nitrification was again present as can be seen in Figure 4D. Influent nitrogen concentrations also increased with the return of students. Figures 4E and 4F show DO concentrations and K_r for the period. DO in the aeration basin is again seen to fluctuate inversely in proportion to the concentrations of influent BOD_5 and COD at anytime. K_r is also shown to vary by a factor of three in Figure 4F at Stations 1 and 2 as the influent organic concentrations increase and DO decreases. Peak K_r is found at station 1, nearest the influent. The wastes are most likely well assimilated before reaching Station 3, keeping K_r low throughout the period.

Figures 5A through 5F present the data for the period of October 11 and 12, 1980. As seen in Figures 5A and 5B, influent flows were lower than the previous study period, yet high influent BOD_5 and COD concentration were maintained throughout. The high influent organic concentration was due to the temporary population increase present to view the homecoming football game. Effluent BOD_5 and COD concentrations were again low and relatively constant throughout the period. Nitrification was occurring as shown in Figure 5D. Small concentrations of NH_3-N, possibly corresponding to the peak organic loads, were noted in the effluent. In Figure 5E aeration basin DO was again seen to be inversely proportional to influent organic concentrations. DO concentrations remained low between the hours of 2 p.m. and 12 p.m. as influent organic concentrations remained high. Figure 4F shows that K_r remained high between the hours of 2 p.m. and 12 p.m., but decreased by a factor of four as dissolved oxygen concentrations increased and influent organic concentrations decreased.

Influent flow rates and organic concentrations were found to be highly variable during each study period. Nitrification was present during each test period. As expected the dissolved oxygen concentration in the aeration basin was found to be inversely proportional to influent oxygen demand. K_r was also found to vary with influent organic and aeration basin DO concentrations by as much as a factor of four, yet no significant changes in effluent quality were seen to occur during any test period.

CONCLUSIONS

Data from three different twenty-four study periods of highly variable influent flow rate and organic loading were analyzed to determine the effects of specific oxygen uptake rate on effluent quality. This was performed to evaluate the reliability of K_r as a control parameter for the activated sludge wastewater treatment process. Based on the results obtained the following conclusions are made:

a) A change in specific oxygen uptake rate for a wastewater of highly variable influent flow and organic loadings does not necessarily produce a significant change in effluent quality as measured by BOD_5 and COD.

b) Specific oxygen uptake rate cannot be used as a control parameter for the activated sludge process when treating a highly variable influent quantity and quality flow.

REFERENCES

1. Benefield, L. D., Randall, C. W. and King, P. H., "Process Control by Oxygen-Uptake and Solids Analysis," Journal Water Pollution Control Federation, 47, 2498 (1975).

2. Young, J. C., Cleasby, J. L and Baumann, E. R., "Flow and Load Variations in Treatment Plant Design," Journal of the Environmental Division, ASCE, 104, 289 (1978).

3. Randtke, S. J. and McCarty, P. L., "Variations in Nitrogen and Organics in Wastewaters," 104, 539, (1977).

4. Duggan, J. B. and Cleasby, J. L., "Effect of Variable Loading on Oxygen Uptake," Journal Water Pollution Control Federation, 43, 540 (1976).

5. Schulze, K. L. and Kooistra, R. D., "Oxygen Demand and Supply in an Activated Sludge Plant," Journal Water Pollution Control Federation, 41, 1763 (1969).

6. Kalinske, A. A., "Effect of Dissolved Oxygen and Substrate Concentration on the Uptake Rate of Microbial Suspensions," Journal Water Pollution Control Federation, 43, 73 (1971).

7. Lawrence, A. W. and Brown, C. G., "Design and Control of Nitrifying Activated Sludge Systems," Journal Water Pollution Control Federation, 40, 1776 (1976).

8. Degyansky, M. E. and Sherrard, J. H., "Effects of Nitrification in the Activated Sludge Process," *Water and Sewage Works*, **124**, 94 (1977).

9. Sherrard, J. H., "Communication: Oxygen Uptake Rate as an Activated Sludge Control Parameter," *Journal Water Pollution Control Federation*, **52**, 2033 (1980).

10. Wilkinson, T. G. and Hamer, G., "The Microbial Oxidation of Mixtures of Methanol, Phenol, Acetone and Isopropanol with Reference to Effluent Purification," *Journal of Chemical Technology and Biotechnology*, **29**, 56 (1979).

11. Haas, C. N., "Oxygen Uptake Rate as an Activated Sludge Control Parameter," *Journal Water Pollution Control Federation*, **51**, 938 (1979).

12. Eckenfelder, W. W. and Ford, D. L., *Water Pollution Control*, Pemberton Press: Austin (1970).

13. Giona, A. R. and Annesini, M. C. "Oxygen Uptake in the Activated Sludge Process," *Journal Water Pollution Control Federation*, **51**, 1009 (1979).

14. Khararjian, H. A., "Communication: Oxygen Uptake as a Control Parameter," *Journal Water Pollution Control Federation*, **52**, 823 (1980).

15. *Standard Methods for the Examination of Water and Wastewater*, **14th Edition**, Washington, D.C., American Public Health Association, (1976).

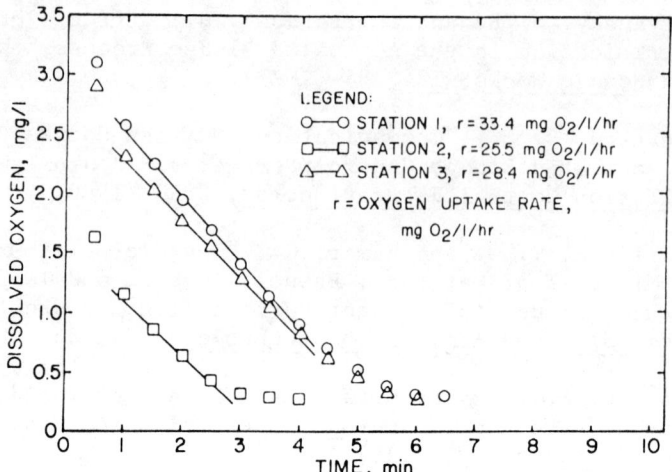

FIGURE 1. OXYGEN UPTAKE RATES, 4 PM, OCT. 11, 1980.

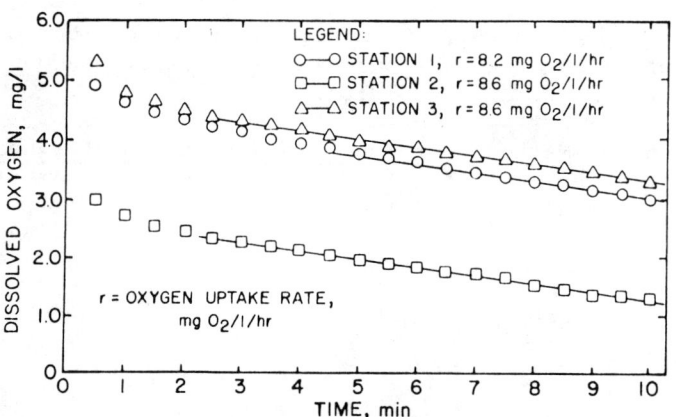

FIGURE 2. OXYGEN UPTAKE RATES, 4 AM, OCT. 12, 1980.

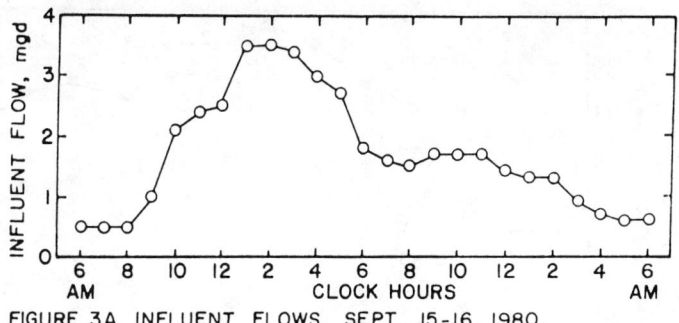

FIGURE 3A. INFLUENT FLOWS, SEPT. 15-16, 1980.

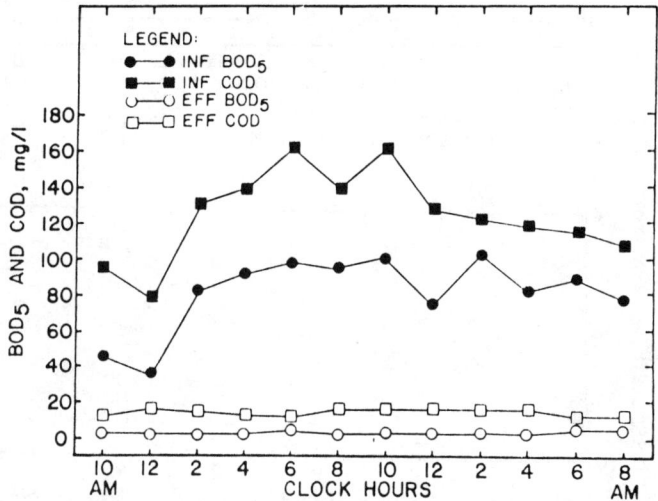

FIGURE 3B. INFLUENT AND EFFLUENT BOD_5 AND COD CONCENTRATIONS, SEPT. 15-16, 1980.

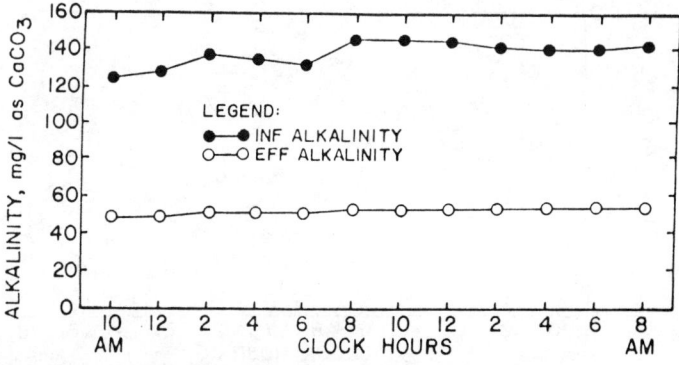

FIGURE 3C. INFLUENT AND EFFLUENT ALKALINITY CONCENTRATIONS, SEPT. 15-16, 1980.

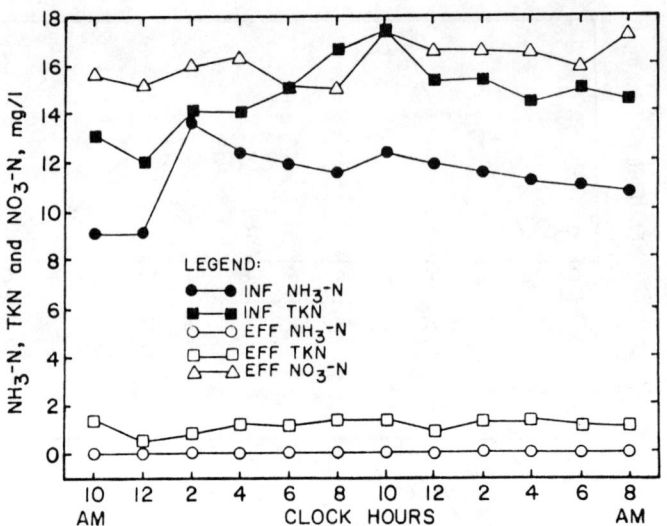

FIGURE 3D. INFLUENT AND EFFLUENT NITROGEN CONCENTRATIONS, SEPT. 15-16, 1980.

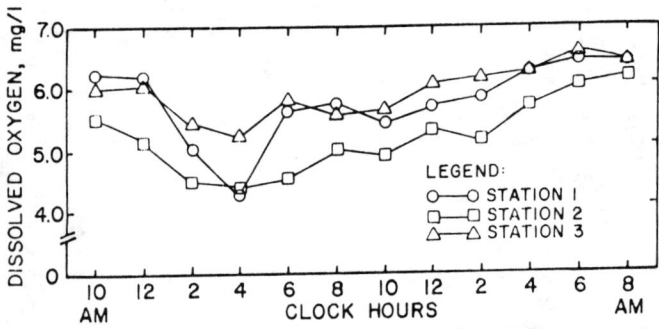

FIGURE 3E. DISSOLVED OXYGEN CONCENTRATIONS, SEPT. 15-16, 1980.

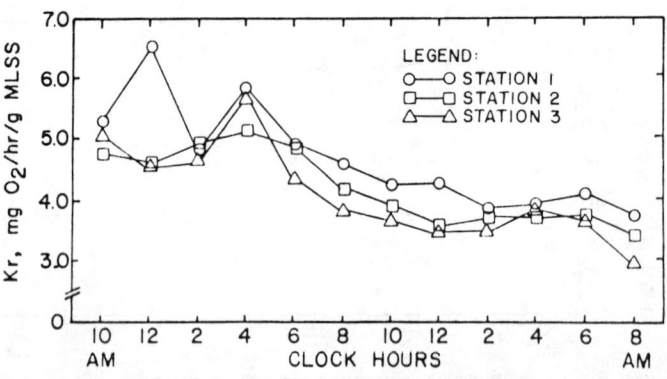

FIGURE 3F. SPECIFIC OXYGEN UPTAKE RATES, K_r, SEPT. 15-16, 1980.

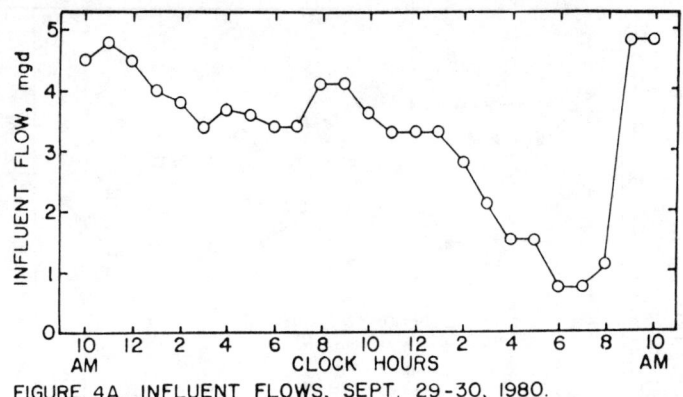

FIGURE 4A. INFLUENT FLOWS, SEPT. 29-30, 1980.

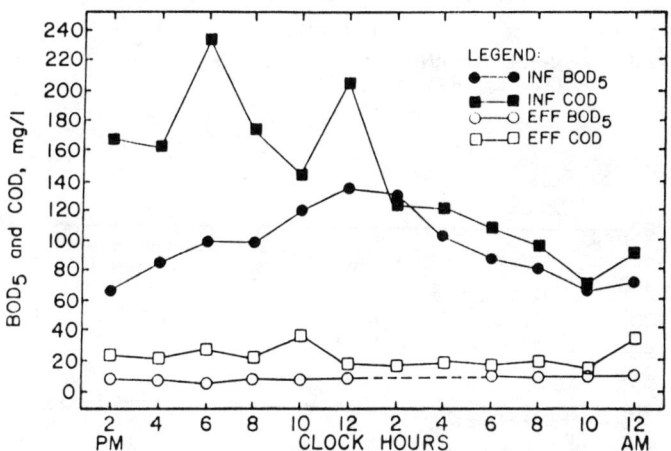

FIGURE 4B. INFLUENT AND EFFLUENT BOD_5 AND COD CONCENTRATIONS, OCT. 2-3, 1980.

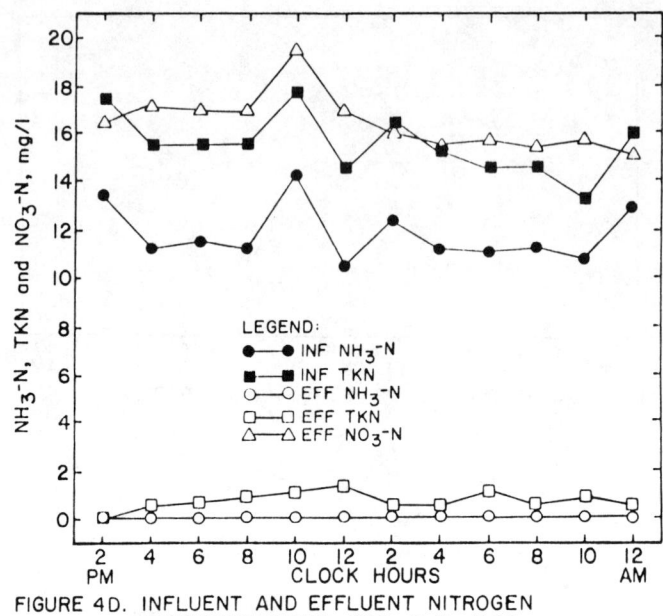

FIGURE 4D. INFLUENT AND EFFLUENT NITROGEN CONCENTRATIONS, OCT. 2-3, 1980.

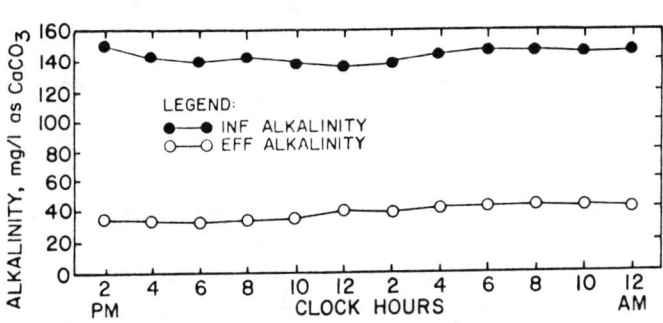

FIGURE 4C. INFLUENT AND EFFLUENT ALKALINITY CONCENTRATIONS, OCT. 2-3, 1980.

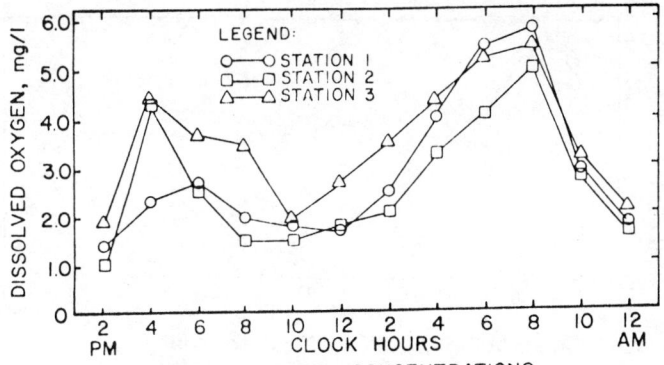

FIGURE 4E. DISSOLVED OXYGEN CONCENTRATIONS, OCT. 2-3, 1980.

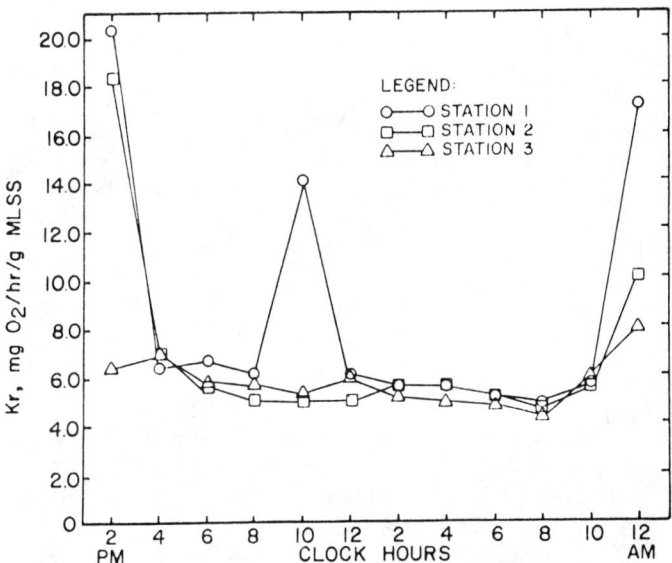

FIGURE 4F. SPECIFIC OXYGEN UPTAKE RATES, K_r, OCT. 2-3, 1980.

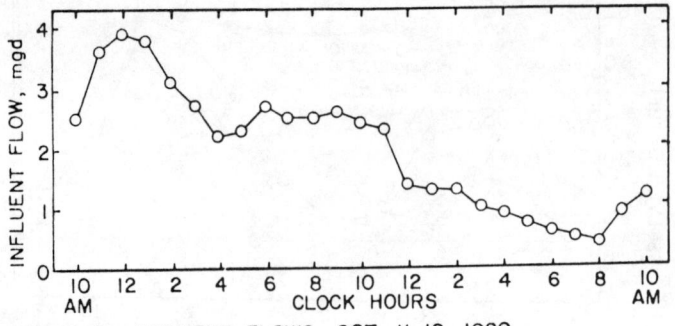

FIGURE 5A. INFLUENT FLOWS, OCT. 11-12, 1980.

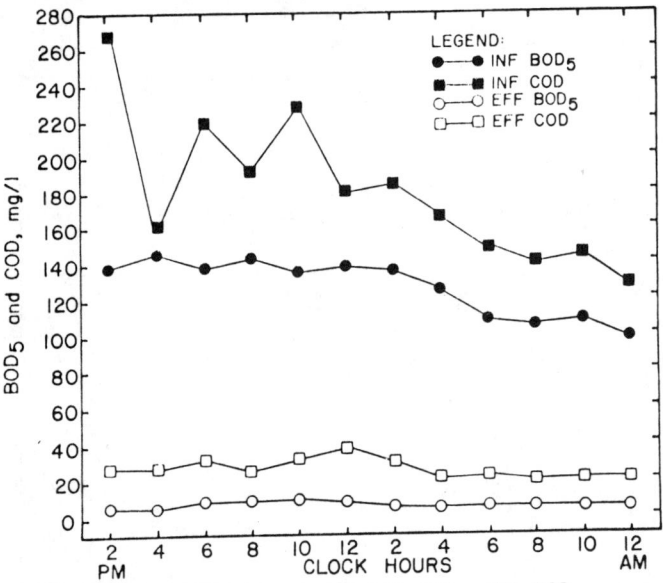

FIGURE 5B. INFLUENT AND EFFLUENT BOD_5 AND COD CONCENTRATIONS, OCT. 11-12, 1980.

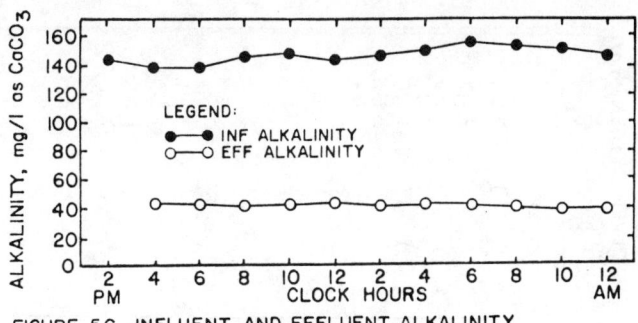

FIGURE 5C. INFLUENT AND EFFLUENT ALKALINITY CONCENTRATIONS, OCT. 11-12, 1980.

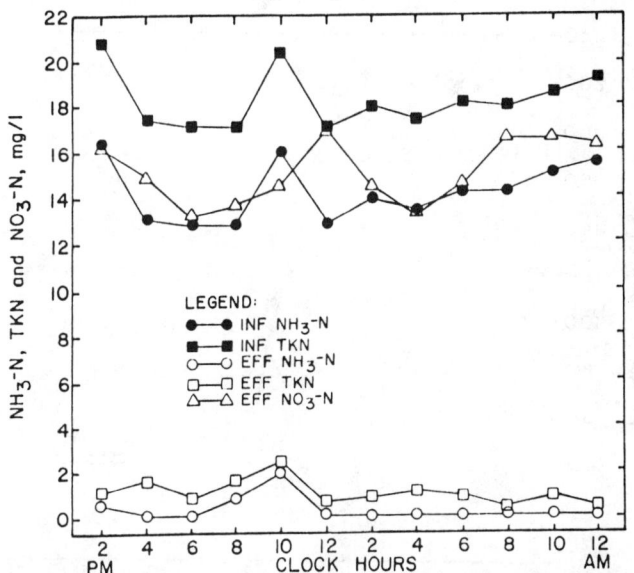

FIGURE 5D. INFLUENT AND EFFLUENT NITROGEN CONCENTRATIONS, OCT. 11-12, 1980.

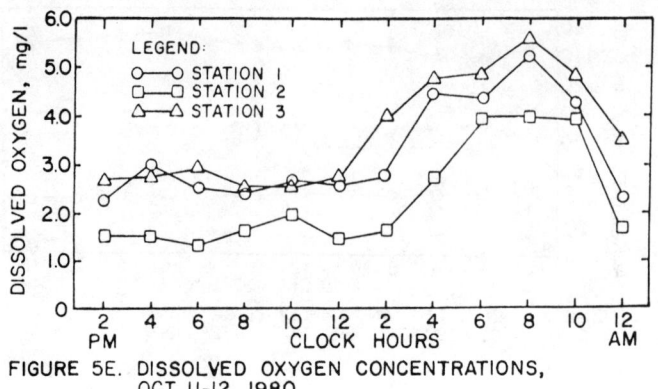

FIGURE 5E. DISSOLVED OXYGEN CONCENTRATIONS, OCT. 11-12, 1980.

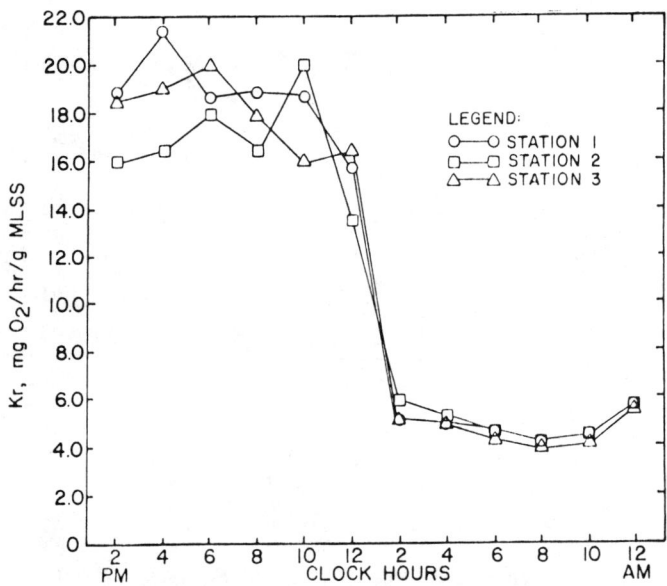

FIGURE 5F. SPECIFIC OXYGEN UPTAKE RATES, Kr, OCT. 11-12, 1980.

EXAMINATION OF THE PERFORMANCE OF ACTIVATED
SLUDGE PLANTS USING OXYGEN TO TREAT PULP
AND PAPER MILL WASTEWATERS

David B. Buckley. National Council for Air and
Stream Improvement, Tufts University, Medford,
Massachusetts

James J. McKeown. National Council for Air and
Stream Improvement, Tufts University, Medford,
Massachusetts

INTRODUCTION

　　The first activated sludge treatment process in the pulp
and paper industry was installed at a Southern bleached kraft
mill in 1955 (1). In the following twenty-five years, approximately 75 to 100 activated sludge systems have been
placed in operation to process wastewater resulting from the
wide spectrum of production categories. Numerous forms of
the activated sludge process are used involving such modifications as conventional plug-flow, complete mix, step aeration, contact stabilization, extended aeration, "hybrid" activated, and two-stage activated sludge. The "hybrid" activated sludge process is an attempt to provide the process
stability of the aerated stabilization basin (ASB) process
with the flexibility of activated sludge by providing mechanical secondary clarification for the return of settled sludge
to the aerated basin. Aeration time for this "hybrid" version is normally two to four days. The two-stage activated
sludge process, although not new to the industry, (2) has received interest in recent years under the name Zurn-Attisholz
process. Fundamentally, the process involves two activated
sludge systems in series.
　　In the late 1960's and early 1970's, the use of pure oxygen in activated sludge received renewed attention in the
form of research and demonstration programs which resulted in
the development of both covered and open reactor processes
(3-6). The covered reactor represents the predominant application and uses aeration tanks (under slight pressure) in
series where the wastewater and pure oxygen travel cocurrently.

Advantages associated with the use of pure oxygen in the activated sludge process are:

1) High settling velocities of the activated sludge.
2) Good compaction of the biological floc in secondary clarification.
3) Less biological sludge production as compared with the air activated sludge processes.
4) The ability to maintain high mixed liquor dissolved oxygen and suspended solids levels to increase the capability to accept swings in process loads.
5) Improved dewaterability of the biological sludge.

Approximately twenty years after the construction of the first activated (air) sludge system in the pulp and paper industry, the first pure oxygen activated sludge process was installed (7) at a bleached kraft mill in the same state as the first air system. Since 1974, at least fifteen additional paper industry locations have employed the pure oxygen activated sludge process and the pulp and paper industry is reported to be the largest industrial user of the oxygen activated sludge process (6). As a result of limited information existing on the full-scale operation of this modification of the activated sludge process, a review was made of the oxygen process and where possible, a comparison with air activated sludge is provided.

PERFORMANCE DATA SOURCE

Ten activated sludge systems were involved in the performance comparison, five utilizing conventional mechanical aeration and five using pure oxygen for mixing and oxygen

Table I. Production Categories and Activated Sludge Process Used

Mill No.	Category	Prod. (tpd)	AST[1] Type
1	Bl. Kraft	500	Air-Conv. Comp. Mix
2	Bl. Kraft	1150	Air-Conv. Plug Flow
3	Bl. Kraft	650	Air-Contact Stabilization
4	Unbl. Kraft	1800	Air-Conv. Comp. Mix
5	NSSC	650	Air-"Ext."[2] Aeration
6	Bl. Kraft	470	Oxygen-4 Stages
	Sulfite	450	
7	Mixed Kraft	1900	Oxygen-3 Stages
	NSSC	250	
8	Unbl. Kraft	850	Oxygen-3 Stages
9	Unbl. Kraft (pulp mill only)	200	Oxygen-3 Stages
10	Chemi-Mechanical	400	Oxygen-3 Stages

1) AST-Activated Sludge Treatment
2) "Ext." Aeration = Hybrid AST

transfer. An approximate twelve months of data were analyzed at each mill location with raw waste loads to biological treatment and effluent quality characterized in terms of annual average BOD_5 and suspended solids. Table I indicates the production categories and type of activated sludge at each of the ten mill locations.

INFLUENT QUALITY TO ACTIVATED SLUDGE TREATMENT (AST)

It is unusual when comparing process performance on a full-scale basis to find identical conditions in influent loads to the processes under examination. Both air and oxygen AST processes are treating wastes resulting from integrated mill operations where chemical pulping is used. Exceptions to this are Mill No. 9 where only pulp mill effluent is treated and Mill No. 10 which uses chemi-mechanical pulping. The predominant waste treated is kraft mill effluent (bleached and unbleached), followed by NSSC, sulfite (in combination with bleached kraft) and chem-mechanical. The waste loads to biological treatment from the ten mills are summarized in Table II.

Table II. Influent Quality[1] To Activated Sludge Treatment

Mill No.	Category	Water Usage (kgal/ton)	BOD_5 (lbs/ton)	(mg/l)	SS (lbs/ton)	(mg/l)
1	Bl. Kraft	30.7	76.0	255	31.2	105
2	Bl. Kraft	17.2	30.7	205	50.9	345
3	Bl. Kraft	21.2	44.0	255	32.6	190
4	Unbl. Kraft	13.7	24.0	205	10.1	85
5	NSSC	11.5	46.2	595	10.8	140
6	Bl. Kraft Sulfite	27.0	77.2	345	25.8	115
7	Mixed Kraft NSSC	28.1	25.5	110	20.7	90
8	Unbl. Kraft	11.4	31.9	335	15.6	165
9	Unbl. Kraft	16.9	43.2	310	13.3	95
10	Chemi-Mech.	14.5	32.7	270	6.8	55

1) Annual Average Basis

The range in water use at Mills 1-5 providing air activated sludge treatment and Mills 6-10 which provide oxygen activated sludge, when considered as two distinct groups, is quite similar ranging from 12 to 31 kgal/ton for the air mills while the oxygen mills ranged from 11 to 28 kgal/ton of production. The same condition exists for BOD_5 losses to biological treatment, with the air systems receiving 24 to 76 lbs BOD_5/ton and the oxygen systems receiving 32 to 77 lbs BOD_5/ton of production. However, the suspended solids quality, ranging from 10 to 50 lbs SS/ton, in the influent to the air activated sludge treatment group of mills was somewhat higher than losses, 7 to 26 lbs SS/ton, from those mills providing oxygen activated sludge treatment.

OPERATING CONDITIONS FOR AIR AND OXYGEN ACTIVATED SLUDGE SYSTEMS

Table III summarizes the operating conditions of biooxidation and secondary clarification provided by each of the ten activated sludge systems.

Table III. Operating Conditions - Air and Oxygen AST

Mill No.	AST Type-Modif.	Organic Loading ($\#BOD_A/\#MLSS$)	MLSS (mg/l)	Residence Time (hrs)	2° Clarification O'Flow Rate (gpd/sq ft)	2° Clarification U'Flow Consist (%)
1	Air-Conv/CM	0.19	3,500	9.1	375	1.0-1.5
2	Air-Conv/PF	0.18	4,500	6.2	675	1.5
3	Air-Cont. Stab	1) { 1.1 CB-1 / 0.29 CB-2	2,000	2.7 CB-1 / 2.1 CB-2 / 2.5 Stab	500	0.8-1.0
4	Air-Conv./CM	0.39	2,000	6.2	450	0.7
5	Air-"Ext." Aer	0.05	2,000	115	450	1.0
6	Oxygen-4 Stage	0.29	7,600	3.7	300	1.5-2.5
7	Oxygen-3 Stage	0.25	4,700	2.2	525	0.9-2.0
8	Oxygen-3 Stage	0.35	5,200	4.4	350	1.3-3.7
9	Oxygen-3 Stage	0.40	2,950	5.8	275	-
10	Oxygen-3 Stage	0.67	3,725	2.6	550	1.0-1.9

1) Composite organic loading reflecting combined stabilization Volume = 0.63 $\#BOD_{applied}/\#MLSS$

Some general observations regarding the two types of activated sludge involved in this review are:

1) The air activated sludge processes operate at a slightly lower organic loading expressed as lbs BOD_5 applied per lb mixed liquor suspended solids (MLSS).

2) Oxygen activated sludge systems maintain higher levels of mixed liquor suspended solids than air systems, generally 2000-4000 mg/l greater than air systems.

3) Residence times (under aeration) in air systems are approximately two to six hours greater than pure oxygen activated sludge.

4) Secondary clarification overflow rates in the oxygen process (range 275 to 550 gpd/sq ft) are somewhat lower than the overflow rates in air systems (range 375 to 675 gpd/sq ft).

5) Settled biological sludge in the oxygen activated sludge systems is approximately 1.0 to 1.5 percent consistency greater than the air systems examined

in this review.

No data were provided on the levels of dissolved oxygen (DO) in the mixed liquor of the air or oxygen processes. However, reported experience (8) with air activated sludge process operation indicates the norm for dissolved oxygen levels in the aeration basin ranges from 1 to 2 mg/l. In addition, operational dissolved oxygen goals have been cited (9) for the air activated sludge process at Mill No. 3 as 1) DO = 1.5-2.5 mg/l for contact basin No. 1,2) DO = 3.0-4.0 mg/l for contact basin No. 2 and 3) DO = 1.0-2.5 mg/l for the stabilization basin.

Dissolved oxygen levels in the reactors of the oxygen activated sludge systems have been reported (7,10,11) to range from 5 mg/l to 10 mg/l.

EFFLUENT QUALITY PRODUCED BY AIR AND OXYGEN ACTIVATED SLUDGE TREATMENT

Table IV summarizes the long-term performance in effluent quality of the five air and five oxygen activated sludge treatment processes.

Table IV. Effluent Quality[1] From Activated Sludge Treatment

Mill No.	AST Type	Prod. Cat.	Water Usage (kgal/ton)	BOD (#/ton)	BOD (mg/l)	SS (#/ton)	SS (mg/l)	Process Eff.(%)[2] (BOD)	Process Eff.(%)[2] (SS)
1	Air-Conv./CM	BK	30.7	6.7	25	21.5	75	91	31
2	Air-Conv./PF	BK	17.2	9.8	65	21.0	140	68	59
3	Air-Cont./Stab.	BK	21.2	1.5	9	4.0	25	97	85
4	Air-Conv./CM	UBK	13.7	2.4	20	6.0	50	90	42
5	Air-"Ext." AER	NSSC	11.5	0.8	10	3.1	40	98	76
6	Oxygen-4 Stage	BK/SUL.	27.0	4.2	20	15.6	70	95	40
7	Oxygen-3 Stage	MIX.KR/NSSC	28.1	3.1	15	15.4	65	88	26
8	Oxygen-3 Stage	UBK	11.4	4.6	50	9.8	105	85	37
9	Oxygen-3 Stage	UBK	16.9	3.9	30	10.0	80	89	5
10	Oxygen-3 Stage	CH.MECH.	14.5	8.4	65	6.4	50	74	5

1) Annual Average Basis
2) Across Biological Treatment, lbs/day Basis

The BOD_5 effluent quality (concentration basis) is quite similar for both the air and oxygen activated sludge processes, ranging from 10 to 65 mg/l, with the air group providing a somewhat better average BOD_5 of approximately 25 mg/l, versus 35 mg/l for the oxygen group.

The range in effluent suspended solids quality is somewhat broader for the air activated sludge (25 to 140 mg/l) than for the pure oxygen systems (50 to 105 mg/l); however, the air group provided slightly better average performance (65 mg/l) than the oxygen systems (75 mg/l).

The process efficiency of both types of activated sludge in reducing the level of BOD_5 is quite similar with the air systems providing a slightly greater efficiency, 89%, than the oxygen activated sludge, 86%. It is in the area of suspended solids capture that the air systems in this review outperformed oxygen activated sludge. The air group provided a suspended solids capture of approximately 60% compared with 25% for the oxygen processes.

SETTLING CHARACTERISTICS OF OXYGEN ACTIVATED SLUDGE

Two of the oxygen activated sludge systems measured the settling velocity of the biological sludge under aeration. The data appearing in Table V represent the average conditions of approximately 12 months of operation at both locations. In addition, data appearing in the literature (10,12) on the settleability of oxygen activated sludge is also included. It should be noted that the sludge velocity presented in Table V is that velocity indicated for the clarification function (zone settling or initial settling velocity) of a secondary clarifier and not the overflow rate specified for the thickening function that also occurs in such clarifiers.

Table V. Oxygen Activated Sludge Velocity

Mill No.	Prod. Cat.	Organic Loading (#BOD/#MLSS)	MLSS (mg/l)	Sludge Velocity[1] (ft/hr)	Equiv. O'Flow Rate (gpd/sq ft)	2° Clarifier O'Flow Rate (gpd/sq ft)	U'Flow Consist. (%)	Eff. SS Quality (mg/l)
8	UBK	0.35	5,250	3.4	610	350(57%)[2]	1.3-3.7	105
10	Ch. Mech.	0.67	3,725	6.2	1120	550(49%)	1.0-1.9	50
Ref. 10	Bl. Kraft	0.8[4]	4,500	5.5	990	450(45%)	2.0[3]	40
		0.9[5]	4,000	3.5	630	450(71%)	1.1[3]	75
Ref. 12	Ch. Mech.	0.56[6]	4,800	5.7	1025	595(58%)	1.6	25
		0.77[7]	4,000	6.3	1135	610(54%)	1.2	76
		0.68[8]	5,100	5.0	900	540(60%)	1.6	44

1) Zone (or initial) settling velocity
2) The percentage figure represents the fraction of the hydraulic capacity of the secondary clarifier utilized by the activated sludge settleability or velocity
3) Estimated from sludge volume index data
4) Characterized by mill as nonfilamentous operating conditions
5) Characterized by mill as filamentous operating conditions
6) Characterized by mill as good operating conditions
7) Characterized by mill as poor operating conditions
8) Characterized by mill as typical operating conditions

The settling velocity of the two sludges at Mill No. 8 and Mill No. 10 differ by an approximate factor of two, 3.4 ft/hr for the unbleached kraft location and 6.2 ft/hr for the chemi-mechanical mill. The equivalent overflow rates to reflect such settling velocity are approximately 600 gpd/sq ft and 1100 gpd/sq ft, respectively. The secondary clarifiers at these two oxygen systems can accommodate activated sludges which settle at lower settling rates than the measured sludge velocities. At Mill No. 8, secondary clarification occurs at an overflow rate of 350 gpd/sq ft which represents 57 percent of the settling velocity of the activated sludge. Mill No. 10 presents a similar situation in that secondary clarifiers provide an overflow rate, 550 gpd/sq ft, which is 49% of the sludge settling capability of the activated sludge. In this particular case, the magnitude of the sludge settling velocities appears to parallel the quality of clarifier overflow suspended solids in that the sludge with the higher settling velocity produces a lower concentration of effluent suspended solids.

The data provided by Reference 10 appear to reinforce this observation. At this particular location the settleability of the activated sludge has been classified as that occurring in the absence and presence of filamentous organisms in the activated sludge. The reduction in settling velocity occurred at a time when there was a slight increase in organic load to the process.

The performance cited in Reference 12 indicated in Table V does not completely support the observation that an activated sludge with a high settling velocity will produce an effluent suspended solids of better quality than a sludge exhibiting a lower settling velocity. The activated sludge process at this mill has been categorized as operating under "good", "poor", and "typical" operating conditions. In this specific case, the transition from "good" performance to "typical" (or somewhat lower) performance paralleled a reduction in settling velocity. However, under "poor" operating conditions, the activated sludge settling velocity actually increased over the levels observed under "good" and "typical" conditions. The "poor" operating conditions were also associated with the highest operating temperature $103^\circ F$ and the highest activated sludge organic loading although the operating temperature was only $3^\circ F$ higher than the $100^\circ F$ temperature associated with "good" performance. It would appear that the combined factors of increased temperature and organic loading experienced during poor performance conditions more than offset the effect of increased settling velocity during this same period.

The four processes cited in Table V provided secondary clarifier overflow rates which more than adequately reflected the initial settling velocity of the oxygen activated sludges. The settling velocities observed represented a fraction of the overflow rates provided by secondary clarification, ranging from 45% to 71% of the hydraulic loading. While it is difficult to generalize in the performance of secondary clarification due to the complex and interacting factors in the activated sludge process, it would appear that when the initial settling velocity of an activated sludge exceeds 50% of the secondary clarification overflow rate some deterioration in effluent suspended solids quality can be expected to occur.

Of particular interest is that while one of the reported advantages of pure oxygen activated sludge is the development of a sludge which settles faster than air activated sludge, secondary clarification rates for the oxygen systems in this review are less than the air activated sludge group. This may reflect 1) an unwillingness to take advantage of the higher settling characteristics of the oxygen sludge by designing smaller secondary clarifiers or (2) a provision to use a conservative, "factor of safety" approach in secondary clarification.

PRODUCTION OF WASTE ACTIVATED SLUDGE

Four of the five oxygen activated sludge locations provided data on wasting of activated sludge. No data were provided by the air activated sludge systems examined in the review; however, data appearing in summaries of two NCASI activated sludge process operation and maintenance technical sessions conducted on the West Coast in 1978 and 1980 were drawn upon for comparative purposes (13,14). Table VI contains the waste activated sludge experience for both air and oxygen processes. These data do not reflect a balance around the biotreatment process but the actual sludge wasted from the process. Influent to biotreatment and effluent from secondary clarification should be taken in account if actual biological sludge production is under consideration. Data in Table VI indicate that the pure oxygen activated sludge processes reviewed in this report appear to waste less biological sludge than air systems.

Table VI. Estimate of Waste Bio-Sludge Production

Mill No.	AST Type	Prod. Cat.	Waste Sludge Production ($\#SS/BOD_R$)	
			Avg.	Range
Ref. 13-14	Air	TMP	-	0.3-0.6
"	"	BK & GW	-	0.3-0.5
"	"	Sulfite	-	0.2-0.5
"	"	Dis. Sulf.	-	0.2-0.4
6	Oxygen	BK & Sulf.	0.40	0.24-0.63
7	"	Mix Kr & NSSC	0.21	0.11-0.33
8	"	UBK	0.17	0.10-0.33
10	"	Ch.-Mech.	0.39	0.26-0.59

HORSEPOWER UTILIZATION IN AIR AND OXYGEN ACTIVATED SLUDGE PROCESSES

The horsepower levels (nameplate basis) involved in oxygen transfer and mixing in the ten activated sludge systems are indicated in Table VII.

Table VII. Aeration Energy Associated with Air and Oxygen Activated Sludge

Mill No.	AST Type	Prod. Cat.	HP[1]	BOD_5 Removed (lbs/hp-day)	
1	Air-Conv/CM	BK	1200	28.3	
2	Air-Conv/PF	BK	1800	13.2	
3	Air-Cont.Stab.	BK	1065	25.8	Avg.=19.5[2]
4	Air-Conv/CM	UBK	2400	16.1	
5	Air-"Ext." Aer.	NSSC	1200	21.3	
6	Oxy.-4 stage	BK/SUL.	2850	23.6	
7	Oxy.-3 stage	MIX. KR/NSSC	1875	20.4	
8	Oxy.-3 stage	UBK	1000	23.2	Avg.=20.7[2]
9	Oxy.-3 stage	UBK	680	12.1	
10	Oxy.-3 stage	Ch. Mech.	680	14.3	

1) Nameplate Basis
2) Based upon the Total Energy & BOD_{REM} in Air and Oxygen Process

Surface aerators provide the aeration energy in the air systems with one exception. In the contact stabilization process at Mill No. 3, submerged static aerators were used in the stabilization basin. The blower horsepower was added to the existing surface aerator capacity in the contact basins to provide the total horsepower level. The aeration capacity for the oxygen systems represents the horsepower used to provide the pure oxygen and the mixers in each stage of the process.

The BOD_5 removed by each of the ten processes was then related to the installed horsepower to provide a unit removal capability in lbs of BOD_5 removed per horsepower day. As shown in Table VII, the averaged performance of the air and oxygen processes is quite similar with the air systems indicating 19.5 lbs BOD_5 removed per horsepower day while the oxygen units provided a slightly higher capability of 20.7 lbs of BOD_5 removed per horsepower day. The highest removal capability was observed at one of the air systems while the lowest performance occurred at a pure oxygen location.

It should be noted that the installed horsepower was used at each location and the BOD_5 removal capability per HP-day does not reflect possible mill specific controls directed towards reducing aeration capacity during periods of low BOD_5 levels to treatment.

SUMMARY

The performance of five oxygen activated sludge processes has been reviewed and, where possible, a comparison made with air activated sludge processes. Duplicate conditions, such as would be encountered in a controlled study under laboratory conditions of influent loads and production categories treated, were not experienced in this review; however, the range in influent loads to both the air and oxygen processes was not dramatically dissimilar. Observations made concerning the operation and performance of the ten activated sludge systems are as follows:

1) The air activated sludge processes operated at a lower organic loading than the pure oxygen systems.

2) The mixed liquor suspended solids in the oxygen processes are generally 2000-4000 mg/l greater than in air activated sludge.

3) As a group, the air activated sludge treatment systems performed at slightly higher secondary clarification overflow rates than the oxygen systems.

4) Effluent quality, characterized in terms BOD_5 and suspended solids (concentration basis), from the two forms of activated sludge differed to a small degree. Collectively, the air activated sludge systems produced slightly lower levels of effluent BOD_5 and suspended solids, approximately 10 mg/l less.

5) The oxygen processes developed an activated sludge which compacted in secondary clarification to a consistency approximately 1-1.5% greater than the air systems.

6) The capture of suspended solids, as indicated by the removal efficiency of influent suspended solids to the activated sludge process, was somewhat greater for the air systems.

7) The pure oxygen systems appear to waste less biological sludge from the process.

8) Data at four oxygen locations indicate that the settling velocity of the oxygen activated sludges represents 50-60 percent of the clarification capacity of the secondary clarifiers.

9) The energy associated with mixing and oxygen transfer in the two processes is quite similar with the pure oxygen systems as a group being slightly more efficient in removing BOD per unit of energy expended.

ACKNOWLEDGEMENT

Appreciation is expressed to the mills providing data for this review. A majority of the mills provided data through their participation in the National Council's program on the performance of wastewater treatment practices in the pulp and paper industry while the remaining locations responded to solicitation for performance data.

REFERENCES

1) Burns, O.B. Jr., "Recent Improvement Studies at the West Virginia Pulp and Paper Waste Treatment Plant in Covington, Virginia," Presented at NCASI South Central Regional Meeting, August 21, 1963.
2) "Biological Waste Treatment Case Histories in the Pulp and Paper Industry," NCASI Stream Improvement Technical Bulletin No. 220, November, 1968.
3) "Use of Molecular Oxygen in Treating Semi-Chemical Pulp Mill Wastes," NCASI Stream Improvement Technical Bulletin No. 149, September, 1961.
4) Albertson, J. G. et al, "Investigation of the Use of High Purity Oxygen Aeration in the Conventional Activated Sludge Process," Federal Water Quality Administration Department of the Interior, Report No. 17050-DNW 05/70, Washington, D. C. 1970.
5) Pearlman, S. R. and Fullerton, D. G., "Full-Scale Demonstration of Open Tank Oxygen Activated Sludge Treatment." U.S. Environmental Protection Agency, Municipal Environmental Research Laboratory, Office of Research and Development, Cincinnati, Ohio, EPA-600/2-79-012, May 1979.
6) Brenner, R. C., "Status of Oxygen Activated Sludge Wastewater Treatment," U. S. Environmental Protection Agency, Environmental Research Information Center, Technology Transfer, Cincinnati, Ohio, EPA 625/4-77-003a, October 1977.
7) Branko, D. et al, "High Purity Oxygen Application at the Chesapeake Corporation of Virginia," Presented at 1975 Tappi Annual Meeting, New York, New York, February 24-26, 1975.
8) NCASI Technical Conference, "Detection Prevention and Solution of Activated Sludge Treatment System Operational Problems in the Paper Industry," Chicago, Illinois, November 11-12, 1976.
9) Metzger, L. R., "A Comprehensive Program of Water and Waste Management at P. H. Glatfelter Company," Water Pollution Control Association of Pennsylvania Magazine, November-December 1979.
10) Nichols, W. E., "Start-Up and Operation of UNOX Secondary Waste Treatment System," Presented at 1978 Tappi Environmental Conference, Washington, D.C., April 12-14, 1978.
11) Peterson, G. et al, "Operating Experience at Longview Fibre Company's Pure Oxygen Activated Sludge System," Presented at 1978 Tappi Environmental Conference, Washington, D.C., April 12-14, 1978.

12) Gilbreath, K. R., "Chesapeake Corporation Operating Experience with a High Purity Oxygen Activated Sludge Waste Treatment System," Presented at 1976 Tappi Environmental Conference, Atlanta, Georgia, April 26-28, 1976.
13) NCASI Special Technical Session, "Activated Sludge Process Operation and Maintenance," Portland, Oregon, March 30, 1978.
14) NCASI Special Technical Session, "Activated Sludge Process Operation and Maintenance," Renton, Washington, February 19, 1980.

LEAST COST DESIGN AND DESIGN SENSITIVITY OF AN ACTIVATED SLUDGE TREATMENT SYSTEM

Robert B. Paterson* and Morton M. Denn. Department of Chemical Engineering, University of Delaware, Newark, Delaware.

INTRODUCTION

The basic goal of a wastewater treatment system design engineer is to design plants that reliably produce an effluent of specified quality at the lowest possible cost. This paper describes an interactive computer program that designs a complete activated sludge wastewater treatment system, specifying sizes and operating conditions of the various units and construction and operating costs. The application of the program to least cost design is also discussed.

UNIT PROCESS MODELS

This section presents a brief description of each of the unit process models. A detailed discussion is provided elsewhere [1].
Preliminary treatment consists of screening, grit removal, and flow measurement. The design program calculates the capital and operation and maintenance (O&M) costs of these components on the basis of influent flow rate.
The design of the primary clarifier is based on data presented in a Water Pollution Control Federation design manual [2]. The model for this unit is entirely empirical, and relates the fractional removal of primary suspended solids (R) to the primary overflow rate and the total suspended solids (TSS) in the influent. The equation also contains a "polymer factor," based on the data of Voshel and Sak [3], which is used to describe primary clarification with chemical addition.

*Presently with Radian Corporation, McLean, Virginia.

The aeration basin design equations are based on Monod kinetics [4, 5] in which the specific growth rate of the carbonaceous bacteria is specified as a unique function of the dissolved BOD present in the aeration basin mixed liquor. The sludge age (θ_c) is the inverse of the specific growth rate; thus, specifying the sludge age fixes the effluent dissolved BOD_5. The aeration basin volume is determined from a mass balance on BOD_5. The design equation requires that a mixed liquor volatile suspended solids (MLVSS) concentration be selected by the designer. The aeration basin model also includes nitrification kinetics.

Proper design of the secondary settler requires that both clarification and thickening requirements be considered [6]. The design program considers clarification requirements first, using a correlation developed by Rex Chainbelt [7] to compute the secondary settler overflow rate as a function of effluent TSS and aeration basin MLVSS. The value of effluent TSS is specified by the design engineer on the basis of the appropriate regulatory TSS limitation.

It is necessary to check that the area specified for clarification will also be adequate for thickening. Thickening requirements are predicted using the batch flux method as developed by Kynch [6, 8], using an empirical form of the batch settling curve suggested by Vesilind [9]. The settling theory, together with the overall material balance equations relating the secondary settler to the aeration basin, then leads to five nonlinear equations for the settler zone concentrations and the recycle and sludge waste rates. If the solution is such that the area determined by clarification is insufficient and thickening constraints dominate the settler design (see [1]), then the design program automatically increases the settler area until thickening requirements are met.

Aeration requirements are calculated from consideration of the oxygen requirements of BOD_5 consumption, endogenous respiration, and ammonia conversion, using the mass transfer coefficient data of Schmit et al., [10].

The chlorine contact basin is designed to provide 15 minutes residence time at the peak daily flow. The chlorine dose is 5 g/m^3.

The sludge thickener area is specified on the basis of solids loading rates of 98 $kgTSS/m^2/day$ for primary sludge, and from 20 to 88 $kgTSS/m^2/day$ for waste activated sludge, depending on the θ_c specified by the designer.

The program designs a two-stage anaerobic digester to treat the combined primary and waste activated sludge from the sludge thickener. The digester is designed to provide 60 percent volatile suspended solids reduction.

Design of the solids centrifuge is based on the optimal conditions specified by Zenz et al. [11] in pilot scale studies of centrifugation: 93.2 percent solids capture, and a final sludge of 20 percent solids.

ESTIMATION OF COST AND ENERGY REQUIREMENTS

For each unit process, the installed capital cost and annual costs of labor, maintenance materials, power and chemicals are calculated using the data of Patterson and Banker [12]. Both capital and O&M costs were updated to September 1978 using the Chemical Engineering plant cost index [13]. The design program can be easily modified to update all cost equations to the present by using the current value of this index.

There are three degrees of freedom left to the designer when effluent standards and material balance constraints are taken into account. These are taken to be: (1) percent removal of TSS in the primary clarifier (R), (2) aeration basin MLVSS, and (3) sludge age (θ_c). These three variables must be selected by the user as inputs to the design program. The program output is a total present system cost (TPSC), which is the sum of the total construction cost and the product of annual O&M costs with the present worth factor. The program also provides detailed cost breakdown and size specifications, and material balances on each unit. Cost matrices showing the cost as a function of any two of the design variables, with the third fixed, are easily and rapidly generated automatically.

EXAMPLE DESIGN OPTIMIZATION

A plant to treat 20,000 m^3/day (5.3 MGD) of medium strength domestic wastewater was selected for the purpose of conducting an example design optimization. The influent wastewater has a total BOD_5 concentration of 200 $gBOD_5/m^3$ and a TSS concentration of 200 $gTSS/m^3$. Values of all kinetic constants assumed are provided elsewhere [1]. An interest rate of 10 percent and a system evaluation period of 20 years were assumed.

Optimization of the Primary Clarifier

The total present system cost as a function of primary TSS removal (R) is shown in Figure 1 for a plant in which the MLVSS is fixed at 2000 $gVSS/m^3$ and θ_c is fixed at 10 days. Curves are shown for strong and weak wastewaters as well as the medium strength base case. The results indicate that the optimal plant design has no primary clarifier. This same result was found to hold for all θ_c and MLVSS and for primary clarifiers with and without the addition of polymer. Increasing the value of R reduces the BOD_5 loading to the aeration basin, and thus decreases the cost of the aeration basin and air supply system. At the same time however, the cost of the primary clarifier, sludge thickener, anaerobic digester, and solids centrifuge are increased. The savings in the cost of the aeration basin and air supply system are more than offset by the increased cost of the primary clarifier and the sludge handling system. This result is quite striking since primary clarifiers are routinely included in the design of domestic wastewater activated sludge treatment plants. However, the same conclusion has been reached by several other authors [14-19].

It is important to note here that this result is based on the assumption that suspended and dissolved BOD_5 are degraded in the aeration basin at the same rate. Some studies have concluded that the rate of consumption of suspended waste is much slower than that of dissolved waste [20, 21], but the data are inadequate to justify a more detailed model. Thus, two major conclusions can be drawn: (1) when conventional Monod kinetics are assumed, the optimal size of the primary clarifier is always zero, and (2) the present lack of data in the literature concerning the relative rates of suspended and dissolved waste consumption is seriously

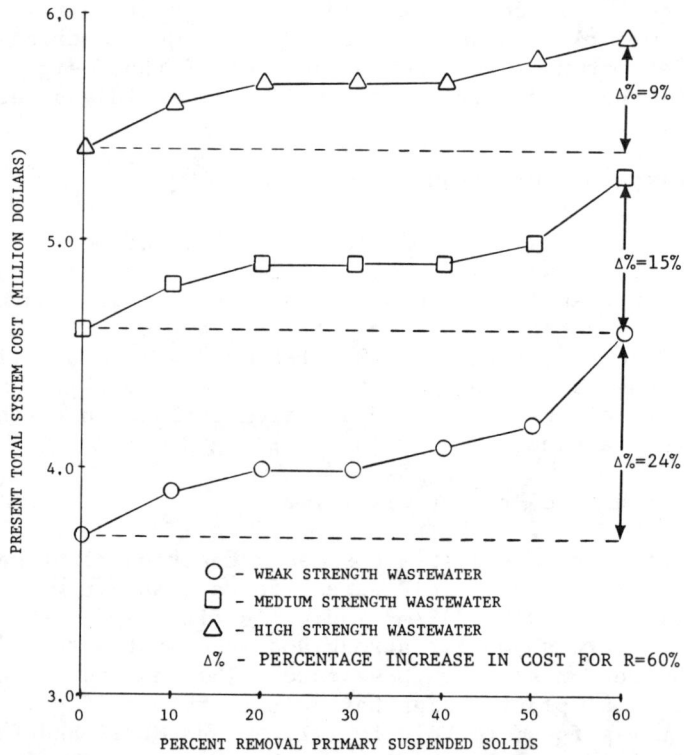

Figure 1. TPSC as a function of R.

limiting the state-of-the-art of activated sludge treatment system optimization.

Effect of Various Sludge Settling Assumptions on Optimum Plant Design

Work by Bisogni and Lawrence [22] and Ford and Eckenfelder [23] has shown that the settling characteristics of activiated sludge improve with increasing θ_c in the range of 2 to 15 days. At some point beyond 15 days, the settling velocity decreases with additional increases in θ_c. One objective of this study was to determine the sensitivity of plant design to uncertainties regarding the change of settling velocity with θ_c. The results are presented in Figure 2. The total present system cost (TPSC) is plotted versus θ_c for the three alternative assumptions concerning the change in settling velocity with θ_c shown schematically in the figure. The detailed data on which the settling curves are based are discussed elsewhere [1].

Settling assumption A, that V_o is constant regardless of θ_c, represents the assumption most commonly applied to system design. The assumed settling characteristics were typical of θ_c of about 10 days, so the results for the smallest values of θ_c are not reliable. For θ_c values greater than 6 days, TPSC is shown to be almost constant. A similar result was found by Middleton and Lawrence [24]. Higher values of θ_c require that more sludge be recycled to the aeration basin, and the costs of the aeration basin, air supply system, secondary settler, and recycle pump thus increase with sludge age. Less sludge is wasted, however,

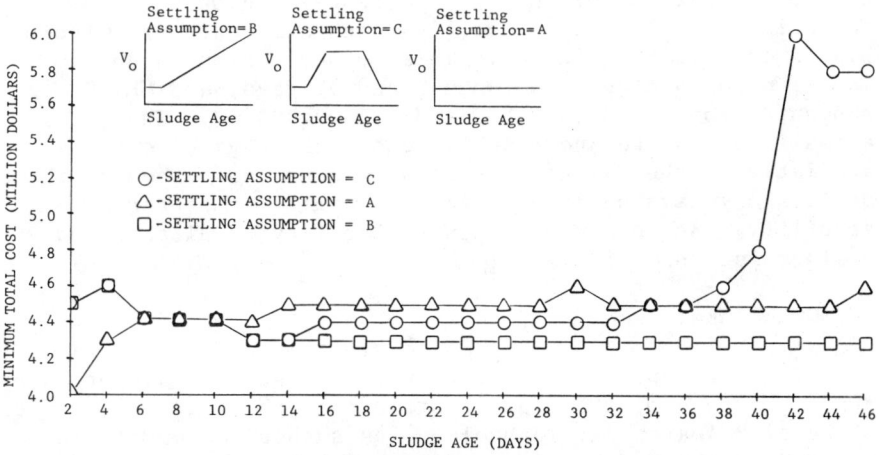

Figure 2. TPSC (minimum) versus sludge age.

and the cost of the sludge handling system is reduced. These two effects tend to balance each other.

Settling assumption B, that V_o is constant for θ_c up to 3 days and then increases linearly with θ_c up to 46 days, is highly unrealistic beyond θ_c of about 15 days. Beyond θ_c equal to 12 days, TPSC is constant. Improved settling characteristics lead to higher solids fluxes through the settler and a high underflow solids concentration, which allows the underflow and recycle flow rates to be reduced. At high V_o, however, clarification tends to dominate settler design, and settler area is not specified on the basis of thickening constraints; thus, additional improvements in settling characteristics do not offer cost savings.

Settling assumption C, is believed to be the most realistic, based upon the available data. It is shown to lead to a relatively flat cost curve over a wide range of θ_c with an optimum design having a θ_c in the range of 12 to 14 days.

In summary, although TPSC is not highly sensitive to θ_c, different assumptions regarding the change in settling characteristics with θ_c lead to different optimal designs. Because of the relative insensitivity of the minimum cost to the settling characteristics, design can be carried out efficiently using conservative estimates of the actual settling characteristics.

Effect of Reactor Solids on Plant Optimization

Total present system cost is shown as a function of both θ_c and MLVSS in Figure 3 for a plant with no primary settler. Three distinct regions have been denoted in the figure. The center region represents the lowest cost, 4.3 million dollars. Surrounding the center region on both sides and below is a second region, in which all costs are within 2.5 percent of the minimum. This second region is quite large, encompassing θ_c between 8 and 32 days, and MLVSS concentrations between 2,500 and 4,500 gVSS/m^3. Thus, for a small percentage increase in cost, the range of options available to the designer is greatly expanded. Some of these designs may have superior effluent quality or better dynamic stability. An analysis of controllability of alternative designs is contained in [1].

ACKNOWLEDGMENT

The work upon which this paper is based was supported by funds provided by the U.S. Department of the Interior, Office of Water Research and Technology, as authorized under the Water Research and Development Act of 1978, P.L. 95-467.

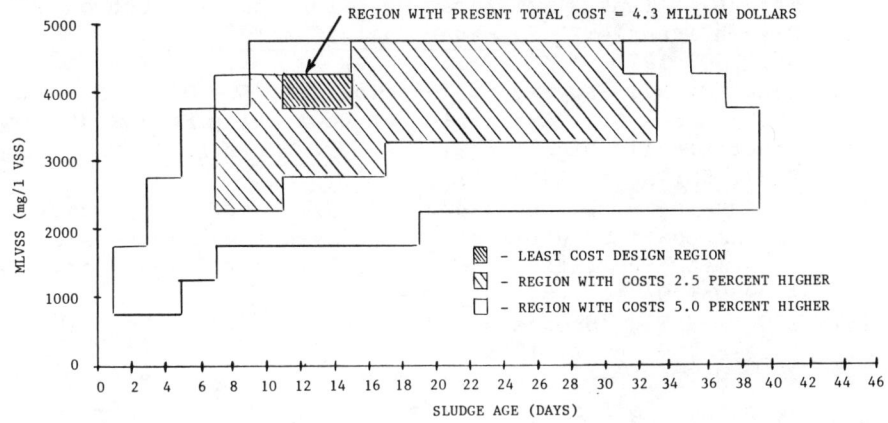

Figure 3. TPSC as a function of sludge age and MLVSS.

REFERENCES

1. Paterson, R. B. and Denn, M. M., "Design and Control of an Activated Sludge Process for Municipal Wastewater Treatment," Technical Completion Report for Project A-041-DEL, Water Resources Center, University of Delaware, Newark, Delaware.
2. Water Pollution Control Federation, "Sewage Treatment Plant Design Manual of Practice #8," 1959.
3. Voshel, D. and Sak, J. G., "Effect of Primary Effluent Suspended Solids on BOD and Activated Sludge Production," J. Water Poll. Control Fed., 40, 5, p. R203, 1968.
4. Metcalf and Eddy, Inc., Wastewater Engineering, McGraw-Hill, 1972.
5. Monod, J., "The Growth of Bacterial Cultures," Annual Review of Microbiology, 3, p. 317, 1949.
6. Dick, R. I., "Role of Activated Sludge Final Settling Tanks," J. Sanit. Eng. Div. ASCE, 96, SA2, p. 423, 1970.
7. Rex Chainbelt, Inc., "A Mathematical Model of a Final Clarifier," EPA Project No. 17090FJW, U.S. Government Printing Office, 1972.
8. Kynch, G. J., "A Theory of Sedimentation," Trans. Faraday Soc., 48, 166, 1952.
9. Vesilind, P. A., "Design of Prototype Thickeners from Batch Settling Tests," Water Sewage Works, 115, p. 303, 1968.

10. Schmit, F. L., Wren, J. D., and Redman, D. T., "The Effect of Tank Dimensions and Diffuser Placement on Oxygen Transfer," J. Water Poll. Control Fed., 50, p. 1750, 1978.
11. Zenz, D. R., Sawyer, B., Watkins, R., Lue-Hing, C., and Richardson, G., "Evaluation of Dewatering Equipment for Anaerobically Digested Sludge," J. Water Poll. Control Fed., 50, 8, p. 1965, 1978.
12. Patterson, W. L. and Banker, R. F., "Estimating Costs and Manpower Requirements for Conventional Wastewater Treatment Facilities," Water Poll. Control Res. Ser. No. 17090DAN10/71, U.S. EPA, 1971.
13. McGraw-Hill, "Economic Indicators," Chemical Engineering, 86, 2, p. 7, 1979.
14. Kuo, M. T., Fan, L. T., and Erikson, L. E., "Effects of Suspended Solids on Primary Clarifier Size Optimization," J. Water Poll. Control Fed., 46, p. 2521, 1974.
15. Mishra, P. N., et al., "Simulation Studies of the Activated Sludge System Including a Primary Clarifier," Paper Presented at the Silver Jubilee of the Indian Institute of Chem. Eng., New Delhi, 1972.
16. Sedzikowski, T., "Influence of the Presettling Tank Size on Dimensioning and Cost of a Sewage Treatment with Activiated Sludge," Water Res., 6, p. 341, 1972.
17. Parkin, G. F. and Dague, R. R., "Optimal Design of Wastewater Treatment Systems by Enumeration," San. Eng. Div. ASCE, 98, p. 833, 1972.
18. Fan, L. T., Kuo, M. T., and Erickson, L. E., "Effect of Suspended Wastes on System Design," J. Env. Eng. Div. ASCE, 100, p. 1231, 1974.
19. Von der Emde, "To What Extent are Primary Tanks Required?" Water Res., 6, p. 395, 1972.
20. Takahashi, S., et al., "Metabolism of Suspended Matter in Activated Sludge Treatment," in Adv. in Wat. Poll. Research, Proc. 4th Int. Conf., Prague (Ed. Jenkins, S. H.), Pergamon Press, p. 341, 1969.
21. McKinney, R. E., Microbiology for Sanitary Engineers, The McGraw-Hill Book Co., New York, 1962.
22. Bisogni, J. J. and Lawrence, A. W., "Relationships Between Biological Solids Retention Time and Settling Characteristics of Activated Sludge." Water Research, 5, p. 753, 1971.
23. Ford, D. L. and Eckenfelder, W. W., "Effect of Process Variables on Sludge Floc Formation and Settling Characteristics," J. Water Poll. Control Fed., 39, 11, p. 1850, 1967.
24. Middleton, A. C. and Lawrence, A. W., "Least Cost Design of Activated Sludge Systems," J. Water Poll. Control Fed., 48, p. 889, 1976.

TREATMENT OF TANNERY BEAMHOUSE WASTE WITH A BENCH-SCALE
ANAEROBIC REACTOR

M. H. Tunick, A. A. Friedman[1], D. G. Bailey, Eastern Regional
Research Center, Agricultural Research, Science and Education
Administration, U.S. Department of Agriculture, Philadelphia,
Pennsylvania.

INTRODUCTION

Effluent streams from the leather tanning and finishing industry present a difficult challenge for economical waste treatment. These wastewaters contain large quantities of BOD consisting primarily of dissolved and suspended hair proteins from the unhairing step in which the hides are prepared for tanning. The suspended and dissolved solids are high due to the use of large quantities of salt and lime in various steps of leather manufacture. Sodium sulfide, used in the unhairing process, adds to the chemical oxygen demand (COD) and is a potential toxicant in a biological treatment system.

Conventional aerobic treatment, even after dilution and high rates of aeration, has not been shown to achieve the proposed standards set for the tanning industry [1]. Further treatment, particularly for direct discharge, is necessary. Several aerobic systems for tanning wastes have been described [2,3]. It is possible to acclimate activated sludge to this waste but the cost of energy for aeration remains high [3].

Anaerobic treatment of a variety of industrial wastes has demonstrated the potential of this method for treating difficult wastes. Pharmaceutical [4] and polymer synthesis byproduct [5] are two of many examples. A. A. Friedman et al. presented work last year at this conference on the use of a bench-scale anaerobic filter (AF) to treat tannery beamhouse waste [6]. Pretreatment, in which the pH was lowered, liberated sulfide as a gas and removed part of the dissolved protein by precipitation. Removal of up to 70% of

---

[1]Department of Civil Engineering, Syracuse University, Syracuse, N.Y. 13210.

the remaining COD was then achieved by the AF. A combination of the AF plus a rotating biological contactor removed over 98% of the 5-day BOD applied to the dual system.

This paper describes a 35 l bench-scale suspended anaerobic system which has treated beamhouse waste and achieved COD removal comparable to that of the AF. In this work, the only pretreatment was dilution. Aerobic followup treatment was not included in this study but would be necessary following an anaerobic treatment process. The unexpected finding of methane generation as well as COD reductions—in spite of sulfide levels in excess of 1000 mg/l—improves considerably the potential for practical application of this process for removing COD from tannery effluent.

MATERIALS AND METHODS

Reactor

The bench-scale anaerobic reactor consists of two cylindrical tanks (Figure 1). The anaerobic reactor is 12 inches ID and 18 inches high. It has an airtight cover of plexiglass, through which provision is made for stirring, heating, feeding, and sampling. A horizontal tube 1/2 inches ID and 15 inches from the bottom connects the reactor to the second cylinder which is 6 inches ID and 18 inches high. This serves as a settling tank from which settled sludge is constantly recirculated to the reactor and effluent leaves the system through an overflow port.

Suspension of the anaerobic sludge is maintained by a stainless steel stirrer, driven at 60 RPM by a 0.05 HP electric motor. The impeller consists of two 2-inch square blades pitched at 45° and suspended 4-1/2 inches from the bottom of the reactor. The mother liquor is maintained at 37 C by a hotfinger located in a silicone-filled test tube suspended from the lid. Two similarly held test tubes in the lid are used for monitoring temperature and for a thermistor providing feedback for the temperature control. Gas produced in the reactor and the settling tank is collected, and volume is measured by water displacement. Samples for determination of gas composition are obtained by a hypodermic syringe through a septum mounted in the top of the reactor. Two stainless steel tubes extending 12 inches are mounted in the lid. One of these is used for feeding and the other for withdrawing liquid samples.

As the effluent leaves the reactor it flows into the settling tank and down through a stainless steel tube to within 4 inches of the bottom. The bottom is cone-shaped to collect settleable sludge which is recirculated at 70 ml/min back to the reactor. The final settled effluent exits through an opening opposite the entrance port and passes through an inverted siphon to prevent escape of gas.

The tannery influent consisted of beamhouse unhairing waste diluted to the desired COD. The feed was continuously added to the reactor with a peristaltic pump at a rate of 10-15 l/day.

Chemical Analyses

Representative samples of feed, mother liquor, and effluent were taken each work day. Total chemical oxygen demand (TCOD) was measured by the culture tube method of Knechtel [7]. Correction for sulfide was made by taking twice the sulfide concentration and subtracting this number from the TCOD concentration. Gas composition was measured with a Hewlitt-Packard model 5734A gas chromatograph, with a 6 ft. x 3/16 inch aluminum molecular sieve column and a thermal conductivity detector (carrier: 20 ml He/min; sample volume: 0.5 ml; temperature programming: 85 to 200 C at 32 C/min). A typical chromatogram is shown in Figure 2. Volatile fatty acids (VFA) were analyzed on the same instrument with flame ionization detection and a 6 ft. x 1/4 inch stainless steel column consisting of 15% SP-1220/1% H_3PO_3 on Chromosorb W [8] (carrier: 20 ml He/min; sample: 10 µl; temperature: 150 C). A chromatogram of this type is shown in Figure 3. Suspended solids (SS) and volatile suspended solids (VSS) were determined by procedures in Standard Methods, 14th edition [9]. Sulfide (S^-) was analyzed titrimetrically by oxidation with $K_3Fe(CN)_6$ [10]. Total Kjeldahl nitrogen (TKN) and ammonia levels were measured on a Technicon Auto Analyzer [11].

Start-up

The reactor vessel was filled with sludge from an existing anaerobic pilot plant and fed with diluted tannery waste at a rate of up to 72 g TCOD/day. After 16 weeks of operation, reactor supernate was removed and additional anaerobic sludge was added. Sludge was never wasted during the course of the experiment. Four weeks later the VFA concentration in the reactor was increased by the addition of glucose to the feed. The glucose loading was maintained at 35 g/day for 2 weeks; after that time, the feed was again made up to consist solely of diluted tannery waste. Monitoring of gas production began 7 weeks after glucose addition was stopped. At this time the experiments and data collection reported below were begun.

RESULTS AND DISCUSSION

TCOD Removals

An average reduction in influent TCOD during the 4-month period reported was 62% after correction for sulfide (Figure 4). This average was maintained over a volumetric TCOD loading range of from 0.92 to 4.92 kg/m^3/day, and a feed concentration of 2000 to 15,000 mg/l COD. At the highest levels of TCOD applied there seemed to be a small decrease in efficiency. Sulfide increased in the reactor and contributed a significant portion of the TCOD measured. Since this would be rapidly removed by subsequent aerobic treatment, the corrected TCOD is a more accurate means of evaluating the effectiveness of the anaerobic treatment.

The VSS content of the anaerobic reactor varied between 500 to 2000 mg/l over the test period. It fluctuated widely but did not appear to be increasing. When the TCOD removed was plotted vs the TCOD applied per gram of VSS, a graph similar to Figure 4, suggesting a removal of about 60%, resulted. However, in this case the scatter of data points around the line was greater. VSS ranged from 30 to 70% of the total solids in the feed. Hydraulic detention time of the reactor was generally 60 hr.

The TCOD applied/day and TCOD in the effluent/day are shown in Figure 5. Accurate sampling of the feed and the effluent from these experiments is difficult, as indicated by the scatter. However, the distinct changes in level of feed loading rates are apparent. The average removal efficiency over the period from January 8 to February 11 was 62% at a loading rate of 1.27 kg TCOD/m^3/day. During the rest of the period in February and March it was 62% while the applied TCOD averaged 3.89 kg TCOD/m^3/day.

Other Parameters

The pH of the reactor varied from 7.4 to 8.2 but changed very slowly. The system rapidly developed a level of 600 mg/l of VFA; although it ranged up to 1800 mg/l, it never dropped below 600 mg/l. This is an indication that the microorganism population was always changing but never fell below a certain level. The stability of the pH, in spite of VFA variability, is probably due to the high alkalinity of the feed from the unhairing waste.

Kjeldahl nitrogen and ammonia analyses were performed on samples from the feed, reactor, and effluent. TKN across the system remained constant. However, 32% was converted to free ammonia which appeared in the effluent.

Gas Production

Perhaps the most interesting finding of the research deals with methane production in this system. Figure 6 shows the concentrations of sulfide ions and volatile fatty acids in the reactor and the volume of methane produced each day. It is generally accepted that sulfide, which reached levels as high as 1000 mg/l in the reactor, inhibits methane generation. As expected, the reactor initially did not produce gas, but shortly after start-up gas was generated at a rate of 4-5 l/day. This is equivalent to nearly 0.2 l CH_4/g COD destroyed/day. Even at this level, however, roughly 30 times less gas was generated per unit of COD removed than normally produced in municipal digesters. Over the period of these experiments, gas production slowed to low levels on two occasions. In both cases the decrease corresponded to a rapid increase in TCOD loading on the reactor. The pH of the reactor, as indicated before, was quite stable and changes in gas production did not correlate with a change in pH.

It appears from Figure 6 that concentration of VFA, concentration of sulfide, and methane generation all reacted to the TCOD change shown in Figure 5 rather than to each other. The gas production reacted most dramatically and was the slowest to recover. At precisely the point where feed concentrations were increased all of these parameters dropped and all ultimately recovered.

Figure 7 shows the percent of TCOD removed daily. Comparison of this with the loading pattern shown in Figure 5 clearly shows that the TCOD removal was not affected by a rapid change in TCOD applied. On one occasion, a return sludge line broke and the sludge in the system spilled onto the laboratory bench. In the morning it was scooped back into the settling tank, and by the next sampling time the TCOD removal had completely recovered. This indicates that the system is not acutely poisoned by oxygen and is quite stable under anaerobic conditions.

A second feature on this figure demonstrates how rapidly the system recovers from a spill. These interactions are consistent with the generally accepted thesis that TCOD removal and methane formation are two different systems in the anaerobic process. Since the volatile fatty acids reacted and recovered at different rates than the methane production, perhaps there are even three systems.

Scale Up

The practicality of an anaerobic system for tannery waste depends on the size of the system required and the energy costs associated primarily with maintaining the necessary temperature.

The size of the system depends on the concentration of waste which can be treated, the removal reaction rate, the mean cell retention time, and the hydraulic detention time. The bench-scale system would permit treatment of waste with COD as high as 15,000 mg/l. The hydraulic retention time required is 60 hr. Equalized waste from the entire tannery has a much lower COD (approx. 3-4000 mg/l). The ability of this system to treat very concentrated waste might not be used to full advantage on the total effluent.

In addition, the anaerobic system must be run at a temperature of 37 C. In a municipal digestor the methane generated can be burned to produce sufficient energy to maintain that temperature. The yields of gas in this reactor have not been nearly that good. In part, this is due to the fact that this waste is largely proteinaceous. Through the use of heat exchange between the treated effluent and the influent, a significant gain in efficiency could be obtained.

Anaerobic treatment of tannery waste might be efficient only for specific concentrated streams; the smaller amount of liquid would require less heat and reactor size. The effluent from this reactor would then be combined with the rest of the streams for further aerobic treatment.

CONCLUSIONS

Experimental results obtained on the treatment of tanning unhairing effluent with a bench scale anaerobic reactor can be summarized as follows: TCOD reductions of over 60% were achieved on this waste with loadings as high as 4.92 Kg/m^3/day (15,000 mg/l), with only a 60-hr hydraulic retention time. In this series of experiments, the removal appeared to be independent of the loading over the range tested. Production of methane, sulfide, and volatile fatty acids was strongly affected by a rapid change in the feed concentration. Concentration of each dropped quickly when TCOD increased. Recovery of sulfide and VFA was much more rapid than that of methane generation. Methane was generated in this system in spite of the presence of sulfide at levels greater than 1000 mg/l. Sulfide levels continually increased during the test period. It was presumably formed from the disulfide linkages present in the dissolved hair protein, keratin. Approximately 32% of the Kjeldahl nitrogen in the feed was converted to ammonia over the same time period. Periodic tubing connector problems resulted in spills of sludge from the reactor, which, however, recovered rapidly.

REFERENCES

1. U.S. Environmental Protection Agency, "Leather Tanning and Finishing Point Source Category Effluent Limitations

Guidelines, Pretreatment Standards, and New Source Performance Standards," *Federal Register*, 44, 38746 (July 2, 1979).

2. Bailey, D. G., Tunick, M. H., Cooper, J. E., Friedman, A. A. and Chang, I. L., "Aerobic Suspended Fixed Film Treatment of High-Strength Tannery Beamhouse Waste," *12th Mid-Atlantic Industrial Waste Conference Proceedings*, Bucknell University, Lewisburg, Pa., 142 (1980).

3. Cooper, J. E., Happich, W. F., Mellon, E. F. and Naghski, J., "Acclimatization of Activated Sludge to Lime-Sulfide Unhairing Effluents," *JALCA*, 71, 1 (1976).

4. Jennett, J. C. and Dennis, N. D., "Anaerobic Filter Treatment of Pharmaceutical Waste," *JWPCF*, 47, 104 (1975).

5. Witt, E. R., Humphrey, W. J. and Roberts, T. E, "Full-Scale Anaerobic Filter Treats High Strength Wastes," *34th Industrial Waste Conference Proceedings*, Purdue University, Lafayette, Indiana, 229 (1979).

6. Young, K. S., Friedman, A. A. and Bailey, D. G., "Pretreatment of Tannery Beamhouse Wastewater Using an Anaerobic Filter: Preliminary Results," *12th Mid-Atlantic Industrial Waste Conference Proceedings*, Bucknell University, Lewisburg, Pa., 102 (1980).

7. Knechtel, J. R., "A More Economical Determination of Chemical Oxygen Demands," *Water and Poll. Control*, (May-June 1978), 25.

8. Supelco, Inc., *Analysis of VFAs from Anaerobic Fermentation*, Bulletin 748E, Bellefonte, Pa. (1975).

9. *Standard Methods for the Examination of Water and Wastewater* (Washington, DC: APHA, 1976), 14th edition.

10. Booth, H., "Atmospheric Oxidation of Alkali Sulphide Solutions," *J. Soc. Leather Trades Chemists*, 40, 238 (1956).

11. Technicon Industrial Systems, *Industrial Method No. 34474 W/B*, Tarrytown, N.Y. (1977).

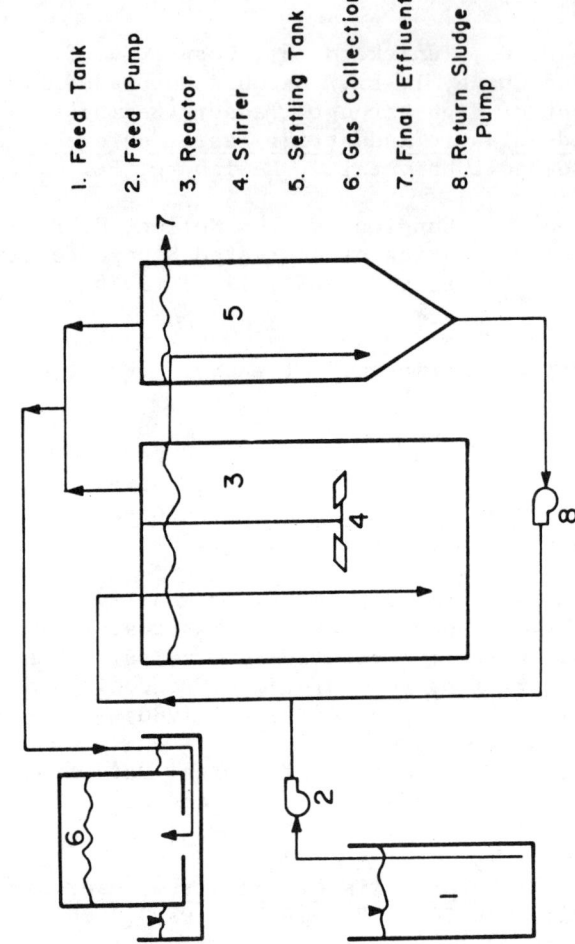

Figure 1. Schematic diagram of bench-scale reactor.

1. Feed Tank
2. Feed Pump
3. Reactor
4. Stirrer
5. Settling Tank
6. Gas Collection
7. Final Effluent
8. Return Sludge Pump

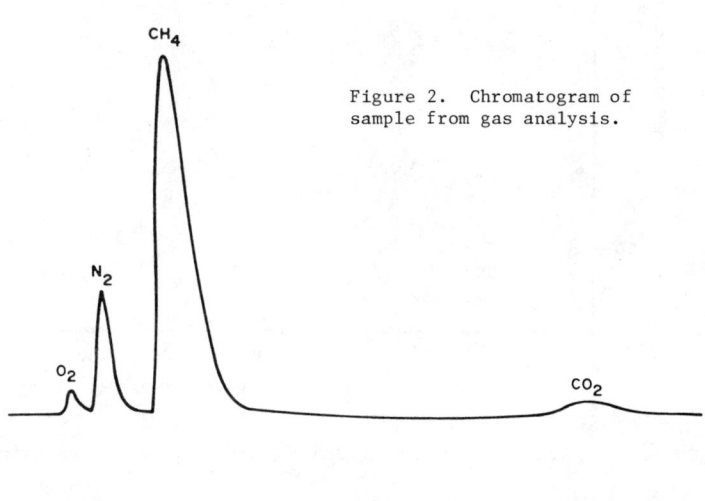

Figure 2. Chromatogram of sample from gas analysis.

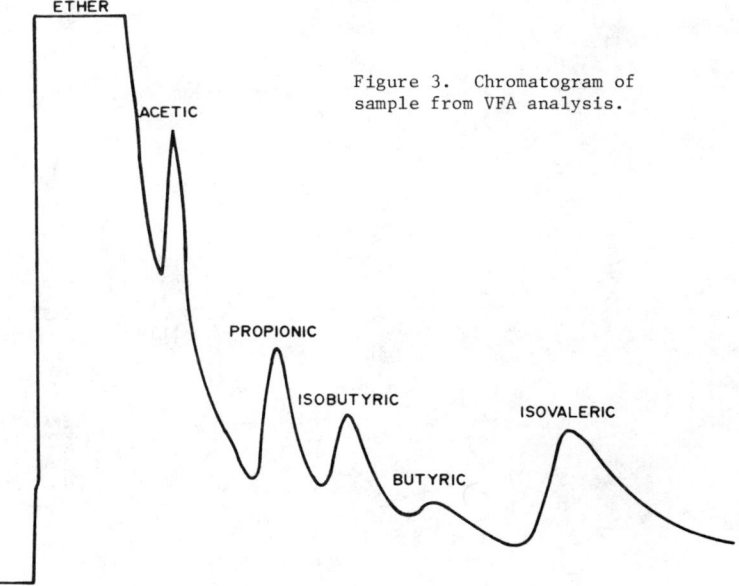

Figure 3. Chromatogram of sample from VFA analysis.

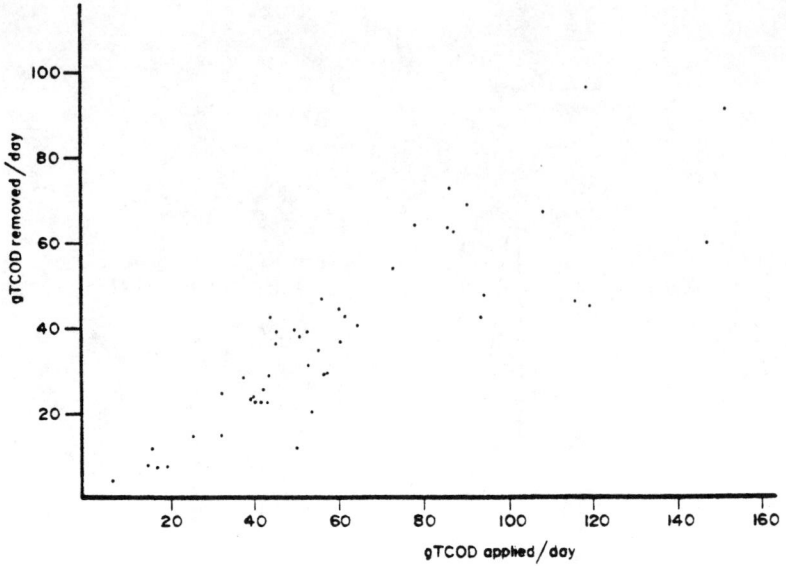

Figure 4. Application and removal of tannery beamhouse waste

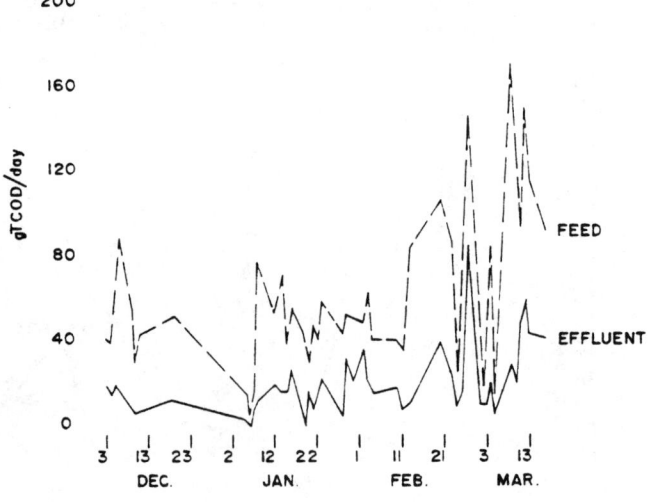

Figure 5. Chronological variation of TCOD in feed and effluent

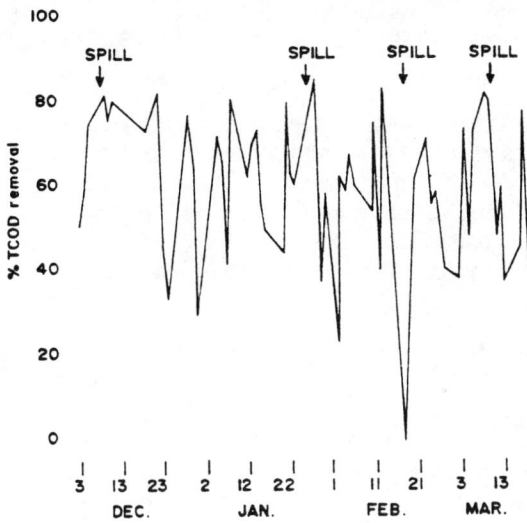

Figure 6. Chronological variation of TCOD removal

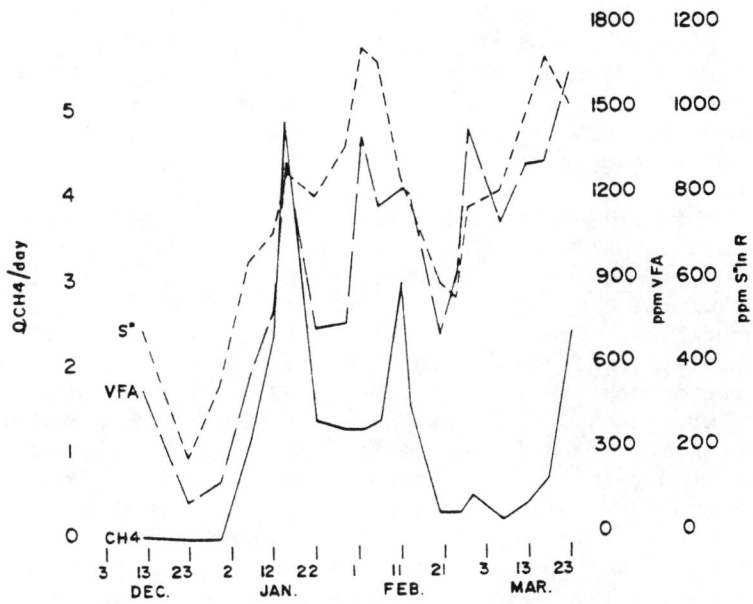

Figure 7. Comparison of reactor CH_4 production, VFA, and $S^=$

COMPARATIVE ANALYSIS OF CELLULOSIC SUBSTRATE
UTILIZATION IN A METHANE FERMENTATION PROCESS

By: Jan Allen and Katherine Ruetenik
Purdue University, 1979
School of Civil Engineering

INTRODUCTION

The area of waste disposal and alternative use is currentl
in a period of rapid development. This is partly due to changing economic conditions concerning resource scarcity and growing environmental concern. With the passage of the Federal Water Pollution Control Act of 1972 (PL-92-500), Congress has helped to develop interest in far-reaching programs to conserve both energy and the environment. More emphasis is on small scale conservation to cut national demands than on a few large scale projects.

New technologies in waste-treatment offer some very exciting energy saving, and resource-producing possibilities. Considering animal waste as a potential source of energy, through treatment, is one example. The idea is not new but still exist in the infancy stage of development. Many ideas have yet to be pursued.

This project investigates a potential solution to waste disposal problems in the swine industry. Swine raising results in the production of two waste products; manure and corn cobs. Both of these wastes are eventual food sources for bacterial communities. However, if this bacterial process could be controlled, one could produce usable methane as a by-product in the process of waste disposal. To take this idea one step further, research has proven that there is a relationship between surface area and certain bacterial growth rates. Present day anaerobic digesters are typically a hollow, unpartitioned tank in which digestion noturally occurs. This is due to a high retention of methane-fermenting bacteria which seem to prefer an attached existence on a digester packing. Short detention times have caused wash-out problems on such slow growth species. Research shows that if surface area is increased, so is the efficiency of the digester. The increase in gas production and digestion rate can be dramatic.

Applying these principles to the swine industry, one has all the elements to test this theory. Corn cobs could be put into existing digesters to increase surface area. This gives the potential for much more rapid waste stabilization. More rapid stabilization spells increased processing copability (i.e. more units of waste per day). Creating a dynamic or constant mixing condition within the digester could insure optimum contact time between organisms and substrate. This would further enhance the reaction rate. These concepts and ideas will be explored and tested in depth within the following report.

The objectives of this report were twofold. After constructing a series of three identical, 7 liter anaerobic digesters a comparative analysis could be performed. One objective was to determine an ultimate substrate loading rate for a corncob packed recirculating system. A feeding schedule designed to incrementally increase the substrate concentration until failure occurred was used. The second objective was to compare the effects of different detention times under identical substrate loading conditions between 2 reactors. The third system was used as a control. It was operated with only the cellulosic corncobs as substrate. This represented the background or secondary substrate for the other 2 digesters. It was used as an indicator of external effects on all 3 digesters and gave the amount of gas produced due to the substrate packing alone.

BIOCHEMICAL PROCESS *
Acid Fermentation *
Methane Fermentation *

FACTORS AFFECTING METABOLIC REACTION RATES *
Temperature and Pressure *
pH *
Alkalinity *
Volatile Acids *
Ammonia *
Toxic of Inhibitory Substances *
Nutrient Deficiencies *

EFFECTS OF CIRCULATION RATES AND SURFACE AREA *
Circulation Rates *
Effects of Surface Area *
Bacterial Habitat *

PROCEDURES AND APPARATUS

System Set-up and Equalization Period

The objectives of the project called for the design of a single stage, batch fed, chemostat. The energy required to provide circulation was kept at a minimum, so that a maximum amount of "net" energy could be produced (in the form of methane gas). The actual system incorporated an upflow packed column and an in-line reservoir. For high rate digestion a constant recirculation rate is necessary. This maintains a homogeneous medium and optimum contact of substrate and the bacterial community.

Once the three identical systems were constructed, the microbial community had to be established. Supernatant from the West Lafayette, Indiana Sewage Treatment Anaerobic Digester was used. This was a community producing domestic sewage which is similar to swine waste. The primary difference lies in the cellulose content. The digesters were charged and circulated with only the cellulosic packing as a substrate. The digesters were operated for several months without any substrate addition. This allowed the methane-formers sufficient time to become well established. While these methane-formers were slowly growing, the acid producing bacteria quickly became adjusted to the new conditions. During the first several weeks of system operation the digesters behaved unpredictably. The pH fluctuated daily, necessitating the addition of bicarbonate on several occasions. There were several leaks where a substantial portion of digester liquor was lost. This was particularly harmful to the attached anaerobic growth. These organisms are extremely sensitive to oxygen toxicity. Exposure of the digester packing to the atmosphere should be avoided at all costs. The problems mentioned are typical of a start-up procedure on an anaerobic digester. No unusual problems of substrate toxicity or nutrient deficiency were experienced. Most of the problems involved pH fluctuations due to insufficient alkaline buffering capacity in the digesters.

All three digesters behaved similarly. The systems all seemed to establish bacterial communities quickly. The systems were operated for 72 days before the feeding schedule began. System one was designated the control because of its displayed average behavior. There was little that could be done to encourage equal gas production between the digesters. The closer they were to each other (comparing liters produced per day), the greater the level of confidence in the final results would be. After the initial 72 days of start up time, the digesters underwent a feeding schedule from day 73 to day 113 (41 days). At this point the systems had reached their physical limitation of medium circulation.

Apparatus

The system that was developed for the objectives mentioned consisted of a column with fragmented solid cellulosic substrate, a reservoir to create more liquid capacity and better circulation, a pump for circulation and a wet test meter to record the amount of gas produced. The column provided food and a habitat for the microorganisms. The liquor was pumped from the top of the column into the top of the reservoir. The liquor then flowed by gravity from the bottom of the reservoir through the bottom of the column. This insured a homogeneity between the liquor and the substrate throughout the system. Tygon tubing was attached to the top of the column and the reservoir and then connected to the gas meter to record gas production. The system was designed as a closed loop system. Samples were taken from the reservoir regularly for testing.

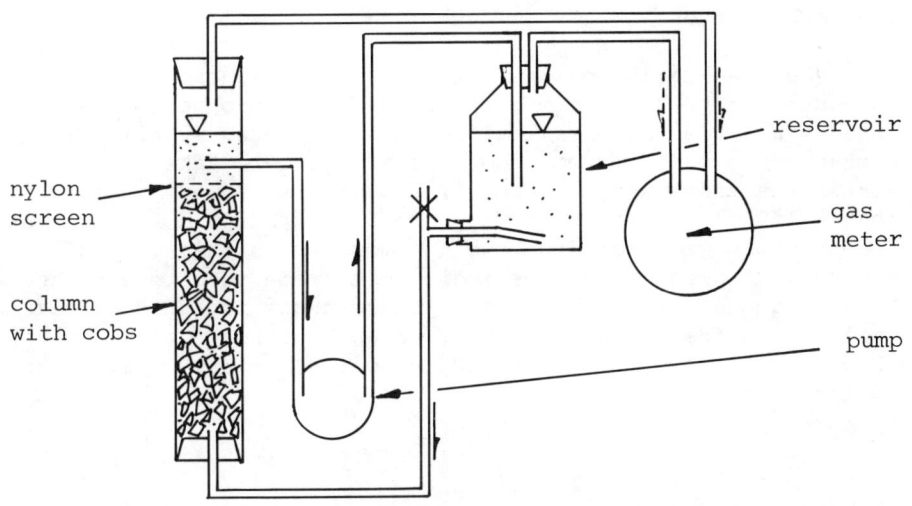

DIAGRAM OF EXPERIMENTAL REACTOR SYSTEM

Besides the samples withdrawn, liquor was removed to be replaced by diluted swine waste during the feeding phase of the experiment.

Materials *
Leaks *

Substrate Loading Rate

Substrate loading rates were designed to progress from an initial loading of .025 lbs v.s./ft^3/day to a maximum of .500 lbs v.s./ft^3/day. In order to achieve this in the alotted time the loading rate would have to be increased incrementally each week. Each week, the loading rate was increased by a factor of 1.65. This resulted in an increase from .025 to .500 in a six week period. As it turned out, this was a very optimistic schedule. Do to the gradual accumulation of solids in the systems, circulation became impossible and the ultimate loading wasn't reached. These loadings were the same for both systems 2 and 3. The difference was in the detention times. System 2 had a 14 day detention time and system 3 had a 7 day detention time. Detention times and loadings were established by removing a specific volume of each digester (2 and 3 only), daily and replacing it with diluted substrate to its original level. The volume replaced each day determined the detention time. System 2 had 500 ml replaced (1/14 of the system), each day. System 3 had 1000 ml replaced daily (1/7 of the system).

Data Collection

Several daily measurements were made. These included gas production, pH, and room temperature. Gas production was measured via a Precision Scientific Wet Test Meter (30 liter), attached to each system. Meter readings were taken at 24 hour intervals and reset to zero. The pH was measured by siphoning out a 150 ml sample which was tested and returned to the system Temperature was read from the thermometer on the room wall. There were also several periodic tests made to determine changing parameters in each system. These tests included alkalikity volatile acids, total mitrogen, organic nitrogen, COD, **soluble** COD, total solids, total volatile solids, soluble solids, and soluble volatile solids.

DATA PRESENTATION *
System Startup *
Operation and Sampling Methods *
Notes to Data Presentation *
Discussion of Data *

CONCLUSIONS

Drawing conclusions from the data collected proved to be interesting. The effect of the two detention times on gas production did not behave as expected. Total solids and total volatile solids concentrations increased sharply when loadings were applied to digesters 2 and 3. Solids concentrations however, suffered a drop during an intermediate period and then

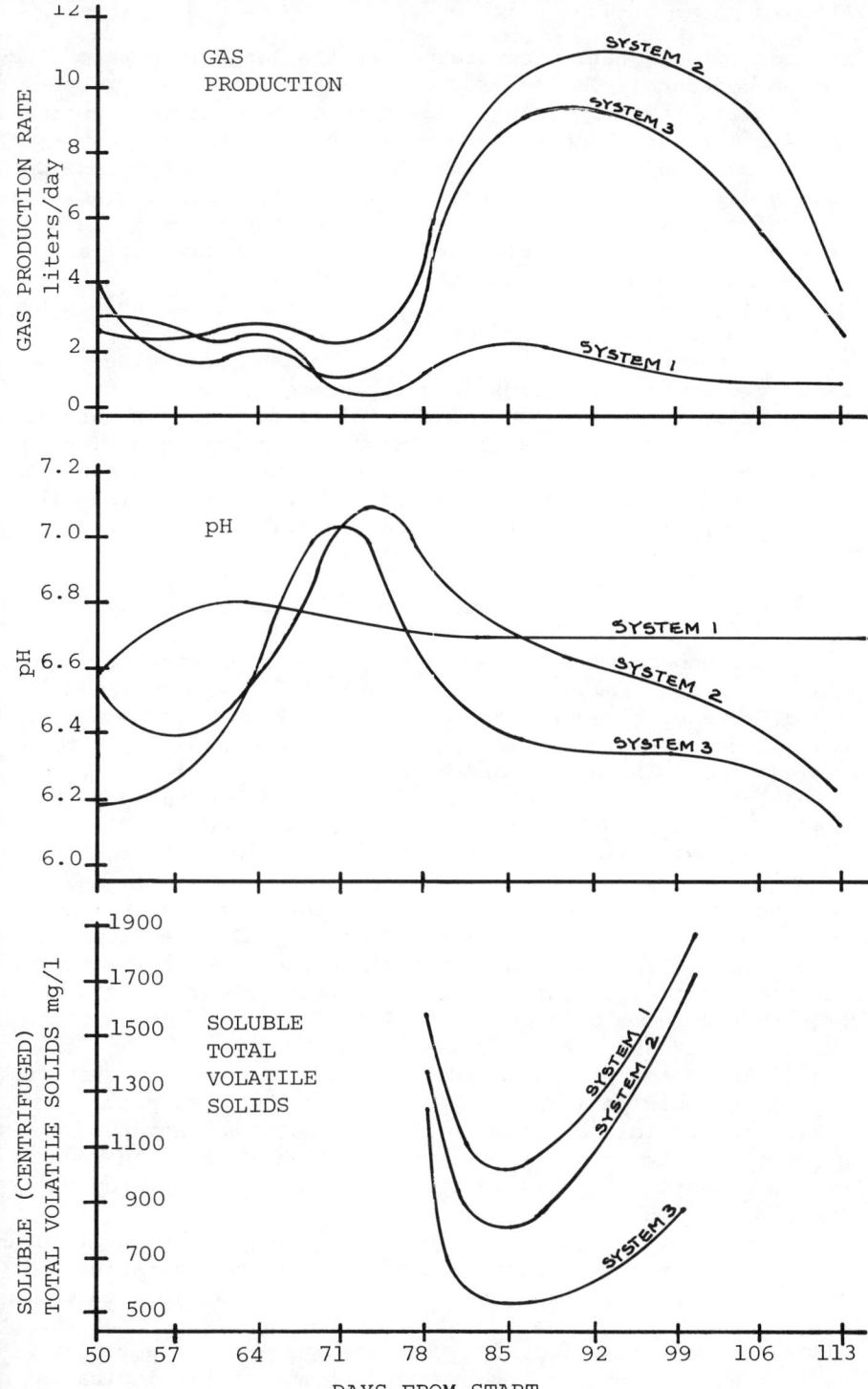

began to rise again. Temperature in the laboratory room fluctuated continually. Gas production rates changed correspondingly. Unlike the temperature, the pH had no consistent correspondence with the amount of gas produced.

Conclusions will be drawn based on the solids data, COD data and pH measurement. Regretfully, alkalinity and volatile acids levels were not monitored. These measurements would help give a much more informative overall picture of what was happening in each digester.

Solids levels were obviously increasing as witnessed by the solids buildup in the systems. A more important parameter however, was soluble solids values. The soluble values indicated the amount of solids actually being hydrolyzed to volatile organic-acids. The soluble solids data gave an indication of the rate of volatile acid production in each digester. There should be a correlation between soluble solids and soluble COD. Both are measures of the organic substrate that feeds the methane-fermentation stage. Soluble concentrations of TS, TVS, and COD did show comparative trends. As expected from the 7 and 14 day detention times, the ratio of respective concentrations was approximately 2 to 1. This was particularly true for soluble TS and COD.

Correspondingly, gas production for the 14 day detention time (system 2), was higher than for the 7 day detention time. The difference however, was not near the 2 to 1 ratio of system concentrations. The average ratio of gas production ranged from 1.15 to 1.50 and averabed about 1.3 to 1. Theoretically, the ratio should have been 2 to 1. This suggests there are other factors that must be considered.

It should also be noted that a most important conclusion is that a shorter detention time does not necessarily mean less efficiency. Reactor 3 produced approximately 79% the volume of gas that reactor 2 produced, at half the detention time. This illustrates an important point. Detention time was not the limiting factor on gas production rates. It can be seen that both soluble COD and soluble volatile solids concentrations increased dramatically between October 21, and the end of the feeding program (November 14). After observing these soluble organic ratios, one would assume that gas production would also exhibit a dramatic difference. The fact that gas production rates did not compare as expected left the results open to question. There were several theories explored to explain this phenomenon.

One possible explanation might have been that the methanogenic subcommunity was overloaded by high concentrations of substrate. This theory has several flaws as a probable explanation. It is not likely that the methanogens are that susceptible to substrate overload. Bacteria grow much faster in a culture medium than in concentrated sewage. This indicates that bacteria can grow on a "super" substrate at higher organic concentrations than swine waste could possibly produce.

Another argument against this explanation can be seen when comparing the ratios of gas production at various loading rates stayed constant between the 7 and 14 day detention times. Because this ratio remained constant as the substrate concentration increased the possibility that the system's (14 day, #2) gas production rate would have decreased instead of increased.

Another consideration is the salt content in each system. High salt concentration will cause cell dehydration. It acts as a biostatic agent. Salts are inhibitory, but not necessarily toxic. However, salt concentrations would also result in high alkalinity values. Alkalinity did not reach dangerous levels in the periodic samples taken. This was not thought to be the problem because of the constant gas ratios at increasing loading rates. Salt concentrations would increase with substrate concentrations and cause a decrease in gas production. The performances did not support this possible explanation.

For this same reason, one can rule out the possibility of the presence of a toxic or inhibitory cation or heavy metal in the system liquor. As mentioned in the introductory discussion on constituent toxicity, copper, zinc, and nickel are especially toxic. The swine were being fed a premix containing significant levels of zinc and copper. Both of these are fed to the swine in oxide form. Assuming that a substantial fraction of these compounds reach the digester, a problem could exist. However, the iron and calcium sulfates are also found in the premix. Sulfates and oxides would be reduced and sulfides would react to form insoluble percipitates with both zinc and copper, renduring them harmless. In light of these probable chemical reactions and no noticed effect due to higher substrate concentrations, we can neglect this as a likely explanation.

High concentrations of ammonia or selenium (also found in swine feed premix), should also be investigated as being inhibitory. No tests were done for ammonia, selenium or heavy metals at this stage in the program. Obtaining data on these constituent levels would quickly determine if they were the cause of the gas ratio problem.

The last and most likely explanation explored involved the relative concentrations of volatile organic acids and alkalinity. It has been found that volatile acids present in a methane-fermentation process can have an inhibitory effect on substrate utilization rates. The high acids levels tend to suppress methanogenic activity even if the pH is maintained. In this particular case, the pH would not drop to critical level because of the strong buffering capacity of high alkalinity concentrations. Therefore, there would appear to be no problem. Because the operating pH of the system was somewhat below optimum, one might suspect a problem but not that it was serious. If this was the case, it would explain the difference between reactors 2 and 3. Regretfully, alkalinity and volatile acids were not monitored closely enough to support this hypothesis. However, based on the data available, this seems the most likely explanation.

Further experimentation and testing is necessary to actually determine the cause of the digesters behaviors. But, given the information available, the probable cause of the gas production ratio imbalance can be traced to excessively high concentrations of volatile acids and alkalinity. However, both reactors continued to operate, even when the systems had become completely plugged with accumulated solids.

Closing Remarks *

Note to the Reader: This paper has been condensed for conference presentation. Omitted sections will be indicated by an asterisk following each such heading. For further information, Contact: Jan Allen; Gray & Osborne, Inc. P.S.; 4055 21st Ave. West; Seattle, WA 98199.

PACT TREATMENT OF SYNFUEL RETORT WATERS. I. INFLUENCE OF PROCESS WATER COMPONENTS ON OXYGEN UPTAKE RATES

David H. Foster. Department of Civil & Architectural Engineering, University of Wyoming, Laramie, Wyoming

Richard D. Noble. Department of Chemical Engineering, University of Wyoming, Laramie, Wyoming

Oluafemmi T. Abudu. Department of Chemical Engineering, University College of London, London, England

INTRODUCTION

The production of fuels from in situ combustion of tar sands, oil shale, and coal has definite environmental advantages. Since the liquid and gaseous fuels are produced underground and brought to the surface, little disturbance to the surface environment occurs at the surface. In situ processing permits energy extraction from deposits that are not amenable to traditional ex situ techniques. For example, underground gasification of coal (UCG) in thick, deep deposits will frequently recover more energy than subsurface mining methods. Yet these in situ processes are not without liabilities. Water is brought to the surface along with the fuels. The water arises in part from combustion of hydrocarbons in the extraction process and represents a potential environmental hazard if disposed of improperly. Water also arises from dehydration of minerals in the burn zone and from groundwater intrusion into the burn area. The in situ treatment of oil shale is representative of synfuel processing of the shale kerogen. The shale is fractured or rubbleized in the formation to be used. Air is pumped at high pressure into a central injection well until communication with peripheral production wells is achieved. Combustion is then initiated in the center well with the combustion zone moving towards the production wells. Volatilized oil, gases, and some water are withdrawn at the production wells. These gases condensed and oil and water are separated following condensation. In shale oil production approximately one barrel of water is produced for each barrel of oil. Tar sands treatment and underground coal gasification also result in significant quantites of water produced with the synfuels. These retort waters are potent pollutants, containing a wide range of organic and inorganic contaminants (Table I). Priority pollutants, toxic metals, reducing compounds, and high levels of acids or bases all make these waters difficult to treat by conventional methods. Yet pollution control is a must if environmental damage is to be prevented.

TABLE I SELECTED CHARACTERSITICS OF SYNFUELS PROCESS WATERS [1]

Characteristic (mg/l except where noted)	Synfuel Process		
	Oil Shale	UCG	Tar Sands
Chromium	0.0.5	–	<0.1
Copper	0.019	–	<0.1
Zinc	0.28	0.02	0.15
Dissolved Solids	6800	2150	3840
Ammonia-N	7000	1895	375
COD	18500	2350	2950
Sulfate	1400	684	699
TOC	2770	982	838
pH (pH units)	8.7	8.7	5.4

The activated sludge process is commonly used to treat dilute organic wastes. An activated sludge process is an aerobic biological process which utilizes the metabolic reactions of microorganisms to attain an acceptable effluent quality by removing substances exerting an oxygen demand. The process by itself is an efficient means of removing dissolved biodegradable organic matter. However, toxic or otherwise biologically resistant materials may be poorly removed, or cause upset or failure of the activated sludge process. Modification of the conventional process is necessary if it is to be successful in treating synfuel retort waters.

One modification of activated sludge that shows promise is the powdered activated carbon-activated sludge process (PACT) developed by DuPont [2]. PAC technology has been applied to petrochemical wasted and mixtures of industrial and domestic wastes. Grieves [3] felt that PACT is a promising technique for meeting best available technology standards for oil refineries. The addition of PAC greatly increases the surface area in suspension in the aeration tank, promoting adsorption. The large liquid-solid interface tends to concentrate organics and oxygen in a localized area where microorganisms can continuously metabolize the organics. Since the kinetics of the oxidation reactions appear to be concentration dependent, the localization of reactants on the carbon surface serves to drive the reaction further towards completion resulting in improved BOD removals [4]. The adsorption process immobilizes pollutants on the carbon surface and greatly increases the degradation of slowly metabolized organics. A number of other benefits have been claimed for the PACT process [2, 5].

EXPERIMENTAL METHODS

This paper reports initial results of an ongoing study of control technology for treatment of pollutants from synthetic fuel processing. The basic thrust of the work was to determine the response of activated sludge to shock loads of components of the whole retort waters and to raw retort waters. The oxygen uptake rate (OUR) as determined by a Warburg respirometer was used as a measure of biological response. Scaramelli and DiGiano [6] compared OUR in PAC treated and untreated activated sludge using a membrane systems and found no significant increases in uptake rates when the PAC was added to the activated sludge system. The Warburg respirometer is a more sensitive measuring device for determine OUR's than electrode systems. To be of value in studies of biological systems, devices for measuring oxygen uptakes rates must be sensitive as well as accurate. Respirometric methods appear to meet this need.

The basic experimental units used in this study are a pilot wastewater treatment plant and Warburg apparatus.

The pilot plant, a 250 gpd system, was used primarily to generate sludge for biological response studies and to provide pilot scale data on the efficiency of the PACT process. The pilot plant consists of primary and secondary clairfiers an aeration unit, pumps and a control panel with accessories. The aeration unit was divided into two independent parts in order to conduct symmetric twin plant test for comparative experiments. One was used as control while the other with PAC addition was used as the test system. The powdered carbon used was Hydrodarco H. A carbon dosage of 100 mg/l was fed to the influent flow. The waste fed to the two units was identical in composition and flow rate. The units were fed from a holding tank containing a 24 hour supply of a synthetic wastewater consisting of powdered milk and instant breakfast drink, and nutrients all adjusted to give a COD of 275 mg/l. This synthetic waste was similar in its chemical characteristics to municipal wastewater and provided the necessary nutrients for the biological process. The raw feed was mostly soluble and averaged 250 mg/l of COD.

The Warburg respirometer device used in this study (Figure 1) allows the measurement of volumetric gas changes. This device consists of a Warburg flask with a sidearm for toxic additives attached to the inner arm of the manometer. Brodie Solution (specific gravity = 1.033) was used as the

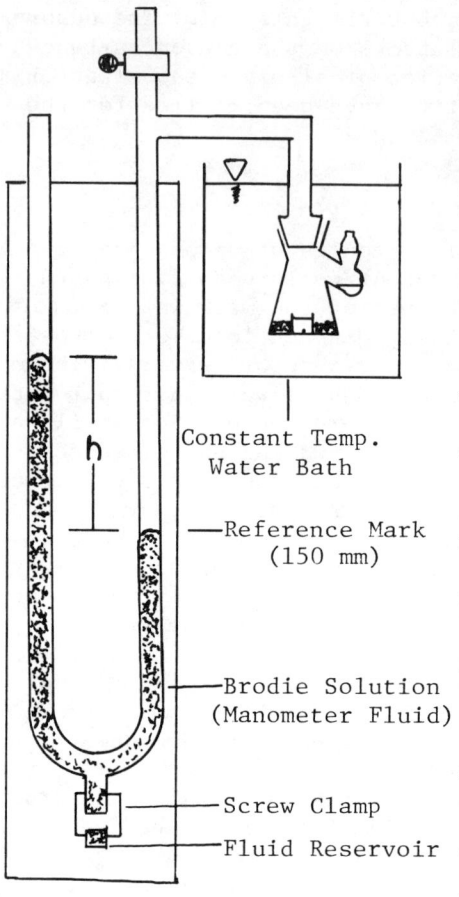

Figure 1. Constant Volume Warburg Respirometer

manometer fluid using the manometric technique in Umbreit where details of the Warburg procedure maybe found [7]. Five ml. of sludge was placed into the bottom of each prepared Warburg flask at the beginning of the experiment and the flask was immersed in a constant temperature water bath with the manometer stopcock open to the atmosphere for 20-30 minutes. The flask was agitated at a fixed rate in the bath to bring it to equilibrium and the stopcock was closed. Readings were taken at regular intervals on the manometers over a 2.5 hour period while agitation to bring about gas transfer continued. Toxic materials or retort waters, placed in the sidearm of the flask at the beginning of the experiment, were tipped in to the flask contents at 60 minutes after the stopcock was closed. This time was used since the contents of the flask had reached steady state as judged by nearly constant oxygen uptake rate. One flask and manometer out of each set in an experiment was used as a thermobarometer to correct for changes in atmospheric temperature of pressure occurring during the experiment. The cells used in these experiments were in an endogenous condition. No substrate was fed during the experiment and changes that did occur in OUR compared to controls attributable to the inhibiting substances added.

Activated sludge samples were analyzed for cell concentration in terms of mixed liquor suspended solids (MLSS), mixed liquor volatile suspended solids (MLVSS), and organic concentration in terms of biochemical oxygen demand (BOD) and chemical oxygen demand (COD). Oxygen utilized was measured with respect to time by noting the decrease in pressure of the system at constant volume. OUR's provide a direct measurement of cellular activity without causing drastic changes in the intracellular environment. BOD of domestic and industrial wastewaters is the amount of molecular oxygen required to stablize the biodegradable matter present in a water by aerobic biochemical action. BOD and COD changes occurring during the course of the experiment were determined on filtrates obtained from the flasks contents. COD is a measure of the amount of organic matter that can be chemically oxidized. MLSS, MLVSS, BOD, and COD were all determined according to <u>Standard Method for the Examination of Water and Wastewater</u>, Fourteenth Edition [8].

EXPERIMENTAL RESULTS AND DISCUSSIONS

Dawson and Jenkins [9], using manometric techniques, found that the cadmium, chromium, copper, nickel and zinc are among the most toxic substances to activated sludge. Other studies [10, 11] have used manometric techniques to monitor toxic effects in continuous feed and shock loaded systems.

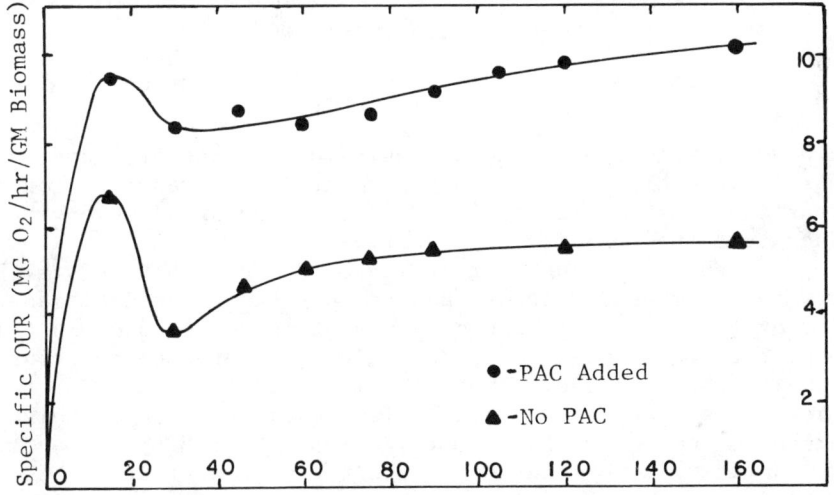

Figure 2. Specific Oxygen Uptake at 25° C and 50 mg/l Cu ++

Figure 2 shows a typical response of activated sludge organisms in terms of oxygen uptake to the addition of a toxic metal, in this case, copper. There is an initial sharp increase in uptake as the system comes to a steady state. The copper was added at 60 minutes and appeared to cause no response by the biomass. McDermott et al [12] reported that a four hour slug dose of copper sulfate in doses greater than 50 mg/l had severe effects on the efficiency of unacclimated systems. Others [13, 14] have reported toxicity from copper in levels ranging from 0.1 mg/l to 2.5 mg/l. Figure 3 shows the oxygen uptake rate for various copper concentrations. The decline in OUR over the range of 50 to 250 mg/l Cu ++ is only slight.

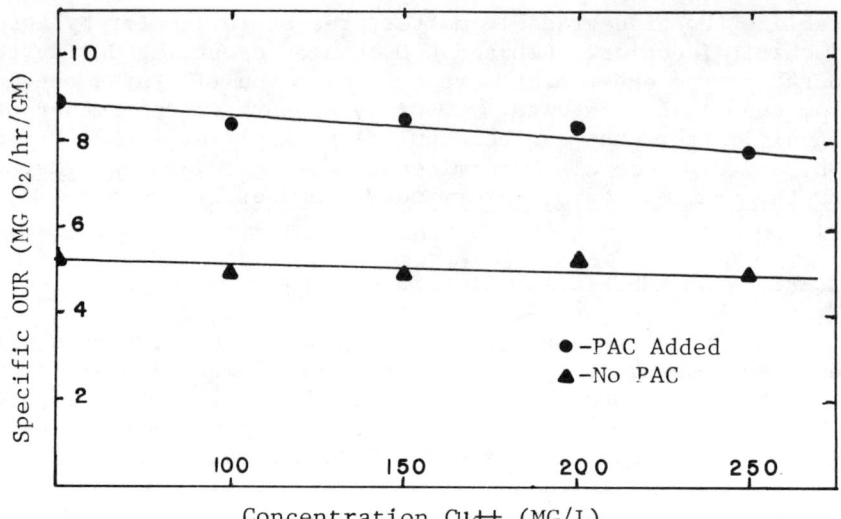

Figure 3. Effect of Cu ++ Concentration on Oxygen Uptake Rates

However, when copper cyanide was used in place of copper sulfate, OUR declined by over 35 percent in the range of concentrations of 45 to 120 mg/l. The accompanying anion obviously increased the toxicity of copper.

It should be noted from Figures 2 and 3 that OUR for PAC treated sludges is higher than for non PAC treated biomass. The oxygen uptake of virgin carbon at the doses used in this experiment was found to be negligible compared to that for studies using sludge. Thus, the increase in OUR in PAC treated systems must be due to stimulation of the microorganisms, either by concentration of nutrients on the carbon surface, or by providing a physical substrate for microbial growth and activity, or by a combination of the two. DiGiano and Scaramelli [6] found no difference in OUR in PAC and non-PAC treated sludges when an electrodes system was used for monitoring. In this study, however, the pattern of significantly increased OUR in the presence of PAC was a consistent finding.

Chromium was examined in both trivalent and hexavalent forms. There is confusing and conflicting data published on the influence of these metals on activated sludges. The large biomass present here (around 2500 mg/l) reduces the amount of chromium available per cell. This factor may explain some of the variance seen in the published literature. Ingols [15] noted that toxicity of chromium decreased as organic matter increased. Under aerobic conditions the chromic ion was reported to be more toxic than the chromate ion. The hexavalent ion was found to be more toxic under anaerobic conditions. In this study, even high doses of chromium (300 mg/l) resulted in higher uptakes rates in PAC enhanced systems. Chromium will adsorb onto the carbon surface [16] and it is possible that the carbon in the sludge partially detoxifies the chrome by adsorption.

Results for zinc are similar to those for the other metals. At concentrations below 250 to 300 mg/l virtually no toxic effects were observed. Reported toxicity of zinc is low and thus this result is not overly surprising. Response was temperature dependent. At 25° C virtually no change in OUR was observed. At 35° C, however, OUR decreased by 33.5% over the range of 50 to 500 mg/l.

Organic constituents in retort waters represent a wide range of chemical types and concentrations. Phenol was chosen as representative of potentially toxic components present in both underground coal gasification and oil shale retort waters. The complex matrix of synfuel retort waters makes much of the typical pure compound research rather difficult to interpret. However, in this preliminary study it was felt that exposure of sludge to phenol might provide initial insight into the response of sludges to shock loads of organic toxicants. Phenol is known to have a strong affinity for the carbon surface [17] and has been reported to be the controlling toxicant in a significant number of oil refinery waste discharges [18]. The range of phenol studied was 15 to 1000 mg/l. When uptake of phenol spiked sludges is compared to other rates observed in this study, it was generally seen that uptake rates were higher than for sludges not exposed to phenol (Figure 4).

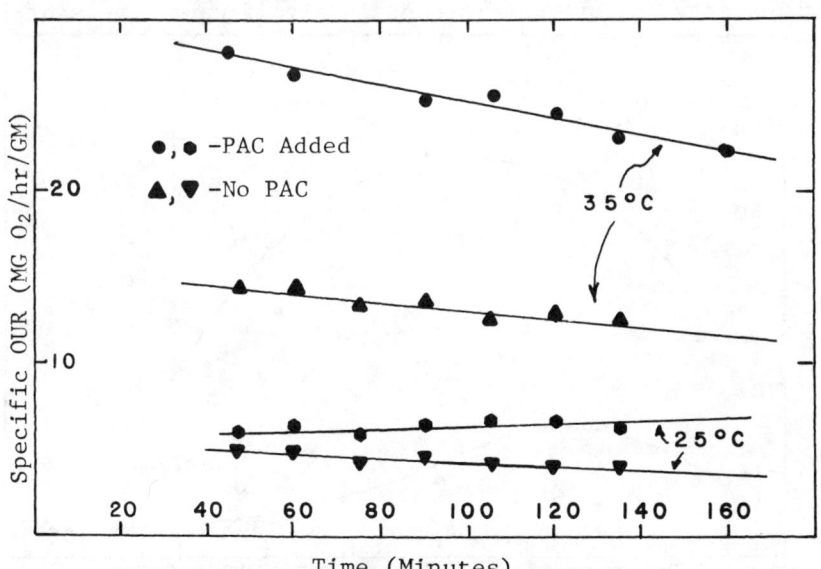

Figure 4. Specific Oxygen Uptake at 25° and 35° C for 50 mg/l Phenol

This was particularly true at 35° C where rates for phenol exposed sludges were nearly 50 percent higher than controls. The sludges were in an endogenous condition prior to contact with the phenol and appeared to be stimulated rather than inhibited by the phenol. Exposures were short (3 hours or less) and acclimation would not have occurred to any great extent in this time. Nor were cumulative effects possible. Longer term exposures showed a higher level of OUR maintained for 6 to 8 hours with a subsequent decline. It would appear that the phenol is acting as a nutrient and OUR will decline only when the source of exogenous substrate is used up. OUR did not vary significantly over the concentration range studies. Thus the toxicity of phenol in the presence of high levels of biological solids and PAC appeared to be quite low and perhaps nonexistent.

Omega 9 oil retort water is a somewhat nonrepresentative synfuel waters of this type. Derived from an in situ experiment at Laramie Energy Technology Center, Site 9 at Rock Springs, Wyoming, the water was considerably affected by groundwater intrusion into the burn area. To provide a range experimental concentrations, Omega 9 was diluted with distilled water in this study. When PAC treated sludges were exposed to retort water a significant increase in uptake was observed compared to control sludges (Figure 5). The rate does not increase over the entire concentration range. Above 50 percent the toxic components of the retort water appear to lead to a decline in OUR. Ongoing studies at Wyoming indicate, however, that sludges may be slowly acclimated to oil shale retort waters. Retort waters contain significant levels of re-

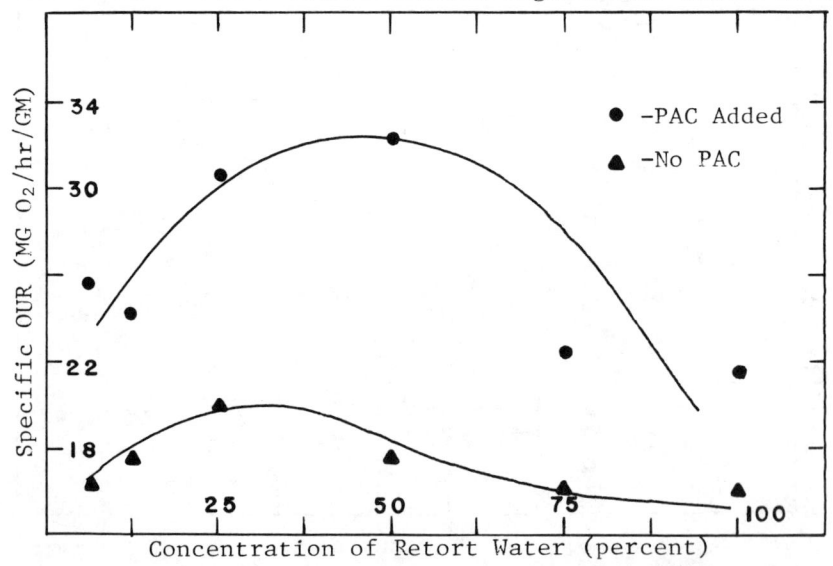

Figure 5. Response of Sludge Organisms to Omega-9 Retort Water (35° C)

ducing compounds, particularly thiosulfate (1000 mg/l or more). Since the uptake could be explained simply by oxidation of these compounds, retort waters were agitated in the absence of biological sludge in the Warburg apparatus. Within the time frame of the experiments, less than 10 percent of the uptake could be accounted for by chemical oxidations alone. It would appear that the sludges metabolized the retort water and that PAC addition enhanced the process. Even though the composition of oil shale retort waters would not promote the notion that biological systems would lead to successful treatment, the OUR studies are encouraging.

CONCLUSIONS

1. Warburg respirometry proved to be a useful and sensitive tool in evaluating sludge response to toxicants.
2. Toxicity to metals varied with type being slight for copper, chrome, but more pronounced for zinc and copper cyanide. Temperature effects were noted.
3. Phenol and Omega 9 oil shale retort waters acted as substrates for the biomass.
4. PAC reduced toxicity where present and stimulated OUR in organic systems.

ACKNOWLEDGEMENTS

The helpful assistance of Dr. Richard Poulson of the Laramie Energy Technology Center, Department of Energy, is gratefully acknowledged.

REFERENCES

1. DiBenedetto, P. J., "A Combined Powdered Activated Carbon - Activated Sludge Wastewater Treatment Process for the Treatment of Alternative Fuel Process Waters", M.S. Thesis, University of Wyoming, Laramie, (1981).
2. Adams, A. D., "Improving Activated Sludge Treatment with Powdered Activated Carbon", <u>Proceedings 28th Purdue Industrial Waste Conf.</u>, Purdue University, (1973).
3. Grieves, C. G., Stenstrom, M. K., Walk, J. D., and Grutsch, J. F., "Powdered Carbon Improves Activated Sludge Treatment", <u>Hydrocarbon Processing</u>, (Oct.1977)
4. Lawrence, A. W., McCarty, P. L., "Unified Basis for Biological Treatment Design and Operation",

Jour. Sani. Eng. Div., ASCE, 95, 757 (1979).

5. Adams, A. D., "Powdered Carbon: Is It Really That Good?", Water and Wastes Engineering, 11, B7, (1974).

6. Scaramelli, A. B., DiGiano, F. A., "Upgrading the Activated Sludge System by Addition of Powdered Carbon", Water and Sewage Works, 90, 94, (1973).

7. Umbreit, W. W., Burris, R. H., Stauffer, J. F., Manometric and Biochemical Techniques, 5th Edition, Burgess Publishing Co., (1972).

8. Standard Methods for the Examination of Water and Wastewater, 14th Edition, American Public Health Association, Inc., New York, (1975).

9. Dawson, P. S. S., Jenkins, S. H., "The Oxygen Requirements of Activated Sludge Determined by Manometric Methods, II. Chemical Factors Affecting Oxygen Uptake", Sewage and Industrial Wastes, 22, 490, (1950).

10. Dawson, P. S. S., Jenkins, S. H., "The Oxygen Requirements of Activated Sludge Determined by Manometric Methods", Sewage Works Journal, 21, 463, (1949)

11. Ayers, K. C., Shumate, K. S., and Hanna, G. P., "Toxicity of Copper to Activated Sludge", Proceedings of the 20th Industrial Waste Conference, Purdue University, 516, (1965)

12. McDermott, G. N., Moore, W. A., Post, M. A., Ettinger, M. B., "Effects of Copper on Aerobic Biological Sewage Treatment", Journal of the Water Pollution Control Federation, 35, 227, (1963).

13. Kalabine, M., "Effect of Copper and Lead Bearing Wastes on the Purification of Sewage", Water and Sewage Works, 93, 30, (1946).

14. Gellman, I., Heukelekian, H., "Studies of the Biochemical Oxidation by Direct Methods III. Oxidation and Purification of Industrial Wastes by Activated Sludge", Sewage and Industrial Wastes, 25, 1196, (1953).

15. Ingols, R. S., "The Toxicity of Chromium", Proceedings of the Eighth Industrial Waste Conference, Purdue University, 86, (1953).

16. Hwang, C. P., "Chemical Interactions Between Inorganics and Activated Carbon", Carbon Adsorption Handbook, Cheremisinoff, P. N., Ellerbusch, F., Eds., Ann Arbor Science, (1978).

17. Snoeyink, V. L., "Sorption of Phenol and Nitrophenol by Active Carbon", Environmental Science and Technology, 3, 918, (1969).

18. Grieves, C. G., "Effluent Quality Improvement by Powdered Activated Carbon in Refinery Activated Sludge Processes", American Petroleum Institute, Division of Refining, Proceedings 42nd Mid-Year Meeting, Chicago, 56, 9012, (1977).

ASSESSMENT OF ACTIVATED SLUDGE PROCESS IN TREATING
SOLVENT-EXTRACTED COAL GASIFICATION WASTEWATERS

Yung-Tse Hung and Guilford O. Fossum. University of
North Dakota, Grand Forks, North Dakota 58202

Leland E. Paulson and Warrack G. Willson. U.S.
Department of Energy, Grand Forks Energy Technology
Center, Grand Forks, North Dakota 58202

ABSTRACT

In this investigation, a treatability study was performed on wastewater derived from operation of the Grand Forks Energy Technology Center's (GFETC) slagging fixed-bed gasifier (SFBG) using North Dakota lignite.
Wastewater was subjected to lime precipitation followed by ammonia stripping and solvent extraction prior to activated sludge treatment. Di-Isopropyl ether, methyl isobutyl ketone, octyl alcohol, 2-octanone and N-butyl acetate were investigated as potential process solvents for phenol extraction. Results indicated that over 95% of the phenolics and up to 80% of the chemical oxygen demand (COD) were removed from wastewater after three stages of extraction using a 1:25 solvent-to-gas liquor ratio.
Four bench-scale completely-mixed activated sludge reactors were operated to evaluate the applicability of biological oxidation for treatment of solvent-extracted wastewater. The reactor influent had average COD, TOC, phenol, and ammonia concentrations of 4000, 2100, 140, and 100 mg/l, respectively. Increasing the aeration tank hydraulic detention time from 0.7 to 9.7 days improved COD, TOC, and phenol removal efficiencies from 32 to 63, 36 to 64, and 45 to 79 percent, respectively.

INTRODUCTION

Coal gasification produces a wastewater that contains ammonia, dissolved organics, cyanides and other contaminants which have a high pollution potential if released to the environment or are highly corrosive and fouling if not treated before reuse [1]. A number of processes, that are commercially available, appear to be applicable in treating the liquid effluent stream from a coal gasification plant. These include tar/water separation, lime precipitation, flocculation/coagulation, acid gas stripping, dilution, solvent extraction, carbon adsorption, softening, anaerobic and aerobic biological treatment, and ozonation. Each treatment plant will be an individually planned combination of appropriately designed process units that meet the specific requirements of the gasification technology, the coal feedstocks and the imposed environmental and economic constraints for that particular installation. Existing coal-specific data on the nature and amounts of contaminants and effectiveness of treatment processes for wastewater streams are not adequate to support reliable process designs.

The Civil Engineering Department of the University of North Dakota has conducted studies on treating wastewater from the Slagging Fixed-Bed Gasification (SFBG) process operated by the Grand Forks Energy Technology Center (GFETC). This method of gasification is a modification of the Lurgi fixed-bed process with the difference being in the method of ash removal. In the Lurgi process excess steam is introduced to the gasifier to prevent melting of the ash material, while in the slagging process only stoichiometric quantities of steam are introduced to react with the coal and the ash is removed as a molten slag. In the Lurgi process the excess steam is condensed and becomes part of the wastewater stream while in the slagging fixed-bed process the components of the wastewater are the coal moisture and the water produced during the carbonization reaction. A description of the slagging fixed-bed process at GFETC has previously been published [2].

Previous investigators have demonstrated that biological treatment can reduce organic contaminants in coal conversion wastewaters after pretreatment by lime precipitation, ammonia stripping and dilution [3]. Published results have shown that biological treatment can reduce total organic carbon (TOC), chemical oxygen demand (COD), biochemical oxygen demand (BOD), and phenols by 80, 80, 93, and 99 %, respectively [4].

The objective of this investigation is to demonstrate that coal gasification wastewater can be biologically treated at full strength after pretreatment by lime precipitation, ammonia stripping and phenol extraction.

CHARACTERISTICS OF WASTEWATER

Wastewater studied came from two different gasification tests on lignite from the Indianhead mine (Mercer County, North Dakota). The first wastewater had a phenol concentration of 6,000 mg/l and a COD of 44,000 mg/l, and was used for determining suitability of solvent for phenol extraction. The second wastewater was used for a biological treatability study after lime precipitation, ammonia stripping and solvent extraction. Analysis of raw and processed wastewater II is shown in Table I.

Table I. Chemical Summary of Wastewater II

Parameter	Raw Gas Liquor	Lime precipitated and NH_3-stripped	Solvent extracted
TSS	110	1,420	
VSS	110	360	
Phenol	9,500	8,100	140
COD	27,900	14,900	4,000
NH_3-N	1,810	250	105
Alkalinity	11,900	2,900	715
pH	8.2	9.2	7.1

All values in mg/l except pH.

The raw wastewater II contained 27,900 mg/l COD, 1810 mg/l ammonia, 11,900 mg/l alkalinity and had a pH of 8.2. Lime precipitation and ammonia stripping reduced the ammonia to 250 mg/l and the alkalinity to 2,900 mg/l. Solvent extraction reduced COD to 4000 mg/l and phenol to 140 mg/l.

The pretreated wastewater did not contain sufficient nutrients for biological treatment. Nitrogen and phosphorus as Na_2HPO_4, $NaH_2PO_4 \cdot H_2O$ and NH_4Cl were added to give a BOD_5 : N : P ratio of 100:5:1.

PROCEDURES

Solvent Extraction

Suitable solvents for extraction of wastewater were selected based on their distribution coefficients for water. The solvents chosen were methyl isobutyl ketone (MIBK), 2-octanone, octyl alcohol, di-isopropyl ether and n-butyl acetate with distribution coefficients of 115, 110, 48, 42, and 61, respectively. A summary of solvent characteristics is shown in Table II (adapted from Luecke) [5].

The extraction process consisted of combining one part solvent and twenty five parts wastewater in a large separatory funnel. After agitation for mixing, followed by phase separation, the solvent was decanted. The wastewater was aerated for several minutes to remove residual solvent before the procedure was repeated using fresh solvent.

Table II. Solvent Characteristics

Solvent	Boiling point °C	Solubility in H_2O (%)	Distribution Coefficient	Approx. Cost ($/lb)
Methyl Isobutyl Ketone	118	2.0	115	1.51
2-Octanone	173	insoluble	110	7.50
Octyl Alcohol	98	0.05	48	5.98
di-Isopropyl Ether	68	0.20	42	1.57
n-Butyl Acetate	126	0.70	61	1.18

Activated Sludge Treatment

Four bench-scale completely mixed activated sludge reactors, as shown in Figure 1, were used to determine their applicability in treating full strength solvent extracted wastewater. Each reactor consisted of an aeration tank and

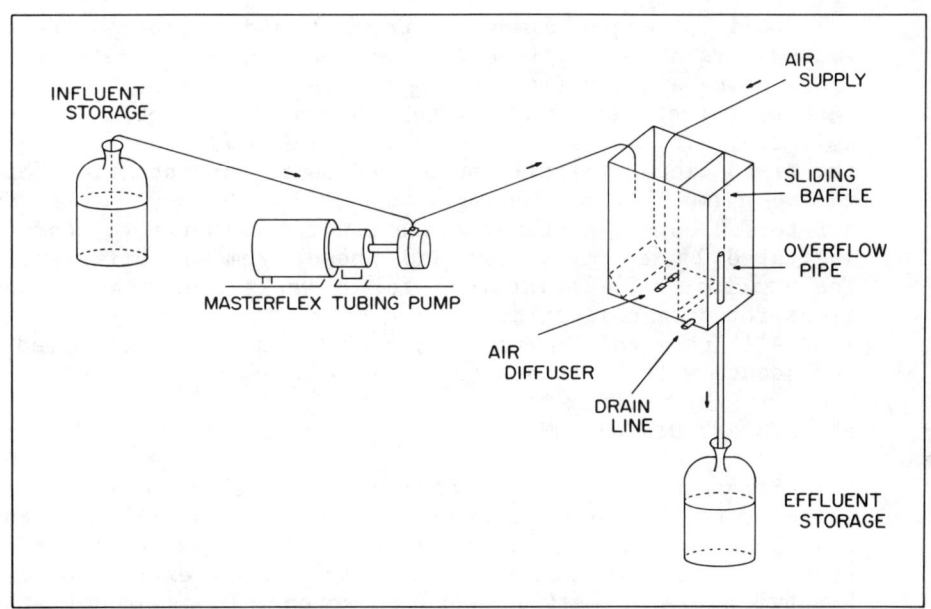

FIGURE 1. - Laboratory scale activated sludge reactor

a settling tank separated by a baffle. Aerobic conditions in the aeration tanks were maintained using filtered, compressed air which, in addition to providing oxygen, also provided agitation to insure thorough mixing of substrate and microorganisms. A biological culture acclimated with pretreated gas liquor was used as seed for all reactors. A summary of operating conditions is shown in Table III.

Table III. Activated Sludge Reactor Operating Conditions

Parameter	Reactor			
	1	2	3	4
Aeration Tank Vol. (l)	0.7	2.6	4.6	9.6
Settling Tank Vol. (l)	0.3	0.4	0.4	0.4
Hydraulic Detention Time (day)	0.7	2.6	4.6	9.7
Influent Flow Rate (l/day)	1.0	1.0	1.0	1.0
Sludge Age (day)	5.2	5.7	6.4	7.2
F/M Ratio (day^{-1})	2.8	1.7	1.8	1.3
pH (reactor tank)	8.0	8.0	7.9	7.9
MLSS (mg/l)	3350	1330	704	446
MLVSS (mg/l)	1960	916	495	320
Effl. TSS (mg/l)	180	230	180	160
Effl. VSS (mg/l)	130	170	130	120

Full strength ammonia stripped and solvent extracted wastewaters were supplied to the reactors at a rate of 1.0 l/day using a Masterflex tubing pump. Reactors 1, 2, 3, and 4 were operated with aeration tank hydraulic detention times of 0.7, 2.6, 4.6, and 9.7 days, respectively. In the test series, the mixed liquor volatile suspended solid concentration (MLVSS) varied from 320 to 1960 mg/l and the sludge age varied from 5.2 to 7.2 days. A steady state condition in the reactors was indicated by constant COD and phenol removal efficiencies. The reactors were maintained for 8 weeks under these conditions for data collection.

All chemical analytical procedures were performed in accordance with "Standard Methods" [6].

RESULTS AND DISCUSSION

Screening tests were conducted to select the preferred solvent to extract dissolved organics from GFETC lignite gasification wastewater. Figures 2 and 3 show phenol and COD concentrations at each stage of a five stage extraction using N-butyl acetate, methyl isobutyl ketone, Di-isopropyl ether, octyl alcohol and 2-octanone as solvents. All solvents removed significantly greater amounts of COD than that which can be accounted for by reduction of phenolics alone, indicating that solvent extraction removes other organics in

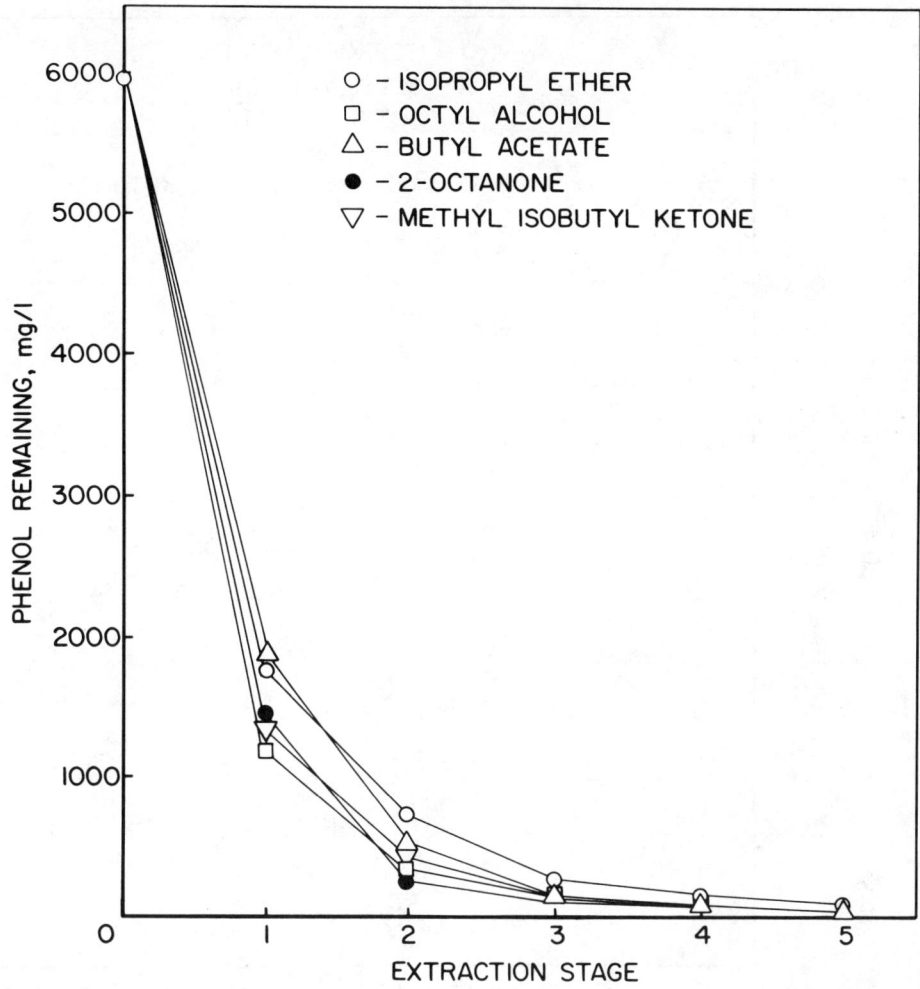

FIGURE 2. - Removal of phenol from raw, filtered gasification wastewater by solvent extraction

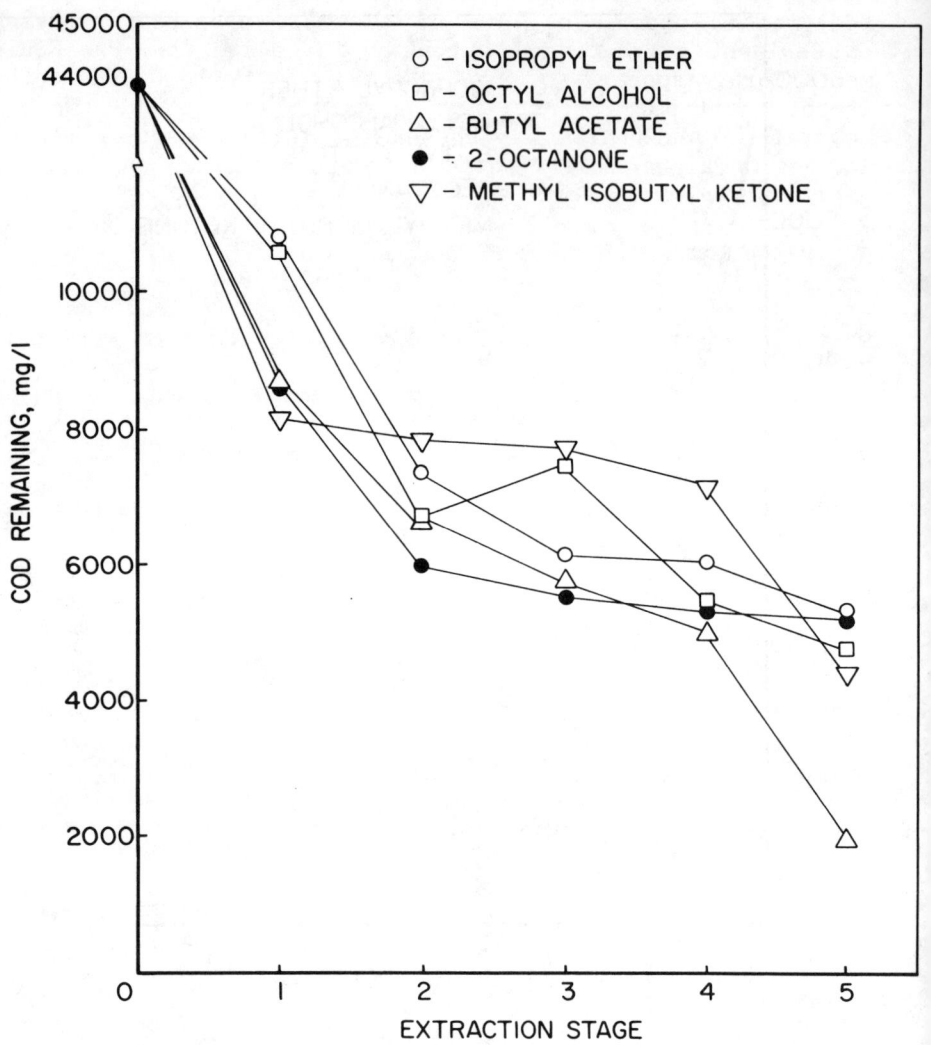

FIGURE 3. - Removal of COD from raw, filtered gasification wastewater by solvent extraction

addition to phenolics. Two of the more prominent processes for solvent extraction licensed by Lurgi and Chem-Pro Equipment Corp. appear to use di-isopropyl ether and methyl isobutyl ketone, respectively as solvents. Of the two commercial solvents MIBK was chosen for further testing because it has a higher distribution coefficient for phenolics in gasification wastewater.

Activated Sludge Performance

The average influent-effluent concentration data of selected parameters for the activated sludge treatment reactors are shown in Table IV.

Reactor influent concentrations of COD, TOC, and phenol averaged 4000, 2100, and 140 mg/l, respectively. Ammonia and alkalinity averaged 100 and 700 mg/l, respectively. Increasing hydraulic detention time from 0.7 to 9.7 days decreased effluent COD from 2700 mg/l to 1480 mg/l, TOC from 1310 mg/l to 740 mg/l, phenol from 77 mg/l to 29 mg/l, ammonia from 130 mg/l to 70 mg/l, and alkalinity from 656 mg/l to 189 mg/l. With increasing hydraulic detention time, from .7 to 9.7 days, COD removal increased from 32 to 63 pct, TOC removal increased from 36 to 64 pct, phenol removal increased from 45 to 79 pct and alkalinity removal increased from 8 to 74 pct. Nitrification may have been inhibited as only 33 pct of the ammonia nitrogen concentration was removed. Oxygen uptake rate decreased from 930 mg/l-day to 81 mg/l-day with increasing hydraulic detention time. Figure 4 graphically presents the percent removals of COD, TOC, and phenol as a function of hydraulic detention time. The most significant change in rate of removal for these parameters occurred between hydraulic detention times of 2.6 and 4.6 days. Increasing detention time above 5 days did not significantly improve the effluent quality.

CONCLUSIONS

1. Full strength coal gasification wastewater can be treated by an activated sludge system if it is properly pretreated by lime precipitation, ammonia-stripping and solvent extraction.
2. The activated sludge process with an aeration tank detention time of 5 days and a sludge age of 6.4 days removed 70% phenol, 62% COD and 61% TOC from the solvent extracted wastewaters.
3. Increasing the aeration tank hydraulic detention time beyond 5 days did not significantly improve the removal efficiencies of phenol, COD and TOC.

Table IV. Average Performance Data of Activated Sludge Reactors

Reactor	Hydraulic detention time (day)	COD			TOC			Phenol		
		Infl.	Effl.	% Removal	Infl.	Effl.	% Removal	Infl.	Effl.	% Removal
1	0.7	4000	2700	32	2100	1310	36	140	77	45
2	2.6	4000	2500	37	2100	1310	36	140	75	47
3	4.6	4000	1520	62	2100	790	61	140	43	70
4	9.7	4000	1480	63	2100	740	64	140	29	79

Reactor	Hydraulic detention time (day)	NH_3-N			Alkalinity			Oxygen uptake (mg/l-day)
		Infl.	Effl.	% Removal	Infl.	Effl.	% Removal	
1	0.7	105	130	-	715	656	8	930
2	2.6	105	138	-	715	591	17	300
3	4.6	105	71	32	715	285	64	180
4	9.7	105	70	32	715	189	74	81

* All values in mg/l unless stated otherwise.

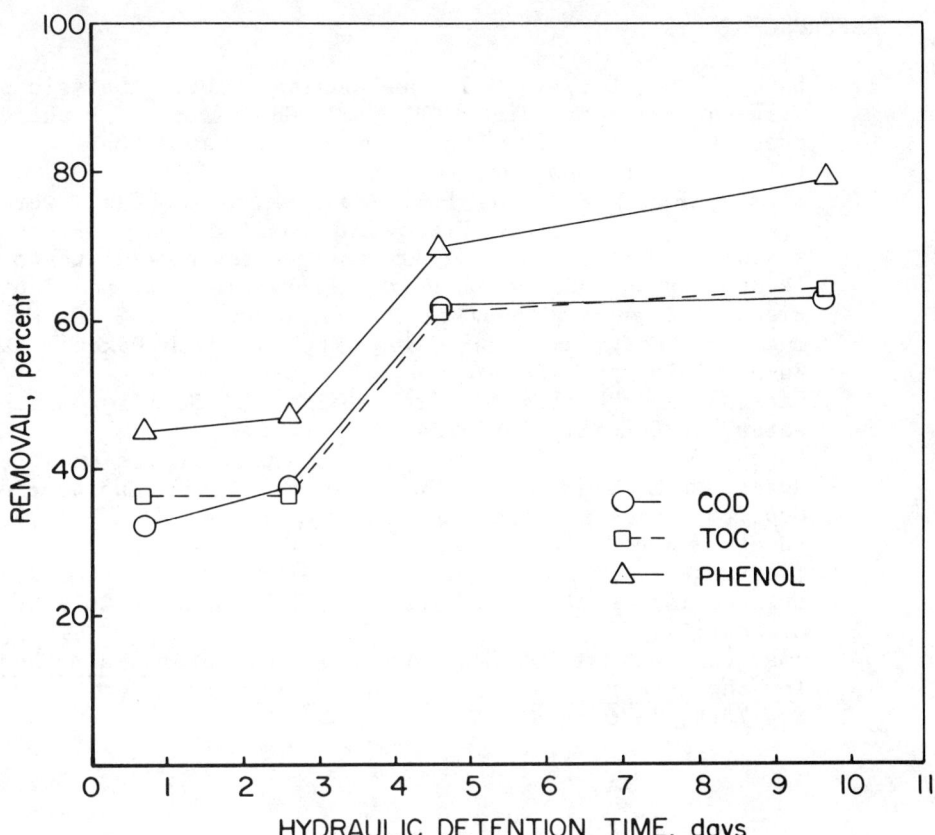

FIGURE 4. - Removal efficiency of COD, TOC, and phenol as a function of hydraulic detention time of activated sludge reactor

REFERENCES

1. Luthy, R.G., Massey, M.J., and Dunlap, R.W., "Analysis of Wastewaters From High BTU Coal Gasification Plants," Presented at the 32nd Purdue Industrial Waste Conference, Lafayette, Indiana, May (1977).
2. Ellman, R.C., Paulson, L.E., Hajicek, D.R., and Towers, T.R. (1979) Slagging Fixed Bed Gasification: Project status at the Grand Forks Energy Technology Center; Technology and Use of Lignite, Proceedings of the Tenth Biennial Lignite Symposium sponsored by the U.S. Department of Energy and the University of North Dakota, May 30-31, 1979, pp. 236-277.
3. Sack, W.A., "Biological Treatability of Gasifier Wastewater," USDOE METC/CR-79/24, June (1979).
4. Luthy, R.G., and Tallon, J.T., "Biological Treatment of Hygas Coal Gasification Wastewater," Civil Engineering Dept., Carnegie-Mellon University, FE - 2496, (1978).
5. Luecke, R.H., "Assessment of Solvent Extraction for Treatment of Coal Gasifier Wastewater," Chemical Engineering Dept., University of Missouri, Columbia, Missouri, (1979).
6. American Public Health Association, Standard Methods for the Examination of Water and Wastewater, 14th ed., New York, 1976.

AEROBIC TREATMENT PROCESS WITH BY-PRODUCT UTILIZATION
FOR WASTEWATER CONTAINING HIGH ORGANIC SOLIDS

P. Y. Yang. Department of Agricultural Engineering,
University of Hawaii at Manoa, Honolulu, Hawaii.

J. K. Lin. Department of Agricultural Engineering,
University of Hawaii at Manoa, Honolulu, Hawaii.

SUMMARY

Aerobic treatment of a waste containing high organic solids with by-product utilization was investigated. Spent fruit flies medium (SFFM) was used as a substrate. Semi-continuous flow operation (solid retention time of 5 days) was conducted for a period of 190 days and 100 days under 20°C and 27°C, respectively. A steady state performance of the removal of filtrate COD and suspended solid concentration can be achieved. Organic loading rate of 20-40 grams SFFM per liter of liquid volume per day, temperature of 20°C and airflow rate (without mechanical mixing) of 0.8 liter per liter of liquid volume per minute should be provided. Under these conditions, 70-80% of filtrate COD is removed and the system runs with a mixed-liquor suspended solids (MLSS) concentration of 20,000 mg/L. The protein content of the MLSS is 25.7%. Suspended solid recovered from the presently suggested process can offer a high degree of potential aquaculture or animal feed supplement.

INTRODUCTION

Agricultural waste management has been receiving increased attention from scientists, engineers, and administrators interested in the enhancement of the environment. Animal and food processing wastes are the two major agricultural wastes being considered. Generally, these two types of organic waste products contain high organic solids. Therefore, land disposal and anaerobic fermentation processes generating methane fuel have been generally studied and used. Application of agricultural wastes to land has been a traditional method of disposal and remains the best approach in most locations. However, high organic concentrations of animal and food processing wastes in small areas may cause surface runoff, contamination of ground water, soil clogging, and odor problems. These conditions commonly prevail in Hawaii. Other treatment alternatives via the methane fermentation process have been studied for the past 3 years in our department [1, 2, 3]. It has been found that pH control is required for certain kinds of food processing waste and longer detention times should be provided for both types of waste products.

A conventional aerobic treatment process (e.g., oxidation ditch) for high organic solid wastewater is not economical because of longer aeration time (power consumption) and higher content of NO_2-N and/or NO_3-N (toxicity for animal feed). From our previous batch studies on biological treatment of food processing wastes [4, 5], it was concluded that the aerobic removal of soluble COD and recovery of suspended solid for animal feed for a waste containing high organic solid is the most promising one for minimizing power consumption for aeration, achieving organic pollution reduction, recovery of protein-rich by-product, and elimination of presence of NO_3-N and/or NO_2-N. In the present study, semi-continuous flow operation of aerobic treatment of spent fruit flies medium (SFFM) and utilization of recovered by-product were further investigated.

MATERIALS AND METHODS

Spent fruit flies medium (SFFM) was used as substrate for the present study because of its high content of organic solids The SFFM is the waste product of the result of mass rearing operation of fruit flies for biological protection of tropical fruits in Hawaii. It is expected that in the near future, the amount of 30 tons per week must be disposed. An analysis of the components of fruit fly medium used in Hawaii shows that mostly wheat mill, grain, sugar and torula yeast are provided. Therefore, it is expected that the SFFM should contain a major portion of the unused organic matter. This waste product (85% of moisture content) was diluted with tap water and maintained the desired loading rate (6.05-40 grams SFFM per liter per day) used in the present study.

Two 208-liter metal drums were used as a reactor (liquid volume - 150 liters) with introduced-diffused air as shown in Figure 1. For one of these reactors, a heating device was installed on the air line in order to increase the temperature as required. Two levels of temperature were maintained, 20°C and 27°C. A solid retention time (SRT) of 5 days was used for all experiments. The SRT is defined as the amount of total solid contents in the reactor divided by the amount of solid removal per day. In other words, a certain volume of mixed liquor was withdrawn per day with addition of an equal volume

Figure 1. Overview of total operational setup - aerators and centrifuge.

of fresh spent fruit fly medium per day. It was found to be homogeneous within the present range of air supply (0.66-0.95 liter per liter of liquid volume per minute) used in the system. Chemical oxygen demand, pH, suspended solid, nitrogen in the solids and solution, were measured following the Standard Methods [6]. Dissolved oxygen was measured by an oxygen meter (YSI Model 54A). Operational parameters, such as organic loading rate, detention time, oxygen requirement, removal of chemical oxygen demand and protein content in the solid product, were investigated.

The mixed liquor withdrawn from the reactor was centrifuged by using a Sharples Superspeed Centrifuge as shown in Figure 1. The solid product separated from the centrifugation process was fed to Malaysian prawns. Twelve rectangular tanks (30 x 60 x 30 cm^3) with a 26°C temperature and dissolved oxygen of 10 mg/liter were used to determine the best utilization of this product. Four types of feed (regular or commercial, aerobically processed, half regular and half processed, and raw SFFM or non-aerobically processed) were divided into these 12 prawn growing units. Percent of weight increase and survival was measured every month for 3 months. Regular or commercial prawn feed contains 24% protein, 7% fiber, 2.5% fat, and 7% ash. Non-aerobically processed SFFM contains 18.6% protein (12.7% solid protein and 5.9% soluble protein or other type of nitrogen compound), 4.9% fat, 9.8% fiber, and 4.4% ash. Every month the number of prawn and weight per prawn were measured for 3 months. Thus, percent of survival and weight increase of prawn related to these four different types of feeding materials can be determined.

RESULTS AND DISCUSSIONS

1. Operational Characteristics and Performance

Three different loading rates, 40, 20, and 6.05 grams spent fruit fly medium (85% moisture) per liter per day (6, 3, and 1.5 grams Total Solids per liter per day, respectively) at 20°C were conducted. Part of these results are shown in Figure 2. The results are summarized in Table I. As shown in Figure 2, influent total COD and filtrate COD are 25,000 to 35,000 mg/L and 5,000 to 15,000 mg/L, respectively. With 190 days of operation, the effluent filtrate COD is reduced to 1,500 to 2,500 mg/L for this long period of operation. Suspended solids concentration in the effluent is 10,000 to 25,000 mg/L. Protein content of the effluent suspended solid is about 25.7% (original solid protein is 12.7%).

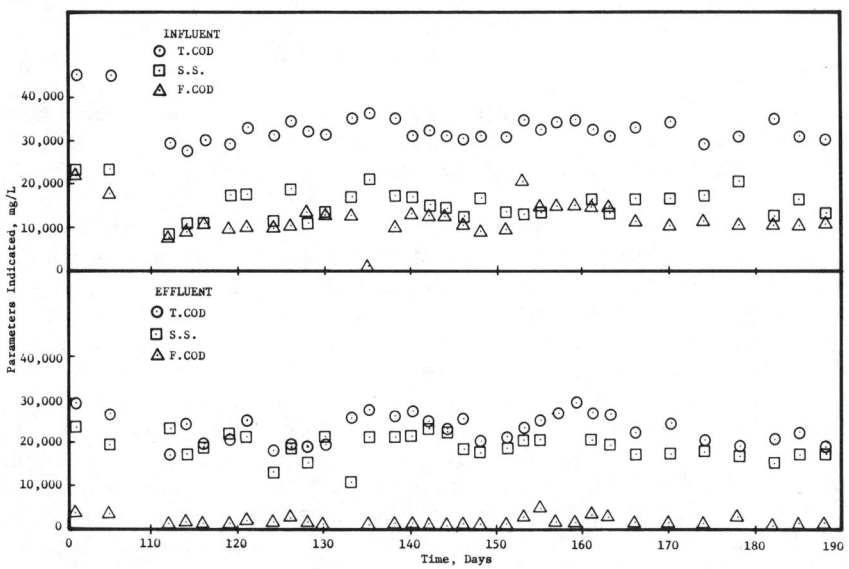

Figure 2. Operational characteristics for a loading rate of SFFM 20 g/L/day (20°C).

In order to investigate the effect of increasing temperature on the operational performance, the heating device was installed prior to the inlet of the air diffuser as shown in Figure 1. With this device, the reaction temperature was maintained at 27°C throughout the experimental period. Two different organic loading rates, 20 and 6.05 grams SFFM/liter/day, were operated. The resulting operational characteristics and performances are shown in Figure 3. Air supply was

242

Table I. Effect of Loading Rate and Temperature on the Operational Performance

Loading Rate, Temperature	Operational Performance			
	T.COD Removal, %	F.COD Removal, %	Protein Content in Suspended Solid, %	Residual F.COD, mg/L
40 gram SFFM/liter/day (20°C)	10-60	70-85	27.1-30.7	2000-3000
20 gram SFFM/liter/day (20°C)	10-60	70-85	25.7	1500-2500
6.05 gram SFFM/liter/day (20°C)	10-60	70-85	21-27.8	450-900
20 gram SFFM/liter/day (27°C)	12-62	85-92	35.6	1000-1400
6.05 gram SFFM/liter/day (27°C)	10-66	57-79	21-26	400-1000

purposely reduced to 60% of the original level between days 17 to 54. As shown in Figure 3, influent total COD and filtrate COD for 20 grams SFFM/liter/day are 27,000 to 38,000 mg/L and 7,000 to 15,000 mg/L, respectively, during this period of operation. When the air supply is returned to the original level (0.946 liter per liter of liquid volume per minute) from day 55, the filtrate COD in the effluent is back to the original level of 1,000 to 1,400 mg/L. With this long period of semi-continuous operation (100 days) by using an SRT of 5 days and a flow rate of 0.946 liter air flow/liter/day, the filtrate COD removal efficiency at steady state can be maintained at 85 to 92% and provides effluent filtrate COD with the range of 1,000 to 1,400 mg/L. Apparently, an inappropriate supply of air will affect the COD removal. Suspended solids concentration in the effluent is maintained in the range of 12,000 to 19,000 mg/L with a protein content of about 35.6%.

Based on the above findings, the operational characteristics and performance at 20°C and 27°C are summarized and shown in Table I.

From Table I it can be seen that there is no effect of loading rates (within present range of study) on the percent of

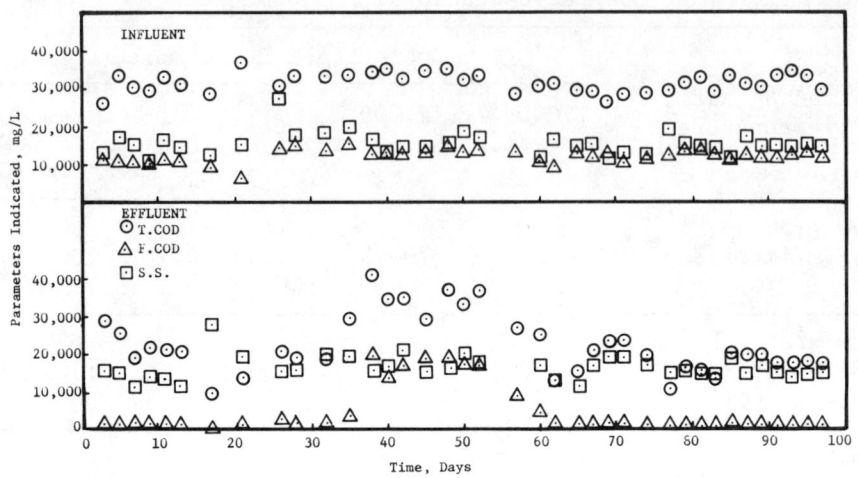

Figure 3. Operational characteristics for a loading rate of SFFM 20 g/L/day (27°C).

COD removal and protein content in the suspended solid at 20°C. However, at 27°C a loading rate of 20 gram SFFM/liter/day provides higher filtrate COD removal efficiency and protein content in the suspended solids than a loading rate of 6.05. Also, the lower loading rate gives lower concentration of remaining filtrate COD at both levels of temperature. For loading rate of 20 gram SFFM/liter/day, the remaining filtrate COD is much lower at 27°C than at 20°C. Therefore, it seems that a loading rate of 20 gram SFFM/liter/day and 27°C can be adopted as possible design criteria for higher filtrate COD removal efficiency, higher protein content in the suspended solid and lower remaining filtrate COD content in the effluent. However, these design criteria should be considered together with the oxygen transfer efficiency and oxygen uptake under different loading rates and temperatures used.

2. Oxygen Transfer Efficiency and Oxygen Uptake Rate

Oxygen transfer efficiency of the aerobically biological treatment process will be dependent on waste characteristics, temperature, airflow rate, mixing and organic loading rate. In the present study, oxygen transfer coefficient ($K_L a$) and oxygen uptake rate (R_r) were measured after 24 hours of each organic loading without external mixing. Determinations of these values followed procedures specified by Baker, et al. [7] and are summarized in Table II.

Values of R_r presented in Table II can be considered as the minimum required oxygen rate to meet the microbial growth. Measurement of $K_L a$ values will reflect the capacity of aeration equipment required. In the present study, $K_L a$ values range from 3.84 to 11.7 hr^{-1}. In the conventional biological wastewater treatment system, $K_L a$ values range from 10 to 30 hr^{-1} between 15°C and 24°C by using diffused air without mechanical mixing. Apparently, the air transfer system used in the present study should be improved when an actual plant is planned for construction. This requires additional research.

With airflow rate of 0.686 L/L/min., 40 and 6.05 gram SFFM/L/day of organic loading, and 20°C, oxygen uptake rates were measured every 2 or 3 hours until they returned to the original oxygen uptake (i.e., prior to daily loading) levels. Part of the performances at such oxygen uptake rates is shown in Figure 4. For 40 gram SFFM/L·day, it takes 24 hours of aeration to return to the original level, 40 mg/L·hr. The highest oxygen uptake is about 260 mg/L·hr. For 6.05 gram SFFM/L·day, it takes about 12 hours of aeration to return to the original level, 2 mg/L·hr. The highest oxygen uptake rate

Table II. Effect of Loading Rate and Temperature on Oxygen Transfer Coefficient ($K_L a$) and Oxygen Uptake Rate (R_r) after 24-hour Loading

Loading Rate	$K_L a$ and R_r	
	$K_L a$, min^{-1}	R_r, mg/L·min
0.30 g T.COD/g.ss-day (20°C, airflow rate, 0.9 L/L·min)	0.196	0.228
0.32 g T.COD/g.ss-day (20°C, airflow rate, 0.68 L/L·min)	0.108	0.354
0.23 g T.COD/g.ss-day (20°C, airflow rate, 0.66 L/L·min)	0.096	0.071
0.41 g T.COD/g.ss-day (27°C, airflow rate, 0.946 L/L·min)	0.112	0.11
0.59 g T.COD/g.ss-day (27°C, airflow rate, 0.686 L/L·min)	0.064	0.201

is about 17 mg/L·hr. When the airflow rate is reduced from 0.686 L/L·min to 0.359 L/L·min with 6.05 gram SFFM per liter per day, the time to reach the highest peak of oxygen uptake is delayed. This is due to the decrease of oxygen transfer efficiency by the reduction of airflow rate.

Performances with respect to oxygen uptake rate were also observed at increasing the temperature from 20°C to 27°C. With loading rate of 20 gram SFFM/L·day and airflow rate of 0.686 L/L·min, it takes 14.5 hours to recover the original oxygen uptake rate, 20 mg/L·hr. Its highest oxygen uptake rate is at 240 mg/L·hr. For loading of 6.05 gram SFFM/L·day, it takes 13 hours to return to the original level, 2 mg/L·hr. The highest oxygen uptake rate is about 125 mg/L·hr. The overall performance is summarized and shown in Table III. The required oxygen uptake rate can be estimated approximately by using the highest value of oxygen uptake rate corresponding to the organic loading rate applied and operating temperature. Oxygen transfer efficiency can also be improved by increasing $K_L a$ values, which is related to airflow rate, temperature and mixing device.

If the highest oxygen uptake rate is selected as the oxygen supply rate for different organic loading rate at 20°C and 27°C, the values of gram oxygen required by 1.0 gram of

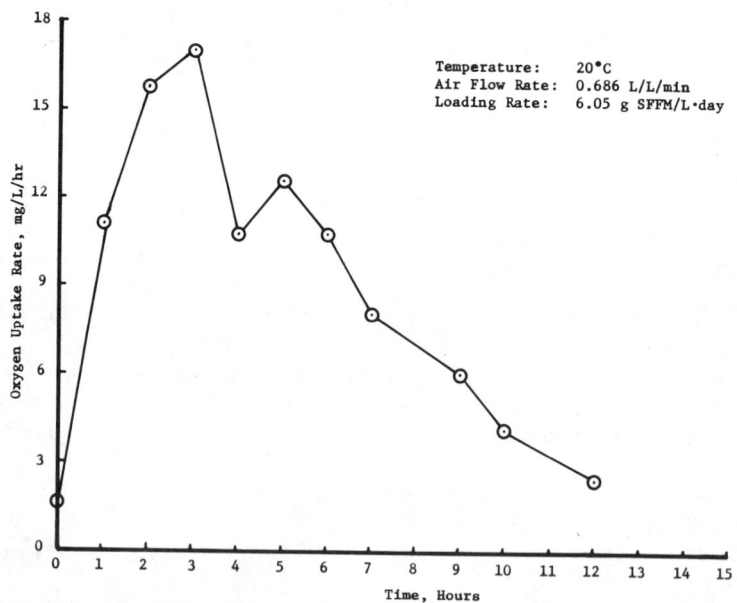

Figure 4. Performance of oxygen uptake rate after organic loading (7/27/79).

Table III. Effect of Loading Rate and Temperature on Recovery Time of O_2 Uptake Rate and Highest O_2 Uptake

Loading Rate (g SFFM/L·day)	Temperature (°C)	Recovery Time of O_2 Uptake Rate (hrs)	Highest O_2 Uptake Rate (mg/L·hr)
6.05	20	12	17
40	20	24	260
20	27	14.5	240
6.05	27	13	125

SFFM applied will be 0.067-0.156 for 20°C and 0.288-0.495 for 27°C. Based on the difference found between these two levels of temperature, it is suggested that a temperature of 20°C should be selected because of the lower value of oxygen required for processing SFFM.

3. Utilization of By-products

The solid product recovered from the semi-continuous flow operations of aerobic treatment of SFFM was fed to a population of prawns and compared with the regular prawn feed. Results are shown in Figure 5. The weight increase of prawn in the mixed feed (half regular and half processed feed) unit is the highest after the 3-month period. Compared to the regular feed, the mixed prawn feed shows a tendency for a continued rapid growth. With regard to percent of weight increase and a total weight increase in gram per prawn, the mixed feed is comparable to the regular prawn feed. However, more significantly, the mixed feed shows a higher survival rate compared to the regular feed. Furthermore, the mixed prawn feed provided a more homogeneous increase in body weight. Therefore, solids recovered from the presently suggested process can offer a high degree of potential for use as an aquacultural or animal feed. Detailed components of processed and non-processed SFFM were presented in a previous report [5].

CONCLUSIONS

Following the long-term study of the semi-continuous flow system for SFFM, it can be concluded that:
- Organic loading rate of 20 to 40 grams SFFM/liter/day is suggested which can provide 70-85% of filtrate COD removal.
- Temperature of 20°C is suggested because of less oxygen requirement (0.067 to 0.156 per gram of SFFM used) compared to 27°C (0.288 to 0.495).

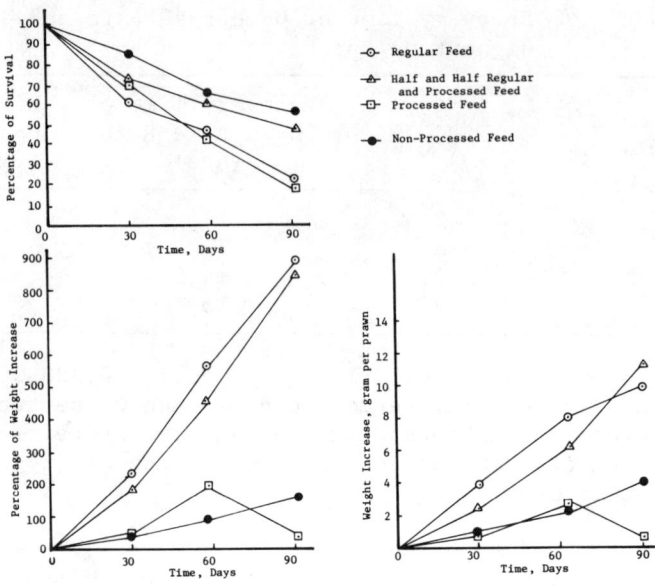

Figure 5. Effect of types of feed on survival and weight increase of prawn.

- Airflow rate of 0.8 liter/liter of liquid volume/minute is required when mechanical mixing device is not provided.
- Processed solid should be separated with highly effective solid-liquid separation.

The processed solid part can offer a high degree of potential for aquaculture or animal feed supplement. The liquid part should go through aerobic treatment process (e.g., extended aeration activated sludge process or aerated lagoon system) to produce the desired effluent COD of 300 mg/L which is permissible for discharge into the city sewer system (in Hawaii), if necessary. This treated effluent can be considered as irrigation water if crop land is available around the treatment system, or part of the treated effluent can be further reused in preparation for the aerobic treatment process.

ACKNOWLEDGMENTS

This work was partly supported by a research grant from SEA, USDA, and approved for publication by the Director of Hawaii Institute of Tropical Agriculture and Human Resources as Journal Series No. 2505.

REFERENCES

1. Yang, P.Y. and Chan, K.K., "Anaerobic Digestion of Poultry Waste with and Without Acid Hydrolysis Pretreatment," <u>Food, Fertilizer and Agricultural Residues</u>, Ann Arbor Science, 423-435 (1977).

2. Yang, P.Y. and Huang, C.J., "Enhanced Energy Production and Water Reuse for Swine Operation in Hawaii," presented at the 1979 ASAE and CSAE Summer Meeting, Winnipeg, Canada, Paper No. 79-4063.

3. Yang, P.Y., Huang, C.J. and Gitlin, H.M., "Kinetic Aspects of the Anaerobic Fermentation of Swine Waste with and without Algal Biomass," Hawaii Institute of Tropical Agriculture and Human Resources, Journal Series No. 2492.

4. Yang, P.Y., Myers, A.L. and Tanaka, N., "Management and Utilization of Waste from Mass Rearing of Fruit Flies," presented at the 1979 ASAE and CSAE Summer Meeting, Winnipeg, Canada, Paper No. 79-6019.

5. Yang, P.Y., Lin, J.K. and Nagano, S.Y., "Biological Treatment of Wastes from Mass Rearing of Fruit Flies and Pineapple Centrifuged Underflow," <u>Proc., 34th Annual Purdue</u> Industrial Waste Conference, West Lafayette, Indiana, 354-359 (1979).

6. <u>Standard Methods for the Examination of Water and Wastewater</u>, 14th ed., Amer. Public Health Assoc., Washington, D.C. (1975).

7. Baker, D.R., Loehr, R.C. and Anthonisen, A.C., "Oxygen Transfer at High Solids Concentration," <u>Jour. Environ. Eng. Div., ASCE</u>, 759-774 (1975).

PERFORMANCE OF THREE TYPES OF ACTIVATED SLUDGE PROCESSES
UNDER VARIABLE ORGANIC AND PESTICIDE LOADINGS

John W. Smith, Department of Civil Engineering, Memphis
State University, Memphis, Tennessee.

Hraj Khararjian, Department of Civil Engineering,
Memphis State University, Memphis, Tennessee.

George Harvell, Department of Civil Engineering, Memphis
State University, Memphis, Tennessee.

INTRODUCTION

In recent years, especially since the Love Canal incident, the concern of environmentalists and many others has shifted towards the fate, management and treatment of toxic wastes. Environmental control agencies and the public are becoming alarmed with the problem resulting in large sums of money being appropriated for the proper disposal of toxic materials. Many municipal and industrial wastewater treatment plants are faced with serious problems in terms of treating wastewater that in many instances contains very minute concentrations of toxic substances. This situation is especially true in municipalities where industries are allowed to discharge untreated industrial waste into the domestic sewer system. Such a case is the North Treatment Plant in Memphis, Tennessee. The wastewater entering the plant is composed of combined industrial and domestic waste. The major contributors include pulp paper, chemical and oil products industries. Identified toxic waste discharged from the chemical industry includes pesticides and cyanides. Several spill incidents have occurred that have resulted in upsets of the existing treatment plant.

The North Wastewater Treatment Plant was designed and is operated under the contact stabilization mode. The wastewater flows through coarse screens at the wet well and then is pumped to fine bar screens followed by gravity flow grit chambers. The grit chamber effluent flows by gravity to the main biological treatment and is mixed with return sludge from the stabilization basin. The mixture flows to the contact basins with short hydraulic detention times (generally less than 2 hours). The biological mass is then separated in the secondary clarifiers with the treated effluent passing through chlorine contact chambers, even though chlorination is not presently utilized, followed by submerged discharge into the Mississippi River. A portion of the settled sludge is pumped to the stabilization basins with the wasted biomass flowing to aerobic digestors followed by thickeners and then to temporary lagoons.

The present organic loading on the plant is already at its design maximum. The plant performance in the past has been inconsistent. During the fall and spring of 1979-1980, research and pilot plant studies were conducted at the MSU Environmental Engineering laboratories. The following were observed:

1. Fixed film reactors are capable of withstanding shock organic loads and toxic wastes with faster recovery compared to conventional and Powdered Activated Carbon (PAC) activated sludge systems.

2. There was no significant difference between conventional activated sludge and PAC units in terms of toxic wastes removals at the levels present in N.T.P.

During the summer of 1980, laboratory studies were conducted with the three systems under various pesticide loading conditions to determine the removal capabilities of the system. Also, an in-depth bacteriological study was conducted to better understand the biological processes. The procedures followed and the results obtained during these studies are discussed in this paper.

MATERIALS AND METHODS

Laboratory experiments were conducted at the Environmental Engineering laboratories of Memphis State University to partially evaluate the treatment capabilities of each system. The following units were operated in parallel:

1. Conventional completely mixed activated sludge. (CMAS)
2. Complete mix activated sludge with powdered activated carbon added in the influent. (CMPAC)
3. Fixed film rotating discs in activated biofilter mode. (RBCA)

A description of the units and operation procedures is presented in the following paragraphs.

Bench Scale Laboratory Units

The conventional CMAS unit and the CMPAC unit were similar to those recommended by Eckenfelder. The units are shown schematically in Figure 1. Rotating biological contactors were designed and constructed with wood as a media for growth. The RBCA units were connected in series followed by an aeration basin and a final clarifier. A schematic diagram of the RBCA units is shown in Figure 1. The basins where the RBCA was housed were constructed of plexiglass having the dimensions of 6"X6"X5". The RBC media were constructed from two plywood plates, 0.5" thick and 5.5" in diameter, connected by a central shaft connected to a 1 rpm motor with 1/4" diameter dows as the media for growth. Each unit contained 108 dows patterned as shown in Figure 1.

The RBCA units were initially operated with diffused air. It was thought that rotating wheels and air would provide enough mixing to keep the solids suspended. However, after operating the unit for three weeks, dead corners and spots were apparent. The diffused air also caused excessive sloughing on specific areas on the wood media. The basins were then modified so that a circular insert was installed at 0.5" clearance from the wheels. This allowed the movement of the disk to provide sufficient agitation and the diffused air was discontinued in the basins.

Operation and Control of the Unit

The units were operated on a continuous 7-day a week basis. Daily wastewater was collected on a batch basis from the North Wastewater Treatment Plant (NTP). The RBCA and the conventional units were fed from a 50 liter container with an electric mixer providing agitation of the wastewater. The PAC unit was fed from a 30 liter container with an air driven mixer to provide agitation. The wastewater was pumped through a Harvard peristaltic pump using 3/8" I.D. Tygon tubing. The same pump was also used to pump the return sludge from the clarifier of the RBCA unit. The containers and the feed tubings were cleaned daily with tap water to prevent excessive biological growth. The effluents were collected in 20 liter containers. The CMAS and CMPAC units were controlled by mean cell residence time (8 days) and MLSS limits of 5000 mg/l and 7500 mg/l in CMAS and CMPAC, respectively. The control of RBCA units was based on maintaining the MLSS in the aeration basin between 1000-1500 mg/l. The wastage rate was varied to maintain these limits. Flow measurements were performed daily using a 50 ml

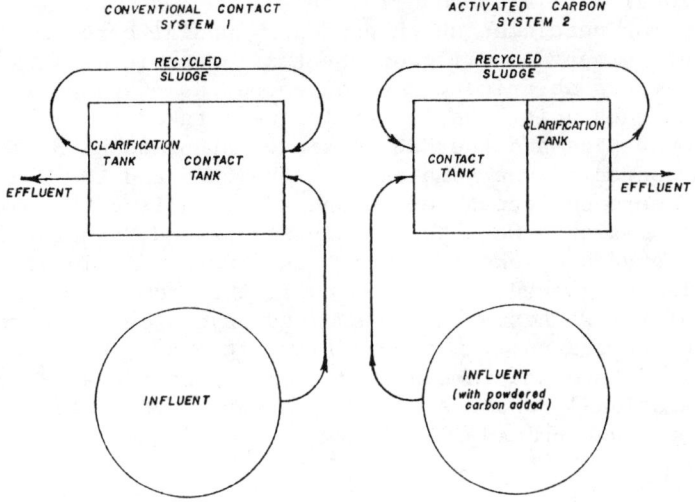

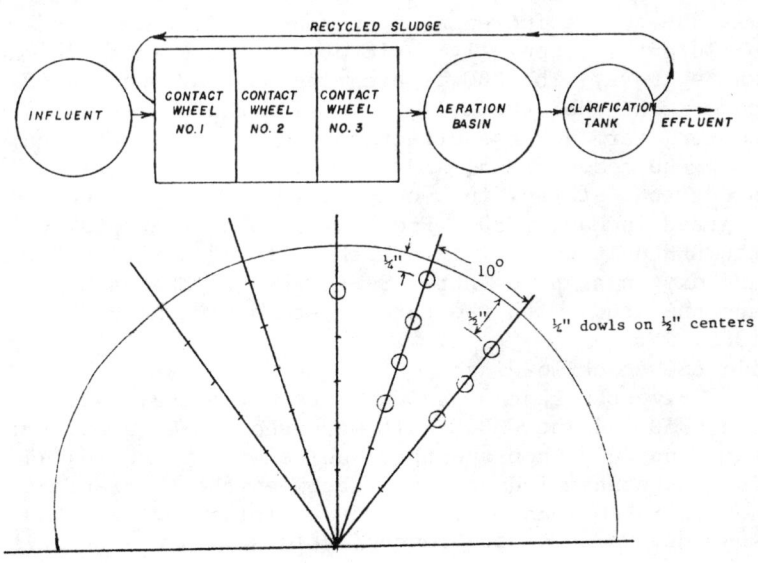

Figure 1

Schematic of Laboratory Test Units

graduated cylinder and a stop watch. Typical daily operational data are shown in Table I.

During pesticide shock periods, special care was taken to insure proper handling of the toxic material. Known quantities of chlorinated carbon compounds (CCC) were added to each unit. Daily samples were taken from the aeration basins and the RBCA basins. MLSS and MLVSS levels were measured in the samples. In the CMAS and CMPAC unit, samples were collected before and after pulling the baffle allowing mixing. Once the baffle was pulled, samples for SVI and oxygen uptake rate were also measured. These tests were also conducted on samples collected from the aeration basin of the RBCA system. Dissolved oxygen and pH were also measured before pulling the baffles with the D.O. level controlled to maintain concentrations greater than 2 mg/l. The operational parameters were tested 5 days a week, i.e. Monday through Friday.

RESULTS

The principle purpose of this investigation was to study and compare the effect of CCC shock load on three types of biological treatment systems. In recent years, powdered activated carbon and RBC systems have become popular for biological treatment. Much of the interest in PAC systems was due to the increase of awareness for removal of micro-pollutants, especially toxic organic materials, in many wastewaters. The RBC is becoming popular because of lower power costs and the ability to withstand shock loads with no long term adverse effect on the process operation.

The conventional and PAC units were a continuation of a study that started in December, 1979. The RBC unit was started in May, 1980. From May to June the total volume of the RBC unit was utilized with diffused aeration for mixing and oxygen supply. After June, the RBC basins were modified and the study with the three systems continued until August, 1980.

1. CCC shock loading:

Five CCC shock loadings were attempted, one in May when the CMAS and the CMPAC unit were shock loaded and twice in June and July when the three units were shock loaded. There is a substantial discrepency between the theoretical dose concentrations and the analyzed samples. Because the absolute value of the concentrations is not as relevant when comparing between these systems, the results obtained are sufficient to develop conclusions. The three systems were able to substantially remove the CCC from the wastewater. The carbon unit performed more efficiently than was expected, but the difference was not significant.

Table I

Typical Daily Operational Data

2. BOD$_5$ and COD Removal:

Under normal operating conditions, i.e. without CCC and/or organic shock load, the three systems removed most of the BOD (soluble and total) and COD. If one considers the soluble BOD or COD removal as a measure of treatability, the removal efficiencies of greater than 90% were common in all three systems as shown in Figure 2. The RBC unit was in general less efficient in terms of soluble BOD and/or COD removal. This finding could be explained by

    a. Anaerobic conditions in the RBC were excessive during the period when the basins were not aerated.

    b. The optimum operating conditions were not determined, i.e. rotational speed, MLSS, sludge retention time, etc.

Not considering the microbial mass on the growing media, i.e. wood, and considering the total mass in carbon unit, the following equations were used to calculate the F/M ratio of each system

CMAS & CMPAC

$$F/M = \frac{So}{Xt}$$

RBCA

$$F/M = \frac{QSo}{V_R(X_T + X_m + X_b) + V_A X_A}$$

where
F/M - Food to Microorganism Ration, Day -1
So - Influent BOD, mg/l
X - Mixed liquor suspended solids, mg/l
X_T, X_M, X_b, X_A - Mixed liquor suspended solids in top, middle, bottom and the aeration basin of the RBC system, mg/l.
t - hydraulic detention time, days
Q - the flow, liters/day
$V_{RT} V_A$ - Volume of RBCA basin and the aeration basin - liters

The calculated F/M ratio values with time are shown in Figure 3. As can be seen from an analysis of the data, the organic loading on the RBC system was very high when compared with conventional and PAC units.

When a significant increase in the influent BOD or COD occurred, the three systems were effected in a similar manner. This was illustrated when the BOD and COD on June 13 and July 7 significantly increased, a substantial decrease in the removal efficiency was observed as shown in Figure 2. The exact nature of the high load was not known; however, North Treatment Plant personnel speculated that sun tan oil was discharged by one of the industries in the system.

The CCC shock load did adversely effect all three units. The effect was immediate and removal efficiencies decreased significantly as shown in Figure 2.

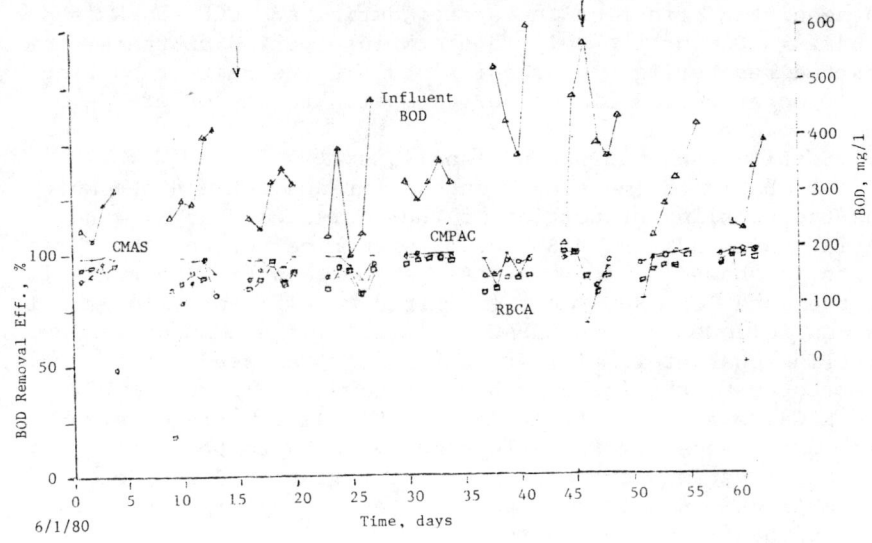

Figure 2

System Performance Efficiency

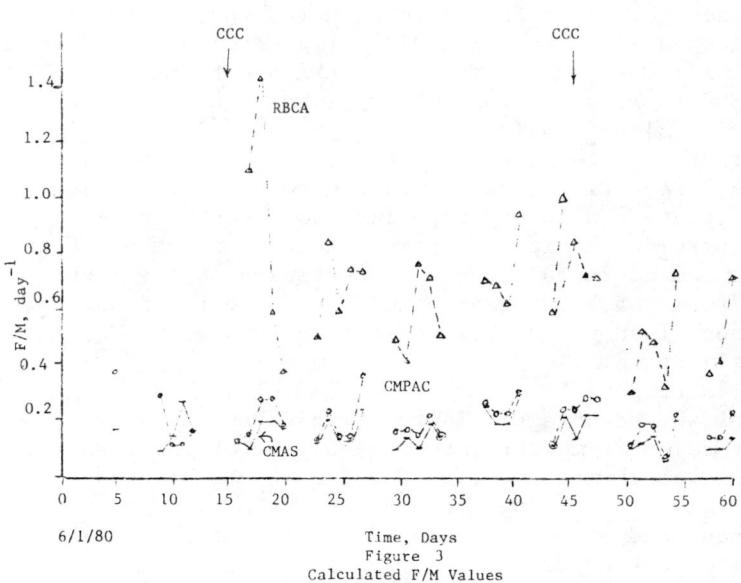

Figure 3
Calculated F/M Values

The strength of the wastewater seems to have an antagonistic effect on the system during the CCC shock load studies. During the July 15 shock load period as the system started recovering, an organic load hit the units resulting in a longer period of recovery compared to the other CCC shock load periods.

3. Settling and Microscopic Analysis:

Probably the most apparent difference between the units was the settling properties of the activated sludge. The variation in SVI and MLSS values with time between the three units is shown in Figure 4. As can be seen from an analysis of the data, the RBCA unit exhibited significantly lower SVI values followed by the CMPAC and finally the CMAS unit. The settling characteristics of the RBCA system seems to be uneffected by the influent BOD as shown in Figure 4. With the RBCA, small variations in the SVI values were observed with wide ranges of F/M values as compared to the CMPAC and CMAS system where a significant variation did exist. The CMAS unit was least capable of accepting organic shock load without adversely effecting the SVI. Both carbon and conventional units exhibited an abundance of bulking filamentous growth especially after an increase of organic load was observed, whereas the RBCA unit did not exhibit an abundance of the bulking filamentous growth but a mold type microorganism. This observation could explain the difference in SVI values between the RBCA and the mixed fluidized activated sludge. It should also be noted that the MLSS in the aeration basin of the RBCA did not generally exceed more than 1500 mg/l whereas the CMPAC and CMAS unit MLSS range was 3000-8000 mg/l and 3000-5000 mg/l, respectively.

4. Bacteriological Studies:

In order to determine the effects of shock loading on the biomass, bacteriological studies were conducted. The results of plate counts and viability do not show any significant relationship or meaningful interpretation of the effects on the microbial flora. The results on the counts of higher forms of life seem to indicate a decrease during the CCC shock load period but this effect cannot be conclusively drawn.

5. Color, Nitrogen:

One of the advantages claimed with CMPAC units is removal of color from wastewater. Analysis of the results of the study confirmed that the CMPAC unit exhibited better color removal when compared to CMAS and RBCA systems.

Influent and effluent ammonia/nitrate nitrogen determinations were performed during this study. Consistent nitrification was achieved in the CMAS and CMPAC units. When NH_3-N level in the influent increased above 20 mg/l, the nitrification efficiency of both the systems was reduced. The RBCA system exhibited a lower nitrification rate when

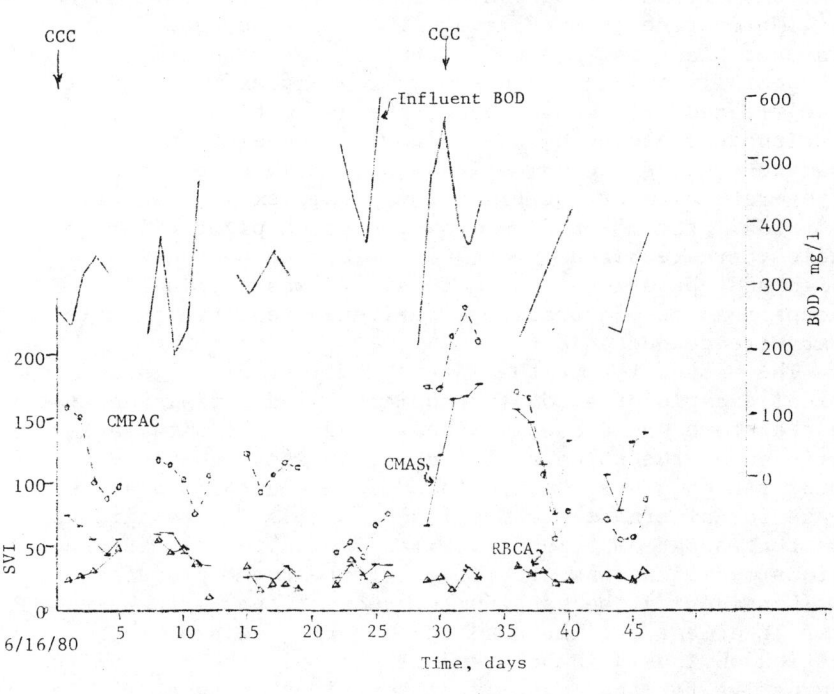

Figure 4

Settling Characteristics of Laboratory Units

compared to the other systems. The lower nitrification rates in the RBCA unit may be due to lack of oxygen in the RBC basin. The literature clearly indicates that a D.O. level of 2.0 mg/l is required for proper nitrification. Another phenomena observed in the RBC unit was the release of ammonia, i.e. effluent NH_3-N concentration greater than influent NH_3-N concentration. This appeared to occur during pesticide shock loading although it was not consistent.

DISCUSSION AND ENGINEERING APPLICATION

In deciding the best alternative treatment for a given wastewater, many questions need to be answered before proper decisions are made. Many historical instances have shown that improper design and management were the cause of treatment plant failure to meet discharge requirements. This is especially evident in cases where complex wastewaters are to be treated. In these cases, the assumption that similar experiences could be used as a tool to develop design parameters has shown to be inadequate. To properly design and operate a treatment plant for a complex wastewater, laboratory studies, and in many instances pilot plant studies, are required to adequately define and develop design and operational parameters. It was within this concept that the laboratory experiments reported in this paper were conducted.

The wastewater at the North Treatment Plant is a typical example of a complex municipal/industrial wastewater and therefore was chosen for this study. The simultaneous operation of the three biological processes (completely mixed, PAC completely mixed and RBC in the ABF mode) provided a base for alternative comparisons. It is interesting to note that such comparisons are lacking in the literature for various modifications of biological treatment processes.

Considering the past performance of the North Wastewater Treatment facility and the future loadings, the question obviously to be asked is which is the best process alternative for the North Treatment Plant wastewater? To answer this question, the feasibility of an alternative in terms of economics, technical, and local constraints should be analyzed. It was not within the scope of this study to perform an economic analysis although references to qualitative cost terms are possible from the data developed.

The three processes exhibited significant BOD, COD, and TSS removals under normal organic loading conditions. In terms of meeting 30 mg/l BOD/30 mg/l TSS standards, all three units consistently obtained lower than standard values. When the process was not under organic or pesticide shock loads, the CMPAC unit provided the best effluent while the CMAS and RBCA followed in sequence of effluent quality. The RBCA was in general less efficient; however, before any conclusions are made, it is necessary to point out that the RBCA unit as operated during this period of the study was not at its optimum conditions when compared with the other two processes. An anaerobic condition did exist in the RBCA basins and the 1 rpm does not appear to be an optimum value. Plans are being made to operate the RBCA unit at 2 rpm. The results at 2 rpm should provide a better understanding of the optimum operational speed for the RBC units.

If one considers the present level of treatment and comparisons made between the three systems, the removal in the RBCA unit would be very comparable to the other two units. The RBCA unit averaged approximately one half the hydraulic detention time of the two other units. This would be a considerable monetary saving in a full scale system and may be sufficient to compensate for the incremental lower efficiency of the RBCA. The sludge settling characteristics were unusually good with the RBCA, whereas the CMPAC and CMAS system developed predominate filamentous growth under high organic loading, which resulted in poor settling. Under similar conditions, the RBCA process operated without any adverse effect on the settling characteristics as expressed by the SVI. This factor has been overlooked in the design of many biological treatment plants. The NTP and the laboratory units received slug organic loads throughout this study. A process which can maintain a stable microbial population without significantly effecting the population dynamics is preferred.

The effluent quality in the three processes deteriorated significantly under the organic shock loads. This was surprising since the increase in soluble BOD was only 1.5-2 times the average and, based on published data, the CMPAC and CMAS systems should have been able to adequately treat the additional load without a significant change in the effluent quality. The range of MLSS in CMPAC and CMAS units was 5000-8000 and 3000-5000 mg/l, respectively, a range commonly recommended for the operation of full scale treatment plants. The maximum F/M ratio in both systems did not exceed 0.3 whereas the RBCA was operated in the range of 0.8 and in many instances greater than 1.0. The argument could be made that not all attached growth is considered in calculating the F/M ratio; however, if it is assumed that the microbial mass on the growing media accounted 50% of the mass the F/M ratio of the RBCA would be reduced to one half. This would still result in higher F/M ratios than the other two systems.

All three systems were adversely effected by chlorinated carbon compound shock loads. The effluent BOD and COD values increased significantly. All three systems effectively removed the CCC. It was difficult to judge which unit was more efficient. It was expected the CMPAC would be more efficient in toxic substance removal and also less susceptible to shock loads. Analysis of the results clearly indicate that the assumption was not substantiated except for more rapid COD and BOD removal recovery efficiency in some instances.

The question is often asked as to the removal mechanism of CCC in a biological process. To partially answer this question, a material balance of the CCC around the system indicated that both adsorption and biodegredation occurred within the system. Results of pure culture studies indicates that 2 to 4 microorganisms in the three systems are capable of breaking down the CCC utilized in this study. Pure culture studies to further define this phenomena are inconclusive at this time and more are planned. Applying these results to the City of Memphis NTP, a deterioration of plant effluent quality could be expected when the total CCC concentrations exceed 0.5 mg/l. At concentrations of greater than 1 mg/l, the effluent quality will be significantly reduced and in general 3 days will be required for the process to recover completely.

Direct changes in the microbial flora within the floc was not evident. The only significant change observed was the apparent decrease in the higher forms of microorganisms, i.e. protozoan, rotifers, etc. This could have significant long term effect if persistent CCC shock loads occurred. Since a desired activated sludge biota contains a diverse population, the need for the higher forms of microorganisms is essential. Published data clearly indicates a correlation between the presence of these microorganisms and good effluent quality. A continuous monitoring of the microbial flora at NTP could indicate the presence of CCC if a decrease in these types of microorganisms is observed. It was also observed that agar plate counts could be eliminated since there seemed to be no apparent correlation or significant interpretation from the plate count data.

An important parameter which counts for a big portion of the operational cost in biological treatment is the oxygen demand. A significantly lower oxygen uptake per unit of MLSS was noted in the RBC process. When an energy balance is made, it is evident that the energy required to rotate the disks is much lower than that required to supply diffused air.

Another aspect of biological treatment that is important in many treatment facilities, but not at NTP is nitrification. The nitrification rate increased in the following order: RBCA, CMAS and CMPAC. Even though nitrification was complete in the range of 5-15 mg/l NH_3-N except in very few cases, the rate decreased when NH_3-N was over 15 mg/l. The RBCA unit did show much lower rates. This could be the result of low D.O. in the RBCA basins. CCC shock loads did not seem to have any effect on nitrification rates. One unexpected phenomena which was noted was the release of NH_3-N in the RBCA unit. A definitive answer could not be given, but low D.O. could result in a deamination process developing in the RBC unit. Additional studies in this area are needed to further enhance our understanding especially where such systems are promoted for nitrogen removal.

The RBCA process would substantially reduce the operational cost and provide large energy savings at a facility such as the Memphis North Wastewater Treatment Plant. The process would provide treatment stability under shock loading as illustrated by the settling properties of the floc. If the present and planned future studies indicate that the 2 rpm will provide the incremental increase in the RBC removal efficiencies, the RBC process would appear to be the most feasible alternative of the three evaluated for NTP.

CONCLUSIONS

1. The CCC (Hex BCH, Hex VCl, Heptachlor, Intermediate Heptachlor and Isodrin) in the concentrations greater than 0.5 mg/l have significant effect on the BOD, COD removals, while in concentrations less than 0.1 mg/l the effect is less apparent.
2. The activated RBC, completely mix conventional activated sludge, and PAC activated sludge systems removed significant quantities of CCC from complex wastewaters.
3. The activated RBC process is a more stable system under organic shock loading conditions as indicated by the capability to maintain good settling characteristics. The operation of the completely mixed and CMPAC activated sludge units was disrupted due to bulking filamentous growth under similar organic shock loads.
4. Higher organic loadings, i.e. F/M ratios, are possible in the RBCA process without adverse effects on removal efficiencies.
5. Lower oxygen supply was required in the RBCA process. This lower power requirement in the rotation of RBC disks would result in a significant economic savings.
6. Better effluent quality could be obtained with CMPAC activated sludge; however, incremental increases in the quality may not justify the additional cost.

COMPOSTING OF INDUSTRIAL RESIDUES

Willibald A. Lutz, Dipl.Ing.Dr., Andritz-Ruthner
Industrieanlagen AG., Vienna, Austria

INTRODUCTION

In Austria and other european countries there are standards and definitions of waste. In general waste could be solid waste or special waste. Solid waste consists of 4 groups, namely household, commercial and industrial, bulky and street sweeping waste. Special waste can be solid like hazardous and toxic waste or liquid like sludge generated in waste water treatment plants. Because of its origin, size and composition the special waste needs separate collecting, hauling and disposal systems. Industrial waste could belong either to the solid waste or to the special waste and can be toxic or non-toxic. The composting process as a method of waste disposal is dealing with all kinds of waste which is non-toxic and primarily organic.

The composting of wastewater sludges and solid residues as a disposal method is being elevated to the forefront of waste transformation technology. This re-evaluation of composting has been caused by several reasons, such as, the change in cost and availability of fossil fuels, the reduction of available disposal alternatives due to environmental restrictions on land, ocean and air disposal, and an increased awareness by the soil scientist, the farmer, and the public of the depletion of valuable soil organics in the urban and rural soils. Compost application, under controlled conditions, can enrich and replenish soil with valuable organic matter, trace minerals and nutrients and renovate the physical and microbiological characteristics for long term growth requirements.

REQUIREMENTS OF COMPOSTING

1. Basic Material

Industrial and commercial waste is basically generated in non residential areas. Depending on the origin there is waste of household type like residues from kitchen and restaurants or it is specific industrial waste related to the production program of the industrial facility like waste oil, waste paper, textiles, leather, etc.

Composting is an aerobic fermentation process based on the activity of microorganism as bacterias, protozoas, actinomycetes and fungus. The utilization of refuses and sewage sludge by composting is a process being in accordance with exact and calculable biological rules. The process mainly consists of microbial decomposition and transformation of organic waste into products of higher quality. From the biological-chemical point of view carbohydrate, sugar, starch, fat, etc. are split up or oxidized by a continuous input of oxygen. The result of a controlled system is the exact know-how and control of this process and the fact that thereby a fully hygienized endproduct is obtained that may be utilized in agriculture or other profitable utilization.

Because of the microorganism the most important limitation of compostible waste is the toxicity. Nevertheless toxic organic and anorganic components are compostible up to a certain concentration. Some microorganisms are adaptable to destroy cellulose, hydrocarbons, fats, nitrogen, lignine, synthetic materials, wax, solvents, phenols and other organic materials. The concentration of these basically fermentable materials will influence the growth and occurance of specific adaptable microorganisms which are mainly actinomycetes. The adaption period varies between days and weeks depending on the kind of toxic element and concentration.

2. Composting parameters

There are certain parameters which have to be achieved to obtain a well running composting process and a good endproduct. One of the most important facts is the sufficient oxygen supply for the microorganism. The amount of oxygen respectively air depends on the basic material, grain size and texture, penetration rate, moisture, age of pile and air pressure. To avoid uncontrolled odour emissions and to increase the dewatering usually air is sucked through the pile or through the enclosed composting vessel. The exhausted air is cleaned in deodorizing filters like activated carbon or in biocompost filters.

The moisture in the raw material depends on the water content of the waste and should be between 40% and 65% by weight. Below 40% the material is too dry to achieve the optimal microorganisms' growth. A moisture content above 65% causes problems with the air supply because of clogging.

The carbon-nitrogen rate is important for the build-up of cell structure of the microorganism. The C/N ratio in the cell is 5:1. The percentage of volatile organic carbon varies in the range of 20% to 50% in the raw material. Therefore the C/N ratio should be approx. 20:1. Household refuse has an average C/N ratio of 40:1 and sludge from municipal waste water treatment plants of 10:1. When mixing both substrates an optimum ratio can be achieved.

The pH value and the salt concentration is of minor importance for the composting process but has a great influence on the compost quality. Acid or alkaline pH is more or less stabilized in the composting process to a neutral value in the range of 6,5 to 7,5 in the endproduct.

The temperature is rising as a result of the exothermic conversion from C into CO_2 by means of microorganisms. The temperature ranges from 50° to 70° C. The pathogene reduction is a function of the temperature and time. In general all pathogene germs are killed during a thermophilic period of 8 to 14 days. The kind of rotting organism also depends on the temperature. Thermophile bacterias prefer higher temperatures while fungus are occuring in lower temperature ranges. Some actinomycetes are producing antibiotics which support the pathogene kill.

3. Quality requirements

According to the present standard in Austria and the european common market there are 4 quality standards, namely
- basic compost material which is mechanically processed solid and liquid waste before decomposition and pathogene reduction,
- fresh compost which requires a rotting period of a minimum of 5 days at a temperature of 65° C. The degree of decomposition of organic substance is more than 20%. The moisture content is also reduced during decomposition in the range of 10%. At the same

time the C/N ratio is lowered.
- cured compost which requires a further decomposition of organic substances of more than 60%. The temperature is still high during a curing period of 6 to 16 weeks depending on the basic compost material and the particular composting system. At the end of the process the compost is decomposed and the temperature will not rise more than $5°$ C when stored in large piles. The moisture will be below 40%. The grain sized vary from 8 mm in fine compost to 20 mm in medium compost and maximum 45 mm in coarse compost. Impurities are only tolerated if they do not cause damage to soil, plants, animals or people. The content of soluble salts should be in the range of less than 3% to avoid damages to soil and plants.
- special compost. For certain fields of application compost has to be processed further and can be used as pig pen soil, filtering media or sound absorbent media. Compost can also be mixed with other materials, e.g. mineral fertilizer, peatmoss, etc.

Concerning heavy metals there are recommendations and regulations in Europe. According to the following table (I) there is a higher and a lower limiting value for compost application. Compost within the range of the lower value can be used without any control. Compost within the higher level of heavy metals is restricted to controlled applications. Furthermore the heavy metal concentration in the soil shall not exceed certain levels within a time period of 50 to 100 years when applying an annual rate of 30 t/ha compost.

Table I. Limiting values for heavy metals in compost, sludge and soil in ppm = mg/kg dry substance

Element		higher value	lower value	soil, air dried
Cadmium	Cd	30	10	3
Lead	Pb	1200	600	100
Chrome	Cr	1200	600	100
Copper	Cu	1200	800	100
Nickel	Ni	200	100	50
Mercury	Hg	25	10	2
Zinc	Zn	3000	2000	200

RUTHNER COMPOSTING SYSTEM

The Ruthner Composting System is a fully mechanised, enclosed, multistep, indoor process. The system consists basically of 4 main steps, namely the preparation, dynamic fermentation in a rotating drum, static fermentation in an aerated pile and the fine compost preparation.

During the preparation phase the waste is ground by special crushing-machines to proper size, thus pulverizing and/or transforming by the heavy impact of hammers hard materials as glass, stone etc. into a gravel-like condition. Iron metals (which may be reused for metallurgic purposes) will be separated by magnetic separators. If required, even nonferrous metals, plastics, and glass may be separated for a possible recycling process.

The pretreated refuses having by means of sludge or water reached a certain degree of humidity, undergo a rapid fermentation process in slowly rotating drums. During this step, regulated by aeration, the temperature increase up to 65° C (i.e. 149° F) because of the exothermic reaction, thus destroying the majority pathogenic germs.

In order to completely mature the compost and enrich it with humic matters, the fresh compost is tranferred to an aerated ripening plate. During this curing in accordance with the RUTHNER System the fresh compost will be aerated following an exact schedule. During the subsequent fungoid growth all pathogenic germs are finally eliminated by arising antibiotic inhibitors as well as by a further increased temperatur

The fine compost preparation is based on screenin milling, mixing and packing. The screening overflow mainly consists of plastic, rubber, leather, paper, ceramics, etc. and has a high BTU value. In larger plants the screening overflow can be utilized in an incineration plant to generate steam and electricity. The screened compost can be sold as compost of minor quality requirements. To achieve higher quality the screened compost is ground in a special compost mill to a small grain size where glass pieces are not visible. For special purposes the milled compost is mixed with mineral fertilizer, peatmoss, etc. to get the highest compost quality. Luxury compost quality is packed in 5 or 10 gallon plastic bags and sold to retailers.

The Ruthner composting system was originally designed to treat solid waste and sludge in a co-composting plant. The largest facility was installed in

Siggerwiesen/Salzburg, Austria, to process 400 t household and commercial refuse and 150 t sludge per day. The plant is working since 1978 and presently produces 60.000 t compost per year. The composition of the Siggerwiesen compost is shown in table (II).

Table II. Compost analysis at Siggerwiesen

Water content	%	34
Volume weight	g/l	600
pH-value (KCL)		7,3
Salt content	% in D.S.*	1,4
Incandescent residue	%	51,5
Total organic substances	%	48,5
Volatile organic substances	%	23,7
C/N ratio		10:1
Main nutrients:	% in D.S.*	
Total nitrogen	N	0,9
Phosphate	P_2O_5	0,6
Potash	K_2O	0,4
Magnesium	MgO	0,9
Calcium	CaO	1,6
Humus matter	%	7,9
Heavy metals in ppm dry substance:		
Arsenic	AS	5
Lead	Pb	250
Borone	B	30
Cadmium	Cd	7
Chrome	Cr	350
Copper	Cu	500
Manganese	Mn	630
Nickel	Ni	45
Mercury	Hg	2
Zinc	Zn	830

* % in D.S. = percentage by weight in the dry substance

Presently Siggerwiesen is producing 14 different kinds of compost. The main advantage of compost application is the improvement of soil fertility and increase of plant yield. The compost is used in the agriculture, horticulture, vegetable gardening, forestry, recultivation, reclamation and for technical purposes as filter media or in sound barriers.

Concerning economic aspects the operation and maintenance costs are in the range of US$ 8,- to 12,- per ton of waste. Approximately one third of the overall operation costs are personnel costs, one third are energy costs and the remaining third are expenses for

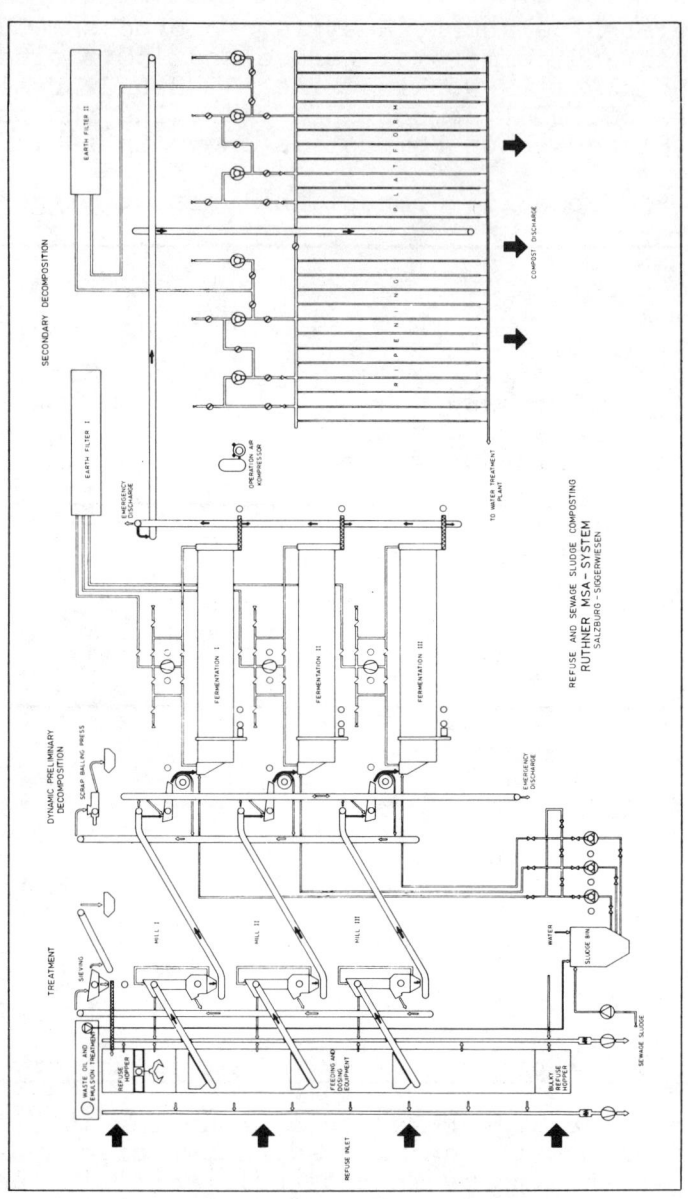

spare and wear parts and supplementary costs. The rate of personnel and energy consumption depends on the standard of automatization. When the Siggerwiesen co-composting plant was started up in 1978 the power consumption amounted to 40 kWh/t refuse. After two years improvement of the operation procedure, installment of additional temperature and CO_2 measuring instruments and a computer system the energy consumption decrease to 22 kWh/t refuse in 1980.

The revenues from the compost selling depend on the local market and the quality. Generally compost is competing with peatmoss and other organic soil conditioners. The most important fact in compost marketing is the guaranteed compost quality and the declaration of application. When following the recommended application rate compost is achieving better soil conditions and higher yields. Evaluating the content of nutrients in compost made of household refuse and sludge the price equivalent would be approximately US$ 10,-/t. Considering additional adding of mineral fertilizers and other improving substances the compost price may exceed $ 5,- per 10-gallon plastic bag.

SUMMARY

Composting is a suitable method of disposal of household, commercial and industrial waste as well as sludge generated in waste water treatment plants. The experiences in several test series and in the large scaled co-composting plant Siggerwiesen/Salzburg indicate that a relatively great number of industrial residues can be composted. Good results were achieved with adding waste oil, emulsions, hydrocarbons, solvents etc. to municipal soild waste and sludge. The amount of industrial and hazardous waste depends on the concentration of toxic elements and the adaption period of the microorganisms in the fermentation process.

To achieve a controlled and marketable endproduct the composting process has to be mechanised and controlled. The various compost products are tested in the laboratory and in field tests. According to the austrian and other european standards and recommendations the content of heavy metals and toxines in the compost has to be controlled and monitored. Considering the application recommendations compost is an excellent soil conditioner and organic fertilizer.

SESSION III

CASE STUDIES

INTEGRATED HAZARDOUS WASTE MANAGEMENT AT A SPECIALTY
ORGANIC CHEMICALS PLANT: A CASE STUDY

Joseph J. Kulowiec, P.E. Manager of Engineering
Consulting Environmental Engineers, Inc., Hartford, Conn.

INTRODUCTION

The proper management of wastewater and hazardous waste has historically been a concern and responsibility of the chemical processing industry (CPI). In the last decade, this concern has been magnified and accelerated markedly by the federal/state regulatory implementation of NPDES permit, POTW pretreatment and hazardous waste management (RCRA) programs. King Industries, Inc., located in Norwalk, Connecticut is a manufacturer of corrosion inhibitors based upon alkyl naphthalene sulfonate formulations. This case study is a summary of the wastewater and hazardous waste management efforts made by the company over the last six years. These efforts have been highlighted by an emphasis on: optimizing in-process raw material/product loss control, waste segregation, removal of materials from wastewater for proper disposal or recovery and the use of treatment technologies that minimize the generation of difficult to manage "hazardous" residues. These technologies include liquid-liquid extraction, activated carbon adsorption and acid neutralization. Wastewater characterization, laboratory/pilot scale process development studies and process design development will be presented and discussed.

WASTEWATER SURVEY AND CHARACTERIZATION

The company produces a series of products based upon dinonylnaphthalene sulfonic acid (DNNSA). The manufacturing process consists of sequential steps: alkylation of the naphthalene with nonene to form dinonylnaphthalene (DNN), sulfonation of purified DNN with oleum to form DNNSA, purification of DNNSA, and finally preparation of purified products, incorporating DNNSA, its salts or by-products sulfonic acids. Purification steps separate reaction mixtures and generate by-products or waste streams (Figure 1).

Figure 1 GENERATION OF PROCESS WASTE STREAMS

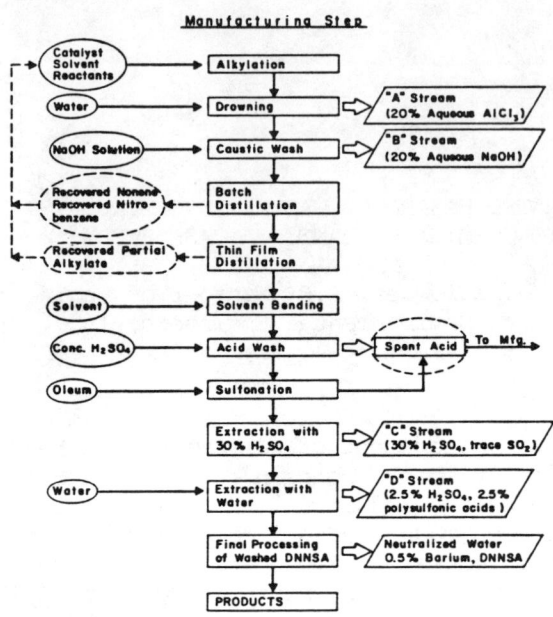

These wastewater and by-product sources were extensively surveyed and characterized. Pertinent parameters are summarized in Table I. BOD_5 testing was not performed on the raw wastewaters because of low pH, presence of inhibitory materials (nitrobenzene) and the suspected bio-resistant nature of sulfonic acids.

Table I. Summary of Wastewater and By-Product Characterization

Parameters	Wastewater or By-Product				
	A	B	C	D	Neutr. H_2O
Flow (gals./week)	2400	2100	2000	25,000	8000
$AlCl_3$ (%)	20	–	–	–	–
NaOH (%)	–	20	–	–	–
H_2SO_4 (%)	–	–	30	2.5	–
Sulfonic Acids (%)	–	–	–	2.5	trace

Nitrobenzene(%)	0.1	0.1	-	-	-
TOC(g/L)	4-5^2	3-4	-	18-20	4-5
Barium(g/L)	-	-	-	-	1-6
pH	1.0-1.5	11	1	1.0-1.5	7-9
Specific Gravity	1.15	1.22	1.22	1.01	1.01

1 - Sulfonic acids of higher alkylation products of naphthalene, referred to as TNNSA

2 - Chloroaniline (0.3 - 0.8 g/L) included

Although the wastewater volumes appear small relative to dilute waste streams from larger CPI plants, the historical processing and disposal problems were substantial. Lime neutralization of "D" stream, followed by vacuum filtration produced approximately 21 cubic yards weekly of 30 percent filter cake and a filtrate with a high TOC level. Combined neutralization of stream "A" and "B" produced approximately 25,000 gallons weekly of 7-8% Al(OH)$_3$ slurry, also with a high TOC level. These practices, along with seepage/evaporation lagooning, were questionable at best. The implementation of more stringent environmental regulations in the 1970's made the development of alternative solutions environmentally and economically expedient.

PROCESS DEVELOPMENT

In order to devise environmentally sound, technically feasible and economical processes, a number of quite practical considerations were used:

- optimizing manufacturing process raw material and product recovery

- extracting valuable and/or toxic components from wastewater streams for internal/external re-use or disposal

- cost-effective end-of-pipe treatment or pretreatment prior to final disposal

- minimizing treatment by-products or "waste sludge" generation

A review of the various laboratory and pilot scale process development evaluations performed on selective wastewater streams follows.

Liquid-Liquid Extraction of Organics

Liquid extraction has been used extensively in laboratory work and since the 1930's in large scale industrial applications as a seperation technique (1). In summary the process is based upon the removal or "extraction" of a component or *"solute"* from a feed stream by interaction with a *solvent* stream. The *solvent* stream is usually chosen such that it is immiscible or only partially miscible in the *feed* stream. The *solute's* relative concentrations in the two phases will depend upon its relative affinity for the *feed* and *solvent*. By definition and convention:

- extract(Y)-desired *solute* in *solvent* product

- raffinate(X)-remaining feed stream after extraction

For a given component or *solute* A, the equilibrium concentration of A in *extract* y of A and raffinate x is called a distribution coefficient D:

$$D = \frac{Y_A}{X_A}$$

Two general categories of solvent extraction processes can be distinguished based upon *solute-solvent* interaction:

- non-specific - physical interaction resulting from hydrogen bonding or polarity difference

- specific - chemical interaction, somewhat analogous to "ion exchange"

In order to minimize waste management problems, liquid-liquid extraction was evaluated for recovery of nitrobenzene and alkylale from streams "A" and "B" for recycle back to the alkylation manufacturing step. Laboratory scale decanting test demonstrated that a secondary decant step, following the primary alkylate separations, could result in additional DNN recovery.

Laboratory scale "shake" tests of streams "A" and "B", a "first cut" approximation of liquid-liquid extraction, with nonene (the process alkylate) showed promising recovery rates (90% +) for nitrobenzene the alkylation reaction solvent.

Stream "D" was also evaluated for liquid-liquid extraction of mixed acids (TNNSA). The company recognized potential uses for these materials, one of which was as a metal coating additive. Laboratory and pilot scale testing and evaluations resulted in the development of a liquid-liquid extraction method consisting of the following steps:

- extraction of sulfonic acids from stream "D" using isobutyl alcohol (IBA) as an extractant

- water washing of the sulfonic acid rich IBA extract to decrease water content

- thermal stripping of IBA from the stream "D" raffinate for recovery/recycle

- thermal concentration of sulfonic acid extract with IBA recovery

The resulting extracted stream "D" could be expected to contain a range of 0.1 to 0.5% sulfonic acids (80-90% removal) and 0.1 to 0.2% IBA. The suitability of this wastewater for sanitary sewer discharge after acid neutralization had to be established. The results of preliminary biodegradability testing is summarized in Table II. Based on these results, the necessary regulatory approvals were obtained to permit sanitary sewer discharge of extraction/neutralization stream "D" after pretreatment.

Table III. Extracted Stream "D" Biodegradability Testing

% DSA[1]	% IBA[2]	BOD_5 Mg/L	Weight %	% BOD_5[3] DSA
2.25	0.09	2100	9	3.33
0.45	0.09	2300	43	21
0.50	0.10	1790	30	4.8
2.25	0.18	5200	21	1.11
0.45	0.18	6300	100	80
0.09	0.18	3200	120	56
0.00	0.20	3900	195	—

1 - Sulfonic acids
2 - isobutonal alcohol
3 - based upon reported BOD_5 of isobutanol of 150-160% by weight

Granular Activated Carbon Adsorption of Organics

Although liquid-liquid extraction recovered valuable raw materials and reaction products in streams "A" and "B", these wastewaters still contained relatively high levels of soluble TOC, 3 to 6 g/L. A laboratory scale testing program was developed and implemented for stream "A" to determine the feasibility of activated carbon (A/C) adsorption of the soluble TOC. This program included:

- A/C screening
- time contact or "kinetic" evaluation
- A/C isotherms
- spent A/C washing

A/C screening showed that high powdered A/C dosages, in excess of 20 g/L were necessary to achieve significant TOC reductions. Two granular carbons (A-Filtrasorb 300, and B-Witcarb 950) were used in this screening with the results reported in Table III.

A/C contact time requirements were determined using a batch laboratory test with the results reported in Table IV.

Powered A/C isotherm tests were then conducted. Representative isotherm data are presented in Table V.

In order to determine if A/C regeneration was feasible, spent A/C washing studies were performed. Adsorbed TOC leaching losses and chloride removal averaged 7 and 50 percent respectively with a 20 percent wash.

Table III. Granular A/C Screening (48 hr. contact time)

Carbon:	None	A	B
Dosage (g/L):	0	50	50
Filtered TOC (mg/L):	3700	200	110
Chloroaniline (mg/L):	3620	96	96

Table IV. A/C Contact Time Testing

Time of Contact (hrs.):	untreated	0.5	1	2	3	4
TOC (mg/L):	3,600	210	190	180	165	160

Note: Powdered Filtrasorb 300, used at 50 g/L dosage: all TOC values are for filtered samples

Table V. Stream "A" Isotherm Test Results

	Carbon						
M(A/C dosage, g/L):	A,B	0	15	25	50	75	100
Filtered TOC(g/L):	A	3.1	1.4	.88	.28	.14	.10
	B	6.0	4.6	3.8	.79	.09	.05
X/M(TOC basis):	A	—	.113	.089	.056	.039	.03
	B	—	.093	.088	.104	.079	.059

Pilot scale continuous flow A/C column evaluations followed the encouraging laboratory results. A glass column, 3 inches I.D. with a 32 inch packed granular A/C section was used with carbon B. Operating parameters were as follows:

Feed TOC: 550 mg/L

Contact Time: 2 hrs. (Based on total empty column volume)

Feed Rate: 1.1 - 1.2 L/hr.

Hydraulic Loading: 3.5 L/min/m^2

Flow Regime: upflow

Test results are reported in Table VI.

Table VI. Pilot Scale Column Test Results

Cumulative Throughput Volume (L):	5	14.7	24.5	32	39
Effluent (mg/L):	120	100	120	260	1300

Based upon the laboratory isotherm test data and these pilot column results, a X/M TOC loading at saturation of 0.14-0.15 was extrapolated. Because of the specific gravity of stream 'A', a downflow mode was considered the most practical in any subsequent full/scale design. With stream 'A' thus purged of organic contaminants, the stream became a saleable by-product to recyclers of AlCl$_3$ solutions. Marketing efforts were focused in that direction with good success.

Extracted Wastewater Neutralization and Disposal

With troublesome organics removed, the problem of neutralizing the various wastewater streams prior to ultimate disposal was addressed. Stream 'B', after extraction with nonene, was found to be reusable in the akylalate caustic wash step. Stream

'C' titration curves were generated for 20 percent NaOH. Chemical requirements were substantial, approximately 1 gallon of 20 percent NaOH per gallon of stream 'C'. Because of this high alkalinity demand, it was concluded that this low volume stream should be contract hauled and processed at a nearby commercial treatment facility (RCRA permitted). The extracted stream 'D' titrated studies produced alkalinity requirements for 20 percent NaOH of approximately 28-30 gallons per 1000 gallons of stream 'D'. The use of caustic for neutralizing these wastewaters was preferred for primarily two reasons: current use in the plant, and total avoidance of processing, handling and disposal of precipitation products ($CaSO_4 \cdot 2H_2O$) from lime neutralization. Equipment maintenance, operator attention and sludge handling costs (capital and operating) also shifted the wastewater economics toward the use of caustic in lieu of hydrated lime ($Ca(OH)_2$). Caustic neutralization of stream 'D' exhibited a predictably strong acid/strong base titration curve. In order to achieve reliable pH adjustment, it appeared that a multi-stage neutralization process would be required for a continuous flow system (2,3). First stage adjustment would be made to a pH of 2.0-2.5, satisfying approximately 85% of the alkalinity demand. pH trimming to achieve an endpoint pH of 7-9 would be done in the second stage.

Liquid-Liquid Extraction of Metals

The use of a specific or "ion-exchange" mechanism as a liquid-liquid extraction process was evaluated for removal of barium (Ba) from the *neutralized water* stream. Liquid ion exchange utilizes an ion exchange reagent (extractant) diluted as a continuous liquid phase in any of a number of water-immiscible solvents (kerosene, heptane or proprietary formulations). Successful initial use of the process occured in the 1940's in the nuclear materials industry (1,4) for separation of uranium from its ore and plutonium from uranium and its fission products in spent reactor fuels. Through extensive product application research and development, the company e lished the effectiveness of its primary product, DNNSA, as a very selective ion-exchange reagent. The recovery of the valuable barium salt of DNNSA and concurrent resolution of an effluent treatment problem appeared practical.

Laboratory scale "shake" tests and pilot scale testing established that a 20% DNNSA solution in heptane was the best extractant formulation in terms of separation clarity, concentration gradient (driving force) and subsequent loaded extractant concentration (to 40% extractant in heptane).

A 3 inch diameter packed column was used to pilot the process (8 feet of ceramic saddle packing). A summary of results is reported in Table VII.

Table VII. Neutralized Water Barium Extraction Study-Pilot Scale Results

Extractant Conc. (%)	Feed: Extractant (Ratio)	Loading (gal/hr-ft^2)	Barium (mg/L) Feed	Raffinate
40[1]	4.4	140	3600	56
	3.75	154	1400	16
	2.0	146	1400	5.4
	4.0	242	500	175
	3.0	430	500	225
	4.0	150	500	2.0
20[2]	2.0	139	6000	62
	3.0	173	3800	3.5
10[3]	2.0	130	4333	205

1-8 ft packing   2-7 ft packing   3-6 ft packing

Extrapolation of this data resulted in an 8 to 9 foot packed column length required to reduce barium from 6000 mg/l to 30 mg/L with a 20 percent extractant at loadings of 150 gals/hr-ft^2.

The entrainment of low levels of extractant in the barium extracted *neutralized water* resulted in a mildly acidic pH and TOC levels of 4-5 g/L. Combined neutralization with the extracted stream "D" appeared necessary prior to sanitary sewer discharge.

Summary

Laboratory and pilot scale evaluation of the various segregated wastewater streams resulted in the formulation of a integrated material recovery/waste disposal plan for the company. In summary:

Stream   Disposition

A	Organic recovery and removal; recycle of aqueous wastewater
B	Same as stream "A"
C	Contract disposal
D	Organic recovery; acid neutralization prior to sanitary sewer discharge
Neutralized Water	Metals removal; acid neutralization prior to sanitary sewer discharge

FULL SCALE SYSTEM DESIGN

System designs were developed from the laboratory and pilot scale study of specific wastewater streams. These system descriptions are limited to generalized schematics and tabular design summaries. The internal recycle designs for streams "A" and "B" are beyond the scope of this not included in this paper. Also included are discussions of various design considerations including process configuration, control, and reliability.

Stream "D" Liquid-Liquid Extraction

Figure 2 schematically depicts the sulfonic acid extraction process for stream "D". The stability of the extraction column (II) operation is maintained by controlling the stream "D" exit rate and extractant, IBA, feed rates. The stream "D"/IBA interface at the top of the column is controlled automatically by a pneumatic interface controller/automatic valve loop throttling the extracted stream "D" column exit rate. Extractant specific gravity, a measure of the decrease in IBA concentration, regulates the extract feed rate to the column. The extractant wash column (W) removes additional water from the IBA, thus concentrating the extracted sulfonic acids (DSA) further. The DSA free streams of IBA/H_2O from columns II and W are combined for subsequent stream stripping of the IBA. The DSA/IBA free stream "D" with 2-3% H_2SO_4 is then discharged to subsequent neutralization treatment. Stripped IBA is collected and recycled to the extraction column. Washed extractant is concentrated in a pre-heated continuously wiped thin film rototherm evaporator. All IBA/H_2O condensates are recycled. The DSA concentrate is either drummed for licensed incineration or sent to by-product blending and storage. Flexibility and reliability for each process step are enhanced by monitoring, alarm and/or control instrumentation loops for all critical parameter including pressure, temperature, level, specific gravity and flow.

The continuous flow columns and rototherm are buffered for start-up/shut down cycles by adequately sized receiver tanks.

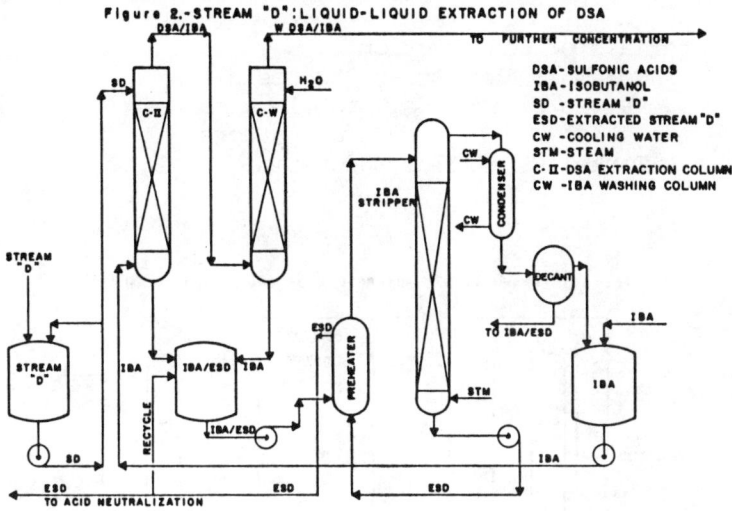

Figure 2.-STREAM "D": LIQUID-LIQUID EXTRACTION OF DSA

DSA- SULFONIC ACIDS
IBA - ISOBUTANOL
SD - STREAM "D"
ESD-EXTRACTED STREAM "D"
CW - COOLING WATER
STM - STEAM
C-II-DSA EXTRACTION COLUMN
CW -IBA WASHING COLUMN

Activated Carbon A/C Adsorption of Organics-Stream "A"

A mutiple column A/C adsorption system was designed for stream "A" organics removal. System design symmary and schematic are presented in Table VIII and Figure 6 respectively. Decanted/extracted stream "A" is pumped at controlled flow to the lead column, C-I, for gravity flow (downflow) thru three columns in a series. This sequence insures a uniform by-product TOC quality along with maximum carbon utilization. When C-I carbon is exhausted, the column is isolated for drainage, washing, exhausted carbon removal and recharging. Piping design allows the recharged C-I to rotate to the final column position with C-II becoming the lead column. The sequence is repeated when columns C-II and C-III become exhausted.

Table VIII. Stream "A" A/C Adsorption System-Design Summary

Volume Processed: 2400 gallons per week
TOC: 4.5 g/L (average)
6.0 g/L (peak)

A/C Loading (X/M) : 0.15 lbs. TOC/lbs. A/C (saturation)
Carbon Consumption: 600 lbs. per week (22-23 cu. ft)

Stream "A" Feed Rate: 2 gpm
Column Dimensions: 24 inches dia. x 10 feet long

Hydraulic Loading: 0.64 gpm/ft^2
Detention Time (empty): 82 minutes per column
Carbon Volume: 22 cu. ft. (2' x 7'6"H)
Head Loss: 2 ft. @ 4 gpm thru 3 columns in series

Materials of Construction
    Columns: FRO (Atlac 382 resin)
    Valves/piping: polypropylene (stream "A" feed)
                 CPVC (all others)
    Tanks: FRP (Atlac 382 resin)

Figure 3.—GRANULAR A/C ADSORPTION OF ORGANICS, STREAM "A"

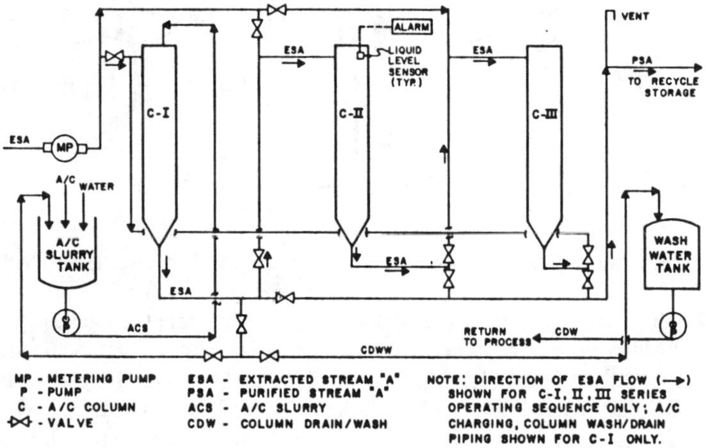

MP - METERING PUMP
P - PUMP
C - A/C COLUMN
⋈ - VALVE
ESA - EXTRACTED STREAM "A"
PSA - PURIFIED STREAM "A"
ACS - A/C SLURRY
CDW - COLUMN DRAIN/WASH

NOTE: DIRECTION OF ESA FLOW (→) SHOWN FOR C-I, II, III SERIES OPERATING SEQUENCE ONLY; A/C CHARGING, COLUMN WASH/DRAIN PIPING SHOWN FOR C-I ONLY.

Materials of construction were selected to insure reliable operation in a corrosive environment. Conventional metal well screen was deemed unsuitable for a column exit assembly. A special assembly was designed and fabricated from FRP and woven polypropylene fabric. A FRP false bottom hopper using polypropylene filter fabric was design to increase exhausted carbon dewatering.

Extracted Stream "D" Acid Neutralization

The process schematic for extracted stream "D" acid neutralization system is presented in Figure 4. Extracted stream "D" is fed continuously to two completely mixed pH adjustment reactors (N-1, N-2) in series. A 20% caustic soda solution is added to each reactor by pneumatically operated needly type control valves. Uniform valve operation is assured by use of a constant back pressure pumping loop. pH is controlled automatically using direct acting proportional process controllers. Neutralized stream "D" is then pumped to a a 10,000 gallon holding tank (N-3). This tank is continuously mixed and equipped with pH monitoring and recording. In the event of process upset in N-1 or N-2, caustic can be added on low pH alarm set point to N-3 to insure a uniform effluent pH. Neutralized stream "D" is then pumped to sanitary sewer. All construction materials selected were corrosion resistant and capable of service at temperatures up to the 180° F experienced in the neutralization reaction.

Figure 4.- EXTRACTED STREAM "D" ACID NEUTRALIZATION

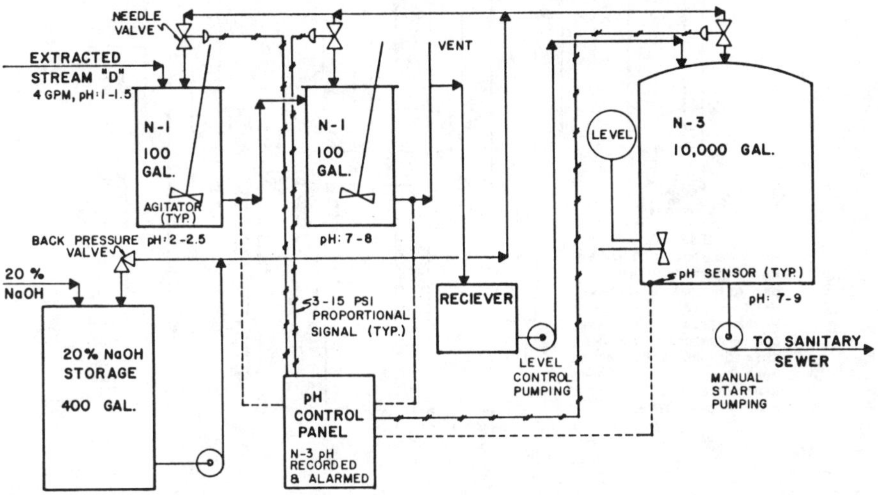

Neutralized Water Liquid-Liquid Extraction

The process schematic for this system is presented in Figure 5. After phase separation and flow equalization (T-1, T-2), *neutralized water* is pumped to a simple packed column, C-I, for countercurrent extraction of barium (Ba) with a 20% solution of sulfonic acid in heptane. The evaperative recovery of heptane, the extractant solvent, is also shown. Extracted *neutralized water* is then pumped to the extracted stream "D" acid neutralization system prior to sanitary sewer discharge.

SUMMARY

The program of wastewater characterization, segregation, by-product recovery and effective pretreatment has resulted in a projected total plant process wastewater discharge of:

    Flow: 13,000 - 20,000 gpd
    pH: 7 - 10
    TOC: 3 - 5 g/L

Ba: 5 mg/L
Nitrobenzene: 0.5 mg/L

The application of these technologies demonstrated the feasibility of developing an effective waste management program integrated with specific, cost-effective recovery at a specialty organic chemical plant.

Figure 5.-NEUTRALIZED WATER:LIQUID-LIQUID EXTRACTION OF BARIUM

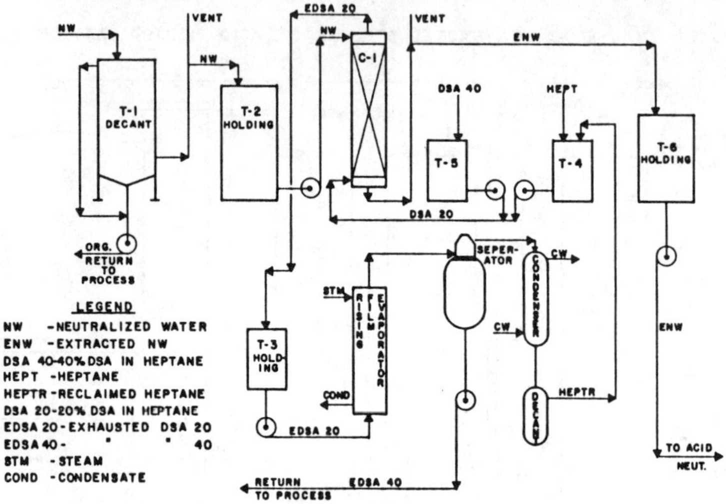
NW
VENT
EDSA 20
NW
VENT
ENW
NW
C-1
DSA 40
HEPT
T-1 DECANT
T-2 HOLDING
T-5
T-4
T-6 HOLDING
DSA 20
ORG. RETURN TO PROCESS
SEPERATOR
CW
CONDENSER
STM
RISING FILM EVAPORATOR
CW
T-3 HOLDING
COND
HEPTR
DECANT
ENW
EDSA 20
TO ACID NEUT.
RETURN TO PROCESS
EDSA 40
LEGEND
NW -NEUTRALIZED WATER
ENW -EXTRACTED NW
DSA 40-40% DSA IN HEPTANE
HEPT -HEPTANE
HEPTR-RECLAIMED HEPTANE
DSA 20-20% DSA IN HEPTANE
EDSA 20-EXHAUSTED DSA 20
EDSA 40- " " 40
STM -STEAM
COND -CONDENSATE

CREDITS

The author wishes to thank Les Chirgwin (Process Engineer), Bob Olson (Plant Engineer), Larry Gallacher (V.P. Research), and Dick King (V.P. Engineering) all of King Industries, Inc., Norwalk, Connecticut, for their assistance and cooperation during the development of the waste management program described.

REFERENCES

1. Bailes, P.J. et al, "Liquid-Liquid Extraction: The Process, the Equipment", Chem. Engr., Jan. 19, 86(1976)

2. Hoyle, D.L., "Designing For pH Control", Chem. Engr., Nov. 8, 121(1976)

3. Hoffman, F., "How to Select a pH Control System For Neutralizing Waste Acids", Chem. Engr., Oct. 30, 105(1972)

4. Bailes, P.J. et al, "Liquid-Liquid Extraction: Metals", Chem. Engr., Aug 30, 86(1976)

THE INDUSTRIAL WASTE COOPERATIVE CONCEPT - POTENTIAL FOR THE METROPOLITAN SEWERAGE DISTRICT, BOSTON, MA.

<u>Michael K. Prescott</u>. Black & Veatch Consulting Engineers, Bethesda, MD.

<u>Noel D. Baratta</u>. Assistant Director, Metropolitan District Commission, Boston, MA.

<u>Wayne T. Grandin</u>. Associate Sanitary Engineer, Metropolitan District Commission, Boston, MA.

INTRODUCTION

In this paper we will present the theory behind the industrial waste cooperative, outline the different types and requirements of a cooperative and then identify some potential cases in the Metropolitan Sewerage District in Boston where a cooperative could work. This particular exercise, although specific to the Boston MSD can be repeated for any other metropolitan area where treatment and disposal of industrial wastes are a problem. The industrial waste cooperative is a relatively new concept with only a few functioning in the United States and other parts of the world. Its need has only come about recently because of new regulations governing the discharge and disposal of industrial wastes and the high costs of compliance with these regulations.

THE INDUSTRIAL WASTE COOPERATIVE CONCEPT

A central cooperative facility for treating industrial wastes from several plants could provide numerous benefits under the right conditions. To be successful, a cooperative treatment facility would need to meet the following requirements:

    o   Close proximity of cooperating industries. Wastes could be piped to the facility or reduced in volume and transported to the facility by truck if the industry was not in the immediate area.

o   An equitable cost recovery system. To pay for capital and operating costs of the treatment facility, a system of user charges would be needed based on the costs incurred in treating the waste. Charges would be assessed to each industry according to volume and strength of their wastes.

o   Compatibility of industrial wastes to be treated. Waste streams are considered compatible under one of two conditions: (1) The different wastes can be <u>treated together</u> using the same process or (2) one waste stream can be <u>used to treat</u> other waste streams. Two examples will illustrate these points: (1) food processing wastes from a cannery and a poultry slaughterhouse, can be <u>treated together</u> using the same biological system; (2) alkaline wastes from scouring operations in textile mills can be <u>used to treat</u> acid wastes from an electroplating operation.

Some of the obvious benefits of a central cooperative facility are:

o   Economics of scale due to one large operation as opposed to several small ones. Equipment and chemical costs, operating and laboratory personnel requirements may be minimized.

o   Combined financial resources to fund the facility.

o   Only one discharge point requiring analysis, permitting, and regulation.

However, it should be noted that there may be other benefits unique to each industry involved.

Management and ownership of such a facility could take a variety of forms. An independent firm, a group of industries, a public agency, or any combination of these could sponsor the project and operate the systems.

APPLICATION OF INDUSTRIAL WASTE COOPERATIVES IN THE METROPOLITAN SEWERAGE DISTRICT

To study the potential for cooperative industrial waste treatment*, one must first look at the industrial population. The industries that discharge large amounts of wastewater requiring pretreatment are scattered throughout the Metropolitan Sewerage District. Even in any one industrial area, the number of industries requiring pretreatment may be far

---

* Actually the correct term would be pretreatment with regard to discharge to the Metropolitan Sewerage System. We shall use this terminology for the rest of the paper.

outnumbered by industries not requiring pretreatment. This makes a cooperative industrial waste treatment facility technically and financially unfeasible in many instances. Remember, what is required for cooperative treatment as defined previously is close proximity of industries discharging compatible wastes. It is also important that the quantity of wastes be large enough to economically pipe or truck to a central facility for treatment.

There are three possible situations which one must look at to determine the potential for an industrial wastewater treatment cooperative.

1. Some of the industrial plants are located close enough to pipe their waste to a central facility while others truck their wastes to the facility.
2. The industrial plants are located close enough to pipe their wastes to a central facility.
3. The industrial plants are located close enough to truck their wastes to a central facility but not close enough to pipe their wastes.

Currently there are several industrial areas located in cities in the Metropolitan Sewerage District where cooperative waste treatement could work without involving relocation of cooperating industries. A description of three particular cases will illustrate the three situations listed above.

Case 1:

An industrial park in one suburban town has several captive and job shop electroplaters, circuit board manufacturers and metal finishers within approximately 1000 feet of each other. These firms at the present time are not pretreating their wastes before discharge to the Metropolitian Sewerage System. Additionally, in the proximity of the industrial park are chemical and electronics manufacturers generating concentrated spent acid, caustic, and other chemical wastes. One treatment plant could service all these industries. The plating and metal finishing companies could pipe their wastes to a central treatment facility employing the best technology available. Consideration should be given to segregating wastes according to the type of treatment required. All cyanide wastes could be piped separately to cyanide destruct units, chromium wastes piped to reduction units and all other wastes piped directly to equalization, neutralization, and precipitation units. Equalization of all three waste streams, cyanide and chromium wastes after oxidation and reduction, and all other wastes before neutralization, could be accomplished in a special equalization tank or within the neutralization unit. Equalization would help smooth out flow and pH fluctuations and would also serve as reserve capacity during treatment plant upsets and servicing. Concentrated acid and caustic wastes from the chemical and electronics manufacturers could be trucked to the treatment plant and used for pH adjustment in the cyanide and chromium treatment systems and for neutralization of the equalized waste streams. This would cut the costs of virgin material required for pH adjustment and also bring in revenue from the other firms as compensation for disposal of their wastes. Of course, analysis of both the acid and caustic wastes would be required prior to utilization to determine compatibility with the treatment processes and to detect the presence of any incompatible pollutants.

Case 2:

A second group pretreatment site is in Boston and involves a job plating shop and two captive plating shops. The job shop is across the street from one of the captive shops and within a block from the other one. Perhaps this group could pipe their wastes to the centrally located job plating shop for combined treatment, using methods similar to those mentioned above in Case 1. Alternately, this group could try other, more advanced treatment methods aimed at recovery of process materials.

Case 3:

The last case involves the printing and photofinishing industries concentrated in Boston as well as in other parts of the Metropolitan Sewerage District. These establishments generate waste photochemicals, but more specifically waste fixer containing high concentrations of silver. The majority of these firms have silver recovery units to extract silver from the fixer before discharge. However, there are still quite a few of these firms, especially the smaller ones, that discharge raw waste fixer to the sewer, dispose of it in the trash, or have it picked up by an independent reclaimer. For these firms, cooperative treatment at a central location might be a worthwhile option. A group of these industries could cooperatively purchase a silver recovery unit and centrally locate it within one of the cooperating firms. The company housing the unit would, of course, be reimbursed for the space by the cooperative.

One specific case in Boston concerns three firms, all in the same building, generating and discharging waste fixer without treatment. For this group, it would be a simple procedure to purchase cooperatively a silver recovery unit to process the three firms' waste fixer.

There are other waste photodeveloping chemicals that may require treatment, particularly the bleaches used in color photofinishing. Some bleaches contain cyanide complexes which may cause problems when discharged to the sewerage system. Some of the larger photofinishers regenerate their bleach in a special expensive process. The smaller color photofinishers can not afford these systems so they discharge the spent bleach solutions. But cooperatively, a group of them could afford a complete bleach regeneration system and probably recover the capital investment in a short time.

The above examples are based on applying the cooperative concept to functioning industrial facilities. If we assume that participating industries are willing to relocate to new facilities, then the possibilities for an industrial waste cooperative will be greatly expanded.

Industries discharging similar wastes, such as food processors and meat packers, or electroplaters and metal finishers, could move into one large complex designed specifically for their individual and group needs. Separate manufacturing areas would be offered, with wastewater connections to the central treatment plant. For example: A group of four or more (the number of participants is limited only by contract negotiating restraints) meat packers could build an industrial complex and treatment plant suited to their processes. Each firm would segregate its wastes according to the pollutants it contains and the type of treatment required, and discharge it to the appropriate line leading to the plant. At the treatment plant, the different waste streams would be equalized separately and treated accordingly.

A second type of industrial waste pretreatment cooperative would treat compatible wastes from different industries. The complex would be designed to accommodate various participating industries according to their manufacturing and wastewater treatment needs. Depending on the types of wastewater generated, there could be several segregated treatment systems for different waste streams; perhaps with a final polishing lagoon before discharge. The central treatment facilities could incorporate neutralization for highly acidic or basic wastewater, precipitation for metals and suspended solids, biological treatment for biodegradeable wastewater, flotation for oil and grease, and any other unit operations and processes needed to meet effluent standards. Because there is only one discharge point, analyzing, monitoring, permitting, and billing procedures are simplified for both the industry and the MDC.

A central facility for treating concentrated wastes brought in by truck or railcar from industrial plants in the general area is a necessary alternative for industries which have well established facilities and don't want to relocate. Under this plan, compatible wastes that are concentrated or can be economically reduced are transported to a centrally located facility for treatment. This facility could be located within an existing cooperative member's plant, or at a new complex built to serve the cooperative's needs. The latter involving a separate new complex case could also serve as a process and treatment facility for some industries willing to relocate. This combined central treatment facility for both piped and trucked in wastes would appear to be the most practical and economical plan. It offers participating industries the option of staying at their present location or relocating to the central facility.

Developing The Industrial Waste Cooperative

The first major steps for developing an industrial waste cooperative are to inform the industrial community of the plan and to obtain financial backing for a feasibility study. The results of the study could then be presented to prospective participants. The initiating group could be a national or regional industrial association, government agency, Chamber of Commerce, independent developer, or group of industries. Then, once the study is completed, and provided the concept is proven viable, the financing negotiations and actual design of the facility can be undertaken. There are various methods of financing and managing a treatment facility and recovering operating costs, but an in-depth discussion is beyond the scope of this paper.

Federal, State or Local government agencies could help provide a major impetus for development of industrial waste cooperatives. Government monies funneled to an agency or consultant could lay the foundation by determining the feasibility of such a concept in industrial areas and in certain industrial categories. Special low interest loans could be made available for studying, designing, and building the actual facility. Tax incentives for industries participating in a cooperative would provide further stimulus.

One possible problem to be dealt with is trying to get competing industries to work together. The philosophy does not at first sound very promising, but there are proven instances where industries have joined efforts. (Two cases that immediately come to mind are the chemical manufacturing and metal finishing industrial associations that have confronted the EPA on effluent discharge standards that they felt were unfair). In Boston, for example, there are several cooperative industrial complexes owned by an association of competing industries. These include the Boston Flower Exchange (coop of flower distributors), New England Seafood Center (fish processors and distributors), New England Produce Center (produce distributors) and the New Boston Food Market (meat and poultry processors and distributors). The latter, the New Boston Food Market, is a functioning form of an industrial pretreatment cooperative. In addition to providing space and utilities to member companies, the complex also provides and maintains grease traps for removing animal fats and greases from process discharges.

Based on the experience of these functioning cooperative and the need for pretreatment by industry, why couldn't there be a New England Plating Center, or Metropolitan Paint Manufacturing Cooperative, or Bay Area Chemical Manufacturers Central Treatment Plant? Certainly the idea has merit and should be studied further for possible implementation in the future.

Summary of Benefits

The benefits derived from institution of industrial waste cooperatives are numerous and affect all parties involved. The regulatory agencies would have only one treatment complex with one discharge point to regulate and surcharge instead of several separate locations. This reduces monitoring and enforcement efforts. The participating industries are able to comply with environmental regulations at reduced capital and operating costs. The business community word enjoy a lower attrition rate and therefore a lower unemployment rate because the economic burden of treatment is reduced. And last but not least, the public would enjoy a cleaner environment.

References

1. Monaghan, Charles A., Simmons, Don, Bayport Centralizes Treatment of Industrial Wastes, Civil Engineering, June, 1973.

2. Gorden, Marsha, et al., Regional Materials Management as an Alternative to In-Plant Treatment of Industrial Residuals, National Conference on Management and Disposal of Residues from the Treatment of Industrial Wastewaters, Information Transfer Inc., 1975.

3. Coulter, Kenneth R., Centralized Treatment of Metal Finishing Wastes, Plating and Surface Finishing, November 1978.

PROCEDURES FOR PROCESSING ATYPICAL INDUSTRIAL WASTES AT
A PUBLICLY OWNED TREATMENT WORKS

Robert T. Mohr. Bureau of Water and Waste Water,
Department of Public Works, City of Baltimore, Maryland.

Paul H. Keenan. Bureau of Water and Waste Water,
Department of Public Works, City of Baltimore, Maryland.

INTRODUCTION

　　The City of Baltimore, Maryland owns and operates two
waste water treatment facilities. The Back River Waste Water
Treatment Plant is a 185 MGD secondary facility employing
activated sludge and standard rate trickling filters. The
plant was originally constructed in 1911 and serves a collection system covering an area of approximately 140 square miles
with an estimated population of 1,385,000. The Patapsco Waste
Water Treatment Plant is in the midst of an expansion program
which will increase the capacity to 70 MGD with treatment by
the oxygen activated sludge process. Together, both plants
serve a population of 1,619,300 contributing to a collection
system covering 178.5 square miles in Baltimore City, as well
as parts of Baltimore, northern Anne Arundel and Howard
Counties.
　　The City of Baltimore has promulgated sewer use regulations under City Ordinance #914 which governs the conditions
by which industries can dispose of industrial wastes to the
municipal sanitary sewerage system. There are those materials
that are prohibited from being discharged to the sanitary
sewerage system as identified in the first section of the
Ordinance, and those substances which shall be regulated and
controlled to specific limitations that are either established
or will be established as contained in the second section of
the Ordinance. The remaining sections of this Ordinance deal
with the surcharge imposed on industries for excessive BOD
and suspended solids, and the penalties for non-compliance
with any of the aforesaid sections.

There are certain conditions, however, which arise by which industries and other users of the system cannot directly discharge the wastes that they have generated due to unusual or unforeseen circumstances. In these instances the City is normally contacted for guidance as to the proper means of disposal. The City has developed procedures by which industries can dispose of these non-routine wastes in an environmentally safe manner. Although there is no legal requirement to provide this type of service to the industries within the service area, the City feels it has a moral obligation to provide a means of protecting the environment and at the same time stem the occurrence of illegal dumping to the municipal sanitary sewerage collection system or to the municipal storm drain system.

These procedures were established for a cooperative program geared towards assisting the industries in the disposal of wastes that they were generating that were not typical of the normal everyday discharges to the collection system.

Procedures

The procedures that were developed, first include notification from the industry of the disposal problem and their request for assistance in solving this problem. All requests for assistance are first referred to the Pollution Control Section of the Waste Water Engineering Division within the Bureau of Water and Waste Water, Department of Public Works. The first thing that is done once a call for assistance is received is to define the nature of the problem. This definition includes the source and quantity of the waste to be treated, a description of the quality and nature of the waste to be disposed, and the urgency of the solution to the problem depending on the industry's ability to store this waste.

After the initial contact has been made by the industry, a Pollution Control Analyst from the City is dispatched to the site in order to obtain necessary information that may not have been transmitted during the original discussion with the industry. The Pollution Control Analyst investigates the site and views the waste to be disposed. He then collects samples which represent the waste to be discharged so that they may be analyzed. Samples are analyzed either at the Laboratory facilities at the treatment plant or at a local private laboratory. The Analyst explains the costs and the procedures for handling this type of waste to the management of the industry requesting assistance. The Analyst then prepares a preliminary report of the problem and his investigation and reports this information to his superiors.

Once the laboratory results are completed they are compared with the sewer use regulations and limitations that have been established (as shown in Table I) in order to reveal whether the industry can discharge without any additional treatment or determine whether additional pretreatment would be required.

Table I. Summary of Allowable Limits[1]

Parameter	Limitation
pH	6.0 - 10.0
Fats, Oils, and Grease	100 mg/l
Hexavalent Chromium	5.0 mg/l
Cadmium	0.2 mg/l
Copper	3.0 mg/l
Nickel	3.0 mg/l
Lead	1.0 mg/l
Zinc	5.0 mg/l
Mercury	0.01 mg/l
Cyanide	0.2 mg/l
Sulfide	10.0 mg/l
Phenol	5.0 mg/l
Temperature	150°F

[1]These limits will be revised to reflect actual treatment plant tolerances. The above limits were derived from limits contained in ordinances from Municipalities comparable to Baltimore.

In addition a consultation with the operating management of the treatment plant is then held in order to assure that the normal functioning of the treatment plant processes are not in any way inhibited or upset due to the acceptance of the waste in question. Based on this consultation it is determined whether the waste can be discharged or whether another means of disposal must be sought.

In the case where it is acceptable for the industry to discharge this waste, the industry is notified of the date, time, and flow rate that the discharge can be made. The discharge is monitored during the entire procedure and a composite sample is collected by City Forces for the purposes of charging the industry for treatment services. Table II indicates the normal parameters used to characterize the waste.

In the case where the industry has a waste which is not amenable to treatment at one of the Waste Water Treatment Plants the industry is notified that it is not feasible to discharge to the POTW.

Table II. Analytical Parameters for Atypical Waste

Parameters	Reason
BOD	Determine strength and treatment cost
TSS	Determine strength and treatment cost
COD	Check on strength (BOD) & Toxicity
pH	Conformance with limitations
F.O.G.	Conformance with limitations
Metals	Conformance with limitations
Organics	Pretreatment data base, compatability
Toxicity	Compatability with treatment system

The industry is then assisted in finding an alternative treatment facility in which the waste would be acceptable in accordance with all Federal, State, and Local regulations. The other regulatory agencies within the State of Maryland are notified as to the status of this waste in order that it may be traced to a proper disposal site. A simplified flow chart of these procedures is shown in Figure 1.

PROCEDURE FLOW CHART

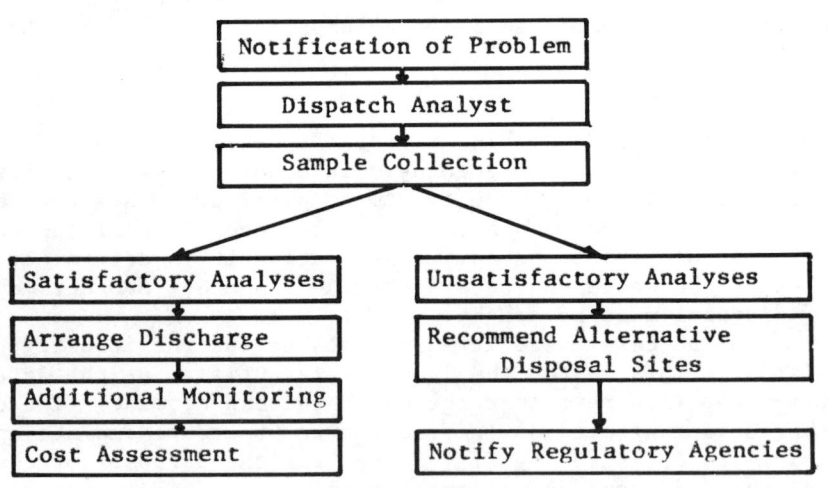

Figure 1. A simplified procedure flow chart indicating steps for processing atypical wastes.

CASE HISTORIES

I. BBL Microbiology Systems

The City was notified by authorities of the Baltimore County Department of Public Works that a county industry within the service area of the Back River Waste Water Treatment Plant had requested permission to discharge a pharmaceutical product that had exceeded its expected shelf life.
The company stated that they desired to dispose of 605 cartons containing ten packages per carton. Each package contained five envelopes with 6.2 grams of barbital buffer in each envelope. This amounted to a total of approximately 188 kilograms of barbital buffer.
Samples of this material were analyzed for toxicity using the Beckman Microtox Model 2055, and permission was granted to discharge 55 cartons per 24 hour period based on results of these analyses. All 605 cartons were discharged without any detrimental effect to the treatment processes.
The company approached Baltimore County personnel a short time later with regard to another material that must be disposed of from time to time. This material was excess culture media consisting of an agar base with blood or sugar nutrient additives.
The main problem with disposal to the sanitary sewerage system was the gelatinous nature of the agar. The company developed a procedure by which the waste culture media would be mixed with chlorox or industrial bleach and could be flushed to the sanitary system with a small volume of water. The company found that approximately 32 pints of bleach was needed for every 30-50 gallons of media to be discharged.
The above method of discharge was approved and the company has discharged this material without causing adverse effects to either the conveyance or treatment processes.

II. Chem Clear of Baltimore, Inc.

The City of Baltimore was contacted by Chem Clear, Inc. The company, which treats industrial wastes to acceptable discharge concentrations, requested permission to build and operate a treatment facility in Baltimore City. The company would be involved mainly in the treatment of waste streams containing heavy metals in concentrations far in excess of the limitations discussed previously.
Chem Clear proposed that they would process these wastes, remove the sludge, and discharge the supernatant to the municipal sanitary sewerage system. The waste water was to be brought in by tank truck, treated by means of pH adjustment and polymer addition, and stored in a holding tank prior to discharge.

The City reviewed this proposal and felt that a formal contract with Chem Clear was necessary due to the toxic nature of the waste streams that would be treated by their facility. This document contained all restrictions, limitations, and procedures that Chem Clear must adhere to in order to be granted permission to discharge to the sanitary sewerage system. After some modifications to this agreement, the document was signed and construction was begun to modify an existing tank farm as the site of the facility.

All wastes treated at Chem Clear are stored in holding tanks which are tested for the standard parameters, heavy metals, and toxicity. All of the initial discharges were analyzed additionally for the priority organic pollutants, using a Finnigan GC/MS/DS.

All discharges are monitored and controlled by the Pollution Control Section utilizing automatic sampling devices and a locking valve system on the holding tanks.

III. A Vinegar Manufacturing Plant

The City was notified in writing by the plant manager of a vinegar manufacturing plant that their company was in possession of 24,000 gallons of vinegar product having 3.4% acidity and denatured alcohol, that was approximately 74 proof. The company reported that the alcohol in this mixture had been contaminated with coal tar oil and a surfactant, and therefore the company could not use it in the manufacture of vinegar. They, therefore, requested permission to allow disposal of this material after neutralization to the municipal sanitary sewerage system. Following notification from the company, a Pollution Control Analyst was assigned to collect samples of this alcohol to determine the BOD and suspended solids' loadings. Following the analytical workup of this waste the company was notified of the charges that would be assessed by the City of Baltimore for discharging this waste to the sanitary sewerage system. This figure was given to the Plant Manager prior to discharge of the waste. The company was informed as to the manner by which they should introduce this waste into the sanitary system in order to avoid possible toxic effects of the alcohol and/or the possibility of a fire hazard. The company discharged this material in accordance with the regulations and paid the treatment costs for said discharge in a timely fashion.

IV. Allegheny Pepsi-Cola Bottling Company

The Allegheny Pepsi-Cola Bottling Company notified the Pollution Control Section that due to flood conditions during Hurricane David, approximately 27,000 gallons of pepsi syrup and 4,000 gallons of liquid sugar had been contaminated by flood waters. It was the intention of the company to discharge this contaminated material to the municipal sanitary sewerage system under the guidance of the Pollution Control Section. Samples were taken of the Pepsi syrup and liquid sugar for analysis. The company was notified of the surcharge costs for the excess BOD and suspended solids that would be created by this highly biodegradeable material.

Samples of the tanks were analyzed for the standard parameters and results of this analysis was reported to both plant management and corporate management of the Pepsi Cola Bottling Company. The liquid sugar had a BOD concentration of approximately 640,000 parts per million and a suspended solids concentration of approximately 240,000 parts per million. The Pepsi syrup had a BOD concentration of approximately 130,000 parts per million and a suspended solids concentration of approximately 14,000 parts per million.

The company was told at what rate to discharge this material and did so in accordance with the procedures provided by the City. Appropriate charges for treatment were assessed and paid by the industry.

CONCLUSION

The purpose of this paper was to document the role of the Publicly Owned Treatment Works in dealing with unusual waste disposal problems that occur within its service area. The goal of the City's program is to cooperate with industry to solve these problems in an environmentally sound manner. Cooperation is the key to the success of the program and a timely solution is the goal. However, at no time is the procedure short circuited simply to benefit an industry or solve its problem. Protection of the operation of the treatment plant must be safe guarded at all times. In the absence of a coordinated service oriented program, illegal or uncontrolled dumping of toxic or hazardous materials into the collection system would probably occur. While the procedures outlined herein have not entirely eliminated this situation, certainly they have been of benefit in reducing the problem of illegal, uncontrolled discharges.

DEVELOPING PARKS, GOLF COURSES AND OTHER RECREATION AREAS
ON COMPLETED LANDFILLS

Edward F. Gilman. Department of Plant Pathology, Cook
College, Rutgers University, New Brunswick, New Jersey.

Franklin B. Flower. Department of Environmental Science,
Cook College, Rutgers University, New Brunswick, New Jersey.

Ida A. Leone. Department of Plant Pathology, Cook College,
Rutgers University, New Brunswick, New Jersey.

INTRODUCTION

As urban population continues to grow and landfills are
closed, we can anticipate greater stimuli for converting former landfill sites into recreational areas. Communities may be
persuaded to turn these former unused wastelands into parks,
golf courses and nature centers. The advantages of converting former landfill sites into attractive recreational areas
are (a) inexpensive land acquisition, (b) proximity to urban
centers, (c) landfills can accommodate rolling landscapes
which are frequently desirable for parks and golf courses,
(d) value of land will increase after reclamation and (e)
could provide a local recreational hunting ground.

The list of disadvantages for converting former landfills
into parks is long. Frequently (a) soil cover is thin, (b)
cover soil is compacted by the soil spreading equipment causing severe restriction of root growth, (c) soil nutrients and
soil moisture levels are quite low, (d) irrigation requirements are greater on landfill areas than on nonlandfill areas,
(e) surface irrigation is difficult because of uneven refuse
settlement, (f) settlement causes breakage of permanent irrigation and other underground service lines, (g) settlement
will cause undulating fairways and greens on golf courses,
(h) landfill gases may migrate into soil cover and kill plants,
(i) soil oxygen may be forced out of the soil atmosphere by

landfill gases and asphyxiate roots, (j) several heavy metals become more available to plants under anaerobic soil conditions and this can lead to poor plant growth and (k) trees (plants) tend to produce shallower roots (1).

Despite the apparent difficulties encountered in vegetating completed landfills, they have been overcome in a few instances within the United States. At these sites, those charged with designing and maintaining the vegetation project were aware of several of the obstacles. However, the majority of reclamation programs were designed by individuals who were unaware of the special problems associated with growing plants on completed dump sites.

This paper will provide guidelines for planting trees and shrubs on a completed landfill. Most recommendations are based on the results of a six-year research grant funded by the Municipal Environmental Research Laboratory of the USEPA. A more comprehensive detailed manuscript is being developed for publication by the Environmental Protection Agency and should be available from EPA in Cincinnati or NTIS by the spring of 1982.

GAS GENERATION

The gases of anaerobic decomposition (primarily methane-CH_4 and carbon dioxide-CO_2) may migrate from the refuse layers into the cover soil and/or into adjacent property (1, 2). Gas contamination of the cover soil will not be uniform over the entire landfill site. Some areas will contain relatively high carbon dioxide (>25%) and methane (>40%) concentration (consequently low oxygen (2%) content); whereas, other areas will be influenced very little by the underlying refuse material. These gases are best controlled by specific activities which should take place soon after the final layer of refuse is spread. It may be best to allow refuse to settle awhile before control measures are installed.

GAS BARRIERS

There is no doubt that combustible gases must be kept from the root system in order for vegetation to survive (2). The provisions which will ultimately lead to a desirable root gas environment are likely to function best and cost less if they are installed prior to spreading the final cover material. These precautions should consist of the selection of a barrier material which is both relatively gas-impermeable, flexible and durable (should last for decades and withstand settlement). This gas barrier material should be placed over the final layer of refuse. This should keep much of the gas away from the cover soil which will be placed over the barrier.

If the gas is prevented from traveling vertically into the cover soil, it is likely to move laterally into adjacent properties. In this case, gas migration control measures should be installed around the periphery of the refuse fill. There is a variety of systems currently available (3).

Impervious liners used to control gas flow include plastic, rubber or similar synthetic membrane, natural clay and asphalt. Polyvinylchloride (PVC) has been the most widely used synthetic membrane, because it can control gases, has a high resistance to deterioration and is relatively inexpensive. However, PVC will reportedly lose its plasticizer (and perhaps its effectiveness) after a few decades. Hypalon^R, chlorinated polyethylene (CPE) and other membranes are more attractive where the synthetic material must maintain its performance standard for fifty-plus years. Natural soil barriers, such as water saturated clay or bentinite lagoon liners, may be highly efficient if kept saturated, otherwise cracks develop in surface cover as the refuse settles. Dry clay or bentinite is not an effective barrier.

There are two approaches to installing barriers. One is to cover the entire landfill with the material and the other is to place the barrier only in areas where deeper rooted vegetation (trees and shrubs) will be planted. Also, settlement is likely to cause cracks or breaks in the seals used in either of these approaches allowing gas to penetrate the cover soil.

An alternate method of gas exclusion may consist of the mounding of soil in specific areas where deeper rooted vegetation is to be planted. The mound may be underlain by a clay layer or a membrane at least 20 mil thick. At least 90 cm of soil should be spread over the barrier in areas where trees will be planted.

LOCATING AREAS UNSUITED FOR VEGETATION

One or more years prior to the time when trees are to be planted, a ground cover should be seeded first to provide protection against erosion as well as to indicate where it would not be advisable to plant deep-rooted plants. Suggested annuals for the eastern United States include weeping lovegrass and annual rye grass. These should be seeded with perennials such as tall fescue, birdsfot trefoil or crown vetch. The ideal time for hydroseeding such disturbed areas is in the early spring or fall. In most cases, if the grass cover with its shallow roots dies or fails to germinate because of the influx of gases from the landfill, one can be certain that other deeper-rooted vegetation will not thrive either at these locations.

The easiest and quickest method of locating areas which are unsuited for plant growth is to observe existing vegetation patterns. The carbon dioxide (CO_2) and CH_4 concentrations are likely to be high and limiting to plant establishment in areas where the soil appears thin or the refuse is exposed. As a general rule, one should avoid planting trees and shrubs in areas which are devoid of ground cover.

Poor soil conditions for tree growth may be encountered even in places where grasses and other ground covers are thriving. Anaerobic soil or soil high in CO_2 or low in O_2 concentration may be present in the tree root zone which usually extends below the root zone of grasses. Table 1 points out several differences between aerobic soil which is likely to support tree growth and anaerobic soil. Even though an area has been chosen based on the above criteria for a planting location, differential settlement and varying gas production rates may cause gases to channel into a portion of the cover soil previously uncontaminated and kill existing plant material.

Table I. Guide for Evaluation of Landfill Soil Gas Problem

Characteristic	Anaerobic soil	Aerobic healthy soil
odor	septic	pleasant
color	darker	lighter
moisture content	higher	lower
friability	poor	good
temperature	higher	lower

Refuse and soil settlement, caused by loss of refuse material through decomposition, will create an undulating surface causing water to pond in low spots during rainy and/or irrigation periods. Vegetation in flooded areas may eventually die if water remains for extended periods; whereas, plants on the unsettled higher areas may suffer from drought. Excessively undulating greens will not be tolerated in most golf courses and therefore, provisions such as installing concrete slabs beneath each green should be made to compensate for settlement.

COVER SOIL

Depth

Numerous investigators (4, 5, 6) report that 60-80% of tree root-volume can be found in the top 20 cm of mineral soil. The remaining portion is located at varying depths (from 30-90 cm+) depending on species and soil characteristics. Although a 30 cm landfill cover soil may accommodate a good portion of the root volume, it would dry out quickly during seasonal dry spells. Because of the excessive cost of covering the entire landfill with deep rich soil, one should consider spreading 90 cm only in those areas where trees are to be planted. This need not be composed entirely of topsoil, however; topsoil should be incorporated into the top 20 cm since this is where most of the feeder roots will be growing. At least 60 cm should be spread in areas where trees will not be planted since this complies with Federal and State regulations.

Texture

The texture of cover material will necessarily vary depending on availability. Generally, if no gas barrier has been installed, it is best to first spread soil with a high clay content and cover with a highly organic soil. The deepest organic soil should be placed in areas where trees will be situated, such as a parking lot island, a clump planting on a golf course, or any other group planting. Where a membrane gas barrier has been installed, sand should first be spread on top of the barrier followed by a loamy textured soil capable of supporting plant growth. The sand will serve as a drain so that water will not accumulate above the membrane and roots growing in this region. The sand also asphyxiate serves as a protective buffer between the tires or tracks of soil spreading equipment and the membrane.

Spreading

One should strive for a soil bulk density of 1.2 to 1.4 g/cc in a loam soil. A higher density can generally be tolerated in a coarser soil. Cover soil is frequently deposited with an earthscraper. A layer of soil loosely spread from the scraper box may be compacted to half its original thickness by the passage over it of the rear wheels of the machine. Bulk densities in the order of 2.0 g/cm^3 in the wheelings are common compared with 1.2 g/cm^3 in the loosely spread soil (7) between the wheels. The end product of such spreading is a series of horizontal layers of more or less loose soil, with intervening compacted zones. These have been shown to restrict penetration by roots and the downward movement of water.

Avoiding or eliminating these compacted layers is vital to good soil reclamation. Soil should be spread only when it is dry, using alternative earth moving machinery to the normal earthscraper. Dragline excavators, bucket-wheel excavators, forward acting shovels and dumper trucks have been proposed but entail costs considerably higher than earthscrapers (8). Until the industry is prepared to accept these greater costs, earthscrapers will continue to be the most common tool used and means will have to be used for overcoming the compaction they create.

If several different soils are to be used in the final 60-90 cm of cover material, they should be mixed together and spread as a unit, not in separate layers. Spreading soil in a thick layer will promote less overall compaction than spreading in several thin layers. Water movement and root growth will also be less restricted in the former method.

If soil must be spread in the conventional manner and bulk densities are above 1.7-1.8 g/cm^3, consider the following procedures designed to promote better root growth and soil water flow thereby ultimately providing a more successful reclamation project.

Preparing Compacted Soil for Plant Growth

The destruction of the compacted layers by means of sub-soiling or deep-tine ripping after the full thickness of soil cover has been replaced is not usually completely effective because the available machinery cannot draw the tines deep enough or close enough together or the operation has been performed at too high a soil moisture (8). There has been some success with ripping each layer after it is spread and with the use of specially designed subsoilers. A vibrating subsoiler and the double-digger may be useful, but they are currently only being tested for their usefulness in aiding landfill reclamation projects (9).

Soil structure can also be improved by establishing a grass or ground cover (e.g. tall fescue, rye grass, crown vetch) for several years before planting trees and shrubs. The processes of freezing and thawing, rainfall, earthworm activity, soil and insect activities, percolation and leaching will also help increase soil porosity and promote more desirable soil physical properties. This process of reestablishing the network of pores and fissures by which the soil drains, may take many years. Thorud and Frissell (10) report that after artificially compacting a sandy loam from 1.14 to 1.45 g/cc bulk density in the 0- 7.5 cm depth, recovery to original bulk density occurred within 8.5 years but no recovery was recorded for deeper layers even after 8.5 years.

Organic amendments are beneficial to the physical, chemical and biological properties of most cover soil. Addition of these materials will reduce the soil strength and increase water infiltration and retention. Some organic materials provide an energy source and improve the environment for beneficial soil microorganisms and other fauna. These organic materials may include humus, peat moss, manure, crop residues, food wastes, logging wastes, industrial organics, leaf compost, composted sewage sludge or refuse compost. Gypsum, perlite or vermiculite may also be disced or chiseled into the existing soil but it is best if they are incorporated into as deep a soil layer as possible, preferably at the time the soil is originally spread on the landfill.

Fertilizing and Liming

Soil tests for pH and the major nutrients: nitrogen, potassium, and phosphorus; soil conductivity and, where possible, for the other macro and micro-nutrients should be made in a number of areas within the proposed tree planting site. Samples should be collected from a 0 to 20 cm deep soil column over a large area in a cross or zig-zag pattern. Heterogeneity in soil can be overcome by collecting several (5) subsamples over an area and pooling these to form a single composite sample. Replicate composite samples should be collected at each location. For routine soil nutrient-level surveys, any of a number of soil auger types may be used to obtain samples. The State University will help to interpret the results and make recommendations for the addition of fertilizer and/or lime. However, where trees and herbaceous species are seeded together, the rapid vigorous growth of the herbaceous vegetation in response to fertilizer may suppress or prevent establishment of seeded trees.

Heavy Metals

Soils which contain high concentrations of zinc, copper, manganese, iron, cadmium or lead should not be used for cover material unless this situation can be corrected by increasing the pH to a level not exceeding 7.0. There are several publications which discuss metal contents of soils in relation to plant growth (11, 12).

Soil Moisture

Soil moisture content of landfill cover soil will be considerably lower than the same soil on a nonlandfill area (13, 2). The factors contributing to this include thin cover soil, low organic matter content, lack of established capillary pore space, extreme micro-climates of most landfills, higher leaf transpiration rates than in nonlandfill areas, and less infiltration of rain and irrigation water due to the highly compacted nature of the soil and in some cases, sloping ground. Drought resistant tree and shrub species frequently seed themselves on completed landfills and can also draw more moisture from the soil at low moisture contents (e.g. 8% moisture content, dry wt. basis for a sandy loam) than drought sensitive species (14, 15).

Many different types of organic matter are listed in the compaction section of this manual which will increase water holding capacity of the soil. Mulching with wood chips, bark, saw dust, grass clippings, plant debris or plastic can help control evapotranspiration by reducing soil temperatures, weed growth and evaporation from the soil surface. More irrigation water will be required to maintain plantings on a landfill than on a nonlandfill area even if these policies are implemented. An above-ground irrigation system will require less maintenance than an in-the-ground system which will require continual maintenance because of breaks caused by settlement.

Erosion Control with Herbaceous Species

Mixtures of annual and perennial species are best suited for quickly stabilizing soil and protecting against erosion on disturbed land. The annual plants provide a quick temporary cover succeeded by a more permanent perennial species. Grasses and legumes should be selected on the basis of local climate and desired end use. Seeding rate recommendations should be followed carefully for the quick cover species because higher rates could produce dense stands that prevent or retard establishment of the permanent species. Contact your State University for specific ground cover species and planting technique recommendations in your area.

SELECTING TREE AND SHRUB MATERIAL

Slow-Growing vs. Rapid-Growing Species

There is evidence that slow growing trees are more tolerant to landfill conditions than rapidly growing species (6). Faster growing trees generally draw more moisture from the soil than slow growers and would, therefore, require more irrigation than the latter in order to maintain growth comparable to that on a nonlandfill area. However, if you are not concerned about how rapidly the trees will grow on the landfill compared to a nonlandfill area, then a faster growing tree may be more desirable, because it will generally provide a vegetative cover more quickly.

Small vs. Large Plant Material

Trees planted when small (1 m tall) showed significantly better growth on the landfill than those which were planted when larger than 2 m tall, regardless of species (6). Our data point out that this is related to the ability of a small tree to adapt its root system to the adverse environment in cover soil by producing roots closer to the surface (i.e. away from the higher landfill gas concentrations deeper in the soil). Roots of larger trees, on the other hand, start at a much deeper level and sometimes cannot develop adequate growth toward the surface before being killed by landfill gases. Our data (6) indicates that by the time the larger tree adjusts to the landfill by producing a shallow root system, the smaller specimens, which started with a shallow root system, can actually grow to the size of the larger trees and in some cases, surpass them. If trees taller than 1.5 m are necessary, plant them on a raised bed as to allow for shallow root development.

Volunteer Species

Although we have not specifically studied early successional volunteer tree species on landfills, they are generally very adaptable to poor soil conditions. Information on volunteer species can be found in several texts (16, 17).

Natural Rooting Depth

Tree and shrub species which enjoy a shallow root system were found in our studies to be significantly more adapted to landfill sites than species requiring a much deeper root system (6). The deeper roots are subjected to higher concentrations of landfill gases and lower concentrations of O_2. Some species can avoid this adverse gas environment by producing a shallow root system, otherspecies grow poorly because they are unable to produce shallow roots. Observations at the South Coast Botanical Garden in Palos Verdes, California, a former 87-acre landfill site, showed that shallow-rooted plants seldom were affected by landfill gases, but on some occasions there had been root damage to the larger deeper-rooted trees and shrubs (1). Several texts showing natural rooting depth of woody species are available (16, 17, 18).

The fact that trees growing on landfills generally develop shallower roots than the same species growing off the landfill emphasizes the need for more frequent irrigation of landfill soils planted with woody vegetation than comparable nonlandfill areas. In our studies at the Edgeboro Landfill in New Jersey, species excavated for extensive root studies on the landfill had a significantly larger portion of their root system in the top soil layers than in the deeper layers. However, roots were much more evenly distributed vertically on trees growing in the nearby nonlandfill control area. Deep-rooted vegetation can grow well on a landfill only if landfill gas does not migrate into the cover soil where trees are growing and provided adequate soil (90 cm or greater) is supplied to the area.

Flood Tolerance

Our field data indicates that the changes produced in landfill covers by gases are similar to those imposed by the flooding of soils; however, the high moisture content is lacking in the landfill cover soil. Therefore, species which are resistant to "wet feet" (flooding) conditions may do well on landfills only if they are supplied with adequate water. Dry site species should be planted if water will not be readily available on the site.

Size of Plants at Maturity

Another factor to consider when selecting species for landfills is size of the tree at maturity. If the cover soil is relatively shallow (30-60 cm), then choose a tree which remains relatively small at maturity or risk tree toppling during high velocity wind storms. If a deeper soil cover is used (>1 m) then the risk of windthrow will be diminished because the root system has adequate soil space to produce anchor roots, provided landfill gases are kept out of the cover soil and the trees are irrigated. There are a number of publications available which contain information on tree height at maturity (17, 18, 19). Local nurseries, landscape architects and the State University can provide information for your particular locality.

Mycorrhizal Fungi

Mycorrhizal fungi in association with plant roots have been shown to greatly increase water and nutrient uptake by the plants (20). This symbiotic association has been successfully used in coal strip-mine reclamation. Mycorrhizae may also aid in successfully establishing vegetation on completed dump sites since landfill cover soil frequently has a poor capacity for holding nutrients and water. Spore and mycelium inoculated soil has been tested for its ability to promote mycelium development on trees in landfill cover soil. Results indicate that both forms of inoculation may be viable alternatives to planting trees in uninoculated soil (21). The roots may be inoculated directly before planting, thus increasing the likelihood of successfully establishing the beneficial mycorrhizal relationship.

Pathological (Disease) Considerations

The selection of trees, shrubs or grasses should always be based on the ability of the species to withstand attack by damaging diseases or insects common to the given area. The State University can frequently provide valuable practical information concerning disease and insect-resistant plant material, optimum planting time, proper fertilization and other soil amendments critical to a disease control program.

PLANTING AND MAINTAINING VEGETATION

Trees, shrubs and grass survive best if planted in early spring or fall. Check with your Cooperative Extension Service County Agent for the planting time that is best for your area. Do not plant in the summer. Plants purchased from a nursery and delivered to the site should be planted as soon as possible. Bare-rooted material can dry out in a matter of hours if left in the sun. Balled and burlapped material can be left for longer but must be irrigated within a day or two, depending on the weather conditions.

A planting-hole about twice as wide as the root mass diameter and up to 15 cm deeper than the deepest root is well suited for trees and shrubs. Care should be taken to avoid compacting the sides of the planting hole, as this may promote prolific root growth inside the original hole but inhibit root penetration into the surrounding soil.

If you find that the soil in this hole is in the anaerobic condition (see Table I), then the plant should not be planted here. Another hole should be dug some distance away from the original location and this soil checked for anaerobiosis. Plant the tree in this area if the soil is aerobic.

Mix some of the original cover soil with some loamy textured material, preferably a highly organic soil, and spread enough of it in the bottom of the hole to make a 15 cm layer. A 50:50 mixture would be a desirable combination.

Hold the main stem of the tree or shrub and fill in around the root system until the hole is half-filled. Gently press the soil down with the sole of your shoe. Do not pack it down. During the backfilling process, be sure not to relocate the roots at a different depth from their original depth in the nursery. This can usually be determined by letting the roots hang freely before planting. Do not compact all the roots at the bottom of the hole. Spread them out as much as possible. At this point, the soil should be watered so that the entire root system has been moistened. This may take one to several gallons. When the water has all soaked into the soil, fill the rest of the hole with the soil mix and gently press the soil down with your foot. Form a circular depression around the stem with a diameter about equal to the extent of the root system. Fill this well with water until it is about to run over the ridge. These simple procedures, if carried out for all plant material, will retain all possible moisture from rainfall and irrigation and help trees survive through the most critical first season.

Irrigation is an extremely important requirement for establishing and maintaining healthy plant material on former sanitary landfills, particularly during the first 2-3 years after planting. After this time, roots may have established a large enough system to withstand moderate drought periods; however, irrigation should be practiced during extended hot, humid weather even for large established trees. This additional watering is recommended since roots on the landfill will be located close to the soil surface where there is little available soil moisture during extended dry periods. We do not recommend constructing an in-ground irrigation system since settlement is likely to cause frequent breaks in the pipes, requiring continual repair, thus increasing maintenance costs. Various above-ground expandable-joint systems are available.

Plants should be treated for disease and insect infestations and protected from damaging animals. Several animals enjoy chewing the bark of certain species during the winter. Check with the county agent and nurseryman to see if the species which were selected are susceptible to such attack. Some species are particularly susceptible to winter desiccation. Your County Agent can recommend procedures for overcoming this problem. There is at least one good text available with excellent coverage of tree maintenance (19).

GAS EXTRACTION

Extracting gas from within the refuse fill by an induced exhaust system, as is currently being practiced in several states and abroad, should aid in plant growth by reducing the quantity and pressure of the landfill gas generated in the refuse. Extraction should be compatible with establishing vegetation on completed landfills. However, commercial extraction operations are likely to end before the generation of gas has ceased, because extraction will become economically unfeasible after a number of years of operation. Even so, it may be desirable to continue running the extraction equipment, since there is likely to be enough biodegradable material left in the refuse to generate gases in concentrations harmful to plant growth. Undoubtedly the plants have established themselves in soil which most likely had a very low landfill gas content during operation of extraction equipment, and if the pumps are turned off, the gas content in the root zone environment may increase sufficiently to cause plant death.

REFERENCES

1. Flower, F.B., Leone, I.A., Gilman, E.F. and Arthur, J.J., "A Study of Vegetation Problems Associated With Refuse Landfills." EPA Publication 600/2-78-094, 130 pp., May 1978.
2. Leone, I.A., Flower, F.B., Gilman, E.F. and Arthur, J.J. "Adapting Woody Species and Planting Techniques to Landfill Conditions: Field and Laboratory Investigations." EPA Publication 600/2-79-128, 130 pp., 1979.
3. Emcon Associates, Methane Gas Generation and Recovery from Landfill. Ann Arbor Science Publishers, Inc., 139 pp., 1980.
4. Billings, D. "The Structure and Development of Old Field Short-leaf Pine Stands and Certain Physical Properties of the Soil." Ecological Monographs, 8:437-499, 1938.
5. Pritchett, W.L. Soil and Roots, Ch. 10, From: Properties and Management of Forest Soils, John Wiley and Sons, N.Y.. pp. 156-172, 1972.
6. Gilman, E.F. Determining the Adaptability of Woody Species Planting Techniques and the Critical Factors for Vegetating Completed Refuse Landfill Sites. Ph.D. thesis, Rutgers University, 1980.
7. Baver, L.D., Gardner, W.H. and Gardner, W.R. Soil Physics, John Wiley and Sons, Inc., New York, 1972.
8. McRae, S.G. "The Agricultural Restoration of Sand and Gravel Quarries in Great Britain." Reclamation Review, 2:133-141, 1979.
9. Berry, C.R. "Subsoiling Improves Growth of Pine on a Georgia Piedmont Site." USDA For. Serv. Res. Note SE-284, 1979.
10. Thorud, D.B. and Frissell, S.S., Jr. "Soil Rejuvenation Following Artificial Compaction in a Minnesota Oak Stand." Minnesota Forestry Research Note 208, 4 pp., 1969.

11. Chaney, R.L. Crop and Food Chain Effects of Toxic Elements in Sludges and Effluents. In Proc. on Recycling Municipal Sludges and Effluents on Land. Nat. Assoc. State Univ. and Land Grant Colleges. EPA and USDA Workshop. Champaign Ill., pp. 129-141, 1973.
12. Proceedings of the Symposium on Using Municipal and Agricultural Waste for the Production of Horticultural Crops. Hort. Science, 15(2):159-178, 1980.
13. Gilman, E.F., Leone, I.A. and Flower, F.B. "Adaptability of 19 Woody Species to a Sanitary Landfill." For. Sci. (in press, 1981).
14. Kozlowski, T.T. _Water Deficits and Plant Growth_. Vol. IV, Academic Press, N.Y., 1976.
15. Brady, N.C. _The Nature and Properties of Soils_. 8th Edition, MacMillan Publishing Co., Inc., N.Y., 1974.
16. Harlow, W.M. and Harrar, E.S. _Textbook of Dendrology_. 5th Edition, McGraw-Hill, Inc., N.Y., 1968.
17. Fowells, H.A. "Silvics of Forest Trees of the United States." USDA Agr. Handbook No. 271, 762 pp., 1965.
18. New Jersey Shade Tree Commissions. "Trees for New Jersey Streets." 2nd Edition, Cook College, New Brunswick, N.J. 1974.
19. Pirone, P.P. _Tree Maintenance_. 5th Edition, Oxford University Press, N.Y., 1978.
20. Marks, G.C. and Kozlowski, T.T., Eds. "Ectomycorrhizae: Their Ecology and Physiology." Acad. Press, N.J., 1973.
21. Sherwood, J.L. and Klarman, W.L. "IAA Involvement in Fungal Protection of Virginia Pine Seedlings Exposed to Methane." For. Sci., 26(1):172-176, 1980.

DEVELOPMENT OF A CHROME RESIDUAL CONTAINMENT
PROJECT AT HAWKINS POINT, BALTIMORE CITY, MARYLAND

Clinton R. Albrecht, Chief of Engineering Services
Maryland Environmental Service, Annapolis, MD

Baltimore, Maryland has been associated with the production of chrome since the mid eighteen hundreds. The Chrome processing facility has not only provided employment and tax receipts, but also has had a direct impact on the satellite chrome using and related service industries. The Baltimore chrome processing plant is owned by Allied Chemical and is located in the Inner Harbor northwest of where Route 895 enters Baltimore on the north side of the Patapsco River. See Figure 1.

A by-product of chromium processing is an odorless, noncombustible refuse. Allied's Baltimore plant produces approximately 93,000 cubic yards of this refuse per year. Refuse disposal is essential to the plant's operation. For years, Allied disposed of its chrome ore waste at an 85-acre land fill adjacent to the Maryland Port Administration's (MPA) Dundalk Marine Terminal near where Route 695 enters Baltimore on the north side of the Patapsco River. In 1967, Allied sold the parcel to MPA with the agreement that Allied would continue dumping up to 6 million cubic yards of refuse on the property. The agreement included financial penalties for Allied's failure to use the MPA landfill. MPA intended to use the filled land for expansion of the marine terminal.

The chrome tailings or refuse resembles a black sand that originally was desirable as a construction fill and was used in many of the harbor areas for construction improvements. The 1970's brought a change in thinking about the chrome tailings. Hexavalent chrome contained in the chrome tailings was identified as a suspected carcinogenic. Regulations now require that chrome tailings be contained

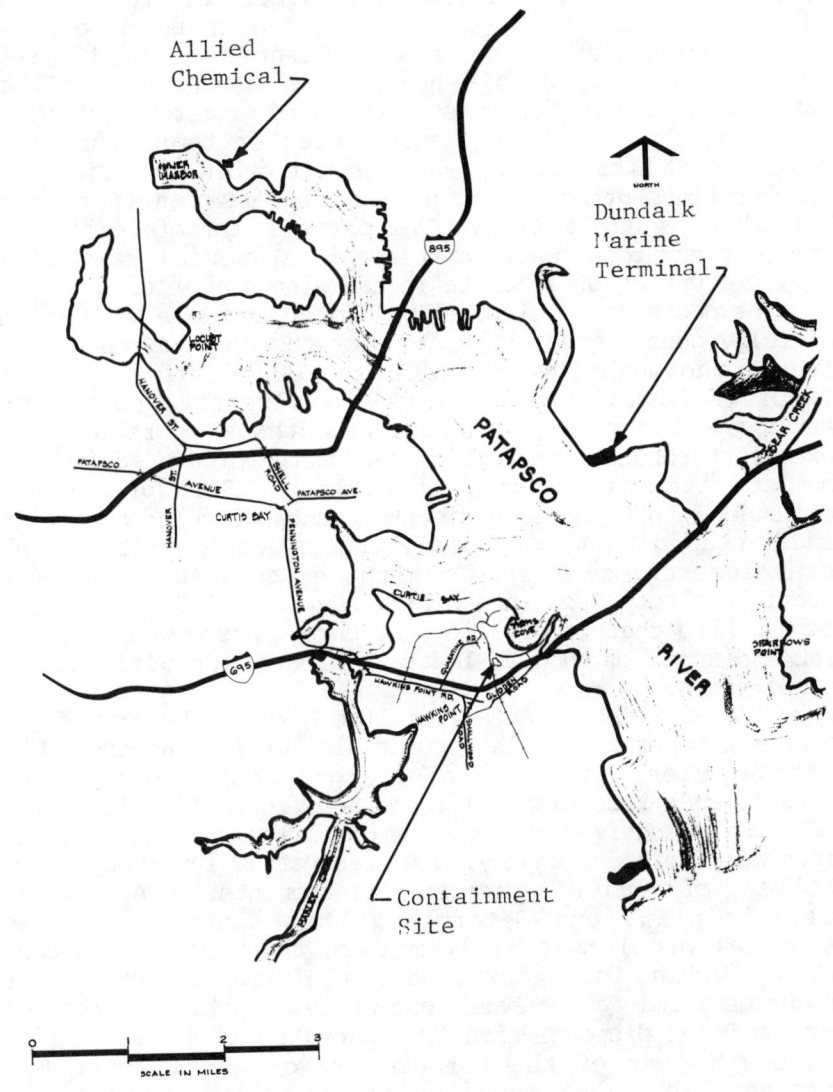

BALTIMORE MARYLAND HARBOR AREA

Figure 1

so as to control the leachate carrying water soluable hexavalent chrome.

The chrome ore waste has the consistency of a moist, heavy sand and contains about 20 percent water by volume. The waste material is principally a mixture of iron oxides, calcium, magnesium and aluminum. The waste material, however, also contains about 0.2 percent water soluble and 1.3 percent acid soluble hexavalent chromium. In the solid, damp state, the waste causes no problem. However, hexavalent chromium concentrations greater than 0.05 milligrams per litre render water unacceptable for drinking. Concentrations greater than 0.1 milligrams per litre are harmful to aquatic life. The chromium therefore is considered a hazardous waste that requires special handling, i.e., prevention of leachate in the disposal site.

The decision to dispose of the chrome ore wastes in carefully constructed cells of impervious clay was made only after a thorough investigation by Allied and the Water Resources Administration (WRA) of alternate means for handling and disposal. A number of methods to reduce the hexavalent chromium content of the waste through additional chemical treatment proved incomplete. Such processing resulted in new and significant amounts of chlorides which still needed land disposal. Ways of recycling the hexavalent chromium into alternate, commercially useful materials also proved infeasible. By June 1975, WRA agreed with Allied that landfilling the chrome waste was the most viable disposal method at least until another alternative became known.

WRA is a department within the Maryland Department of Natural Resources. WRA is responsible for issuing the NPDES surface water, and state discharge permits to all point sources. WRA recognized Allied's need for a landfill site. WRA also recognized a state responsibilty to assist Allied since another state agency, MPA, was partly involved in the problems of Allied's disposal arrangements. WRA reviewed Allied's plans and determined that although certain technical problems still remained, the proposed disposal site at Hawkins Point, appeared eligible for a groundwater discharge permit and a hazardous waste permit. The proposed Hawkins Point disposal site is located near Thoms Cove on the south side of the Patapsco river where Route 695 crosses. WRA also suggested that Allied request the assistance of the Maryland Environmental Service (MES).

The Maryland Environmental Service (MES) was established in 1970 as a non-profit corporate utility enterprise to operate all State owned institutional water and wastewater treatment facilities. MES also operates as a contract agency for local governments and industry in matters concerning waste management and water supply. It may plan, design, finance, construct and operate facilities such as wastewater treatment plants, water treatment and supply facilities and solid waste disposal systems, including resource recovery facilities.

MES is an agency of the Maryland Department of Natural Resources and its utility operations are self-supporting from fees and charges received from State institutions and its contract customers. It is not a regulatory or grant agency and it may enter into contractural arrangements with governments outside of the State. Concern for environmental quality in water, land and air is its central thrust. It seeks improved technology through experimental and innovative programs for the benefit of the citizens of Maryland.

The MES operates as a non-profit enterprise to assist local governments, industries and other state agencies in matters of waste management and water supply. Its purpose is to serve both public and private interests by supporting efforts to achieve environmental quality. MES must conform to all regulations and decisions of other State agencies.

Thus the State of Maryland became involved in the containment of chrome ore tailings as a result of obligations associated with property purchased from Allied Chemical for marine terminal expansion. The project not only involves the Maryland Port Administration as the owner of the property, Allied Chemical as the producers of the chrome tailings and operators of the containment site, the Water Resources Administration as the regulatory agency issuing licenses, the Department of Health and Mental Hygiene for protection of the waters, insurance requirements and bonding requirements including Federal restrictions. The Maryland Environmental Service became involved by invitation and later as the operator of the site for Allied Chemical.

The chrome refuse containment facility is located near Thoms Cove, Hawkins Point, an industrial area of the harbor. Immediately adjacent to the containment site are chemical companies and landfills. The containment site includes 10 acres for the containment of only Chrome Ore Tailings, a Class II Hazardous Material. Approximately 130,000 tons of refuse will be contained in the site. The tailings will be completely encapsulated in clay. A Schematic Cross Section of the site is shown in Figure 2. The cells are approximately 100 feet across and 900 feet long by 16 feet deep. The dark heavy border and the cell dividers indicate the clay liner compacted to over 90 per-

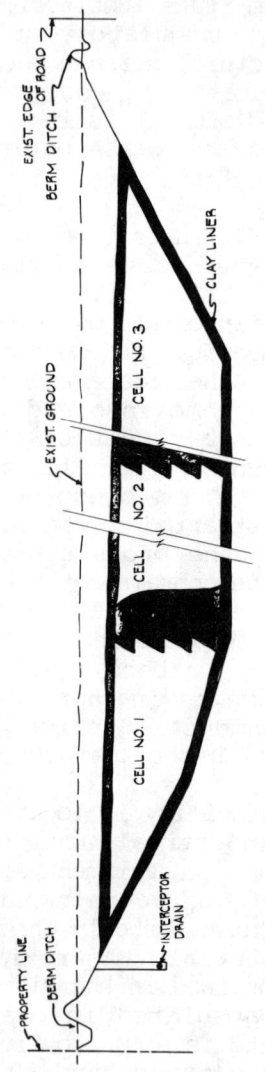

Figure 2

cent compaction. Of the 10 acres, approximately 7 acres have been excavated to an average depth of 16 feet. The containment area is divided into three cells separated by a clay divider. The chrome residual is encapsulated in a minimum two foot thick clay liner compacted to ninety percent of compaction. All surface water is diverted away from the containment area by ditches. Ground water flowing toward the project is intercepted and diverted by an interceptor drain located on the west and north sides of the area. Groundwater quality is monitored by means of five test wells. Any water coming in contact with the chrome tailings is considered contaminated and is caught and pumped to a lined pond for further transport offsite for chemical treatment. Regulatory requirements complicate the operations in that the tailings must be protected from water, i.e. rainfall. Trucks must not track any of the tailings off site.

All of the operation, monitoring and reporting are in accordance with the requirements of the Office of Environmental Programs, Waste Management and Enforcement Programs, Maryland Department of Health and Mental Hygiene Facility Permit, Type A, with special and general conditions. Surface water runoff flows toward a stream on the south side of the containment site and then eastward toward the Patapsco River. Two aquifers exist to the west of the site. An Interceptor Drain is installed on the north and west side of the site to intercept and drain off the water from the aquifers. Monitoring of the ground water surrounding the site is regular. Surface water within the cell being constructed is considered to be contaminated and is pumped to a containment pond for further transport to a treatment facility.

The acquisition of the facility permit required two public meetings with the local citizens groups and several informal meetings with their elected officers. The residents of Hawkins Point and the nearby communities are organized into improvement associations which are strongly oriented against any waste project to operate or to be planned for future operation in the area. However, as a result of these meetings, with continuous reevaluation of the project, and the development of compromises, the permit was issued without legal challenge. The compromises included changes in the haul routes, increased site security and continued contact with the local citizens. The result has been a growing respect between the citizens and the Service.

A number of technical concerns were raised by members of the attending audiences at the hearings. The primary concerns focused on risk of groundwater contamination, surface water runoff from precipitation, nuisances such as dust, noise control, traffic congestion, and safety factors ssociated with hauling seven days a week, and indefinite safeguards for any future site development.

The operating experience to date indicates no change in the groundwater quality, no discharge to surface waters and the project costs are within the estimates. This is based on a total of 50,939 tons placed in Cell No. 1 since the start of operations on August 26, 1980. During this interval 132,000 gallons of runoff from the construction site has been considered contaminated and hauled away for chemical treatment. Cell No. 2 was started on March 13, 1981 and as of the end of March over 4,000 tons of refuse have been placed.

The estimate of the project costs for a period of two years was as follows:

Equipment Rental	$429,840
Equipment Purchase	19,690
Operation & Maintenance	362,621
Other Expenses, Insurance	194,066
Startup, non-recoverable	37,425
Administration	10,436
Total, Estimated	$1,054,078
Cost per Month, Estimated	$ 43,919

The operational costs for the last six months has been $212,209 or an average of $35,368 per month. We still have some large future expenses such as the purchase of suitable clay material for the final cap which will raise the average cost of about $5.00 per ton for operations at the present time to about $8.00 per ton when the site is covered and closed out.

LEACHATE DETECTION IN A SOLE SOURCE AQUIFER:
A CASE STUDY

William F. Graner, P.E., Senior Associate, Charles R Velzy Associates, Inc., Babylon, New York.

Thomas Hroncich, Commissioner, Town of Islip, Department of Environmental Control, Islip, New York.

BACKGROUND

The problem of leachate contamination of groundwater is one that has received little attention until recent years. With increased environmental awareness and recognition of the leachate contamination potential, governmental agencies have instituted comprehensive regulations requiring development of control strategies and monitoring programs.
Deterioration of groundwater by leachate from sanitary landfills on Long Island, New York has become a complex water resource and water supply issue. This concern results from the fact that the vast groundwater reservoir of Long Island is the sole source of freshwater for nearly 2.8 million inhabitants.
The Blydenburgh Landfill, Town of Islip, New York, and the groundwater quality downgradient of the site is of particular interest and concern because of allegations concerning the disposal of hazardous and toxic materials, in addition to municipal solid waste at this location. Furthermore, the landfill is situated at the major groundwater divide and at the prime recharge zone of the groundwater reservoir.
This paper focuses on the hydrogeologic investigations and unique exploratory well drilling program undertaken at the Blydenburgh Landfill to detect and monitor the presence of leachate.

SOLE SOURCE AQUIFER DESIGNATION

After years of comprehensive water supply and wastewater management reports and investigations documenting deterioration of groundwater quality by domestic and industrial wastes, solid waste leachate and other hazardous and toxic materials, the Environmental Protection Agency has designated the two easternmost Counties of Long Island, Nassau and Suffolk, to be a sole source aquifer region.

STUDY LOCATION AND OBJECTIVES

Long Island, New York is a large detached segment of the Atlantic Coastal Plain, Figure 1. This Island is separated from the mainland by the Long Island Sound on the north and by the East River and New York Harbor on the west. The Atlantic Ocean borders the Island on the south and east.

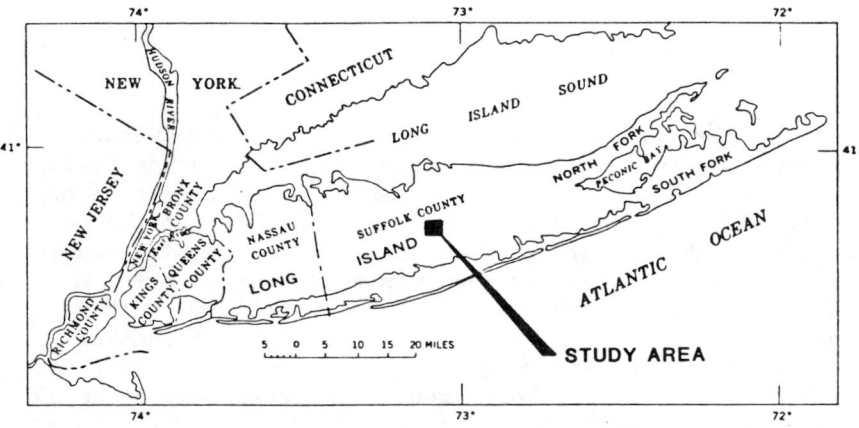

FIGURE 1. - LONG ISLAND AND STUDY AREA

The existing Blydenburgh Landfill is operated by the Town of Islip and situated in the central part of the Island. The landfill has been in operation since 1927 and currently covers an area of approximately sixty-seven (67) acres. This location is currently the primary solid waste land disposal site within the Township and serves a population of about 340,000 persons.

At the Blydenburgh Landfill the potential for leachage generation exists, and an assessment of the leachage problem undertaken to determine what control strategies are required to protect users of the groundwaters. Objectives of the investigations at this particular land disposal site included the following primary tasks:

- Preliminary assessment of hydrogeologic conditions
- Installation of groundwater table observation wells
- Establishing groundwater contours and flow directions in the vicinity of the landfill
- Site selection for installation of permanent monitoring wells
- Exploratory test well drilling program and geotechnical investigations
- Water quality appraisal and implementation of a long-term verification monitoring program

GROUNDWATER FLOW SYSTEMS

A southward-sloping wedge of unconsolidated sediment underlies the entire island and overlies crystalline bedrock. These sediments provide the extensive Long Island fresh groundwater reservoir. A regional groundwater flow system, Figure 2, recharged by precipitation, extends from the groundwater divide near the center of the Island outward toward saltwater, which surrounds the Island. Throughout

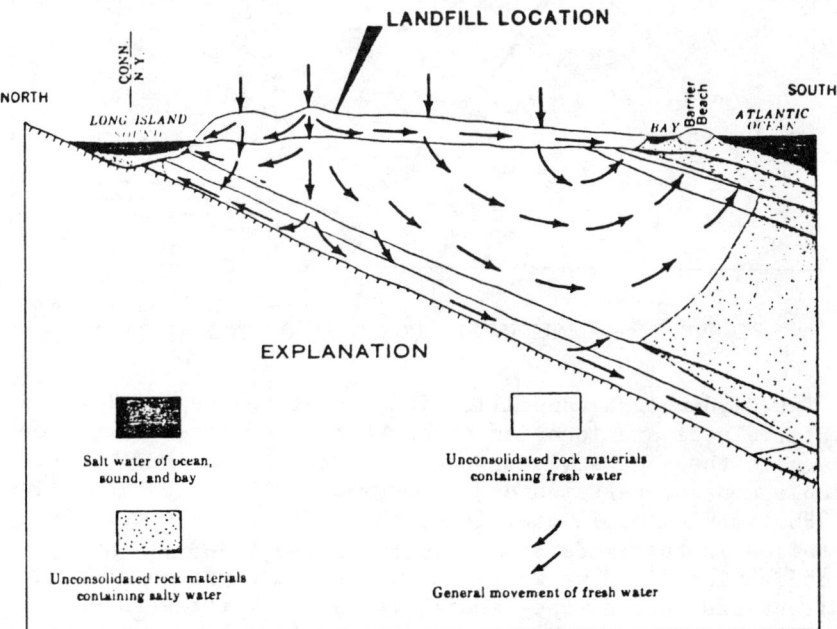

FIGURE 2. - DIAGRAMMATIC SECTION SHOWING GENERALIZED GROUNDWATER FLOW CONDITIONS

the Island, shallow groundwater discharges into nearby stream channels. South of the center of the Island, groundwater in the upper part of the system discharges directly into the Great South Bay, but that in the lower part of the system may discharge by upper leakage into the bays and ocean.

REGIONAL AND LOCALIZED WATER TABLE CONTOURS

For several decades the U. S. Geological Survey in a cooperative program with Suffolk and Nassau Counties has installed an extensive observation well network to detect changes in the water table contour patterns and water quality. A generalized regional water table contour map for Long Island is shown in Figure 3.

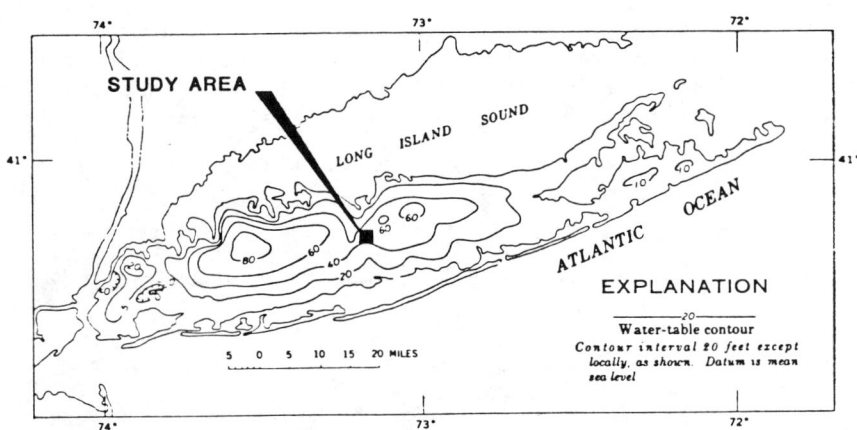

FIGURE 3. - GENERALIZED CONTOURS ON THE WATER TABLE

The Blydenburgh Landfill, situated at the regional groundwater divide and location of highest water level elevations is also the area subject to maximum change in local water table contour patterns and groundwater flow direction. In 1980, supplemental water level observation wells were installed in the immediate vicinity of the landfill to establish localized water table contours and groundwater flow directions through the landfill. Figure 4 reflects the 1980 localized groundwater contour patterns and location of water table observation wells.

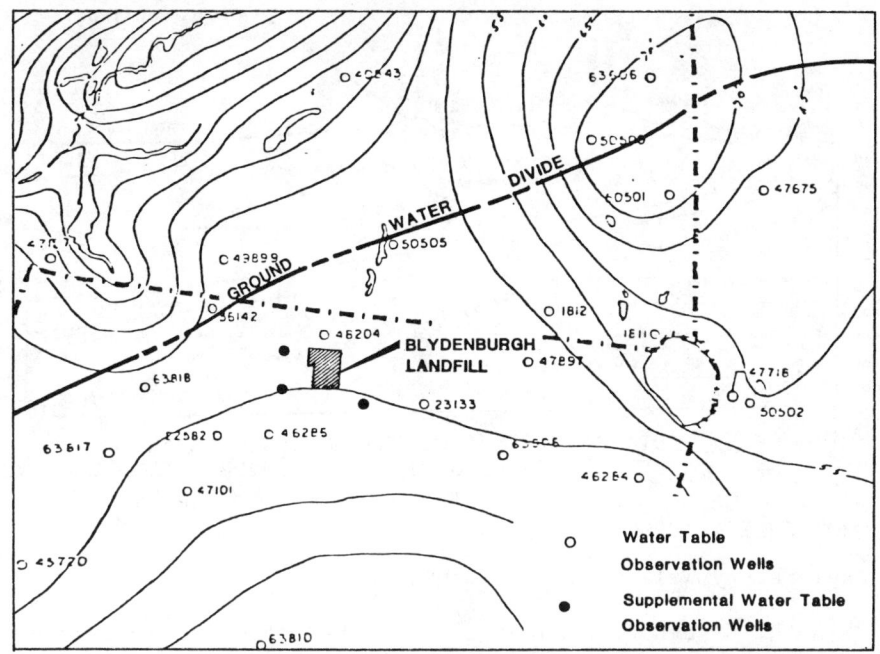

FIGURE 4. - LOCALIZED CONTOURS ON THE WATER TABLE (1980)

HYDROGEOLOGIC SETTING

The materials that constitute Long Island's groundwater reservoir include deposits of gravel, sand, silt, clay and mixtures thereof. These materials can be classified into several hydrogeologic units on the basis of hydraulic properties, relative position, composition, geologic age and other characteristics. A hydrogeologic section through the study location is shown in Figure 5. Pertinent characteristics of the major hydrogeologic units are listed in the table below.

HYDROGEOLOGIC UNIT	GEOLOGIC NAME	WATER-BEARING CHARACTER
Upper glacial aquifer	Upper Pleistocent deposits	Mainly sand and gravel of moderate to high permeability also includes clayey deposits of glacial fill of low permeability.
Gardiners Clay	Gardiners Clay	Clay, silty clay, and a little fine sand of low to very low permeability.
Magothy aquifer	Magothy formation	Coarse to fine sand of moderate permeability; locally contains gravel of high permeability, and abundant silt and clay of low to very low permeability.
Raritan Clay	Clay member of the Raritan Formation	Clay of very low permeability; some silt and fine sand of low permeability.
Lloyd aquifer	Lloyd Sand Member of the Raritan Formation	Sand and gravel of moderate permeability; some clayier material of low permeability.

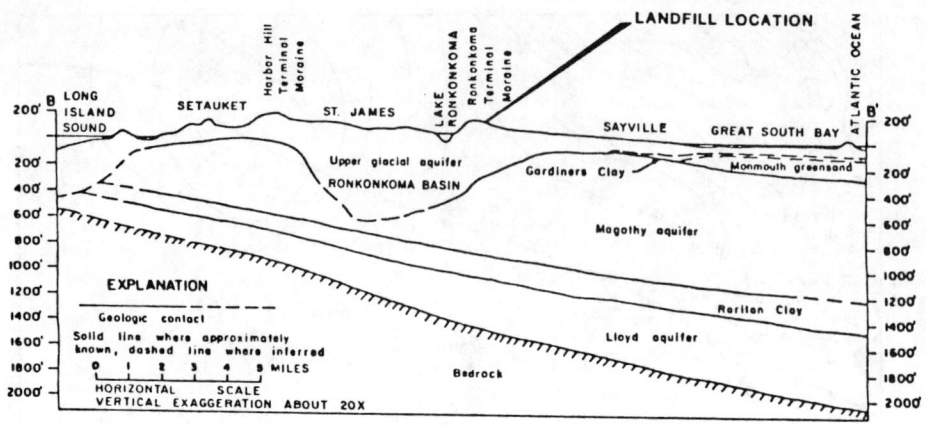

FIGURE 5. - DIAGRAMMATIC SECTION SHOWING GENERAL RELATIONSHIPS OF THE MAJOR HYDROGEOLOGIC UNITS

GEOTECHNICAL PROGRAM

Exploratory Well Drilling Techniques

In the Blydenburgh Landfill area the geologic and hydrogeologic character of the upper glacial formation deposits-Ronkonkoma Basin, has not been studied in detail. Evident in Figure 5, for this area, the upper glacial sedementary materials were deposited on a previously existing and irregular underlying geologic unit. These unique features and expected subsurface characteristics required a geotechnical and drilling program tailored specific to the project location.

Four (4) deep test holes were drilled to depths of between 450 and 610 feet below grade surfaces through the glacial deposits and into the Magothy Aquifer or substantial clay horizon. For each test hole subsurface cores were obtained every 20 feet and at detected changes in formation characteristics.

At the first drilling site a reverse rotary bore hole was installed to a depth of 610 feet below grade. The bore hole was geophysically logged and a 180 foot thick leachate plume detected (265 to 445 feet below grade). A cluster of wells was installed to serve as the monitoring and long-term verification network. Each well was gravel packed and pumped for formation water samples.

After installation of the system at Site No. 1, additional criteria was established which necessitated modification to the drilling sequence and techniques. These exploratory drilling and well installation procedures included:

- Drilling of a reverse rotary hole, Site No. 2, to sampling depth (every 50 feet beyond 200 feet below ground surface), jetting in of a 2 inch temporary well screen through the drill stem and pumping for formation water samples. This procedure was repeated at four (4) separate horizons to a final bore hole depth of 453 feet below grade surface followed by geophysical well logging. A clay formation was encountered between 296 and 353 feet below grade but no indication of leachate. This bore hole was filled and the site abandoned as a monitoring location.

- Drilling of a reverse rotary hole, Site No. 3, to sampling depth with setting and gravel packing a temporary well screen and pumping for formation water samples. This procedure was repeated at 6 horizons to a final bore hole depth of 473 feet below grade, followed by geophysical well logging. A permanent 6 inch well was developed under a 20 foot thick clay lens, at a depth of 386-396 feet below grade surface.

- Drilling of a conventional rotary hole, Site No. 4, to sampling depth with setting and gravel packing a temporary well screen and pumping for formation water samples. This procedure was repeated at 310 and 430 feet below grade surface followed by geophysical logging to a final bore hole depth of 450 feet. A final permanent 4 inch well setting elevation was selected at 430 feet below grade, the bore hole backfilled to required depth, gravel packed, developed and pumped for formation sampling. A second permanent 2 inch well was installed at the unsaturated/saturated zone interface (160 feet below grade) to establish head differentials in the aquifers and to monitor groundwater elevation trends adjacent to the landfill.

GEOPHYSICAL LOGGING

At each permanent well location geophysical logging of the uncased bore hole was performed by the U. S. Geological Survey. The results of this program provided:

- Indirect evidence of subsurface formation characteristics and information supplementing the subsurface soil core data

- Supplemental information and verification of descriptive logging by the well driller

- Data by which to define the vertical distribution of the leachate plume

Electric Logs

Two point electric logging was employed to establish the apparent resistivity of the subsurface formations and plotted in terms of depth below ground surface. This procedure provided the methodology for relating apparent resistivity to characteristics of the subsurface formation and the quality of water contained therein. The degree to which resistivity is lowered partially depends upon the dissolved solids content of the formation water. This characteristic was used to define the vertical distribution of the leachate plume at Site No. 1.

Gamma-Ray Logging

Gamma-Ray Logs are based on measuring the natural radiation of Gamma-Rays from certain radioactive elements that occur in varying amounts in subsurface formations and relate to the emission of Gamma-Rays plotted against depth below grade surface.

Changes in radiation are commonly associated with differences between the types of materials making up successive strata. Characteristically, the log of unconsolidated formations indicate clay lenses when the Gamma Ray intensity is high and generally denote sand strata when the intensity is low.

Variation in water quality has little effect on the Gamma-Ray Log so that the log is more valuable in identifying the position and thickness of clay formations. These continuous logs were also used to assist in the interpretation of subsurface cores and drillers descriptive data.

CONCLUSIONS

The geotechnical and water quality sampling program confirmed the presence of a leachate plume at Site No. 1. This plume had an apparent thickness of about 180 feet and a conductivity at its center of 1000 umho/cm. None of the other test wells showed the presence of leachate and reflected background conductivity levels of between 90 and 125 umho/cm.

For this program the cost for each test well was approximately $40,000. and the overall exploratory program approximately $200,000. To obtain a better definition and extent of the plume would require extensive geotechnical work at a considerable capital cost.

COMPANY SOLVES LATEX WASTE PROBLEM

James E. Kerr. Para-Chem Southern, Inc.,
Simpsonville, South Carolina

Craig H. Lockhart. The Permutit Company, Inc.
Paramus, New Jersey

INTRODUCTION

Para-Chem Southern, Inc. produces chemicals, such as latex compounds, acrylic lattices, and adhesives at their manufacturing facility in Simpsonville, South Carolina. These compounds are custom blended for their customers on a demand basis, and are used primarily by manufacturer's of textiles, furniture and floor and wall coverings.

Waste from this plant is predominantly rinse water which has been used to clean the latex blending vessels. Due to the constant changes in the compounds produced, wide variations occur in the type of latex found in the rinse water and the solids content and pH of the waste stream. Typically, the rinse water may contain between 1.0 and 4.0 percent solids, have a pH range of 6.0 to 11.0 and the raw waste may or may not include some adhesives.

Both latex and adhesive wastes were lagooned on-site until 1976. At that time Para-Chem Southern initiated a program to design and install a waste treatment system which would meet the approval of the South Carolina Department of Health and Environmental Control (DHEC). The DHEC required that a bladeable waste be produced which would be suitable for landfilling. Working with J. L. Rogers and Callcott Engineers of Greenville, South Carolina, The Permutit Company, Inc. developed a system for flocculation, flotation and dewatering of the plant wastes which would meet that requirement.

PROCESS DESCRIPTION

The waste treatment system begins with a rinse water collection system which funnels the waste stream to a flume that carries the waste water by gravity from the manufacturing facility to the waste treatment plant.

At the treatment plant the flume splits into two sections, each discharging into a 26,000 gallon equalization and holding tank. A slide gate allows one tank to be filled at a time. As one tank collects the rinse waters produced in one day, the wastewater collected the previous day is processed through the waste treatment system.

Single-speed agitators mounted over the center of each holding basin keep the waste material mixed. This batch type treatment scheme minimizes the variability of the influent to the waste treatment equipment. Since each batch changes characteristics from day to day, jar tests are conducted each day to determine the proper chemical dosages required to properly flocculate that particular sludge.

A positive displacement pump transfers the sludge from the equalization tank to the first of three 1200 gallon fiberglass contact tanks, where the chemicals are fed and the proper contact time is maintained. Alum (aluminum sulfate) is added in tank 1 to drop the pH to approximately 4.2 to 4.5 to break the emulsion. At the same time alum acts as a coagulant, which assists in the final flocculation of the sludge. The introduction of caustic (sodium hydroxide) in tank 2 raises the pH to approximately 6.5, which puts the waste stream within the plant effluent limits of 6.0 to 8.0 and also, puts the sludge at the optimum pH for polymer addition. In tank 3, the sludge is flocculated using a dual polymer system, introducing both cationic and anionic polyelectrolytes. The total detention time in the three tanks is greater than 60 minutes which insures a good floc which is required for proper operation of the dissolved air flotation unit.

From the contact tanks, the flocculated wastewater is introduced to the Model OB-60 Permutit dissolved air flotation unit. This unit is capable of treating 50 gpm of process flow plus 50 gpm of recycle flow, which is provided from the flotation unit effluent stream. An air compressor and a retention tank (pressure vessel) are provided with the flotation unit which make up the air saturation system, which is essential to the dissolved air flotation process. (Refer to Figure 1)

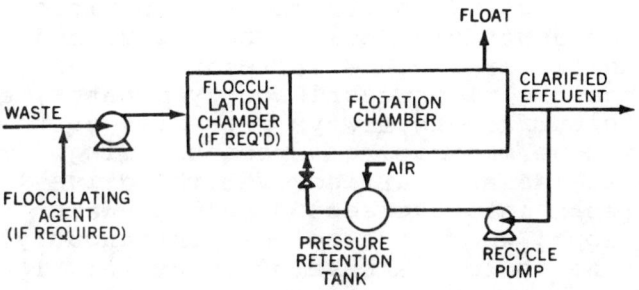

Fig. 1 Dissolved Air Flotation System

The effluent recycle flow is pumped to the retention tank where air is dissolved at approximately 50 to 60 psi. From the retention tank the air saturated liquid flows into the inlet compartment where it mixes intimately with the flocculated process flow. The rapid release of pressure in the flotation unit produces minute bubbles which slowly rise to the surface, carrying with them suspended solids present in the process flow. The attachment of air bubbles to the suspended solids lowers their combined specific gravity below that of water, making the particles buoyant. The buoyant particles rise quickly to the surface forming a "float" which is skimmed and sent to a holding tank from which it is transferred to the sludge dewatering units.

The effluent from the dissolved air flotation unit is lagooned for eventual discharge and contains approximately 30 ppm suspended solids. The float contains approximately 4 to 6 percent solids by weight.

Sludge dewatering of the float is accomplished using Permutit's "Dual Cell Gravity" dewatering unit (DCG-200) and multi-roll press (Model MRP-36). The sludge is pumped from the holding tank to a flocculation tank from which it flows by gravity to the DCG-200. The DCG basically consists of two cells separated by a fine mesh polyester filter cloth. A drive roll and sprocket assembly continuously rotates the filter cloth over front and rear guide wheels. Initial liquid/solids separation takes place in the first or dewatering cell, where liquid drains through the filter cloth at a relatively high rate. Since the cloth continuously rotates, a clean cloth surface is always presented to the influent sludge, minimizing screen blinding. The dewatered solids are carried over the drive roll separator into the second cell or cake formation zone, where the solids continuously roll and mass to form a cake of relatively low moisture content. When the cake grows to a certain size, excess quantities are discharged over the rim of the second cell to a conveyor belt which discharges the sludge cake into the inlet hopper of the MRP-36. (Refer to Figure 2)

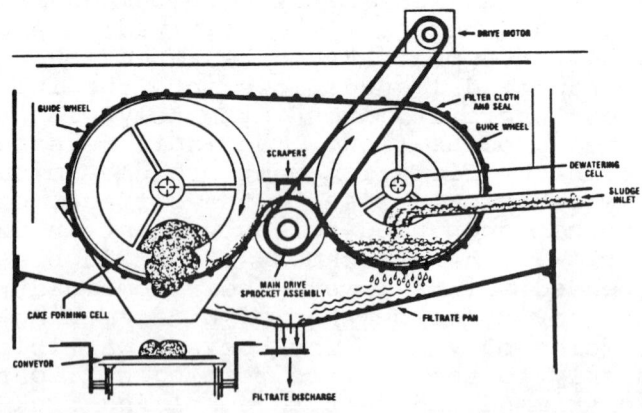

Fig. 2   DCG-200

The MRP presses the sludge between two endless filter belts of monofilament polyester. Space between the belts, through which the sludge travels, is controlled by five rollers which are adjusted by pneumatic regulators to apply a gradually increasing compression. The belts are inclined to assist in the drainage of the filtrate away from the sludge cake. A doctor blade removes the "bladeable" dewatered sludge cake from the belt at the point of discharge. (Refer to Figure 3). The cake is then transferred by means of an inclined conveyor to a dumpster, which is hauled to the landfill site.

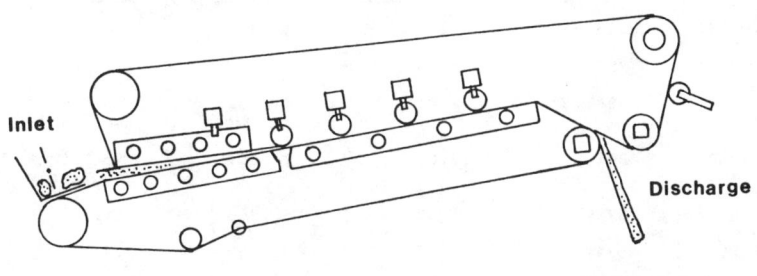

MRP-36

Fig. 3   MRP-36

The cake which is discharged from the DCG contains approximately 10 to 12 percent solids by weight, while the bladeable cake from the MRP contains approximately 20 to 25 percent solids by weight. The combined filtrate from the dewatering systems is recycled back to contact tank #3 where it is mixed with the plant process waste stream for introduction to the dissolved air flotation unit. The flotation effluent is the only stream that is discharged to the lagoons.

OPERATING DATA

Table I provides average monthly data on the influent waste stream for 1980. The influent pH is quite variable on a day by day basis.

However, the monthly averages show that the pH of the wastewater being treated generally falls in the range of 9.5 to 11.0. The average total gallons treated per month during 1980 was 309,645. At an average influent solids content of 1.85 percent, this represents an average of 47,728 pounds of dry solids handled per month.

TABLE I

INFLUENT CHARACTERISTICS OF THE WASTE STREAM FOR ONE YEAR
BY MONTH

MONTH	TOTAL GALLONS TREATED	AVERAGE INFLUENT SOLIDS (W/W%)	POUNDS OF DRY SOLIDS TREATED	AVERAGE INFLUENT pH
January 1980	313,683	1.66	43,497	10.1
February 1980	286,474	1.86	44,491	10.5
March 1980	308,787	1.84	47,442	No Data Available
April 1980	262,000	2.05	44,847	No Data Available
May 1980	378,291	1.93	60,963	11.3
June 1980	244,954	1.66	33,953	10.1
July 1980	293,455	2.14	52,437	10.0
August 1980	306,699	2.25	57,621	10.9
September 1980	359,144	2.00	59,977	10.2
October 1980	348,604	1.60	46,573	9.6
November 1980	290,486	1.68	40,749	10.0
December 1980	323,157	1.49	40,205	9.9
AVERAGE	309,645	1.85	47,728	10.26

During a three year period, March 1978 through March 1981, the discharge cake solids from the MRP fell in the range of 20 to 25 percent solids by weight. Table II provides a random sampling of monthly sludge concentration data, which shows that although the influent solids concentration ranged from 0.9 to 2.5 percent, the discharge cake solids remained very consistent. A bladeable cake was maintained, which was acceptable for landfill.

TABLE II

* AVERAGE MONTHLY SLUDGE CONCENTRATION DATA

MONTH	INFLUENT PERCENT SOLIDS	MRP DISCHARGE CAKE SOLIDS%	CONCENTRATION FACTOR
March 1978	1.70	24.6	14.5
April 1978	0.92	19.7	21.4
May 1978	1.04	22.9	22.0
August 1978	1.78	24.4	13.7
September 1978	1.60	21.2	13.3
January 1979	1.71	20.2	11.8
February 1979	1.59	19.0	11.9
May 1979	1.89	30.9	16.3
June 1979	1.58	28.0	17.7
November 1980	1.68	21.9	13.0
January 1981	2.23	27.9	12.5

* Months Chosen At Random

TABLE III

COST OF TREATMENT

	COST/POUND ($)	COST 1000 GALLONS ($)
50% Alum	0.025 *	4.12
50% NaOH	0.075	0.86
Cationic Polymer (Percol 728)	2.64	5.50
Anionic Polymer (Percol 725)	2.25	<u>1.41</u>
TOTAL		11.89

* Includes Freight

Basis: Six months July 1980 - December 1980
Total Flow = 1,921,545 gallons

Percol polymers are manufactured by Allied Colloids

Table III shows the average chemical costs for treating the waste stream at Para-Chem Southern during the last six months of 1980. The cost per 1000 gallons treated is approximately $11.89, with the bulk of the cost split between the alum and the cationic polymer.

Sludge solids content values are provided in Table IV. Note that at each treatment process stage an increase in solids concentration is realized.

TABLE IV

TYPICAL SLUDGE CONCENTRATIONS

	PERCENT SOLIDS
Influent	0.9 - 2.5
After dissolved air flotation unit	4 - 6
After dual-cell gravity unit	10 - 12
After multi-roll press unit	20 - 25

CONCLUSIONS

Variable waste streams, such as the mixed latex and adhesive waste water found at Para-Chem Southern are normally quite difficult to handle. However, through the joint design efforts of J. L. Rogers and Callcott Engineers and The Permutit Company, Inc., an economical method of treatment has been provided that produces a bladeable waste containing between 20 and 25 percent solids, which is acceptable for sanitary landfill. The clarified plant effluent that is discharged to the lagoons contains only 30 ppm suspended solids.

The installed capital cost for this equipment is approximately $250,000. The operating costs for this plant at the present time are approximately $11.89 per 1000 gallons treated.

The key to obtaining good, consistent operation in a plant of this type is proper chemical treatment to ensure good floc formation. The operators at Para-Chem Southern closely monitor the pH levels in the contact tanks and periodically check the floc formation in the last contact tank to be sure the proper polymer dosages are being added. The operating costs of treating this waste is kept at a minimum by frequent observation of operations and daily analysis and jar testing.

Based on results obtained at the Para-Chem Southern treatment facility, the prospect of using this type of treatment process on similar latex wastes is quite promising.

SESSION IV

PHYSICAL-CHEMICAL TREATMENT
(Solid Residues and Organics)

SELECTION OF BELT PRESSES FOR SLUDGE DEWATERING

<u>K. C. Goel, Ph.D., P.E.</u> President, Goel Associates, Inc., Dover, Delaware.

<u>Harish K. Mital.</u> Project Engineer, Edward H. Richardson Associates, Inc., Newark, Delaware.

<u>Stephen Hsu, P.E.</u> Chief Engineer, City of Wilmington, Wilmington, Delaware.

This paper summarizes the procedure for selection of belt presses recently developed and utilized for the City of Wilmington, Wastewater Treatment Plant, Delaware. The approach consisted of the following steps:

(a) Development of Performance-Based Specifications;

(b) Pre-qualification of manufacturers on the basis of short-term performance test;

(c) Development of mathematical models for obtaining "Total Annual Costs";

(d) Obtaining bids on the unit price basis;

(e) Long-term Side-By-Side Competitive Testing;

(f) Extrapolation of data from Long-Term Test to determine the required number of full-size machine;

(g) Calculation of "Total Annual Cost" on the basis of the Long-Term Test and information supplied with the bid;

(h) Award of contract to the bidder with the lowest "Total Annual Cost";

(i) Manufacture of one full-size machine;

(j) Scale-up Verification Test; and

(k) Adjustment in the number of machines to be purchased if necessary.

At present, primary and secondary sludges are thickened and pumped to high rate anaerobic digesters. The digested sludge is concentrated in lagoons and air dried [1]. The air dried sludge in combination with dirt is used as cover at a landfill. The sludge production on dry weight basis is approximately 40 tons per day. The solid concentration of digested sludge varies from 2% to 4%.

A project called Delaware Reclamation Project (DRP) is under construction for composting of the sludge and shredded garbage. The City is under contract with Delaware Solid Waste Authority to provide the sludge for this project with the following requirements:

Sludge Load on dry weight basis = up to 50 tons per day
Solid Content = 18% or greater
pH = 6.5 to 10.0

The land currently being used for air drying will not be available in the future. Alternative sites for air drying are also not available. The sludge in the lagoon averages 8%. Therefore, in order to obtain the 18% solid concentration required by the Authority, the City needed an alternative solution for sludge dewatering.

An extensive study [2] was conducted to determine the cost-effective method for sludge dewatering. Alternatives considered included the following:

(a) Open Sand Drying Beds;

(b) Covered Sand Drying Beds;

(c) Vacuum Filters;

(d) Centrifuges;

(e) Lagoons;

(f) Belt Filter Presses, and

(g) Pressure Filters including Diaphragm Presses.

The use of Belt Filter Presses with polymer conditioning was found to be the most cost-effective solution in this project. The vacuum filters with polymer conditioning could not produce the required solid contents in the dewatered sludge (cake), while pressure filter with lime and ferric chloride conditioning was not cost-effective due to the high cost of disposal. Vacuum filters and pressure filters were tried with other forms of conditioning, but these alternates were either too costly or could not produce the desired cake dryness. Belt filter presses were able to obtain more than 18% solid in cake with very little increase in the dry weight of sludge and extremely good filtrate quality.

BELT FILTER PRESS

There are several belt press suppliers in the United States. Most of the presses differ from each other in terms of equipment layout, performance, and reliability. Typically, the belt press consists of two endless belts which perform a three stage operation; namely, gravity drainage, low pressure section and high pressure section (shear zone). The sludge is conditioned by mixing with the polymer and by flocculating the mixture so as to produce a free-draining sludge floc. The conditioned sludge is first fed to the belts where dewatering is effected by the force of gravity. The filtrate coming out of gravity zone usually carries 200 to 500 ppm total suspended solids. The sludge from gravity zone passes onto the low pressure zone where light pressure is applied to the sludge by means of rollers in contact with the top belt. Finally, the sludge passes through the high pressure zone where shear forces are used to bring about the final dewatering. These shear forces are created by tensioning of belts while the belts travel over an arrangement of staggered rollers. The filtrate from the pressure zone carries high solid concentration (500 - 2000 ppm). Some belt presses recycle this filtrate by feeding it back at the polymer conditioning tank. Since this filtrate contains polymer, this method reduces the polymer dosage for conditioning. After discharge of dewatered sludge onto a conveyor, the belts are washed by means of high pressure sprays of water. Most belt presses use plant water for cleaning the belts. The filtrate is sent back to the head of plant for reprocessing.

As each belt press manufacturer has a different kind of equipment layout, performance, and cost; a two step selection procedure was adopted.

1. Preliminary Screening Test (Pre-qualification Test)

2. Long-Term Competitive Test for Performance and Evaluation

1. Preliminary Screening Test

   This test was conducted on the belt presses to determine whether they can meet the following minimum requirements:

   (a) Cake Solid: Not less than 18%.

   (b) Suspended Solids Capture Efficiency: Not less than 97%.

   (c) Total Suspended Solids: Not greater than 500 mg/l. In the event the filtrate discharge rate was found to be less than 110% of sludge feed rate, the maximum allowable suspended solids concentration in filtrate was 850 mg/l.

   (d) Filtrate Volume: Not greater than 185% of incoming sludge.

   The first requirement was due to this plant's commitment to the Delaware Solid Waste Authority as discussed earlier. The second requirement was needed so that the existing plant did not get overloaded excessively when the filtrate gets back to the head of plant for reprocessing.

   The supply of potable water at the plant is very limited which necessitated restrictions on volume of filtrate.

   This test was four hours long and the samples of sludge feed, filtrate discharge and sludge cakes were collected every hour. The results of this test are summarized in Table 1.

   As is evident from these results, most of the belt presses were not able to capture enough solids to obtain the 97% efficiency. Only Carter and Euramca were able to pass this preliminary test. The main difference between these two belt presses as compared to others, is that these two belt presses employ filtrate recycle systems, which in turn reduce the filtrate solids concentration and also the volume.

2. Long-Term Test Per Performance and Evaluation

   This test was developed to select the successful bidder. It included observations of the performance of the belt press during a long five day 80-hour operation and generated some important data on the basis of which performance of the belt press could be evaluated. The major requirements of this test were that the cake solid should be at least 18.5% and the solid capture efficiency should not be less than 95% and the belt press must operate successfully for at least 65 hours in

Table 1.

Belt Press Mfg.	Belt Width (inch)	Sludge Feed Rate (GPM)	Feed Solids %	Sludge Cake Solids %	T.S.S. in Filtrate (mg/l)	Filtrate Discharge as % of Sludge Feed Rate	T.S.S. Capture Eff. %
A.	25	25.5	2.93	15.8	1610	97.6	94.1
A.	25	16.7	2.93	15.3	1310	126	94
B.	32	18.7	2.90	19.7	752	92.5	97.4
C.	40	33	3.25	21.3	890	162.7	95.4
D.	40	28.2	2.80	19.7	2300	177	83.9
E.	40	30.6	2.35	16.3	826	179.7	93.3
F.	48	40.6	2.76	19.9	836	285.7	90.9
G.	60	29.4	2.76	23.3	440	98.3	97.5

A. Komline-Sanderson
B. Ralph B. Carter Company
C. Environmental Elements (Koppers)
D. Parkson
E. Envirotech
F. Ashbrook
G. Euramca

this 80-hour test. Other data generated included throughput of the machine, polymer consumption, power consumption, and filtrate quality and quantity. This data along with additional data supplied with the bid was used to calculate the "Total Annual Cost". The basis of selection of the apparent low bidder was the lowest "Total Annual Cost". Apparently, due to their failure to obtain the required bonds, Euramca decided not to submit a bid. Thus, Carter Press Model 2532 was the only one to be tested for long-term performance.

Procedure - This test was conducted for five days, each day 7 a.m. to 11 p.m. or a total of 80 hours. Out of these 80 hours, a minimum of 65 hours of actual operation of the machine was required. This is equivalent to a maximum of 15 hours of downtime. Downtime was charged against the machine whenever the sludge pump was not operating or no sludge cake was produced or the machine was being cleaned or repaired. During the test, samples of sludge feed, sludge cake, and filtrate were collected at approximately every three hours. Filtrate was recycled continuously and filtrate sump was not allowed to be emptied.

Carter's belt press machine was successful in operating more than 65 hours in the 80 hour test. The results of the test are summarized in Table 2.

Table 2.

	Day 1	Day 2	Day 3	Day 4	Day 5	Avg.
Sludge Feed Rate (GPM)	13.83	11.36	11.82	12.05	13.52	12.36
Total Solids in Sludge Feed (%)	3.79	3.67	3.7	3.53	3.62	3.67
Total Solid in Sludge Cake (%)	18.3	20.12	19.42	19.02	18.1	19.13
Solids Capture Efficiency, %	97.76	97.41	98.69	98.6	98.58	98.21

Other data collected during the test were as follows:

1. Polymer Consumption = 11.60 lbs per dry ton of sludge solids.

2. Power Consumption = 16 KWH per dry ton of sludge solids.

3. Total Suspended Solids in Filtrate requiring reprocessing = 34.9 lbs per dry ton of sludge solids.

Additional data provided by the belt press manufacturer as a part of his bid included the following:

1. Equipment Cost (per press) and

2. Equipment Size including total and effective width

EVALUATION

The belt press was evaluated on the basis of the "Total Annual Cost". The following cost items were considered in evaluation:

1. Amortized Equipment Cost;

2. Annual Labor Cost;

3. Amortized Building Cost;

4. Annual Hauling Cost;

5. Annual Polymer Cost;

6. Annual Power Cost, and

7. Annual Filtrate Reprocessing Cost

The machine capacity or throughput is usually expressed in terms of gallons per minute per foot of belt width. Since the effective belt width of belt press tested (24.0") was significantly lower than the belt press which is to be supplied for actual operation (92.5"), a scale-up factor was used to determine the calculated throughput of the full size machine. It was assumed that the throughput of the full size machine per foot of the effective width will be identical to the test machine. The City plans to dewater 50 dry tons of digested sludge per day, and also wishes to operate these machines in two eight hour shifts. The number of full size machines required was then calculated from the calculated throughput of the test machine. These calculations indicated that approximately 13 machines are required for operation. It was decided to purchase two stand-by machines. Therefore, the total number of machines purchased was 15.

Amortized equipment cost was calculated by assuming 20 years as the machine life with no salvage value and a 7-1/8% discount rate. Labor cost was estimated on the assumption that one operator would be required for two belt presses at a time.

Since a minimum clearance is required on all sides around the belt presses, the building size was determined by laying out the required belt presses, control systems, and polymer system. From the building area, the capital cost was calculated on the basis of $55 per square feet of floor area. This cost was amortized the same way as the equipment cost described above.

The sludge cake will be transported from the City wastewater treatment plant to the Delaware Reclamation Project site. The total weight and volume of cake requiring transportation depends directly on the percent solids in the cake. A curve was prepared to estimate the annual hauling cost as a function of the present solids in the cake.

Polymer and power consumption rates were determined during the test and on the basis of those data and by using present polymer cost and electric rates, annual polymer cost and annual power cost were calculated. Total suspended solids in filtrate, which would require reprocessing, was also determined. By assuming a cost of $0.05 per pound of solid for reprocessing, the annual filtrate reprocessing cost was calculated.

CONCLUSIONS AND RECOMMENDATIONS

In view of the wide difference between the claimed performance and the actual performance, prescreening of bidders on the basis of actual tests is highly desirable.

There is a wide variation in the size and throughput of belt presses including throughput per unit of effective width; e.g., a machine with 40" wide belt manufactured by one company may have as high as two times the capacity of another 40" wide machine manufactured by another company. Therefore, comparison based strictly on the number of machines of a specified size seem unfair to the manufacturers of the "high throughput" machines.

There is a trade-off between the cake dryness and filtrate quality; generally, the higher the cake solids content, the poorer the filtrate quality.

In cases where the emphasis is on good filtrate quality, machines with integrated solids recycling feature are more desirable. The solids in the filtrate indicated a tendency to float. Therefore, it may also be possible to improve the filtrate quality of other machines by employing an additional unit process such as dissolved air flotation or a settling tank with a skimming device.

The throughput of machines equipped with solids recycle feature is generally lower than those without the recycle.

The energy requirement for belt presses is significantly lower than either vacuum filters or centrifuges.

The cake obtained from belt presses is generally drier than cakes from vacuum filters or centrifuges.

The belt replacement cost can be very significant. Unfortunately, there is a general lack of operating data on this subject.

SUMMARY

The City of Wilmington wastewater treatment plant selected belt presses for dewatering of the sludge. A comprehensive two stage testing program was developed. The tests indicated that most of the belt presses could not meet the performance standards. Only the machines with integrated solids recycle system qualified for bidding. The performance was also evaluated on the long-term basis and the total annual cost of this machine was estimated. On the basis of the performance evaluation and the lowest life-cycle cost the City has selected Ralph B. Carter Company for supply of 15 Model 2532 belt presses. Initially only one machine will be manufactured. An extended test will be conducted using the first full-size machine to check the scale-up factor assumed earlier. If necessary, the number of machines ordered will then be adjusted upward or downward.

References: [1] Hsu, S. and Wittmer, S., "Air Drying as an Interim Dewatering Method", Tenth Mid-Atlantic Industrial Waste Conference, 1973.

[2] Goel, K. C., Savage, C. S., and Hsu, S., "Cost Effective Selection of Sludge Dewatering Processes", American Public Works Association Solid Waste Management Seminar, 1980.

SLUDGE MANAGEMENT IN THE POULTRY PROCESSING INDUSTRY

William F. Ritter. Agricultural Engineering Department, University of Delaware, Newark, Delaware.

INTRODUCTION

There are over 400 federally inspected poultry processing plants in the United States that perform the functions of slaughtering and evisceration of broilers, turkeys, mature chickens and other classes of poultry. Of these, 218 plants are located in the South Atlantic and South Central regions. These two regions in 1970 accounted for 86.3 percent of total broiler production, 48.1 percent of mature chicken production, and 32.8 percent of turkey production.

In broiler processing, feathers, blood, dirt and viscera are removed from the birds. Large quantities of water (20-57L/bird) are used in both washing and cleaning the poultry during processing. Broiler processing wastes are generally high in BOD and suspended solids with suspended solids production of 21 kg/1000 birds and BOD production of 27 kg/1000 birds.

Approximately 65 percent of the federally inspected processing plants discharge to municipal systems, while 35% percent have their own waste treatment plants. Most of the plants that discharge to municipal systems have some pretreatment to reduce industrial surcharges. Dissolved air flotation, sedimentation and screening are being used as pretreatment methods for reducing the BOD and suspended solids. Dissolved air flotation (DAF) has gained wide acceptability by the poultry industry. It will remove 80 percent of the grease and oil from the wastewater stream. Many processing plants are faced with the problem of disposing of the sludge from DAF units. In some poultry processing plants the grease is reclaimed as a bi-product. The DAF sludge creates operating problems in rendering plants because of the high water content.

Since the poultry processing industry faces a great problem in treating and disposing of high grease content sludges, research was initiated at the University of Delaware to study the disposal of high grease content sludges. The objectives of the project were:

    1. To investigate the effectiveness of commercial enzyme and bacteria cultures in reducing the grease content of the sludge before digestion or land disposal.

    2. To evaluate the change in soil properties and groundwater quality when high grease content poultry processing sludge is applied directly to land.

METHODS AND MATERIALS

Bacteria and Enzyme Cultures

A total of twelve products were tested in laboratory and pilot plant size experiments. The products tested and their manufacturers are given in Table I. Some of the companies recommended rates that should be used in the tests, while several companies had no idea as to what rates to use.

The initial tests were conducted in the laboratory in 3.8L containers. Two L of sludge was used in the laboratory tests at room temperature that are referred to as trials #1 and #2. In both trials #1 and #2 the sludge was aerated for 10 days. After the laboratory trials, three trials (#3, #4, #5) on a pilot plant size scale was conducted in the 113L containers, with 56.5L of sludge.

In trial #3 all of the tests but two were conducted under anaerobic conditions. In trials #4 and #5 all containers were aerated with aquarium pumps. Different concentrations of the products were used in different trials. Some of the rates selected in the experiments were obtained from company literature while rates for some products were selected by trial and error. Dr. Grubbs (1) of Flow Laboratories recommended that approximately 25g of bacteria culture product is used in the 56.5L samples. For many of the products, multiples of this recommended rate were used.

All tests were conducted for a period of 10 days. Before the product was added, a sludge sample was taken and analyzed for chemical oxygen demand (COD), total solids, volatile solids, fixed solids, grease and pH. After 10 days another sample was taken and analyzed for the same parameters. In trial #1, the sludge was not analyzed for grease, since this trial was used to test the experimental techniques more than the products. Some sludge samples were also analyzed for Kjeldhal nitrogen, ammonia, nitrate nitrogen, ortho phosphorus and total phosphorus. The dry bacteria and enzyme culture products were soaked in distilled water for 1 hour before they were added to the sludge.

All sludge containers were stirred by hand for 2 min. while the product was being added.

Table I. Commercial Products Tested

Product	Company Name	Product Classification
DBC Plus Type H-2	Flow Laboratories, Inc. 1601 W. Orangewood Ave. Orange, CA 92668	Digestive Bacteria
LLMO	Gen. Environ. Sci. Corp P.O. Box 22294 Beachwood, OH 44122	Digestive Bacteria
Bi-Chem 1003 FG	Sybron Biochemical Box 808 Salem, VA 24153	Digestive Bacteria
Bi-Chem 1008 SF	Sybron Biochemical	Digestive Bacteria
WD-47C	Wen-Don Corp. P.O. Box 13905 Roanoke, VA	Digestive Bacteria and Enzymes
Enviro-Aid	Binwil, Inc. P.O. Box 75118 Cincinnati, OH 45275	Enzymes and Surface Active Agent
Enviro-zyme	Winston Co. 8511 Gillette Lenexa, KS 66215	Digestive Bacteria and Enzymes
Biozyme	Brulin and Co., Inc. P.O. Box 270-B Indianapolis, IN 46206	Digestive Bacteria Enzymes and Wetting
Replenish	W. Chem. Prod., Inc. Long Island City New York, NY 11101	Enzymes
Formula 90	Harrington-Robb Co. P.O. Box 181 Florence, NJ 08518	Emulsifier
Polybac PDZ-20	Harrington-Robb Co.	Digestive Bacteria
Polybac E	Harrington-Robb Co.	Emulsifier

Sludge for the experiments was obtained from the DAF unit at the Felton plant of Perdue, Inc. A total of four batches of sludge were hauled from Felton to Newark. For trials #1 and #2 sludge was obtained in a 208L container and transferred by hand to the laboratory. For the pilot plant tests, sludge was hauled in a 1135L drum and pumped into 113L containers. Difficulty was encountered in obtaining a uniform concentration of solids in each container. Properties of the four batches of sludge used in the experiments are presented in Table II.

Some of the biological treatment products were also tested in the laboratory in 250 ml flasks. In the first test sludge was dried at 100°C to a constant weight and 4g aloquots were weighed into tared flasks. The sludge was then seeded with bacteria culture and distilled water was added to the 200ml mark along with 0.2g each of nitrogen and phosphorus. The flasks were incubated for 30 days under anaerobic conditions at room temperature and then dried to a constant weight. The percent degradation was determined by weight loss. In the second test only 2g of sludge was used in 100 ml of distilled water and the flasks were aerated for 10 days and then dried to a constant weight.

Table II. Properties of the Four Batches of Sludge Used In the Bacteria and Enzyme Experiment

Parameter	Concentration (mg/L)	
	Mean	Range
COD	60700	43900-88200
TS	44250	22000-64000
FV	8500	4700-12900
VS	35750	17300-51700
Grease	11325	6700-15100
Kjeldhal N	2153	1293- 2540
NH_3-N	1472	1051- 1778
NO_3-N	<10	<10- <10
Total P	925	505- 1618
Ortho P	800	386- 1333
pH	6.5	6.3- 6.9

Land Disposal

Dissolved air flotation sludge from the Felton plant was

applied to plots in July and August, 1979 and in May, 1980 at loading rates of 168, 336, and 1672 kg TKN/ha. The properties of the sludge were similar to the sludge used in the enzyme and bacteria studies. In 1979, soybeans were planted after winter wheat in all plots (Figure 1). Plot 5 did not have sludge applied until shortly after the beans had emerged, which resulted in the beans being killed. In 1980, soybeans were planted in plots 1-4. Before sludge was applied to the plots three groundwater monitoring wells were installed. Two sets of background water samples were taken and analyzed for nitrate nitrogen, ammonia, organic nitrogen, chloride, pH, ortho-phosphorus, total phosphorus, and COD. After sludge was applied in 1979 water samples have been collected at various intervals and analyzed for nitrate nitrogen, ammonia, organic nitrogen, pH, chlorides, COD, ortho phosphorus, total phosphorus, fecal coliform, and total coliform.

Soil samples were taken before the sludge was applied, in August, 1979, April, 1980, July, 1980 and in October, 1980. Soil samples were analyzed for nitrate nitrogen, ammonia, organic nitrogen, pH, phosphorus, organic matter, soluble salts, calcium, magnesium, potassium and cation exchange capacity. Hydraulic conductivity measurements were taken 4 times.

RESULTS AND DISCUSSION

Bacteria and Enzyme Cultures

The volatile solids reduction for different treatments in trial #1 is presented in Table III. The control had the largest volatile solid reduction. Chemical oxygen demand values are not presented because all of the tests showed very little COD reduction (<10%).

Table III. Volatile Solids Reduction for Trial #1

Product	Amount Added (ppm)	Final pH	Percent vs. Reduction
Polybac PDZ-20	1100	6.3	13
Polybac E	5^a	6.4	17
Formula 90	5^a	6.5	15
DBC Plus H-2	1100	6.4	5
Control		6.1	34

^{a}Added 5 ml/L sludge

The results for grease and volatile solids reduction for trial #2 are presented in Table IV. The final pH of the samples was higher in trial #2 than trial #1 and for two tests exceeded the best range for bacteria cultures which should be between 6 and 8. There was considerable variation in the volatile solids reduction. None of the products showed a large grease reduction. Polybac E reduced the volatile solids, but did not reduce the grease, therefore this product is not recommended for wastewater treatment systems. Bi-Chem 1008 SF did not show any grease or volatile solids reduction, possibly due to the high pH, or a representative sample may not have been obtained from the container. Visual observations of trials #1 and #2 indicated there was a lot of floating grease on all containers at the end of the test period.

Table IV. Volatile Solids and Grease Reduction for Trial #2

Product	Amount Added (ppm)	Final pH	Percent Reduction	
			Grease	VS
Polybac PDZ-20	1100	7.0	13	17
Polybac E	5[a]	8.1	0	25
Formula 90	5[a]	6.5	0	22
DBC Plus H-2	1100	7.3	17	22
Bi-Chem 1003 FG	1100	7.3	16	9
Bi-Chem 1008 SF	1100	8.1	0	0
Control		6.5	6	14

[a] Added 5 ml/L sludge

The results for trial #3 are presented in Table V. All of the tests except two for DBC Plus 2-H tests were conducted under anaerobic conditions. All products except the aerated 880 ppm DBC Plus 2-H container and control showed a large decrease in hexane soluble grease content.

After the tests were completed, the sludge was allowed to stand in the drums for two weeks under anaerobic conditions. Before the sludge was dumped, visual observations were taken on all drums. Some floating grease was observed on the Formula 90, control and the aerated 880 ppm DBC Plus 2-H drums, but was not observed on the other drums.

Table V. Grease and Volatile Solids Reduction for Trial #3

Product	Amount Added (ppm)	Final pH	Percent Reduction Grease	VS
DBC Plus 2-H	440	7.4	48	4
DBC Plus 2-H	880	7.3	78	12
DBC Plus 2-H	1760[a]	7.2	94	29
DBC Plus 2-H	880[a]	7.3	0	0
Polybac PDZ-20	880	8.1	88	6
Polybac PDZ-20	1760	7.0	40	3
Formula 90	5[b]	6.0	84	10
Formula 90	2.5[b]	7.3	64	18
Formula 90	1.25[b]	7.3	77	3
Control		7.4	8	5

[a] Drums were aerated
[b] Added 5, 2.5 and 1.25 ml/L sludge respectively.

The results of trial #4 are presented in Table VI. None of the products in trial #4 reduced the grease as much as the products tested in trial #3. The one control test had a grease reduction of 48%, which was greater than any of the products tested. The other control test had only a 4% reduction in grease content. All the drums were aerated in trial #4. Visual observations in trial #4 indicated floating grease was present in all drums except the WD-47-C and Replenish treatments 10 days after the final samples were collected and aeration was stopped.

In trial #5 higher concentrations of some of the products tested in the earlier trials were used. The results of trial #5 are presented in Table VII. All tests showed relatively high reduction of grease. The volatile solids reduction was relatively low in most cases. The batch of sludge used for trial #5 was of an extremely high solids content that contained many large chunks of solids. This made it difficult to obtain a representative sample for volatile solids and grease analysis. Some of the variation in results in Table VII may have been due to sampling error. In trial #5 no floating grease was observed in the Formula 90 treatment. All other drums had some grease floating on the surface after aeration was stopped. The layer of grease did not appear as thick on any of the drums as when the sludge for trial #5 was first pumped into the drums.

Table VI. Grease and Volatile Solids Reduction for Trial #4

Product	Amount Added (ppm)	Final pH	Percent Reduction Grease	VS
Enviro-zyme	880	6.2	42	20
Enviro-zyme	1760	6.3	35	15
Biozyme	880	6.2	13	15
Biozyme	1760	5.9	12	23
WD-47-C	880	6.2	22	16
WD-47-C	1760	6.3	14	12
Replenish	880	7.9	15	2
Replenish	1760	8.3	16	2
Enviro-Aid	8[a]	6.0	14	16
LLMO	10[a]	6.2	18	11
Control		6.3	48	25
Control		6.4	4	6

[a] Added as a solution

Table VII. Grease and Volatile Solids Reduction for Trial #5

Product	Amount Added (ppm)	Final pH	Percent Reduction Grease	VS
Formula 90	10[a]	6.5	70	6
Polybac PDZ-20	8800	5.8	46	0
Polybac PDZ-20	4400	6.4	52	0
DBC Plus 2-H	8800	6.0	51	31
DBC Plus 2-H	4440	6.7	88	13
Bi-Chem 1003 FG	4400	6.0	63	4
Replenish	4400	6.9	56	2
Control		6.6	59	12

[a] Added 10 ml/L sludge.

The laboratory tests in which sludge was dried and placed

in flasks under anaerobic and aerobic conditions are presented in Table VIII. In the anaerobic tests, the products Bi-Chem 1003 FG and Polybac PDZ-20 reduced the weight of the sludge 25 and 15%, respectively. The other three products tested did not cause much of a weight reduction and were similar to the control test. In the aerobic tests, Replenish was the only product that caused a reduction in the sludge weight.

Table VIII. Anaerobic and Aerobic Flask Test Results

Product	Oxygen Condition	Total Percent Solids Reduction
Polybac PDZ-20	Anaerobic	25.0
Bi-Chem 1003 FG	Anaerobic	15.0
Bi-Chem 1008 SF	Anaerobic	8.0
DBC Plus 2-H	Anaerobic	5.5
Biozyme	Anaerobic	5.5
Control	Aerobic	5.9
Replenish	Aerobic	56.4
Polybac PDZ-20	Aerobic	0
DBC Plus 2-H	Aerobic	0
Biozyme	Aerobic	0
LLMO	Aerobic	0
WD-47-C	Aerobic	0
Control	Aerobic	0

In all of the biological degradation tests, no product was completely satisfactory in reducing the grease. In some cases Polybac PDZ-20, Formula 90 and DBC Plus 2-H gave higher grease reduction than other products. There are a number of reasons why some of the products did not function as well as what their manufacturers claimed they would or why some of the products only worked sometimes. For supplemental bacteria cultures to enhance treatment performance, they must become the predominant microflora strain in the system. In some cases this may not have taken place. For some of the aerobic bacteria, the dissolved oxygen content may not have been high enough. In some cases the pH may have affected the performance of the bacteria, since the pH varied in different treatments. There should have been an adequate supply of nutrients for the bacteria, since the sludge was high in nitrogen and phosphorus. The variation in results from the biological degradation tests were similar to some of the livestock odor control tests reported in the literature (2; 3). In some cases the commercial products reduced livestock odors and in other cases they did not reduce odors.

Land Disposal

Average concentrations of ammonia, nitrate nitrogen, organic nitrogen, chlorides, COD, orthophosphorus, total phosphorus and pH for the three monitoring wells are presented in Table IX. Nitrate nitrogen concentrations increased in Well #2 in the fall of 1980 to a maximum of 17.4 mg/L, but decreased again in the winter of 1981 to 11.8 mg/L. Nitrate nitrogen concentrations in Well #1 and Well #3 did not increase after the sludge was applied. The concentrations of the other chemical parameters did not increase to any extent. No total or fecal coliform bacteria were detected in any of the samples.

The sludge did not cause any large increase in the chemical properties measured in the soil. The soil tests indicated no large amounts of nitrogen moved below the top 30 cm of the soil profile.

Table IX. Average Concentrations of Parameters in Monitoring Wells

Parameter	Before Sludge Applied Well			After Sludge Applied Well		
	#1	#2	#3	#1	#2	#3
NH_3-N	.83	.09	<.05	.08	.09	.19
NO_3-N	6.75	10.7	.91	5.48	12.0	.48
Org-N	.73	.12	.49	.56	.95	.56
Cl	22	21	20	8	10	8
COD	15.3	13.0	4.3	9.6	17.8	10.6
Ortho P	.049	<.010	.042	.014	.019	.041
Total P	.18	.10	.20	.13	.13	.19
pH	5.8	5.9	6.2	5.2	5.1	5.7

A summary of the hydraulic conductivity measurements taken on different dates is presented in Table X. There was a large variation in the hydraulic conductivity values. This kind of variability is expected under field conditions. There may have been some surface sealing over the winter on the high rate plot (plot #5), but the permeability increased once the field was tilled. The October 15th measurements indicate the sludge may increase the permeability of the soil as indicated by the higher hydraulic conductivity values. This increase in permeability may be caused by an increase in the number of earthworms. Visual observations indicate there are more earthworms in the soil where sludge was applied.

Table X. Hydraulic Conductivity Measurements of a Horizon

Sample Date	Mean	Range (cm/hr)	Standard Deviation	Coefficient of Variation (%)
5-3-79	8.9	3.8-16.5	4.6	51.1
8-20-79				
Plot #1	11.2	2.8-19.8	8.6	76.5
Plot #2	10.9	9.4-12.2	2.0	18.0
Plot #3	2.5	2.0- 3.0	0.5	22.3
Plot #4	9.9	5.3-14.5	4.6	45.6
Plot #5	6.9	3.0-10.7	3.3	47.8
4-24-80				
Plot #1	15.5	6.6-27.7	10.7	69.6
Plot #2	3.6	0.3- 9.4	5.1	137.4
Plot #3	21.8	0.8-42.9	29.7	136.2
Plot #4	6.6	0.4-17.5	9.4	141.0
Plot #5	1.3	0.5- 3.0	1.4	107.4
10-15-80				
Plot #1	24.1	4.6-49.5	18.8	77.6
Plot #2	24.1	4.1-37.8	14.2	59.4
Plot #3	6.9	2.8-17.0	6.9	98.2
Plot #4	23.3	2.1-43.9	23.1	99.6
Plot #5	8.1	0.4-21.8	9.4	117.0

CONCLUSIONS

1. Commercial bacteria and enzyme products for grease and fat degradation may not perform satisfactorily at all times.
2. DBC Plus 2-H, Polybac PDZ-20 and Formula 90 performed the best of 12 products tested in a majority of the trials.
3. High grease content sludge applied at nitrogen loading rates suitable for crop production will not cause adverse affects on groundwater quality or soil properties.

REFERENCES

1. Grubbs, R. B. Flow Laboratories, Orange, CA. Personal communication. 1979.
2. Miner, J. R. Evaluation of Alternative Approaches to Control Odors from Animal Feedlots. Idaho Research Foundation Report. University of Idaho, Moscow, Idaho. 1979.
3. Ritter, W. F., N. E. Collins and R. P. Eastburn. Chemical Treatment of Liquid Dairy Manure to Reduce Malodors. Third International Symposium on Livestock Wastes Proceedings. ASAE Pub. Proc. 275. pp. 381-384. 1975.

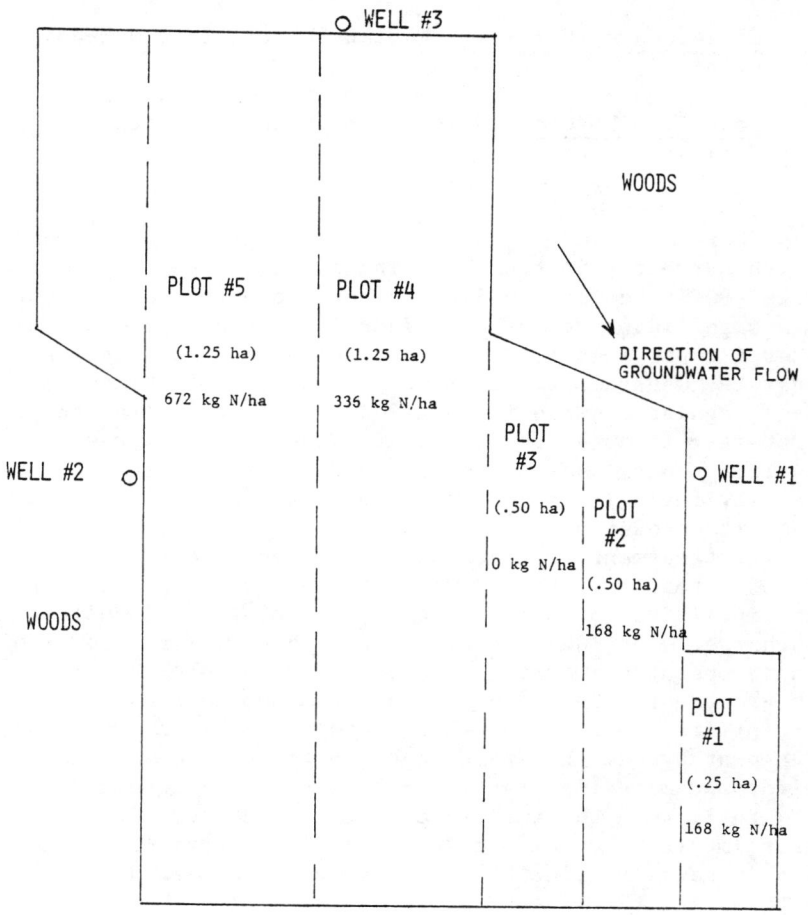

Figure 1. Plots and Monitoring Wells Layout (not to scale).

THE INFLUENCE OF PRETREATMENT ON IMPROVING THE QUALITY OF SLUDGE

Elizabeth F. Gormley. Radian Corporation, McLean, Virginia.

John V. DiLoreto. Washington Suburban Sanitary Commission, Laurel, Maryland.

There are over 27,000 municipal wastewater treatment plants currently in the U.S. These publicly owned treatment works (POTWs) collect and treat both domestic wastes as well as a significant amount of industrial wastes. The Clean Water Act provides for local control over these industries under the General Pretreatment Regulations (40 CFR 403). These regulations require the POTW to direct an industrial discharger to apply on-site pretreatment or otherwise restrict the release of certain constituents which the municipality feels it cannot handle effectively or which could upset its operations.

Pretreatment Regulations are designed to achieve three positive results. First, they will assist in allowing the municipalities to control the quality of their industrial discharger's effluent to the POTW. This will reduce or eliminate operating difficulties and lead to more cost efficient treatment. Secondly, the program should reduce or eliminate priority pollutants which pass through the municipal treatment system and are discharged to surface waters. Third, and more importantly, the regulations should allow the POTW to improve the quality of their generated sludge. This is an important consideration not only in terms of cost but also in terms of flexibility in choosing a method of disposal.

The following is a discussion of the implementation of a pretreatment program in the Washington, D.C. area. The Washington Suburban Sanitary Commission (WSSC) has been providing water supply and wastewater treatment to suburban Maryland since 1918. The Sanitary District now encompasses approximately 1,000 square miles and provides services to 1.2 million people. At the present time, seven wastewater treatment plants process approximatley 50 million gallons of wastewater per day. In addition, the WSSC is part owner of the Blue Plains treatment plant located in the District of Columbia which treats almost 300 million gallons per day of wastewater including approximately 140 million gallons per day generated within the Sanitary District.

One would probably not consider the Washington metropolitan area one of much industry. However, the business of government has actually spawned numerous "hidden" industries. For example, a number of federal agencies conduct laboratory research in the areas of genetics and medicine. Several facilities utilize nuclear reactors in their research and they all discharge to the POTW. It is difficult to see the industry, but the POTWs know it is there. Like other POTWs, there are problems of main line blockages, treatment plant upsets and sewer lines vanishing under the attack of corrosive and hazardous discharge.

The Sanitary District is a showcase of advanced wastewater treatment facilities. The most up-to-date technologies such as advanced microfiltration for the removal of fine suspended solids and nitrification for the reduction of ammonia nitrogen are employed at several of the Commission's treatment plants. For the processing of domestic waste, few POTWs can equal the level of treatment obtained at our facilities. Final effluent discharges often surpass national drinking water standards.

A major sampling program has been established to accurately characterize the POTW. Three consecutive 24 hour composite samples of influent and effluent streams and grab samples of sludge were collected from the three largest Commission operated wastewater treatment plants. An effort was made during sampling to follow, as closely as possible, those procedures established or proposed by the U.S. Environmental Protection Agency. Samples, other than sludge, were taken manually every hour, over a three day period, with flow compositing accomplished in the laboratory. Appropriate preservations were performed in the field at the time of sampling including on-site cooling for all samples. Sludge samples were taken once each day at different times. Analysis for all 129 priority pollutants and oil and grease was planned for each set of samples. Split samples were also taken for verification by a separate laboratory.

Several factors affect the type and quantity of discharge to any given POTW. The type of industry, unit operations, process variables, and pollution control technologies employed may produce an infinite variety of wastewater discharge characteristics.

Table I presents metal analyses which show that several heavy metals of concern appear in the influent, effluent and sludge at elevated levels at all three facilities. Most interesting is that heavy metals such as chromium, copper, lead, and nickel appear at relatively low levels in the influent and at higher levels in the effluent at the Piscataway plant. Potential causes of this phenomenon include introduction of treatment chemicals containing metals or analytical error.

Table I. Heavy Metal Concentrations in WSSC Treatment Plants

	Piscataway			Western Branch			Parkway		
	I	E	S	I	E	S	I	E	S
Cd	.002	.002	.080	.001	<.001	.300	<.001	<.001	.060
Cr	.017	.068	2.96	.014	.011	.970	.018	.011	3.17
Cu	.054	.008	310	.075	.037	384	.093	.035	387
Pb	.029	.064	373	.051	.040	179	.065	.015	142
Ni	.015	.018	68.0	.025	.021	39.0	.015	.010	53.2

NOTES: 1) All units are mg/l
2) I = Influent; E = Effluent; S = Sludge (Dry Solids)

The 37 organic priority pollutants listed in Table II were discovered in influent, effluent, and sludge streams. It is necessary to determine the fate of these organics since they appear so frequently. Other similar studies sponsored by the EPA prove that several organics act in a manner similar to the heavy metals -- that is, they may accumulate in the sludge rather that being treated or passing through the treatment system. This is very important since the WSSC has so many contributors of organic pollutants.

It would seem on the surface that the WSSC has the kind of sludge most municipalities would like to have. Even though it was previously noted that heavy metals and organics are accumulating in the sludge, they seem to be of low enough concentration to allow several options for disposal to be considered. Costwise, POTW sludge processing and disposal can account for as much as 50% of the total wastewater treatment cost. Also, the final disposal of this sludge can easily become not only a concern when disposal options are limited, but also a highly political issue since many residents object to the possibility of disease, groundwater contamination, and odors. The extent of this problem can be better appreciated when one considers that POTWs annually generate an estimated five million metric dry tons of sludge.

Theoretically, an efficiently run pretreatment program would, at a minimum, enhance a POTW's options for the ultimate disposal of sludge.

Table II. Organic Priority Pollutants Found Present in WSSC Treatment Systems

	I	E	S		I	E	S
Volatiles				**Base/Neutrals**			
Benzene	x	x		Bis(2-Ethylhexyl)			
Chloroform	x	x	x	Phthalate	x	x	x
Ethylbenzene	x			Butyl Benzyl Phthalate	x	x	x
Methyl Chloride	x	x		Di-n-Butyl Phthalate	x	x	x
Methylene Chloride	x	x	x	1,2-Dichlorobenzene	x	x	
Tetrachloroethylene	x	x	x	1,4-Dichlorobenzene	x		
Toluene	x	x	x	Diethyl Phthalate	x	x	x
1,2-trans-Dichloro-				Naphthalene	x		
ethylene	x	x		1,3-Dichlorobenzene	x		
1,1,1-Trichloroethane	x			Pyrene			x
Trichloroethylene	x			Anthracene			x
Chlorobenzene	x			Fluoranthene			x
1,1,2,2-Tetrachloro-				Phenanthrene			x
ethane	x						
				Pesticides			
Acid Extractables							
				Gamma-BHC	x		x
2-Chlorophenol	x			Delta-BHC	x		x
Parachlorometacresol	x			Alpha-Endosulfan	x		x
Phenol	x	x	x	Heptachlor	x		
2,4,6-Trichlorophenol	x			Beta-BHC	x		
2,4-Dichlorophenol	x			4,4'-DDD	x		x
				Endrin Aldehyde	x		
				Delta-Endosulfan	x		

NOTES: I = Influent; E = Effluent; S = Sludge

Table III represents the results of a study on the options available to POTWs for the disposal/management of sludge. It is important, however, to take a realistic look at these options and their costs. Primarily because of a change in attitude concerning our environment, stiffer regulations have been written and are being more vigorously enforced. The practice of incineration, for example, has been affected by regulations which control emissions by the Clean Air Act of 1977. Also, its utility as a disposal

Table III. Sludge Quantities by Method of Disposal [1]

Land Application		31%
Food-Chain Application	16%	
Distribution and Marketing	12%	
Non Food-Chain	3%	
Deep Trenching		24%
Incineration		21%
Ocean Dumping		18%
Other		6%

NOTE: This table represents 350 POTWs surveyed, 1.9 million dry metric tons of sludge, and 40% of POTW sludge generated nationwide.

practice has been minimized due to the increased cost of fuel and concerns over our use of energy. Ocean dumping is controlled by the Marine Protection, Research and Sanctuaries Act of 1972 and the EPA has directed municipalities to end ocean dumping of sludges by the end of 1981. Although direct land application of sewage sludge is one of the most cost effective methods of disposal, it does have some drawbacks. Direct land application to forest or agricultural areas pose health, safety, and odor problems. The sludge loads incorporated into an area should be chosen carefully taking into consideration population density, weather, drainage, and prevailing wind direction.

One environmental concern with direct land application is that since this sludge is uncomposted, and is being applied either as a partially dewatered liquid or as a dried solid, it has not achieved pasteurization. Human pathogens and parasites, still exist and pose a potential hazard and health risk. This is especially true when undigested sludge is applied to land. The concentration of heavy metals and industrial organic chemicals must also be monitored and controlled to prevent phytotoxic effects on crops and to limit the potential entry of these pollutants into the food chain or groundwater.

Deep trenching of sewage sludge, prevalent in the Washington Area, is a method being practiced mostly as an interim alternative. Most communities are increasingly reluctant to commit acres of land on a long term basis. This practice costs the WSSC approximately $200 to $250 per dry ton and has been met with stiff opposition from the general public.

These regulations and restraints tend to restrict the options available to municipalities for sludge disposal.

However, the better the quality of sludge, the greater the options for sludge management. Alternatives such as direct land application, composting, and co-disposal with refuse become more feasible with a possible added benefit of marketability. In fact, in Montgomery County in the Washington Metropolitan Area, composting is currently viewed as the favorable alternative. At the present time, it costs approximately $210 per dry ton for composting and dewatering at a facility in New Jersey. This composted material is not currently being marketed, thereby minimizing the cost effectiveness of this sludge management method.

From an economic as well as environmental point of view, direct land application and composting have become the most feasible methods of application as long as sludge quality can be controlled. The main value of sludge application is its possible soil enrichment qualities. Directly applied sludge has about 3 to 4% nitrogen. Composted sludge has only about 1.5% nitrogen and both have small amounts of phosphorus and potash. In general, these values are too low to make the sludge useful as a substitute fertilizer, but rather useful as a soil enhancer.

Table IV presents the limits of application rates pre-proposed by the EPA on May 6th, 1980 for sludges used as fertilizers or as soil conditioners on restricted and unrestricted crops vs. WSSC sludge data. Since maximum values have been used for comparison it would appear that WSSC sludge would be a candidate for direct land application or for composting.

With the preconceived notion of the EPA that the Sanitary District is not an industrial area, let us examine data on heavy metal content of sludge from other municipalities. Data in Table V from current literature shows sludge properties to be similar to other major metropolitan areas of the U.S.

However, Table VI displays U.S. EPA heavy metal content data on 7 other POTW's which indicate sludge quality surpassing the level being achieved by the WSSC. It can be concluded that the level of industry in the Washington Metropolitan Area is significant and that continued monitoring of sludge quality throughout the pretreatment program will be necessary to maintain this quality. If this is true, it clearly establishes the need for an effective Pretreatment program for all regions of the country in order to increase sludge management options.

There are some aspects of sludge quality which must be kept in mind. The first is that the federal pretreatment program has done little to approach the problem of organics in sludge. How many more of these 129 priority pollutants find their way into the sludge? What effect do they have on municipal treatment systems? What effect do they have on

Table IV. Comparison of WSSC Raw Sludge Data With Pre-Proposal Draft Regulation for the Distribution and Marketing of Sewage Sludge Products [1]

	Cadmium ppm	Lead ppm	PCB ppm
WSSC (Max.) Raw Sludge	13.0	480.7	ND
Fertilizer			
1) Unrestricted Use (Total N = 5 - 20%)	7.1 to 154	240 to 4500	ND
2) Restricted Use (Total N = 5 - 20%)	13 to 260	240 to 4500	ND
Sludge Soil Conditioning Products (Compost)			
1) Unrestricted Use			
o Low Cd/Zn, Lime Product (Total N<5%)	10 to 25	250 to 1000	ND up to 2
o Low Cadmium Product (Total N<5%)	2.5 to 5	250 to 1000	ND up to 2
2) Restricted Use	2.5 to 10	250 to 1000	ND up to 2

PCB - Polychlorinated biphenyls
(ND - Not detected)

Table V. Heavy Metals in Sludge - WSSC Compared to Other Metropolitan Areas

	WSSC Min	WSSC Max	Wash., DC [2] Blue Plains	Burlington Wis. [3]	Sommers [2] Min	Sommers [2] Max
Cd	0.5	13.0	19	10.1	3	3,410
Cr	22.0	102.0	---	792	10	99,000
Cu	281.0	418.8	725	415	84	10,400
Pb	58.9	480.7	575	289.5	13	19,730
Ni	38.5	77.7	---	481	2	3,515

NOTES: a) All values are ppm.
b) Sommers data (1977), 200 facilities, 8 states.

Table VI. Heavy Metals in Sludge - WSSC Compared to EPA Data

	WSSC Min	WSSC Max	Fate of Priority Pollutants [4] Plant A	Fate of Priority Pollutants [4] Plant B	7 cities [a] Min	7 cities [a] Max
Cd	0.5	13.0	0.6	0.3	0.04	79.8
Cr	22.0	102.0	17.9	8.1	4.8	73.8
Cu	281.0	418.8	24.2	10.7	2.8	77.4
Pb	58.9	480.7	11.0	7.4	1.5	45.9
Ni	38.5	77.7	3.2	3.1	0.5	20.7

NOTES: a) U.S. EPA data for 7 major metropolitan areas. Testing is still underway for 33 other cities.
b) All values are ppm.

composted sludge? Are they enhanced, concentrated or eliminated?

Another aspect is the administration of the pretreatment program itself. Will it be possible to regulate a high level of sludge quality? This is something which must be considered when developing and implementing a program. To be fair to industry and at the same time maintain a high level of sludge quality may, in reality, be difficult to do.

This, brings politics into the decision-making process, and the major obstacle still remains - a POTW must maintain low levels of pollutants in their collection and treatment systems while not inhibiting regional economic development. This is surely an achievable goal.

REFERENCES

1. Adapted from "Pre-Proposal Draft Regulation - Distribution and Marketing of Sewage Sludge Products," U.S. EPA Office of Water and Waste Management, May 6, 1980.

2. Parr, J. F., Epstein, E., and Willson, G., "Composting Sewage Sludge for Land Application," Agriculture and Environment, Elsevier Scientific Publishing Co., Amsterdam, The Netherlands, 4(1978).

3. Pietila, K. A. and Zacharias, D. R., "Sludge Management in Burlington, Wis.," Sludge, July-August, 1980, p. 25.

4. Feiler, H., "Fate of Priority Pollutants in Publicly Owned Treatment Works," EPA 440/1-79-300 (October 1979).

PERFORMANCE OF FINAL CLARIFIER AS A FUNCTION OF WASTEWATER CHARACTERISTICS

YEUN C. WU. Associate Professor, Dept. of Civil Engineering, Uni. of Pittsburgh, Pittsburgh, PA.

ZAHID MAHMUD. Senior Project Engineer, Rainbow Irrigation Systems, Fitzgerald, Georgia.

The characteristic feature of the activated sludge process is that biological solids are separated from the treated effluent for recycle back into the bioreactor. Industrial wastewaters differ considerably in the manner of generation and their constituents. Many industrial wastewaters are deficit in nitrogen and pose problems in the final clarifier operation due to poor sludge settleability. The effluent of the final clarifier represent the quality of treatment achieved at most treatment plants prior to its discharge into the receiving waters. Therefore the final clarifier has been considered the most important unit process in a secondary waste treatment system. In this paper , performance of the final clarifier is measured by the settleability of the biofloc.

Dick(1) found that ineffective capture of the biological solids increases the suspended BOD in the effluent and ineffective consolidation of the captured solids prior to recycle to the biological phase of the process, results in inefficient use of the aeration tank. The use of conventional design parameters such as SVI is due to that designer is yet incapable of accounting the actual mechanism of the relative degree to which sedimentation proceeds. The concentration to which activated sludge solids can be consolidated in the final settling tank for recycle to the biological portion of the process, commonly has been measured as SVI. But SVI is highly dependent upon the initial suspended solids concentration as well as the process operation. From the above discussion it is clear that the final clarifier design can effect the performance of activated sludge and in turn that will profoundly effect on both effluent clarity and that on recycled sludge concentration.So the settling properties of activated sludge should be taken into serious consideration in the design of the sedimentation tank.

Tenney(2) experimently proved that the degree of biological flocculation of microorganisms or bacterial disperssibility was highly dependent upon the nutrional environment in which they grow. WU(3) stated that this is because the nutrient condition in cultural medium determines the surface structure of bacterial cell that controls the effectiveness of biological flocculation. WU(4) also stated that cell surface charge which effects the bioflocculation,is produced as a result of ionization of some polymeric charge producing substances such as cell protein and polysaccharide compounds making up the surface of bacterial cell. WU(5) found that microorganisms produced under nitrogen limiting growth condition, are high in intercellular polysaccharide in terms of glycogen and low in cell protein. When readily used source of nitrogen is added to the culture medium, the accomulated polysaccharide glycogen is used for protein synthesis. The cell surface charge density changes as growth phase changes.The minimum and maximum bacterial surface charge density occurs at the exponential growth phase and endogenous respiration phase, respectively.

The influence of nutrients and cell growth conditions on the behavior of microbial flocculation and sedimentation is not well defined, although the essential role of carbon and nitrogen in biological wastewater treatment process is understood. Most recently WU(6) has found that charge neutralization and cell flocculation of severely nitrogen-starved sludge organisms is relatively difficult to achieve. The relationship between the structure of exocellular polymers produced by floc forming bacteria like Zoogloea and Pseudomonas and their flocculability was studied and found that some of the isolates possessed a capsular matrix which is composed of exocellular fibrils. Other isolates did not appear to have a capsulate matrix, however this later type did posses exocellular material. The fibrils were susceptable to cellulase although all fibrils did not appear to be identical. It is postulated that the exocellular polymers are responsable for the flocculant growth habit of bacteria, and that the process of bacterial flocculation produced by synthetic polyelectrolytes was essentially the same as that caused by naturally produced exopolymers.

This research was primarily concerned with the cultivation of sludge organisms under conditions where all nutrients other than nitrogen were supplied in excess; and then examine the influence of nitrogen limitation on the effectiveness of sludge flocculation and sedimentation which in turn reflects the performance of the final clarifier.

The sludge used for determining of bioflocculating and settling characteristics were cultivated in a completely mixed, continous flow activated sludge system with cell recycle. The COD:N ratio of the synthetic wastewater ranged from a normal or carbon limiting growth condition to a severe nitrogen limiting growth condition(5.3:1,11:1,21:1,36:1,53:1, 106:1,157:1,and 205:1). The sludge loading rate ranged from 0.126 to 2.27 lb COD/day/lb MLSS. The temperature in the biological reactor was maintained at approximately 29 ± 1° C and pH 7.0. Ten percent tap water was added to the synthetic wastewater to provid the micro nutrients.

The mixed liquor suspended solids withdrawn from the aeration tanks were used for the study of bacterial flocculation. Both mixed liquor suspended solids and sludge volume index tests were conducted according to the procedure presented in STANDARD METHODS. Additionally the sludge flocculability and settleability was determined by varying the initial concentration of suspended solids; therefore the effects of solids concentration on the velocity of sludge settling was determined. The settling velocity of the sludge was calculated by taking the settling data during sludge volume index determination. Anthrone and Buiret tests were used to determine the concentration of carbohydrates and proteins in the activated sludge. Sludge charge and capsule formation was also noted.

DISCUSSION

Results of slude volume index (SVI) and zone settling velocity(ZSV) are presented in Figure 1 through Figure 3. The samples were collected with sludge concentration range of 2300 mg/l to 2700mg/l.

The results shown in Figure 1 through Figure 3 are average of the results of more than fifteen samples.

Well-flocculated good settling sludge (70<SVI<100) was obtained in the nitrogen rich activated sludge system at t_d =8-30 hours, Θ =11.2-26.3 days, and F/M ratio=0.138- 0.585 lb. of COD per day per lb. of MLSS and also in the nitrogen-deficient activated sludge system at t_d = 24 hours, Θ=28 days, and F/M ratio =0.223 lb. of COD per day per lb. of MLSS. In contrast, severe bulking sludge (SVI>200) was found at t_d =2 hours, Θ = 3.8 days and F/M ratio of 2.204 lb of COD per day per lb. of MLSS in the former system and at t_d = 8-18 hours, Θ = 10.9-15.8 days, and F/M ratio = 0.276-0.619 lb. of COD per day per lb. of MLSS in the later system. Generally, an increase in SVI tended to occur at extreme low and/or high end of the loading rate, aeration and solids retention time range in a normal growth activated sludge system. While the activated sludge process was operated under the nitrogen restricted condition, the SVI was lowered sharply at $t_d \gtreqless 18$ hours, $\Theta \gtreqless 18.2$ days, and F/M ratio $\gtreqless 0.276$ lb. of COD per day per lb. of MLSS.

The other important observation obtained during this study seems to be the relationship between SVI and ZSV or effluent clarity. This study consistently shows that the increase in ZSV results in a decrease in SVI, but the magnitude of change between ZSV and SVI is not in a direct proportion. For instance, the ZSV of nitrogen rich activated sludge raised significantly from 6.8 ft/sec to 19.6 ft/sec as a result of decreasing SVI only from 125 to 75. On the contrary although the SVI of nitrogen deficient sludge decreased profoundly from 305 to 140, the ZSV increased only slightly from 0.75 ft/sec to 3.0 ft/sec. speaking of effluent clarity was excellent when SVI was high. In particular , this result became so obvious when dealing with the nitrogen-restricted growth activated sludge system. However, there are many cases also showing that the poorly-flocculated sludge having a high SVI produces effluent in high turbidity.

Effect of mixed liquor suspended solids is shown in Figure 4 for sludges from this experiment study for both nitrogen rich(COD:N= 5.3/1, 21/1, and 36/1) and nitrogen limited (COD:N = 53/1, 106/1, and 157/1) activated sludge. In all but one case (COD:N = 5.3/1, t_d =6.5 hr.), the trend was decreas-

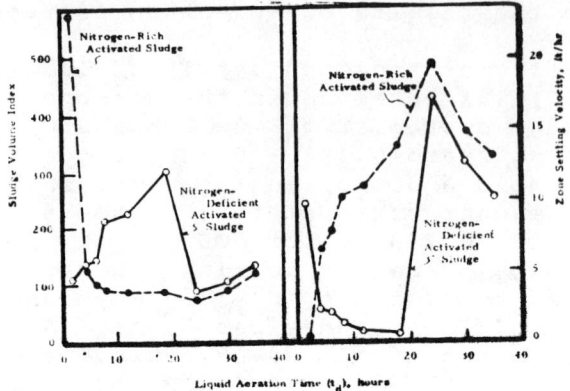

Figure 1. Effect Of Liquid Aeration Time On Sludge Volume Index And Zone Settling Velocity In Both Nitrogen-Rich And Nitrogen-Deficient Activated Sludge Systems

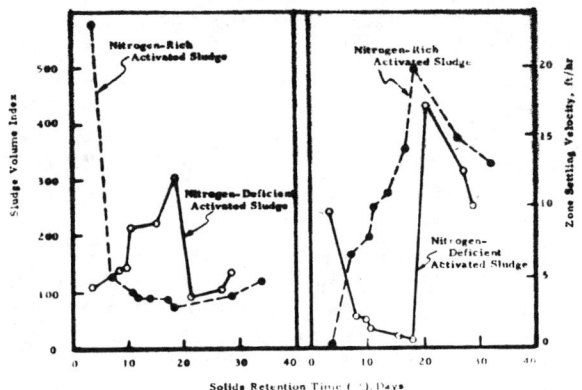

Figure 2. Effect Of Solids Retention Time On Sludge Volume Index And Zone Settling Velocity In Both Nitrogen-Rich And Nitrogen-Deficient Activated Sludge Systems

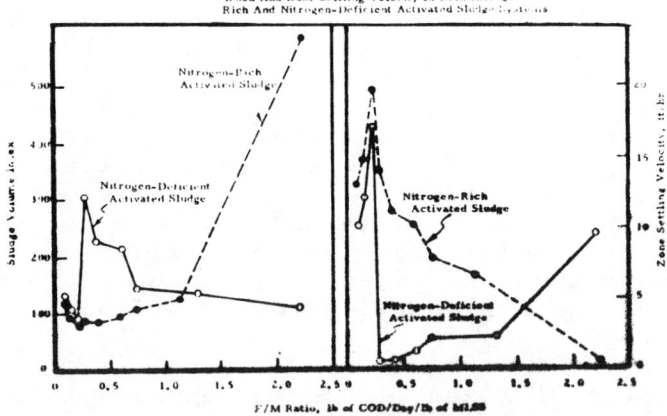

Figure 3. Effect Of F/M Ratio On Sludge Volume Index And Zone Settling Velocity In Both Nitrogen-Rich And Nitrogen-Deficient Activated Sludge Systems

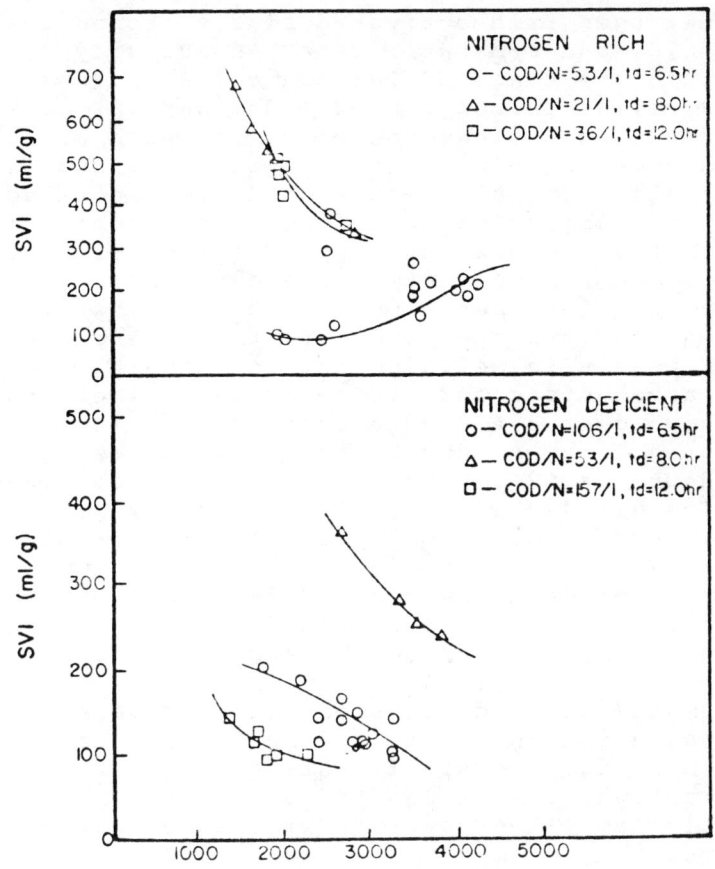

Fig. 4. Effect of MLSS in N- Rich and N- Deficient Act. Sludge System.

ing SVI with increasing MLSS concentration for the MLSS range encountered.

In order to fully understand the major factors other than liquid aeration time, solids retention time, and organic loading which may directly influence the result of sludge flocculation and sludge sedimentation, all of the biological solids generated under various living environments were thoroughly examined to determine the biochemical properties of sludge microorganisms by measuring the synthesis of carbohydrates, proteins, and surface charge.

It was evident from this study that the variation of liquid aeration time, solids retention time, and F/M ratio affects significantly the chemical composition of sludge microorganisms. An investigation of nitrogen rich activated sludge showed that if the t_d, Θ, and F/M ratio were respectively controlled within the range of 2-8 hours, 3.8-11.2 days, and 0.585-2.204 lb. of COD per day per lb. of MLSS, the increase in % sludge carbohydrate with decreasing % sludge protein or, vice versa, was found. Additionally, both sludge carbohydrates and sludge protein on % dry weight basis decreased substantially as soon as these three parameters fell outside the operating ranges mentioned above. The data resulted from the operation of nitrogen restricted activated sludge system indicate that there are two distinct phases in defining the relationship between sludge carbohydrate and sludge protein. The first phase in which the % sludge carbohydrate increased with decreasing % sludge protein or, vice versa, was obtained when t_d=2-6 hours, Θ = 3.4-10.0 days, and F/M ratio= 0.737-2.273 lb. of COD per day per lb. of MLSS. Second phase took place after t_d 6 hours, Θ 10.0 days, and F/M ratio 0.737 lb. of COD per day per lb. of MLSS. Evidently, the decrease in % sludge carbohydrate accompanied by a decrease in % sludge protein was found in second phase.

From the above discussion, it is understood that the flocculation and settling characteristics of activated sludge are highly influenced by the nature of the wastewater, the loading of the sludge as well as the biochemical properties of sludge microorganisms.

During cell morphology study it was observed that the nitrogen restricted activated sludge cultivated at high organic loading usually accumulated a considerable amount of capsulated matter on the cell surface and such substances sustained blue very readily with alcian blue. The degree of capsule formation with nitrogen rich activated sludge was found relatively less except those cells developed under an extremely high organic loading. The bacterial cellular matter in this case stained in red. Inevitably, there is a possibility that the capsulated cells are impossible to be distinguished from the non-capsulated cells if the former group of sludge organisms are entrapped inside the matrix of sludge floc.

The sludge flocculability and settleability are also closely related to the microbial physiology and morphology. Apparently, the sludge settleability was severely impaired by the outgrowth of filamentous organisms which entangled and/ or separated sludge particles from producing a good biological flocculation. So, the sludge floc settled as a small mass and not independently. In general, the present study shows the higher the population of filamentous microorganisms in the sludge floc the higher the level of SVI or the higher the degree of sludge bulking. However, the mechanism which controls the settleability of nitrogen-restricted activated sludge, is slightly different from that concerned with the nitrogen-rich activated sludge. In spite of an excellent growth of filamentous microorganisms, the settleability of the nitrogen-restricted activated sludge increased sharply after the t_d and Θ became less than 18 hours and 18.2 days, respectively, and the F/M ratio became greater than 0.276 lb. of COD per day per lb. of MLSS. This occurrence was not observed in the nitrogen-rich activated system and the result could be explained as to the high production of biopolymeric compounds on the cell surface. This, in turn, leads to an excellent cell flocculation and sedimentation. The non-filamentous bulking was observed on several occasions when the nitrogen-rich activated sludge system was operated at either high or low organic loading. According to the present investigation, the best sludge settleability was found when the flocculant zoogleal microorganisms attached to the short length filaments which served as a backbone for the sludge floc.

REFERENCES

1. Dick,R.I. ,"Role of Activated Sludge Final Settling Tanks", J.of San. Eng. ,ASCE,96, SA2, 423-436 (1970).
2. Tenney,M.W.,"Chemical Conditioning of Biological Sludge for Vacuum Filteration," JWPCF,42, R1-R20 (1970).
3. Wu,Y.C.,"Flocculability and Filterability of Activated Sludge in Response to Growth Conditions,"R.P.N.S.F.,Sep. CE.75-119(1975).
4. Wu,Y.C.,"Some Effects of Nitrogen and Phosphorus Limitation on a Biological-Chemical Waste Treatment Process" Unpublished Doc. Dissertation,Dept. of Civil Eng.,Uni. of Michigan, (1973).
5. Wu, Y.C. ,"Role of Nitrogen in Activated Sludge Process", J.of Env. Eng.,ASCE, V.102 EE5, 897-907 (1976).

APPLICATION OF HYDROPERM® CROSSFLOW FILTRATION SYSTEM
IN REMOVING SUSPENDED SOLIDS

Han-Lieh Liu. Senior Research Scientist,
HYDRONAUTICS, Incorporated, 7210 Pindell School Road,
Laurel, Maryland 20810

INTRODUCTION

   This paper presents the studies of two EPA sponsored research, development and demonstration programs for the application of a HYDROPERM microfiltration system to treat two types of industrial effluents. They are lead-acid battery manufacturing wastewater and electroplating rinse and dragout wastewater. The former demonstration program was successfully completed in 1980, however the performance of the system is still being continuously monitored. The latter program is an on-going project. The evaluation of its field demonstration will be made in September 1981. Nevertheless, the laboratory tests have successfully demonstrated the feasibility of using the HYDROPERM tubes to treat plating wastewater.

   HYDROPERM microfiltration is a new method of suspended solids removal by applying cross-flow filtration principle to thick-walled porous plastic tubes. These tubes, whose wall thickness, inside diameter, porosity, and pore-size distribution can all be closely controlled, can be selected for a particular wastewater so as to obtain optimum performance. Cross-flow microfiltration, which is intended primarily for the removal of suspended solids, is significantly different from membrane ultrafiltration (UF) or hyperfiltration (RO) which remove substances on the molecular level and are primarily for the removal of dissolved solids. They have been, however, used to remove suspended solids in the wastewater in some industries.

   In this paper, the general description of cross-flow

microfiltration will be discussed first, followed by the application case studies on battery wastewater and electroplating wastewater.

CROSS-FLOW MICROFILTRATION

In the treatment of both domestic and industrial wastewater, the removal of suspended solids is a required operation which arises under a variety of circumstances. Two types of filtration processes have been widely used in the removal of suspended solids: (1) the through-flow (e.g., multimedia) filter; and (2) cross-flow filtration (e.g., ultrafiltration and reversed osmosis). In cross-flow filtration, the direction of the feed flow is parallel to the filter surface, so that accumulation of the filtered solids on the filter medium can be minimized by the shearing action of the flow.

The filtration characteristics of HYDROPERM tubes combine both the "in-depth" filtration aspects of multimedia filters and the cross-flow filtration aspects of membrane ultrafilters. For example, in a manner similar to multimedia filters, the HYDROPERM tubes may allow some of the smaller particles in the waste streams to actually penetrate into their wall matrix. However, it should be noted that the pore structure of the HYDROPERM tubes differs from those of membrane ultrafilters in that the pore sizes are of the order of several microns with the "length" of the pores being many times their diameters. The entrapped solids in the wall matrix will not pass through the tube wall because of this long and tortuous path. Chemical solvent can be used to dissolve these entrapped species if necessary. The penetration of these small particles (usually less than a micron) is limited to the formation of a dynamic membrane layer on the inner side of the tube in the cross-flow filtration process. The existence of this layer is due to the establishment of the boundary layer of the flow through the tube. It is this layer which can help in preventing further penetration of small particles into the wall matrix. It is also conjectured that this layer behaves like another filter media. Thus, while the tubes are primarily designed for the separation of suspended solids from the feed, they frequently achieve some levels of dissolved solids separation because of this layer.

The principle element of the HYDROPERM microfiltration system is a thick-walled (~1 mm) hollow tubular filter (6-9 mm I.D.), made of thermoplastic material and containing micron-size pores. Pore-size distributions of two typical tubes are shown in Figure 1. It should be noted that the pore-size distributions shown in Figure 1 are initial values, since the pore-size distribution of the tubes undergoes a change during

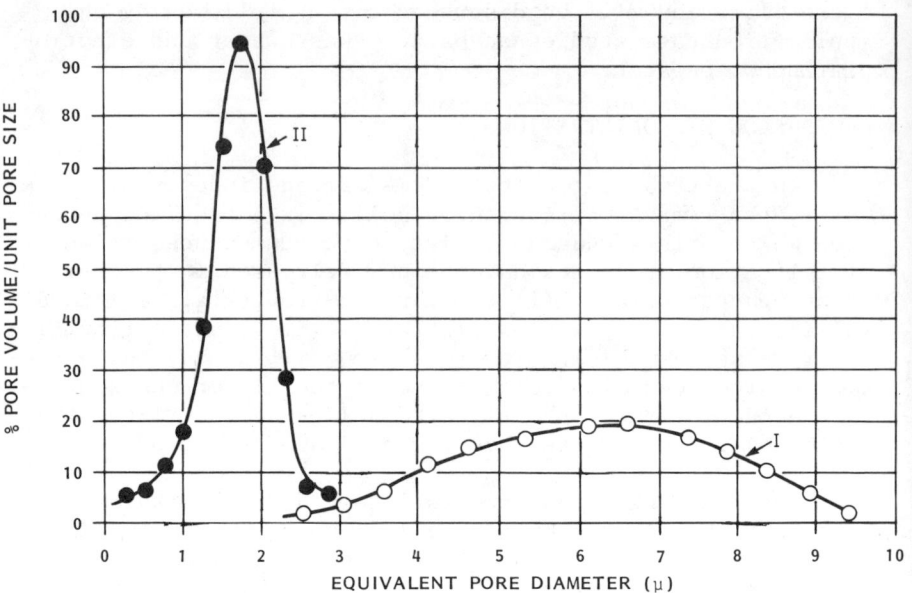

Figure 1. Typical pore size distribution of HYDROPERM™ tube.

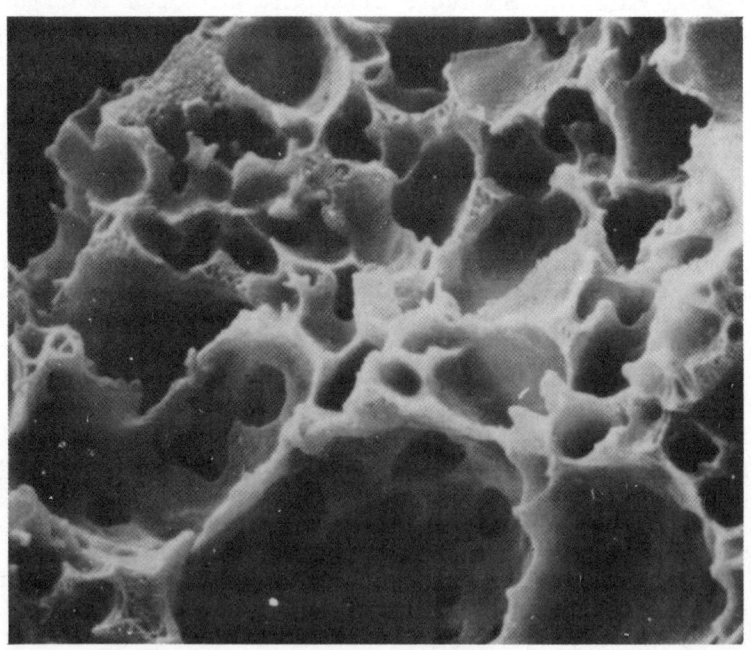

S.E.M. 1000X

Figure 2. Electron microphotographs of HYDROPERM™ tube pore structure - transverse section.

the filtration process because of the penetration of small particles into the tube wall. A typical pore structure is shown in Figure 2. The open cell, reticulated nature of the pore structure can be appreciated.

CASE STUDY I - BATTERY MANUFACTURING WASTEWATER EFFLUENT

The wastewaters from lead-acid battery manufacture are highly acidic (pH ~1), and contain a number of toxic heavy metals, primarily lead, in concentrations varying from 20 to 500 ppm. Other metals in the waste include Cu, Ni, Zn, As and Sb.

Prior to the installation of the HYDROPERM wastewater treatment system at the General Battery Corporation plant in Hamburg, Pennsylvania, acid wastewaters from that plant were treated with lime, and then transported ~20 miles by tank truck 2-3 times/day to the wastewater treatment facility at the General Battery Corporation Reading plant.

Presently the total wastewater treatment system in the Hamburg plant consists of the following unit processes: neutralization by lime; suspended solid separation by the HYDROPERM system; and sludge collection in a settling cone. The design of the HYDROPERM system was reported in Reference 1. In short, about 200 6-mm I.D. nylon tubes were used to make the HYDROPERM module. The module was 6-in. in diameter, 7-ft long, and contained 90-ft^2 filtration surface area. Two such modules, arranged in series, were used for a HYDROPERM treatment system.

The system was operated at 15 psi with a circulation velocity of about 7-ft/sec through the tube. After a year's operation of this unit, the permeate flux values could still be maintained in the range of 400-500 gfd (i.e., 50-60 gpm). The daily cleaning procedure was limited to the permeate flushing of the system. Cleaning with dilute HCl was exercised only on a weekly basis or as needed. Figure 3 shows a typical flux variation of one of the module's performance.

Currently, the permeate from the HYDROPERM treatment system is of a sufficiently high quality that it is discharged without further treatment to a local stream. Table 1 shows the permeate quality analyses at various dates.

The sludge which was concentrated through the HYDROPERM microfiltration process, was collected in the settling cone and was transferred to the landfill site only once every two days instead of 2-3 times every day. It usually contained 25-35% suspended solids by weight (mainly $CaSo_4$). The savings

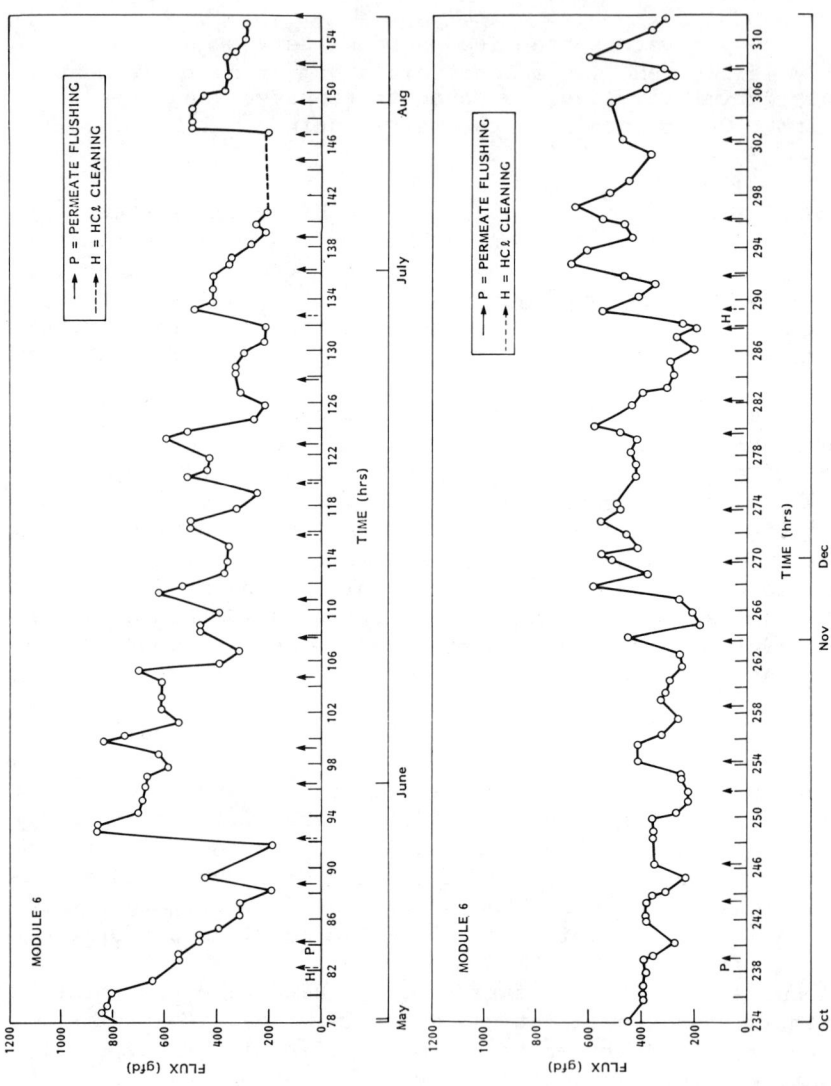

Figure 3. HYDROPERM™ System full-scale flux performance - Module 6.

in sludge transportation are obvious.

CASE STUDY II - ELECTROPLATING WASTEWATER EFFLUENT

The electroplating industry produces a large volume of wastewaters which contain primarily four common toxic heavy metals (i.e., chromium, zinc, copper and nickel), as well as a variety of others such as lead, cadmium, tin, silver, etc. The conventional treatment involves the unit operations of chromium reduction, cyanide oxidation, neutralization, clarification and sludge concentration. The conversion of heavy metals into insoluble hydroxides usually results in difficult to settle solids. The problem is further complicated by the fact that plating wastewaters may contain compounds that interact with dissolved metals and interfere with their precipitation as metal hydroxides. Alternatives such as sulfide precipitation, waste stream segregation and extreme pH treatments, are being studied by many people with some success (Reference 2).

Through the assistance of the American Electroplating Society, plating wastewater samples were received from Alexandria Metal Finisher, Virginia. Wastewater samples were also received from Craftsman Plating and Tinning Corporation and Chrome-Rite Company, in Chicago.

The untreated but detoxicated, samples went through a series of laboratory tests. It consisted of jar tests and filtration tests. The test variables included different precipitants and flocculants used in the industry. In the jar test, the particle formation and its settling characteristics were evaluated. The filterability of the particle laden wastewater was then examined through the HYDROPERM test apparatus. The results of this series of evaluation tests finalized the optimum tube characteristics for the field module tests.

The analytical permeate quality results in the single tube tests were quite satisfactory if pH was well controlled in the test run. Table II shows that all three wastewater sources can be effectively treated with HYDROPERM microfiltration process. The flux values of the wastewater depended upon the chemical used in the various conventional treatment steps. A flux value over 200 gal/ft^2-day could be easily demonstrated.

CONCLUSION

HYDROPERM cross-flow microfiltration is an effective new technology for the treatment of wastewater from the production of lead-acid batteries and electroplating. Two full-scale HYDROPERM units installed at the General Battery Hamburg plant

TABLE I

LABORATORY ANALYSES ON METALS IN THE
BATTERY WASTEWATER EFFLUENTS

Date	Samples	TS	SS	Pb	Cu	Zn	Ni	Sb	As
2-27-80	Feed	162,740	160,000			54.5			
	Permeate	3,621	1.8	0.045	0.017	0.008	0.024	0.141	0.002
4-2-80	Feed	12,364	9,188	319.4					
	Permeate	3,378	1.0	0.029	—	—	—	0.151	<0.002
9-24-80	Feed	9,346	6,804	40.3					
	Permeate	2,773	1.0	0.082	0.023	0.024	0.053	0.058	0.003
11-11-80	Feed	11,932	9,180	55.2					
	Permeate	3,356	6	0.073	0.015	0.016	0.041	0.190	0.001
12-12-80	Feed	7,794	5,144	84.4					
	Permeate	3,358	14	0.064					

TABLE II

LABORATORY ANALYSES ON METALS IN THE
PLATING WASTEWATER EFFLUENTS

	Samples	TS	SS	Cr	Cu	Zn	Cyanide		
1.*	Feed	1860	616	15.4	57.3	32	12		
	Permeate	1320	8	<0.01	36.4	0.38	11		
		TS	SS	Cd	Cu	Pb	Sb		
2.	Feed	4100	721	170	12.5	0.10	0.8		
	Permeate	2860	16	0.3	0.11	<0.01	<0.1		
		TS	SS	Cr	Cd	Cu	Ni	Zn	
3.	Feed	5940	4010	90	11.5	3.85	120	800	
	Permeate	2250	2	<0.01	0.06	0.42	2.16	1.44	

*Note: 1. Alexandria Metal Finisher; 2. Craftsman Plating and Tinning; and 3. Chrome-Rite have been successfully operated for over one year. Each unit has been able to maintain an operational flux value of about 400 gfd (~50 gpm of permeate production). The permeate quality of treated wastewater effluent is very satisfactory. It has been directly discharged into the stream without any violation of regulations.

The laboratory study of using HYDROPERM tubes to treat electroplating wastewater proves the feasibility of this new technology application. Reasonably high flux and good effluent quality have been demonstrated. A full-scale field test of a HYDROPERM system in a plating shop is scheduled in the early summer.

ACKNOWLEDGMENTS

The author wishes to achnowledge the financial support partially provided by the U.S. Environmental Protection Agency in both of the R&D programs. Valuable assistance from the following agencies are also sincerely appreciated: General Battery Corporation, Pennsylvania; American Electroplating Society; Alexandria Metal Finisher, Virginia; Craftsman Plating and Tinning Corporation, Illinois; and Chrome-Rite Corporation, Illinois.

REFERENCES

1. Shapira, N.I., Liu, H-L., Baranski, J., and Kurzweg, D., "Removal of Heavy Metals and Suspended Solids From Battery Wastewaters: Application of HYDROPERM Cross-Flow Microfiltration," HYDRONAUTICS Technical Report 8800, June 1980.

2. EPA Summary Report
    Control and Treatment Technology for the Metal Finishing Industry - Sulfide Precipitation Technology Transfer
    EPA 625/8-80-003, April 1980.

UPGRADING THE DISSOLVED AIR FLOTATION PROCESS

J. Glynn Henry. Department of Civil Engineering, University of Toronto, Toronto, Ontario.

Ronald L. Gehr. Department of Civil Engineering and Applied Mechanics, McGill University, Montreal, Quebec.

INTRODUCTION

Dissolved air flotation (DAF) has been used for the thickening of municipal and industrial sludges and the clarification of dilute suspensions. It has been used extensively in pulp and paper mills, poultry processing, oil, paint, grease, and fibre removal and shrimp cannery wastewater treatment. Although clarification of wastewater and the thickening of organic sludges is emphasized in this paper, the procedures for evaluating performance and the suggestions for upgrading the process apply to flotation of all types of wastewater whether sanitary, municipal or industrial.

The information presented comes from contract work at the Wastewater Technology Centre in Burlington, Ontario (1) and from doctoral research at the University of Toronto (2). Both studies included follow-up work on full-scale DAF units. The objective of the Burlington work was to evaluate DAF as an alternative to sedimentation. Three suspensions [220, 2900, and 4600 mg/L of suspended solids (SS)] derived from municipal wastewater (ie. sanitary and industrial wastes) were employed. The objective of the doctoral research was to investigate the role of polymers in the DAF process. Activated sludge mixed liquor from the Humber Wastewater Treatment Plant in Metropolitan Toronto was used for the waste. Some of the problems encountered in these 2 studies in trying to operate bench-scale and pilot plant units and to correlate the performance of one to the other are described. Modifications to the pilot unit which resulted in improved performance could be beneficial in full-scale systems and make the DAF process more competetive as an alternate method of thickening or clarifying sludges or slurries.

EQUIPMENT

Batch Apparatus

A three cylinder batch flotation appartus similar to that introduced by Wood and Dick (3) was used for batch DAF tests because it simulates full-scale continuous flow units more closely than the traditional "bomb" or two cylinder apparatus does. Because the level of saturation in DAF systems is critical to performance and commonly assumed values of 0.75 to 0.95 had been found to be incorrect (4), the basic unit (see Figure 1) was modified to enable the oxygen and hence the dissolved air to be measured directly. A probe from a Beckman Model 735 DO Analyser was incorporated within the pressure vessel. This batch DAF unit has been described in previous studies by the authors (1, 4, 5, 6) and details will not be given here.

Other minor modifications from the Wood and Dick apparatus concerned the valve and piping arrangements and the use of a larger diameter flotation cylinder (140 mm instead of 63 mm).

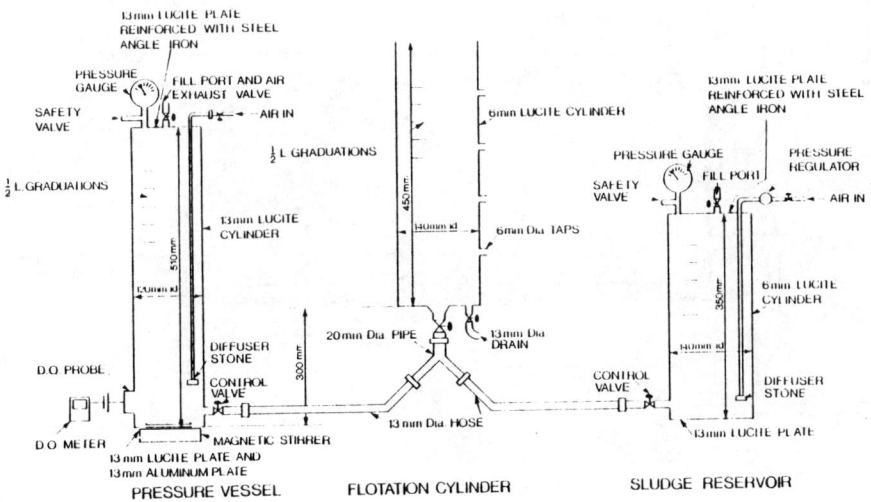

Figure 1. Bench-Scale Batch DAF Unit

Pilot DAF Unit

The continuous-flow pilot DAF unit used was the Komline Sanderson SR 15 unit as shown schematically in Figure 2.

The flotation tank, originally of 1.39 m^2 (15 ft^2) nominal area, was reduced to 0.836 m^2 (9 ft^2) nominal area to allow greater flexibility of hydraulic loading when using the existing limited facilities for the supply of raw sewage and activated sludge mixed liquor. The scraping mechanism to remove floated sludge was a continuous-chain type with three scraper blades; speed was fixed at 1.8 m/min (6 fpm), and control was by automatic timer. The pressure tank had an operating pressure between 330 kPa and 400 kPa (48 psi and 58 psi). Compressed air at about 550 kPa was injected into the recycle feed to the tank, and the water level in the pressure tank was controlled automatically. Air was mixed with the liquid at the elevated pressure by means of the recirculation pump which forced the liquid through an eductor inside the tank, and by venturi action caused a saturation of approximately 60% to be achieved. The recycle flow, measured by a magnetic flow meter, was controlled by a diaphragm valve at the outlet of the pressure tank. The dissolved oxygen in the air-charged liquid was measured by a dissolved oxygen probe attached to a by-pass line over the recycle control valve. Operation of the flotation system was on a 24 hour basis, with continuous runs of up to 2½ weeks.

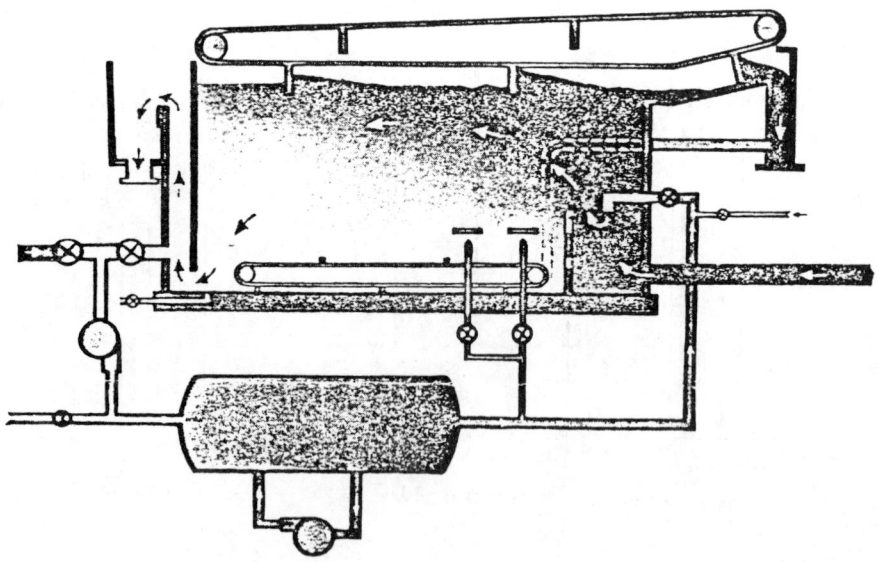

Figure 2. Komline Sanderson SR 15 DAF Unit

PROCESS SCALE UP

Batch tests using the three cylinder apparatus were useful in predicting approximate operating values for the pilot plant. Factors of interest were the air to solids ratio (A/S), polymer type and dosage and the float and effluent solids concentrations. Rise rate as a means for determining maximum hydraulic and solids loadings (and hence tank dimensions) for sludge thickening has been used by many researchers (3, 7, 8, 9) but was found to be impractical when dealing with dilute suspensions. Allowable loading criteria for sizing full scale units should be determined from pilot studies.

A comparison between batch and pilot performance indicated that the float solids concentrations were similar ($\pm$ 10%) and that polymer dosage requirements of the batch unit accurately reflected the optimum percent for the pilot plant. Points of difference between the two systems were the effluent SS which were consistently lower in the pilot plant and the air requirements which ranged from 10 to 40% above that predicted by the batch test (1).

OPERATING PROBLEMS

The major operating problem with continuous operation of the DAF Pilot unit was the air dissolution system. From the outlet of the flotation tank to the point where recycle liquid, charged with air, enters the tank again, three trouble-spots were identified: the level control system in the pressure tank; the eductor and re-aeration system; and the recycle control valve.

It was anticipated that one of the causes of trouble in such an air dissolution system would be fibrous material in the influent, and that a straining system should be employed to remove these offenders. However, it was found to be impractical to use a strainer in the recycle line: when the strainer was used ahead of the pressure tank, it became clogged very rapidly when the recycle liquid quality deteriorated, even if this deterioration were momentary. The strainer chamber was kept however, and this did retain some of the solids. A permanent solution to this problem in the field would involve some form of preliminary screening or settling to remove stringy and fibrous solids.

Other operating problems encountered during this phase included dissolved oxygen measurements, skimmer operation, and sludge accumulations on various parts of the flotation tank and associated mechanical equipment.

Level Control System

The Komline Sanderson air dissolution system relies on a fixed water level being maintained in the pressure tank. This level is monitored in a sight glass, both visually and by a flexible small-bore tube which has one end at the desired level, and the other connected remotely to a liquid sensor. When the water level rises so that the tube end is submerged, the liquid travels by a combination of capillary action and back-pressure to the sensor. This sensor in turn triggers a solenoid valve which forces additional air, at a pressure slightly higher than that in the tank, along the tube, into the sight glass, and thus back to the tank. The water level is thus depressed, and the solenoid valve closes until needed again.

This level control system is subject to numerous problems. Adjustment of the pressure differential between the tank and the forced air feed is critical, as is adjustment of the air flow rate. Impurities and foam in the recycle liquid tend to clog the sight glass, tube and the solenoid valve, and impair the functioning of the liquid sensor. No provision exists for correcting a low-level condition in the tank. As a result of these problems, the water level in the tank is frequently not at the desired level. This impairs eductor operation, and the flotation operation may fail completely. Vigilant attention was required to maintain continuous operation.

A system which uses electrodes in the pressure tank to monitor the level and control the auxiliary air supply would be simpler **and more reliable,** especially on small units where response time is critical.

It should be mentioned, in defense of the level control system used in this study, that it is employed in all full-scale installations at sewage treatment plants in Ontario, and that in these large installations it appears to be successful. The response time in larger units is much longer than in pilot-scale plants, and for this reason the level control system is adequate. Nevertheless, where an option exists for an alternative method, an electrode system might be preferable.

Eductor and Re-Aeration System

In order to obtain adequate levels of air saturation, a re-aeration system is used. Liquid is pumped from the base of the pressure tank, and returned through one or more eductors. The turbulence thus created, and the air forced into the liquid via the eductor mechanism, is responsible for almost all the air which is forced into solution.

In the unit used for this study, only one eductor was

supplied. The focal point of this eductor is an 8 mm (5/16") nozzle through which the liquid is forced. The potential for clogging is high, and once this occurs, air saturation drops and the pressure tank has to be opened, and the eductor removed and cleaned.

A simple method for cleaning the eductor during plant operating was devised. It required only minor pipe changes. The piping configuration and cleaning method is shown in Figure 3.

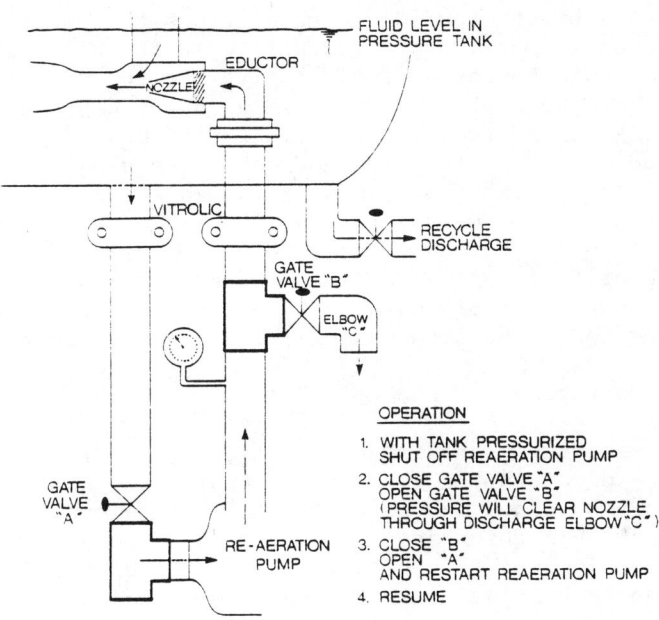

Figure 3. Valve Arrangement for Eductor Cleanout DAF Pilot Plant

Recycle Control Valve

The persistent problem of fibrous materials and other contaminants manifested itself again in the recycle control valve. In order to maintain a steady supply of air available for flotation, it was necessary that the recycle flow remain constant. After many changes, it was found that blockages could be avoided, and a constant flow maintained, by using a diaphragm valve. A 19 mm (3/4") diaphragm valve proved to be more effective than the larger size and provided it was flushed regularly, blockages and flow disturbances were minimized. This small opening had the added advantage of creating a head-loss high enough to ensure sufficient flow past the D.O. membrane located in the line by-passing the control valve. However, recycle flow was thereby limited to 170 L/min (45 gpm).

Dissolved Oxygen Measurement

Dissolved oxygen was measured by means of a Beckman Model 7002 Dissolved Oxygen Monitor, inserted into a flow-through chamber which acted as a by-pass to the recycle control valve. In this way, pressure on the probe membrane was created by the head difference across the control valve, and did not exceed 138 kPa (20 psi). It was hoped that D.O. readings could be taken continuously, but the fragile teflon membrane on the probe was too easily ruptured by the solid impurities in the recycle liquid. Furthermore, it was found that readings would fall if the probe were subjected to the combination of high flow [up to 50 L/min (13 gpm)] and elevated pressure [up to 138 kPa (20 psi)] for longer than about 1/2 hour. Periodically obtaining a value for the mass of air available for flotation was practical and easily applied, but the equipment had to be handled with care, and the probe had to be cleaned and recharged regularly.

Skimmer Operation

The skimmer mechanism had a fixed speed drive, but the frequency of operation could be controlled by an automatic timer. This method for controlling float removal rate was satisfactory in this case, but a variable speed drive would provide additional flexibility. Operation of the skimmer was adequate provided that the chain was correctly tensioned, and that it was regularly lubricated. In addition, the skimmer blades had to be adjusted such that underflow between the blade and the beeching plate was minimized. To this end it was found necessary to raise the beeching plate by the addition of a 6 mm (1/4") PVC sheet because the skimmer blades could not be located closely enough to the original beeching surface.

Accumulations of sludge would occur wherever the skimmer blades changed direction of travel, and also at specific skimmer scraping points. Removal of these accumulations had to form part of the regular maintenance routine.

Skimming problems became more severe with increasing float depth. The continuous flight skimmer is positioned in such a way that the blades do not sweep at the extreme end of the tank (effluent end), and float did tend to accumulate here. If not removed manually, this stored float caused a deterioration in the effluent quality. Furthermore, since float depth was considerably greater than could conveniently be removed by hand scoop, the effective surface area available for flotation was reduced by this dead space, the reduction being about 12%. Theoretically, this reduction should be reflected in the area loading calculations, but this was not done, mainly because the actual area reduction was imprecise. The net effect would be that the actual hydraulic and solids loading were higher than those calculated. The solution to this problem would be to install a baffle, to effectively shorten the tank, and enable the entire surface area to be skimmed.

PROCESS CONTROL

Loading Criteria

As noted earlier hydraulic and solids loading criteria should be obtained from pilot plant results. This is particularly important where clarification of dilute suspensions is intended. Table I summarizes the performance of the DAF pilot unit handling wastewater with SS of 200, 2900 and 4600 mg/L. The manufacturer's recommended hydraulic loading for this unit is 2.19 m/hr (0.9 gpm/ft^2) and a solids loading of 9 Kg/m^2/hr (2 lbs/ft^2/hr). The actual hydraulic (governing) loading exceeded the recommended value in all three cases. In fact with low SS the hydraulic load could be as much as 3 times that recommended. Limited experience with short runs suggests that the hydraulic loads could be increased by at least 50% above the values noted without deterioration in DAF performance.

Table I. Summary of Pilot Plant Performance

	INFLUENT SUSPENDED SOLIDS CONCENTRATION		
	200 mg/L	2900 mg/L	4600 mg/L
Inflow (L/min)	91	68	32
(gpm)	24	18	8.5
Hydraulic Load (m/hr)	6.50[1]	4.88[2]	2.27[3]
(gpm/ft^2)	2.66	2.00	0.93
Solids load (Kg/m^2/hr)	1.44	13.7	10.4
(lbs/ft^2/hr)	0.29	2.81	2.3
Air saturation (%)	52	60	60
A/S	0.30	0.026	0.028
Recycle Ratio	1.2	1.5	2.5
Polymer dose (mg/g)	2.2	1.45	0.9
Effluent SS (mg/L)	40	18	<15
Float T.S. (%)	5.7	6.5	3.8

Note: Comparable Overflow Rates for settling tanks (Ontario Ministry of the Environment) are (1) 0.82 m/hr (2) 0.82 m/hr (3) 0.61 m/hr.

Choice of Recycle Liquid

Batch tests comparing the results when clear water was used for recycle liquid rather than DAF effluent, indicated that the clearer the recycle liquid, the better the performance of the DAF unit (1). The poorer results (higher effluent solids and lower float solids) when DAF effluent was used could have been due to the lower effective polymer dose remaining after some of the polymer had been utilized by the solids in the recycle liquid. In a continuous flow DAF unit, polymer is introduced after pressurization and a polymer loss would not occur. Consequently for a better comparison between batch and continuous flow units clear water should be used for recycle in the batch tests.

The choice of the most suitable liquid for recycle in continuous units is not as clear cut. Gehr (2) has demonstrated that polymer dosage beyond that needed to produce minimum effluent solids will not improve process performance. It appears as residual in the effluent or recycle stream and is still effective as a flotation aid. From the standpoint of polymer usage therefore, it makes no difference theoretically (i.e. where no polymer residual is provided) whether DAF effluent or some other liquid (of similar quality) is used for recycle. However from a practical standpoint a minimum residual level must be maintained.

Measurement of the polymer residual could be used to control polymer dosage. Batch tests indicate the amount of polymer completely taken up by the float, leaving no residual in the recycle liquid. In continuous flow units excess polymer must be routinely applied to cover periods of peak flow or high solids concentration. When recycled DAF effluent serves as the source of pressurized air, it can also provide the residual polymer needed which could in turn be used as the measure to control polymer dosage. If liquid other than DAF recycle were used for dissolving the air, slightly more polymer would have to be added since the residual would be continually lost.

SUMMARY

Extensive work with the DAF process has revealed a number of problems which reduce the effectiveness of this unit operation. The process requires expensive polyelectrolytes and is energy intensive. It is important therefore that it be operated as efficiently as possible

This paper has suggested several ways for upgrading DAF performance. Utilizing a 3 cylinder batch appartus to predict full scale requirements is one way. Improving the air dissolution and air monitoring systems is another. The process could be operated more efficiently if pilot tests were used to establish realistic loading criteria and if polymer residual were used to control polymer dosage. The use of DAF effluent for recycle rather than liquid from an external source would help to reduce polymer dosage.

Incorporation of the improvements suggested could make the DAF process more attractive to industry as an alternative method of thickening or clarifying sludges and slurries.

REFERENCES

1. Henry, J.G. and Gehr, R., "Dissolved Air Flotation for Primary and Secondary Clarification," Report submitted under C.M.H.C. contract no. 2002-01-10/72-25 (1978).

2. Gehr, R., "The Role of Polymers in Dissolved Air Flotation," Ph.D. Thesis, University of Toronto (1980).

3. Wood, R.F., and Dick, R.I., "Factors Influencing Batch Flotation Tests," Jour. Water Poll. Control Fed., 45, 2, pp. 304-315 (1973).

4. Gehr, R. and Henry, J.G., "Laboratory Techniques for Evaluating Batch Flotation of Activated Sludge," Proc. Wat. Poll. Res., 11, pp. 19-33 (1976).

5. Gehr, R., and Henry, J.G., "Measuring and Predicting Flotation Performance," Jour. Water Poll. Control Fed., 50, 2, pp. 203-215 (1978).

6. Gehr, R. and Henry, J.G., "Assessing Flotation Behaviour of Different Types of Sewage Suspensions," Prog. Wat. Tech. IAWPR/Pergamon Press Toronto, Vol. 12 pp. 1-21 (1980).

7. Eckenfelder, W.W. Jr., and Ford, D.L., "Flotation," Chapter 5 in Water Pollution Control, Jenkins Book Pub. Co. (1970).

8. Turner, M.T., "The Use of Dissolved Air Flotation for the Thickening of Waste Activated Sludge," Effl. and Wat. Treat. Jour., 15, 5 (1975).

9. Gulas, V., Lindsey, R., Benefield, L., Randall, C., "Factors Affecting the Design of Dissolved Air Flotation Systems," Jour. Water Poll. Control Fed., 50, pp. 1836-1840, (1978).

EFFECTS OF ACCELERATED DYNAMIC RESPONSE ON SETTLEABILITY
OF BIOLOGICAL SOLIDS IN ACTIVATED SLUDGE SYSTEMS

Mark J. Krupka. Polybac Corporation, Allentown, Pennsylvania.

Jeffrey M. Thomas. Polybac Corporation, Houston, Texas.

INTRODUCTION

Activated sludge systems are designed to perform two key functions. First to convert soluble organics to carbon dioxide, water and biological solids. Second to separate the solids generated from treated process water to provide a stable, uncontaminated effluent. The ability to achieve good solids-liquid separation in the secondary clarifier is critical to the overall performance of the system.

A number of factors have been found to affect the settling characteristics of biological solids. Among these are: concentration and characteristics of polymers; pH and toxic properties; agitation and shear forces; concentration of particles, and solids residence time (SRT). In one study it was clearly shown that an increase in SRT results in an effluent containing fewer suspended solids and a sludge that compacts better(1). The effects of SRT on solids settling characteristics can be attributed to its effects on the physiological state of the bacteria.

Effects of Physiological State on Bioflocculation

During growth in a batch system the physiological state of the cells is a function of the growth phase of the culture. The growth phase is determined by the number and type of bacteria in the system and the amount of substrate available.

Tenney and Verhoff(6) noticed that bioflocculation was also strongly influenced by the physiological state of the microorganisms. (Figure 1). They observed that dispersed growth often occurs during the logarithmic growth phase when the organisms are growing rapidly. Flocculation occurred best during the endogenous phase. Tenney and Stumm(5) attributed the dispersed growth to the fact that biopolymers either were not excreted or were excreted at a rate too low to provide

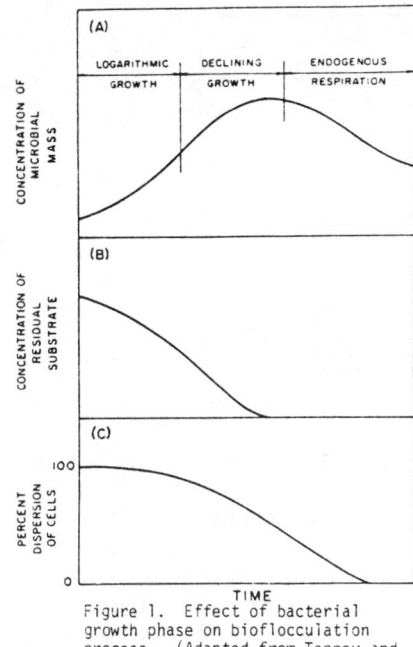

Figure 1. Effect of bacterial growth phase on bioflocculation process. (Adapted from Tenney and Verhoff (6)).

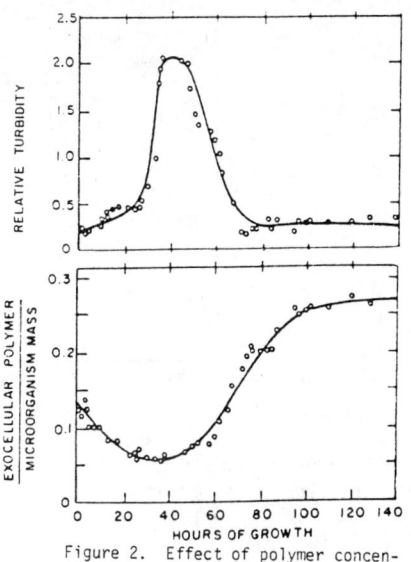

Figure 2. Effect of polymer concentration on bioflocculation process. (Adapted from Tenney and Verhoff (6)).

flocculation of the biomass as it was produced. (Figure 2).

As growth slowed and the physiological state of the cells changed improved flocculation was observed.

Although this information applies to growth in a batch culture, some inferences can be drawn as to how this change in physiological state applies to the operation of a continuous flow bioreactor. For the start-up of a continuous flow system, the dynamics are intially similar to those found in a batch system. However, once the desired equilibrium is reached the culture is maintained in this physiological state by continually feeding substrate and withdrawing mixed liquor.

In an activated sludge system, a form of continuous flow bioreactor, the best effluent quality is achieved by operating just into the zone of endogenous respiration. This zone also provides the optimum slow growth phase conductive to good solid setting.

Accelerating Dynamic Response of Biological Systems

Although the Monod model describes the growth of a pure culture in a biological system, it can be used generally to describe the response of the mixed population in an activated sludge system. According to Monod's model the μ or specific growth rate of an organism is a function of the available substrate concentration to which the bacteria are exposed. (Figure 3). This equilibrium concentration is represented by

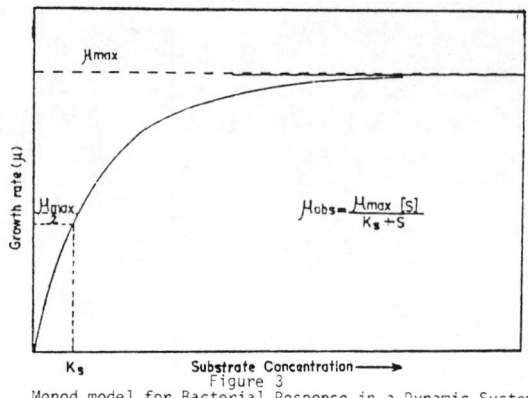
Figure 3
Monod model for Bacterial Response in a Dynamic System

the concentration of organics contained in the effluent from a treatment system.

As transient organic shocks are experienced in a system the F/M normally increases as does the equilibrium concentratio of organics. This results in a shift from the endogenous growth phase to the logarithmic growth phase with its associate dispersed characteristics. Through this natural response the system will increase its biological population to restore steady state operation to the system. The character of the Monod curve indicate the response time or time for the system to return to steady state operation, depending on the severity and duration of the shock.

Laboratory bench trials have indicated that through processes of selective adaptation and mutation superior growth and resistance to inhibition have been demonstrated with comple organic compounds. The ability to accelerate the dynamic response of a biological treatment system through the addition of specialized bacterial cultures was first demonstrated in a full scale side-by-side trial in a refinery treatment system.(7 A hypothetical model to explain this phenomenon has been proposed. (2) The ability to develop cultures with superior biochemical characteristics is common in the fermentation industry. (4) The ability to preserve these cultures for immediate use upon demand has provided several operating advantages in the operation of biological treatment systems.

Case Histories

In two full scale plants application programs for the addition of specialized cultures were engineered to improve the systems performance. In all cases, data defining organic removal efficiency and effluent clarity were compiled. The plants included a large pulp and paper mill and a combined municipal/industrial treatment system. For the pulp and paper waste LIGNOBAC[R], a freeze dried bacterial culture was utilized. FILABAC[tm]-M, a combination of enzyme, surfactants and bacteria formulated for inhibiting growth of filaments and augmenting the performance of the existing biomass were used in the municipal/industrial system.

Results

The data compiled on organic removal efficiency and effluent solid were analyzed for this relative frequency distribution. The plots of these analyses are presented in Figures 4-8.

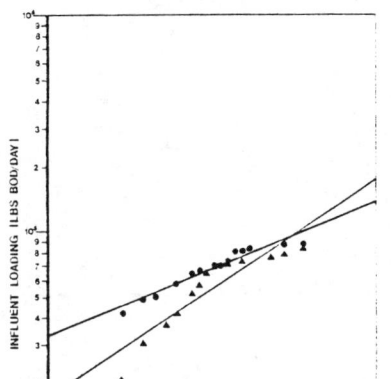

Figure 4
Effects of LIGNOBAC Addition on BOD
Removal-Pulp and Paper Waste

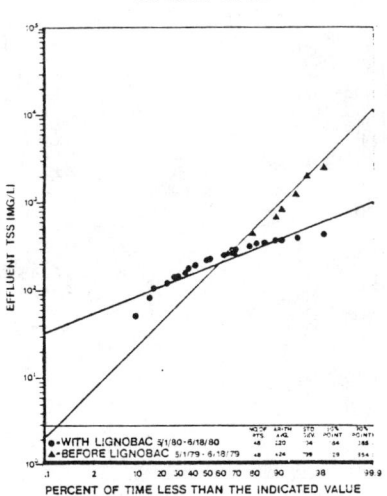

Figure 5
Effects of LIGNOBAC Addition on Effluent
Suspended Solids-Pulp and Paper Waste

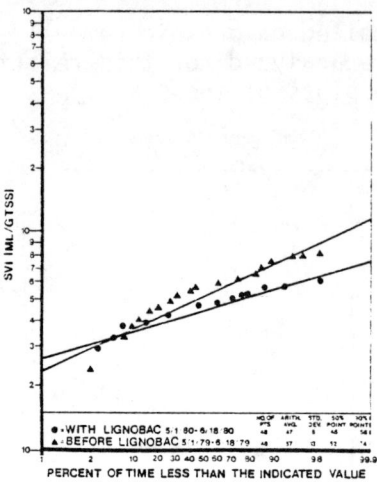

Figure 6
Effects of LIGNOBAC Addition on SVI-
Pulp and Paper Waste

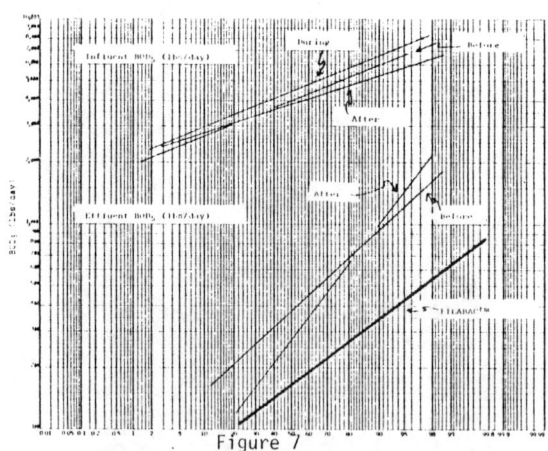

Figure 7
Effects of FILABAC-M Addition on Effluent BOD_5
Municipal/Industrial Waste

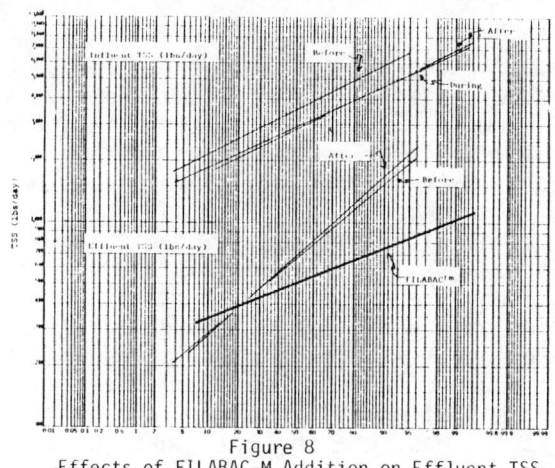
Figure 8
Effects of FILABAC-M Addition on Effluent TSS
Municipal/Industrial Waste

In the system treating the pulp and paper waste results of the performance during the trial were compared with the same operating period one year earlier. During this time no significant changes were made in the operation of the system. The plots show that even under higher loadings the system delivered higher quality effluent, particularly during shocks. This indicates the improved efficiency and stability of the system. The consistency of the settling characteristics were improved as indicated by the effluent TSS and SVI charts.

The results of the municipal/industrial treatment system trial were essentially the same. Improved organic removal efficiency, stability and solids settling compared with the operating periods before and after the trial.

Discussion

The results from these trials indicate the efficiency and stability can be improved through the addition of specialized bacterial cultures capable of accelerating the dynamic response of a biological system. Furthermore, this increased stability has been found to improve the settling characteristics of the biological solid in the system.

In this and other full scale trials the application has proven cost effective through reducing the total polymer consumption in the systems while achieving equivalent or better effluent quality.

SUMMARY

The settling characteristics of biological solids in activated sludge systems is affected by the physiological state of the biomass. Once the desired settling characteristics have been achieved, it is necessary to operate the system as close to steady-state as possible to maintain these properties. Most systems, even those with provisions for equalization experience radical fluctuations in organic loading. These conditions normally alter the steady-state operation of the system and hence, the physiological state and settling characteristics of the biological solids.

By improving the system's dynamic response to the variations in organic loading, more consistent steady-state operation of the system will result. In a number of documented case histories, it has been observed that the dynamic response of an activated sludge system can be improved substantially by the addition of specialized bacterial cultures (McDowell, Purdue 1978). The primary objective of the full-scale trials and stability were observed as well as improvements in effluent suspended solids. These improvements were attributed to the accelerated dynamic response of the systems.

This paper will present data from two full-scale plant trials. A relative frequency distribution analysis was performed on influent and effluent BOD and these results were correlated with effluent suspended solids and/or SVI to correlate the effects of system stability on solid settling.

REFERENCES

1. Bisogni, J. J. and Lawrence, A. W.
"Relationship Between Biological Solids Retention Time and Settling Characteristics of Activated Sludge," Water Research 5, 753-763. (1971).

2. McDowell, C. S. and Zitrides, T. G.
"Accelerating the Dynamic Response of Bacterial Populations in Activated Sludge Systems", 34th Annual Purdue Industrial Waste Conference Purdue University, West Lafayette, Ind. May 8-10, 1979.

3. Palm, J. C., Jenkins, D., and Parker, D. S.
"The Relationship Between Organic Loading, Dissolved Oxygen Concentration and Sludge Settleability in the Completely Mixed Activated Sludge Process," 51st Annual Conference Water Pollution Control Federation Ansheirn, Cal. , (1978).

4. Pelszcn, M. J. and Reivel, R. D.
Microbiology, 3rd Edition, McGraw-Hill, p. 218, 1972.

5. Tenney, M. W. and Stumm, W.
   "Chemical Flocculation of Microorganisms in Biological Waste Treatment," *Journal of Water Pollution Control Federation,* 37, 1370-1388, (1965).

6. Tenney, M. W. and Verhoff, F. H.
   "Chemical Flocculation of Microorganisms in Biological Wastewater Treatment," *Biotechnology and Bioengineering,* 15, 1045-1073, (1973).

7. Tracy, K. D. and Shah, P. S.
   "Application of Bacterial Additives to Refinery Activated Sludge" 50th Annual Meeting, California Water Pollution Control Federation, Sacremento, Cal. April 27, 1978.

SOME ASPECTS OF IRON AND LIME VERSUS POLYELECTROLYTE
SLUDGE CONDITIONING

G. Lee Christensen. Civil Engineering Department,
Villanova University, Villanova, Pennsylvania.

Stephen G. Wavro. Exxon Company USA,
Baytown, Texas.

INTRODUCTION

    The Water Pollution Control Act Ammendments of 1972 and the Clean Water Act of 1977 focused a substantial national effort on the installation of municipal and industrial wastewater treatment facilities. As these facilities comply with more stringent effluent standards, larger sludge volumes are generated, and provisions must be made for their proper processing and disposal. An essential step in the sludge disposal schemes of most intermediate and larger facilities is volume reduction by mechanical dewatering. Vacuum filtration was the preferred process for mechanical dewatering from its introduction early in this century through the sixties. During this same period, lime and iron salts were the preferred chemicals for sludge conditioning prior to vacuum filtration dewatering.
    About two decades ago polyelectrolyte coagulants were introduced as alternatives to iron salts and lime for chemical sludge conditioning. In the last decade new sludge conditioning and dewatering processes and equipment have been introduced in response to the large, national commitment of funds to wastewater treatment facilities. The most significant introductions being a thermal process for sludge conditioning and plate and frame presses, belt presses, and centrifuges for mechanical sludge dewatering. So, at present, designers and operators of sludge conditioning and mechanical dewatering processes can choose from among three conditioning procedures to proceed one of four mechanical dewatering devices.

Thermal conditioning is generally restricted to the larger facilities employing incineration with heat recovery because of the high cost of fuel. The choice of a chemical conditioning system depends, principally, on which mechanical dewatering equipment is selected. Iron and lime are the coagulants of choice for plate and frame presses while polyelectrolytes are the coagulants of choice for belt presses and centrifuges. It is only with vacuum filtration that an initial and continuing decision must be made between inorganic and organic conditioners.

A number of studies have pinpointed the advantages and disadvantages associated with iron and lime or polyelectrolyte conditioning systems since polyelectrolytes were introduced 20 years ago. For example, polyelectrolytes require doses which are two orders of magnitude lower than comparable iron and lime doses (ca. 0.1 percent and ca. 10 percent respectively). These lower doses translate into savings in the cost of storage, preparation, and delivery of the conditioner. On the other hand, iron and lime have substantial side benefits associated with their use including disinfection, odor control, and an ability to immobilize heavy metals. Scaling problems and the generation of chemical solids (iron and calcium precipitates) are often quoted disadvantages of the iron and lime conditioning system.

The purpose of this paper is to report on the results of a laboratory investigation into several aspects of iron-lime and polyelectrolyte sludge conditioning for vacuum filtration including lime requirements for conditioning, sensitivity to dose, dilution of the sludge stream by the coagulant stream, and the significance of chemical solids generation in the iron and lime conditioning process.

EXPERIMENTAL RESULTS

Lime Requirements for Sludge Conditioning

Over the years numerous articles have reported on the iron and lime doses required for effective wastewater sludge conditioning. Ferric chloride doses generally range from 2 - 10% and lime doses from 0 - 30 percent (1). While sludges do vary in their response to lime addition, few biological sludges can be effectively conditioned (specific resistance $\leq 10^{12}$ m/kg) without lime. Usually, iron and lime requirements are higher for waste activated sludges than for primary sludges and higher for digested sludges than for raw sludges (1-3). Experimental results from the present study (not reported herein) are in agreement with the findings of others, but did indicate that lime requirements for satisfactory conditioning are increased much more dramatically by sludge storage than iron requirements.

Polyelectrolyte Requirements for Sludge Conditioning

The selection of an effective polyelectrolyte for a particular sludge conditioning process is a trial and error procedure. While countless conditioning experiments have been conducted with polyelectrolytes over the last two decades, the results usually describe characteristics of the sludge rather than the polyelectrolyte (4). Although a lot of data has been generated on polyelectrolytes, the results of most tests are applicable only to the sludge tested. Studies by Vesilind on polymer usage showed that a polyelectrolyte may be effective on one sludge, but may be completely ineffective on another sludge with seemingly identical characteristics (5). Furthermore, polymers with very similar characteristics will produce varied results on the same sludge.

In order to compare the characteristics of polyelectrolyte conditioning with the characteristics of iron and lime conditioning discussed above, polyelectrolyte conditioning tests were conducted on the same sludge types from the same sources used in the iron and lime conditioning tests. A total of nine polyelectrolytes and three sludge types were investigated. The polyelectrolytes represented two major manufactures, and were all cationic with a range of molecular weights and linear charge densities. The results of two of these tests are presented in Figures 1 and 2.

The data contained in Figures 1 and 2 generally substantiate the rather specific nature of the polyelectrolyte-sludge solids interaction. For example, HF-849 was the most effective polymer on raw mixed primary and waste activated sludge (Figure 1). Nevertheless, HF-849 was the least effec-

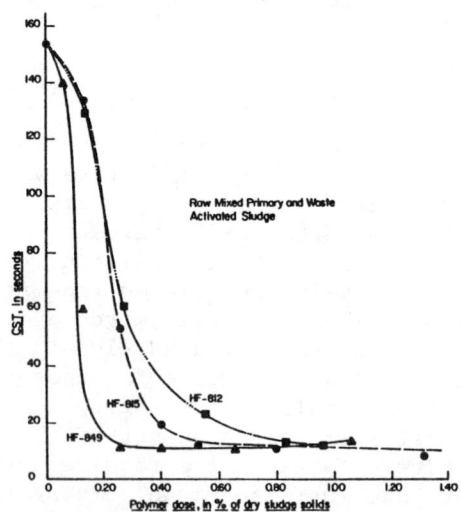

Figure 1. Comparison of Three Polyelectrolytes on a Raw Mixed Primary and Waste Activated Sludge.

tive polymer on the anaerobically digested mixed primary and trickling filter sludge (Figure 2). This rather specific

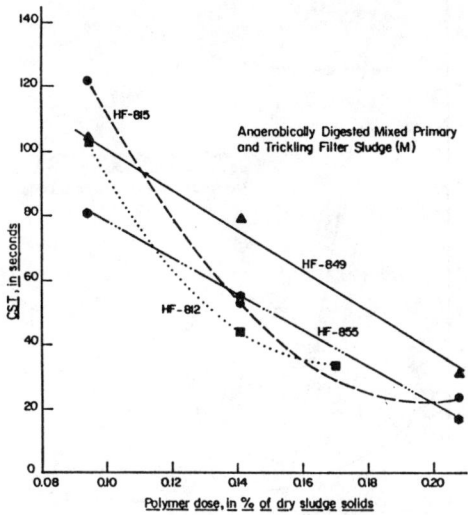

Figure 2. Comparison of Four Polyelectrolytes on an Anaerobically Digested Mixed Primary and Trickling Filter Sludge.

chemical interaction between polyelectrolytes and sludge solids has been noted by others. Cassel and Johnson, for example, were unable to settle on a polyelectrolyte for conditioning at the Blue Plains Plant because no single polyelectrolyte was able to handle the daily variations in sludge quality experienced at that plant (6).

A direct comparison of iron and lime conditioners with a polyelectrolyte conditioner was conducted for an anaerobically digested sludge (see Figure 3). The most obvious difference between iron and lime conditioning and polyelectrolyte conditioning is the order of magnitude of the polyelectrolyte dose required for effective conditioning. While iron and lime doses range from 0 to 10 percent and 0 to 30 percent, respectively, polyelectrolyte doses range from 0.2 to 1.0 percent. The other obvious difference between these two chemical conditioning systems is the sensitivity of conditioning to coagulant dose. Iron and lime conditioning is quite insensitive to dose whereas polyelectrolyte conditioning is quite sensitive to dose. In Figure 3, sludge dewaterability is virtually unaffected by lime doses between 14 and 22 percent. On the other hand, an increase to 0.8 percent or a decrease to 0.4 percent in polyelectrolyte dose reduces the dewaterability of the polyelectrolyte conditioned sludge by a factor of six.

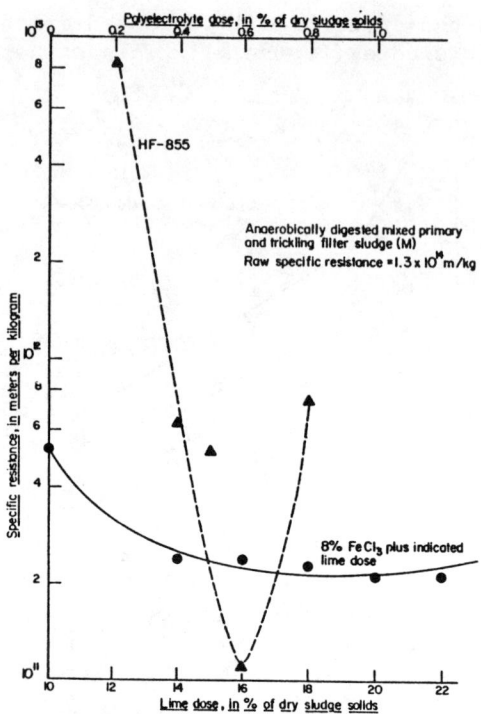

Figure 3. Comparison of Inorganic and Organic Conditioners for an Anaerobically Digested Mixed Primary and Trickling Filter Sludge.

Effect of Conditioner Selection on Sludge Solids Concentration in the Dewatering Process

The effectiveness and cost of mechanical dewatering equipment are a strong function of the feed sludge solids concentration. For example, the thicker the sludge fed to a vacuum filtration unit, the higher the yield and the higher the cake solids concentration. This concept is also applicable to other mechanical dewatering devices including centrifuges, belt presses, and plate and frame presses. All of these processes reduce the sludge volume per unit mass of dry sludge solids by removing water, so reducing the volume of water per unit mass of dry sludge solids fed to the dewatering equipment produces a better dewatering result.

One disadvantage of chemical conditioning is the addition of water to the sludge solids when mixing the coagulant stream with the sludge stream, i.e. the thickened sludge stream is diluted by the coagulant stream (coagulant dissolved and/or suspended in water). The net effect is a reduced sludge solids concentration in the feed to the dewatering process. The amount of sludge solids dilution is a function of the concentration of coagulant in the coagulant stream, the dose of coagulant required (in percent of dry sludge solids), and the concentration of sludge solids in the sludge stream. Although polyelectrolytes require significantly lower doses for effective conditioning than ferric chloride and lime (on a percent of dry solids basis), conditioning with polyelectrolytes produces a more dilute conditioned sludge because polyelectrolyte coagulant streams are much more dilute than iron and lime coagulant streams, e.g. 0.01 to 0.1 percent versus 10 percent respectively. In order to illustrate specifically the effect of coagulant dilution on subsequent dewatering, an analysis of a hypothetical chemical conditioning-vacuum filtration dewatering sequence will be examined.

The effect of the sludge feed solids concentration on vacuum filtration yield can be expresses as (9).

$$Y = \left(\frac{2P\Theta}{\mu r}\right)^{\frac{1}{2}} \cdot \frac{W^m}{t^n} \qquad (1)$$

where

Y = yield, in kg/m^2 sec
P = pressure difference, in N/m^2
μ = absolute viscosity, in kg/m · sec
r = specific resistance, in m/kg
Θ = submergence, unitless
W = mass of sludge solids deposited per unit volume of filtrate, in kg/m^3
t = cycle time, in sec
m, n, = coefficients, unitless

A theoretical development of sludge filtration yield indicates that m should have a value of 0.5. In practice, however, the value of m is closer to 1.0 indicating that yield is directly proportional to the sludge solids concentration in the feed (7). So the drop in yield expected on a vacuum filter due to the dilution associated with coagulant addition will be roughly proportional to the extent of dilution.

In a typical ferric chloride and lime conditioning process, the sludge stream would be approximately 5 percent solids (50kg/m^3), the ferric chloride and lime doses would be 8 and 20 percent on a dry solids basis, respectively, and the coagulant streams would be fed as 10 percent (100 kg/m^3) solutions. These conditions would produce a solids concentration in the feed to the dewatering equipment (neglecting the solids added by iron and calcium precipitates) of 4.4 percent or (44 kg/m^3). If you include the precipitates formed (ferric hydroxide and calcium carbonate), the feed solids concentration would be 5.6 percent (56 kg/m^3).

In the case of polyelectrolyte conditioning, a typical conditioning process might set the coagulant stream at 0.03 percent (0.3 kg/m^3) with a dose of 0.4 percent (8 pounds per ton) on a dry solids basis. Under these conditions with the sludge stream at 5 percent solids, the feed to the dewatering process would have a solids concentration of 3.0 percent (30 kg/m^3). This amounts to a 67% increase in the volume of sludge fed to the dewatering equipment. So, for the same sludge stream the ferric chloride and lime conditioning produces a feed to the dewatering equipment of 4.4 percent neglecting precipitates and 5.6 percent including precipitates while polyelectrolyte conditioning produces a comparable feed of only 3 percent solids. The thicker the sludge stream before conditioning the more pronounced the effect of polyelectrolyte dilution (see Figure 4).

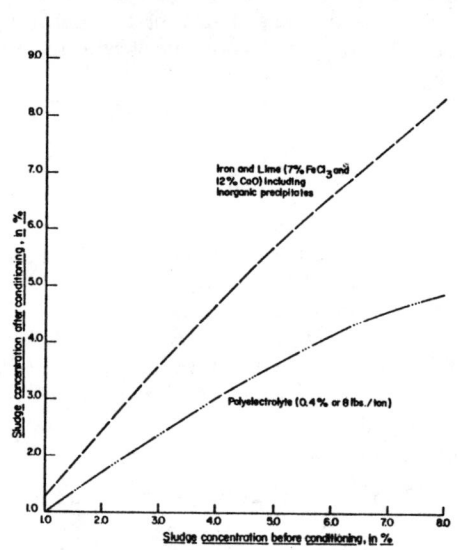

Figure 4. Dilution of Sludge Solids Concentration Associated with Chemical Conditioning

The significance of the dilution associated with polyelectrolyte conditioning for vacuum filtration dewatering can be assessed with the aid of Equation 1. If you take the theoretical value of m=0.5, the ratio of the iron and lime conditioned sludge yield to the polyelectrolyte conditioned sludge yield would be proportional to $(5.6/3.0)^{0.5}$ or 1.36. This factor of 1.36 includes the precipitates for the iron and lime addition, and should be adjusted to strictly sludge solids yield, i.i., 1.36 x (50/64) = 1.06. If m = 1, the yield factor would be proportional to $(5.6/3.0)^{1.0}$ = 1.87 for total yield (including iron and lime precipitates) and proportional to 1.87 x (50/64) = 1.46 on a strictly sludge solids basis.

The significance of the above analysis lies in the presentation of factors which may more than offset an often quoted disadvantage of iron and lime coagulants. The often quoted disadvantage is the generation of dry chemical solids, over and above the dry, biological, sludge solids being coagulated. These iron and calcium precipitates closely approximate, on a dry basis, the amount of ferric chloride and lime (as CaO) added (6). Nevertheless, the significantly higher solids concentration fed to the dewatering process in iron and lime conditioning compared with polyelectrolyte conditioning may more than compensate for the added precipitates in terms of <u>sludge</u> solids yield and final moisture content of the dewatered cake.

Interaction Between Conditioning and Dewatering

Conditioners for a sludge dewatering process are usually selected on the basis of CST and/or specific resistance analyses. After one or more conditioners are selected their performance on specific types of dewatering equipment can be evaluated by appropriate laboratory or pilot scale equipment, e.g. a filter leaf in the case of vacuum filtration. An earlier section of the present discussion reported the results of CST and specific resistance analyses on ferric chloride and lime and a selection of polyelectrolytes on a variety of sludge types. In this section the most effective polyelectrolyte and ferric chloride and lime are employed in several filter leaf analyses on a particular sludge to compare the performance of these conditioners in a vacuum filtration process.

The sludge employed in the filter leaf testing was an anaerobically digested mixed primary and waste activated sludge. The most effective polymer for this sludge was HF-849 (see Figure 5). For the filter leaf testing the optimum doses for HF-849 and iron and lime were employed along with the most effective mixing regime for the respective conditioners. The conditioned sludge volume was set at 2 liters, and two combinations of form and dry time were tested. The results are presented in Table 1.

Even though the polyelectrolyte HF-849 looked more promising than iron and lime after specific resistance testing (see Figure 5, the filter leaf results show that iron and lime produced higher total solids yield (includes ferric hydroxide and calcium carbonate precipitates), higher sludge solids yield (precipitates not included) and a higher cake solids concentration than the same sludge conditioned with an effective polyelectrolyte. The most likely explanation for the poorer performance of the polyelectrolyte (even though the polyelectrolyte looked very promising after specific resistance analyses) is the dilution associated with polyelectrolyte conditioning.

Whenever alternative chemical conditioners are being evaluated it is necessary to understand, and take into account, the interactions which exist between chemical conditioner, dewatering equipment and subsequent disposal schemes, e.g. incineration or land disposal (8). With incineration

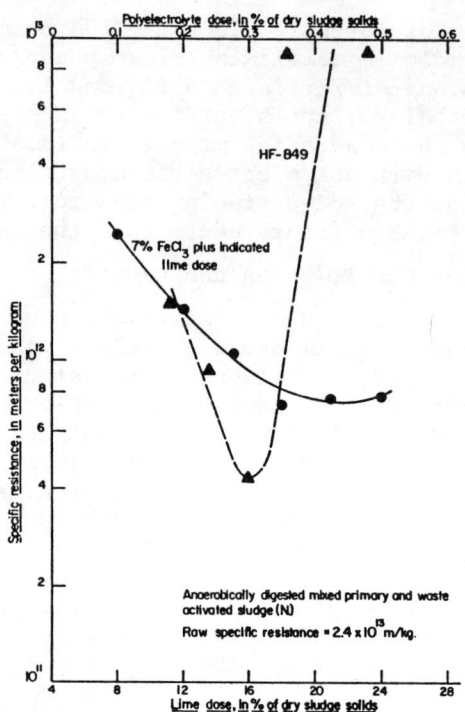

Figure 5. Comparison of Inorganic and Organic Conditioners for an Anaerobically Digested Mixed Primary and Waste Activated Sludge

the cost of auxillary fuel is a major concern associated with this disposal scheme and the quantities of auxillary fuel required are directly affected by the water to sludge solids ratio of the cake feed. In turn, the water to sludge solids ratio in the cake feed depends on the nature and effectiveness of the conditioning and dewatering processes.

Similar interactions exist for land filling or land utilization schemes for sludge disposal. An important consideration in land disposal schemes is trucking costs. Trucking costs will depend directly on the tonnages hauled. For a specific plant, the tonnage of dry sludge solids produced (primary plus secondary) is fixed, but the wet tonnage to be hauled depends on the effectiveness of the conditioning/dewatering sequence. The key parameter in this case is the wet weight of the dewatered sludge per unit weight of dry sludge solids.

Many wastewater treatment plant engineers and operators are aware of the dry chemical solids (ferric hydroxide and calcium carbonate precipitates) which are added to the dry biological sludge solids in an iron and lime conditioning process. Many operators and engineers perceive these precipi-

TABLE I. FILTER LEAF RESULTS - IRON AND LIME VERSUS POLYELECTROLYTES

Result	CYCLE I Form time = 30 sec. Drying time = 120 sec.		CYCLE II Form time = 120 sec. Drying time = 120 sec.	
	Iron and lime[a]	Polyelectrolyte[b]	Iron and lime[a]	Polyelectrolyte[b]
Dry cake, in grams	10.7	7.0	16.5	9.8
Total solids yield, in lb/ft^2/hr	5.7	3.7	5.5	3.2
Sludge solids yield, in lb/ft^2/hr	4.5	3.7	3.7	3.2
Cake solids content, in percent	21.6	15.5	15.3	11.3

[a]Dose = 7% FeCl and 21% on a dry sludge solids basis.
[b]Dose = 0.36% of HF=849 on a dry sludge solids basis.
[c]Sludge was anaerobically digested mixed primary and waste activated sludge.

tates as always placing iron and lime at a disadvantage with respect to polymers which do not add chemical solids to the sludge. While it is true that the chemical precipitates may be a problem in some conditioning/dewatering/disposal schemes, it is not true that these precipitates are a problem in all conditioning/dewatering/disposal schemes. A rational evaluation of alternative chemical conditioners must be based on an analysis of the conditioning/dewatering/disposal interaction described above employing appropriate parameters such as water to sludge solids ratio and wet weight of cake per unit weight of dry sludge solids.

The filter leaf data presented earlier in Table 1 can be used to illustrate the concepts described above. For the 30 seconds form time and 120 seconds drying time, the polyelectrolyte conditioned sludge had a water to dry sludge solids ratio of 5.45 pounds of water per pound of dry sludge solids while the ferric chloride and lime conditioned sludge had a ratio of 4.65 pounds of water per pound of dry sludge solids. For the same set of filter leaf tests, the polyelectrolyte conditioned sludge had a wet weight to dry sludge solids ratio of 6.45 pounds of wet cake per pound of dry sludge solids while the iron and lime conditioned sludge had a comparable ratio of 5.93. These results demonstrate that iron and lime conditioning is capable of producing a more desirable, dewatered cake than polyelectrolyte conditioning. Under different circumstances, e.g. different sludge type, plant, or polyelectrolyte, polyelectrolyte conditioning may produce a more desirable result. The point is that in all cases a rational analysis of the dewatering result with competing conditioners is required.

CONCLUSIONS

The following conclusions on chemical conditioning for vacuum filtration dewatering can be drawn from this study:
1. Lime requirements for sludge conditioning (with ferric chloride) are variable and depend on sludge type and source. In general, however, lime requirements increase with the age of the sludge, e.g. fresh sludges require less lime than stale sludges and raw sludges require less time than anaerobically digested sludges.
2. Sludge conditioning with polyelectrolytes is an inherently less stable process than sludge conditioning with iron and lime. This instability with polyelectrolytes is a result of the significant variability in the surface chemistry and concentration of sludge solids at a given plant along with the (1) specific nature of the polyelectrolyte-sludge solids interaction, and (2) the possibility of process deterioration due to overdosing. Iron and lime conditioning is inherently more stable because changes in the surface chemistry and/or the concentration of the sludge solids does not significantly affect process results.
3. Chemical conditioning with polyelectrolytes substantially reduces the concentration of solids in the sludge fed to subsequent dewatering process. This reduced solids concentration adversely affects the outcome of the mechanical dewatering processes.
4. The formation of chemical solids (ferric hydroxide and calcium carbonate precipitates) associated with ferric chloride and lime conditioning is perceived by many to be an inherent disadvantage of inorganic conditioners. Nevertheless, these chemical solids along with the more concentrated coagulant streams employed with inorganic coagulants assures a more concentrated sludge feed to the dewatering equipment. This more concentrated feed to the dewatering equipment may more than offset any disadvantage associated with chemical solids formation.
5. In the evaluation process for competing chemical conditioners, interactions between conditioning, dewatering, and disposal processes should be examined, e.g. the significance of the chemical solids formed by the inorganic coagulants in the overall solids processing and disposal scheme should be examined. The key parameters for dewatered cake evaluation are water to sludge solids ratio for subsequent cake incinera-

tion and wet weight of cake per unit weight of sludge solids for trucking to land disposal sites.

REFERENCES

1. USEPA, <u>Process Design Manual for Sludge Treatment and Disposal</u>, USEPA-ERIC Technology Transfer, Cincinnati, (1979).
2. Tenney, M.W., et al., "Chemical Conditioning of Biological Sludges for Vacuum Filtration," <u>J. Water Poll. Control Fed.</u>, 42(3), R1 (1970).
3. Harrison, J.R., "Developments in Dewatering Wastewater Sludges," <u>Vol. 1 Sludge Treatment and Disposal</u>, USEPA-ERIC Technology Transfer, Cincinnati, (1978).
4. Novak, J.J., and O'Brien, J.J., "Effect of pH and Mixing on Polymer Conditioning of Chemical Sludge," <u>J. Amer Water Works Assoc.</u>, 69(11), 600 (1977).
5. Vesilind, P.A., "Polymer Usage Gaining for Sludge Dewatering," <u>Water and Wastes Engr.</u>, 50 (1971).
6. Cassel, A.F., and Johnson, B.P., "Evaluation of Dewatering Devices for Producing High-Solids Sludge Cakes," EPA-600/2-79-123, USEPA, (1979).
7. <u>Wastewater Treatment Plant Design</u>, Water Pollution Control Federation M.O.P. No. 8, Lancaster Press, (1977).
8. Christensen, G.L., Elliott, W.R., and Johnson, W.K., "Interactions Between Sludge Conditioning, Vacuum Filtration, and Incineration," <u>J. Water Poll. Control Fed.</u>, 48(8), 1955 (1976).
9. Eckenfelder, W.W., and Ford, D.L., <u>Water Pollution Control</u>, The Pemberton Press, Austin, (1970).

ACKNOWLEDGEMENTS

This work was financially supported by a grant from the National Lime Association. At the time of this work, Stephen G. Wavro was a graduate student in Water Resources Engineering at Villanova University.

TREATMENT OF INDUSTRIAL WASTEWATERS WITH FORMALDEHYDE

Franciszek S.Tużnik. Environmental Protection Department, Institute of Precision Mechanics. Warszawa, Poland.

INTRODUCTION

Formaldehyde /methanol/ HCHO is the simplest aliphatic aldehyde and one of the simplest carbon compound as well. In industry it is used as a 35-40% water solution commonly known as formaline, stabilized with a 4-11% addition of methanol /1/.

HCHO is a highly active chemical compound. It is quite a good reducing agent and can be oxidized to carbon dioxide and water /Fig.1/.

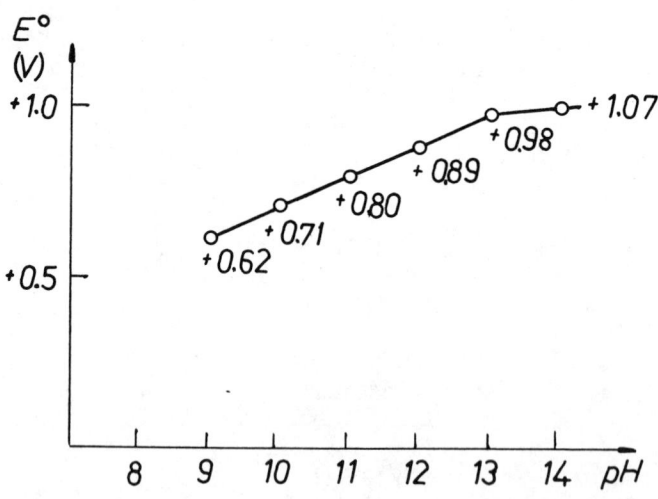

Figure 1. The dependence of the reduction potential of formaldehyde E^o upon the pH level.

With an increase in the pH level the reduction properties of formaldehyde increase almost proportionally.

A cyanohydrin reaction can serve as an example illustrating the properties of the carbonyl group.

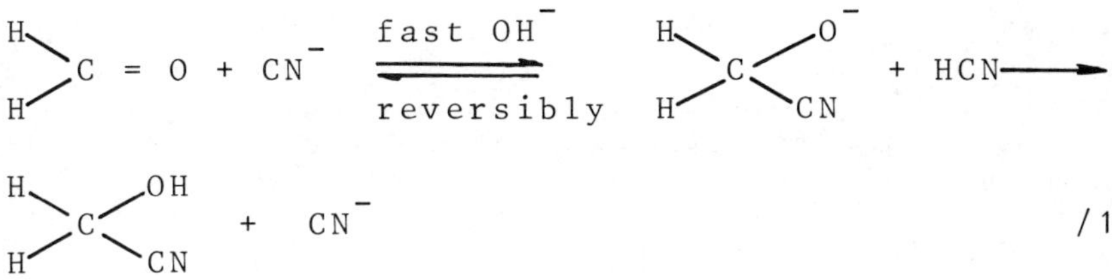

$$\underset{H}{\overset{H}{>}}C=O + CN^- \underset{\text{reversibly}}{\overset{\text{fast } OH^-}{\rightleftharpoons}} \underset{H}{\overset{H}{>}}C\underset{CN}{\overset{O^-}{<}} + HCN \longrightarrow$$

$$\underset{H}{\overset{H}{>}}C\underset{CN}{\overset{OH}{<}} + CN^- \qquad\qquad /1/$$

Alkalies speed up and acids delay the reaction /1/, the second stage determines the reaction rate. This is obvious, since HCN is not a nucleophil, thus only a base catalyst can pull away a proton from it and create a strongly nucleophilic, very active anion CN^-.

Another example illustrating the properties of the carbonyl group is the ammonia attachment.

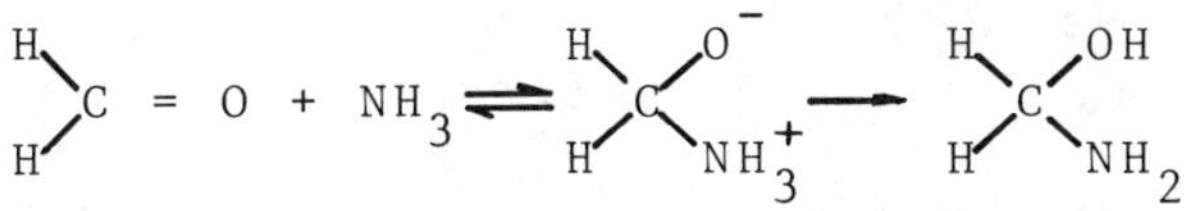

$$\underset{H}{\overset{H}{>}}C=O + NH_3 \rightleftharpoons \underset{H}{\overset{H}{>}}C\underset{NH_3^+}{\overset{O^-}{<}} \longrightarrow \underset{H}{\overset{H}{>}}C\underset{NH_2}{\overset{OH}{<}} \qquad /2/$$

THE POSSIBILITIES OF UTILIZING FORMALDEHYDE IN TREATMENT OF LIQUID WASTES.

Formaldehyde due to its properties, outlined in the introducing, is a very good agent for treating certain industrial liquid wastes and for the recovery of metals from wastewaters.

Thus, HCHO is used for treatment for all kinds of wastewaters containing cyanides, such as:
- electroplating wastes,
- wastes from heat treatment operations,
- blast-furnace wastes,
- wastes from hydrogen cyanide production.

HCHO is used also for treatment of various types of wastewaters containing hexavalent chromium Cr^{+6}, e.g.:
- electroplating wastes,
- tannery wastes.

As an agent for the recovery of metals from was-

tewaters, HCHO is most commonly used in the recovery of silver, copper, cadmium, zinc and chromium from the following types of wastes:
- electroplating wastes,
- wastes from photographic processes.

TREATMENT OF WASTEWATERS CONTAINING CYANIDES

Formaline was first used for treatment of wastewaters containing cyanides. A method has been developed in GDR /2/ for the decomposition of cyanides in hardening shop solutions. The chemical mechanism of the reaction was presented as follows:

a. $NaCN + HCHO \rightleftharpoons CH_2ONaCN$

b. $CH_2ONaCN + H_2O \rightleftharpoons CH_2OHCN + NaOH$

c. $CH_2OHCN + NaOH + H_2O \rightleftharpoons CH_2OHCOONa + NH_3$

d. $CH_2OHCN + NH_3 \rightleftharpoons CH_2NH_2CN + H_2O$ /3/

e. $CH_2NH_2CN + NaOH + H_2O \rightleftharpoons CH_2NH_2COONa + NH_3$

f. $CH_2OHCN + CH_2NH_2CN \rightleftharpoons NH(CH_2CN)_2 + H_2O$

g. $NH(CH_2CN)_2 + 2NaOH + 2H_2O \rightleftharpoons NH(CH_2COONa)_2 + 2NH_3$

h. $NH(CH_2CN)_2 + CH_2OHCN \rightleftharpoons N(CH_2CN)_3 + H_2O$

i. $N(CH_2CN)_3 + 3NaOH + 3H_2O \rightleftharpoons N(CH_2COONa)_3 + 3NH_3$

and totally, where the mole ratio of NaCN to HCHO was 1:1:

$7NaCN + 7HCHO + 8H_2O \rightleftharpoons N(CH_2COONa)_3 +$
$+ CH_2OHCOONa + 3CH_2NH_2COONa + 3NH_3$ /4/

and where the mole ratio of NaCN to HCHO was equal to 1:2.5:

$10HCHO + 4NaCN + 2H_2O \rightleftharpoons (CH_2)_6N_4 + 4CH_2OHCOONa$ /5/

The method for treatment of cyanide wastewaters from a blast-furnace gas cleaning plant developed in Poland was based on the above described reaction chemism, however, the concentration of cyanides used was conciderable lower, approx. 0-500 mg/dm^3.

The US company DuPont de Namours developed the so called Kastone method /4/. This method is based on using a mixture of formaldehyde and peroxides /usually H_2O_2 directly or indirectly/ for treating cyanides in diluted electroplating wastewaters. The mechanism of the reactions which take place was presented as follows:

a. $CN^- + H_2O_2 \rightarrow NCO^- + H_2O$

b. $CN^- + HCHO + H_2O \rightleftharpoons CH_2OHCN + OH^-$

c. $[Zn/CN/_4]^{2-} + 4HCHO + 4H_2O \rightleftharpoons 4CH_2OHCN + 4OH^- + Zn^{2+}$

d. $CH_2OHCN + H_2O \rightarrow CH_2OHCONH_2$

e. $NCO^- + 2H_2O \rightarrow NH_4^+ + CO_3^{2-}$ /6/

f. $CN^- + 3H_2O \rightarrow HCOO^- + NH_4OH$

g. $Zn^{2+} + 2OH^- \rightarrow ZnO\downarrow + H_2O$

h. $CH_2OHCONH_2 + 2H_2O \rightarrow CH_2OHCOOH + NH_4OH$

i. $NH_4^+ + NCO^- \rightleftharpoons NH_2CONH_2$ (carbamide)

It is assumed that reactions d, e and f catalyzed by hydrogen peroxide in an alkaline conditions. So far, this method has been used in the decomposition of cyanides containing Zn^{+2} and Cd^{+2} ions.

The FKJA method of treatment of wastewaters and spent concentrated solutions from the electroplating shops with simultaneous recovery of metals was developed at the Institute of Precision Mechanics. The author of this paper has been instrumental in developing this method. The FKJA method is patented in Poland /5/ and abroad /6/. /Both the method as well as the equipment are patented/.

Formaldehyde is the main component of the FKJ-72 reagent used in the FKJA method for recovery of such metals as: silver, copper, zinc, cadmium, chromium, and for the decomposition of cyanides. The series of types of equipment used to run this process in industrial conditions is called LAFT /7,8,9/.

The assumption behind the FKJA method is the treatment off, and recovery of metals from:
   a. concentrated electroplating baths dragged out on products during chemical rinsing immediately after plating /Fig. 2/.

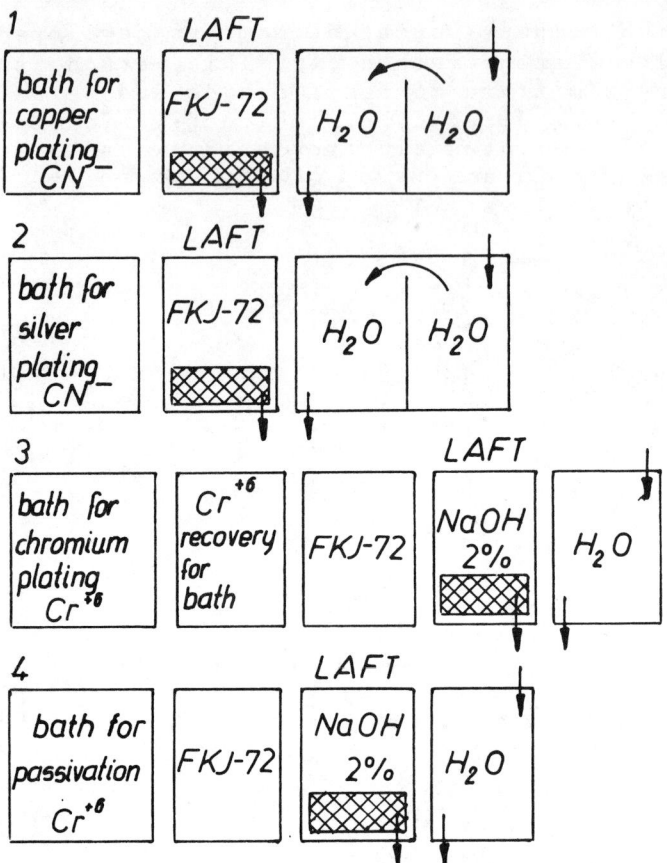

Figure 2. Treatment of wastewaters and recovery of metals using the FKJA method. 1. Recovery of Cu, 2. Recovery of Ag, 3 and 4. Recovery of $Cr(OH)_3$.

b. concentrated spent, electroplating baths periodically dumped which do not have to be diluted.

With reference to /a/ there is always a large excess of formaldehyde in relation to the substance to be decomposed the chemical rinse tank.

With reference to /b/ the reaction is usually run in accordance with stoichiometry; there is a small excess, of the 15-30% order, of formaldehyde in relation to CN^- and the amount of the oxidizing agent, e.g.: Ag^+ and Cu^+ /10/.

Depending on the type of cyanide solution and whether formaldehyde was used in excess /as in the case of chemical rinsing/ or in accordance with the stoichiometry of the reaction, it is possible to suggest a number of variants of the chemical reactions which take place:

Variant: A

a. $3Na_2[Cd(CN)_4] + 14HCHO + 3H_2O \rightarrow 4CH_2OHCOONa +$
   $+ 2CH_2[N(CH_2CN)_2]_2 + 3CdO\downarrow + 2NaOH$

b. $12Na[Ag(CN)_2] + 31HCHO + 2NaOH \rightarrow 4CH_2[N(CH_2CN)_2]_2 +$
   $+ 8CH_2OHCOONa + 3Na_2CO_3 + 12Ag°\downarrow$ /7/

c. $2Na_2[Cu(CN)_3] + 9HCHO + H_2O \rightarrow CH_2[N(CH_2CN)_2]_2 +$
   $+ 2CH_2OHCOONa + 2HCOONa + 2Cu°\downarrow + H_2\uparrow$

Variant: C

a. $3Na_2[Cd(CN)_4] + 18HCHO + 3H_2O \rightarrow 6CH_2OHCOONa +$
   $+ 3CdO\downarrow + 2(CH_2=NCH_2CN)_3$

b. $12Na[Ag(CN)_2] + 39HCHO + 6NaOH \rightarrow 12CH_2OHCOONa +$
   $+ 12Ag°\downarrow + 3Na_2CO_3 + 4(CH_2=NCH_2CN)_3$ /8/

c. $2Na_2[Cu(CN)_3] + 11HCHO + H_2O + NaOH \rightarrow 2Cu°\downarrow +$
   $+ H_2\uparrow + (CH_2=NCH_2CN)_3 + 3CH_2OHCOONa + 2HCOONa$

Variant: D

a. $Na_2[Zn(CN)_4] + 10HCHO + 2NaOH + H_2O \rightarrow ZnO\downarrow +$
   $+ (CH_2)_6N_4 + 4CH_2OHCOONa$

b. $4Na[Ag(CN)_2] + 21HCHO + 6NaOH \rightarrow 4Ag°\downarrow + Na_2CO_3 +$
   $+ 2(CH_2)_6N_4 + 8CH_2OHCOONa$ /9/

c. $4Na_2[Cu(CN)_3] + 34HCHO + 8NaOH + 2H_2O \rightarrow 3(CH_2)_6N_4 +$
   $+ 12CH_2OHCOONa + 4HCOONa + 4Cu°\downarrow + 2H_2\uparrow$

Variant: B

a. $3Na_2[Zn(CN)_4] + 26HCHO + 6NaOH \rightarrow 2CH_2[N(CH_2COONa)_2]_2 +$
$+ 4CH_2OHCOONa + 3ZnO\downarrow + 2(CH_2)_6N_4 + H_2O$ /10/

b. $12Na[Ag(CN)_2] + 55HCHO + 18NaOH \rightarrow 4CH_2[N(CH_2COONa)_2]_2 +$
$+ 8CH_2OHCOONa + 3Na_2CO_3 + 12\ Ag^0\downarrow + 4(CH_2)_6N_4 + 8H_2O$

c. $2Na_2[Cu(CN)_3] + 15HCHO + 4NaOH \rightarrow CH_2[N(CH_2COONa)_2]_2 +$
$+ 2CH_2OHCOONa + 2HCOONa + 2Cu^0\downarrow + H_2\uparrow + (CH_2)_6N_4 + H_2O$

Variants A,C and D represent the relations possible for all types of cyanide solutions found in plating shops. Variant B, on the other hand, shows the course of the reaction of cyanide solutions during the process of chemical rinsing.

As an illustration of this, Fig. 3 presents the dependence of the cyanide concentration in NaCN solution on the reaction time.

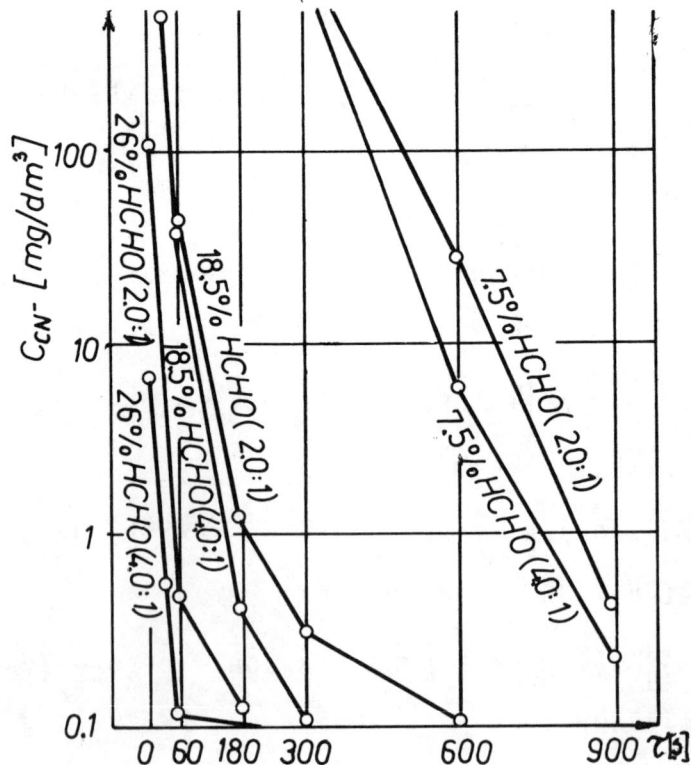

Figure 3. The value of cyanide concentration in solutions in relation to the reaction time for various HCHO concentrations and various HCHO:CN⁻ mole ratios. $C_{oCN^-} \cong 100 g/dm^3$ $t_o = 25°C$, pH≅12.0, 7.5%HCHO≅20% formaldehyde in H_2O, 26%HCHO≅70% formaldehyde in H_2O.

This shows that as the dilution of HCHO increases the rate of the reaction significantly decreases. The doubling of the HCHO:CN⁻ mole ratio from 2:1 to 4:1 has no practical effect on speeding up the decomposition of cyanides.

Fig. 4 shows the dependence of the cyanide in a $[Cu(CN)_3]^{2-}$ solution upon the reaction time at optimal pH condition.

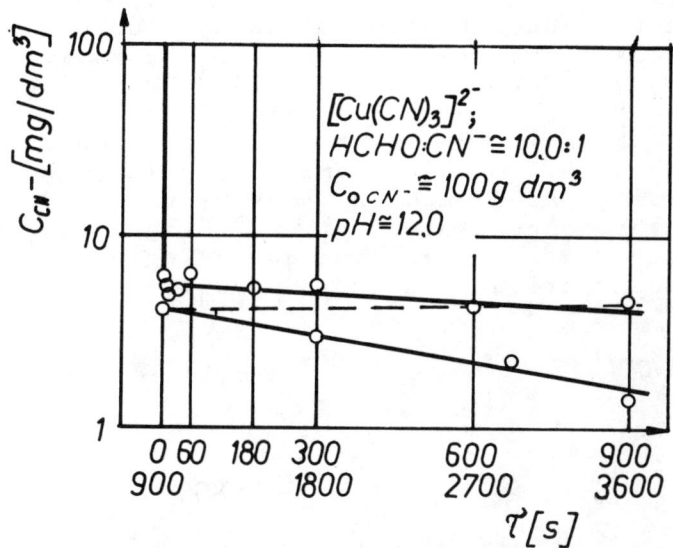

Figure 4. The value of cyanide concentration in solution in relation to the reaction duration for copper.

As the diagram shows /this is important for chemical rinsing conditions/ the concentration of CN⁻ after about 30-40 seconds is 6 mg/dm^3.

TREATMENT OF WASTEWATERS CONTAINING Cr^{+6}

Poland, on base of the FKJA method developed at the Institute of Precision Mechanics was the first country to use formaldehyde to reduce Cr^{+6} to Cr^{+3} in wastewaters.

In a highly acid conditions the reaction runs as follows:

$$8H_3O^+ + 2HCr_2O_7^- + 6HSO_4^- + 3HCHO \rightarrow 19H_2O + 4Cr^{+3} + 6SO_4^{2-} + 3CO_2\uparrow \qquad /11/$$

The stoichiometry of the reaction shows that as a result of a reaction with acid solutions containing Cr^{+6} all organic carbon in the formaldehyde becomes carbon dioxide.

In a highly acid conditions the reaction between concentrated formaline solutions and chromic acid is quite violent; the solution becomes hot and foams due to the rapid evolution of CO_2, which may cause the liquid to overflow.

Since the concentration of bath solutions most often used in chromium plating and chromating is within the range of 100-350g CrO_3/dm^3; and since it is easier to use a properly diluted solution of formaldehyde both in treatment of spent concentrated solutions as well as in chemical rinsing, the concentration of HCHO used in industry varies depending on the concentration of Cr^{+6}. Fig.5 presents a diagram of a safe reaction depending on the initial concentration of HCHO and CrO_3.

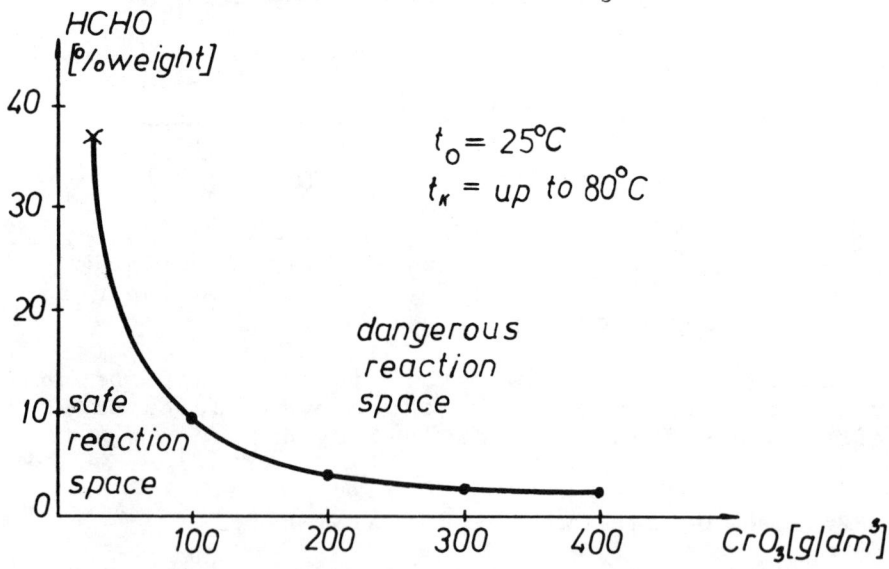

Figure 5. The diagram of maximum HCHO concentrations which allows for a relatively safe process application.

TECHNOLOGICAL AND ECONOMIC ASPECTS OF THE FKJA METHOD

The FKJA has been widly adopted in industry. Over 50 industrial plants throughout Poland are now using this method with very good results. It doesn't only allows for saving of millions zloties yearly, but has significantly contributed to improving the safety and hygienic work and the protection of the environment.

The main difficulties in introducing the FKJA method on a wider scale is the sharp, pungent odour of formaline. It is most oppresive in manually operated electroplating plants during the preparation of fresh and replenishing of already working solutions for chemical rinsing.

Since concentrated solutions of reagents are used, the treatment is usually conducted in very small volume. This diminishes considerably the cubage of the waste treatment plant and also significantly reduces reagent consumption. The reduction of Cr^{+6} to Cr^{+3} and the decomposition of cyanides occur immediately, thus the time required for waste treatment is considerably shorter in comparison with generally used methods.

Using chemical rinsing after cyanide and chromic processes eliminates the necessity of installing a number of wastewater conduits for each type of wastewaters. Only one collector is needed for all wastewaters leaving the plating shop.

As compared to the generally used methods, the FKJA method allows for a significant TDS reduction of wastewaters, since neither hypochlorite nor sodium bisulphite are used.

Formaldehyde vapours are twenty times less toxic than ozone and two times less toxic than chlorine.

REFERENCES

1. "Technological Formaline", Polish Standard PN-74/1-88000.
2. Konitz, G., Ingenieur. für Chemie "Justus von Liebieg", G.D.R., Magdeburg, 1963.
3. Biczysko, J. and Pociecha, Z., Poland IMŻ Report No. N-1447.
4. Lawes, B.C., Fournier, L.B. and Mathre, O.B., "A Peroxygen System for Destroying Cyanide in Zinc and Cadmium Electroplating Rinse Waters", 58 Annual Technical Conference AES, Buffalo,

NY., 6, 1971, Plating, 9, 1973.
5. Polish patents No. 79107, 89567, 89568, 90423, 90800, 74283, 89944, 111516.
6. USA Patent No. 4157942, G.B. patent No. 1556240, French patent No. 2343811.
7. Tużnik, F., "Neutralization of Rinse Waters and Concentrated Silver Plating Baths with Simultaneous Silver Recovery", Powłoki Ochronne IMP, 5, 1974.
8. Tużnik, F. and Lis, A., "Equipment such as the LAFT for Treatment of Concentrated Solutions with Simultaneous Recovery of Metals", Powłoki Ochronne IMP, 6, 1974.
9. Tużnik, F., Kołodko, K., Prusak, J. and Lis, A., "Method of electroplating waste treatment" - Equipment and Technology International 1978-79. Edition USA.
10. Tużnik, F., "Testing the Possibilities of Simultaneous Neutralization of Chromium and Cyanides in Electroplating Wastewaters". IMP-doctoral thesis. Warsaw, 1980.

UV PHOTODECOMPOSITION OF COLOR IN DYEING WASTEWATER

Calvin P.C. Poon. Department of Civil and Environmental Engineering, University of Rhode Island, Kingston, R.I.

Bruno M. Vittimberga. Department of Chemistry, University of Rhode Island, Kingston, R.I.

INTRODUCTION

The total U.S. commercial color production amounts to 0.5 billion pounds annually. Practically every industry uses dyes and pigments to color its products. The textile industry discharges approximately 10 million pounds of dye annually to waste streams. The textile industry concentrates in the mid to south Atlantic coast regions although there is still a significant number of textile dyeing and finishing mills operating in Rhode Island, Connecticut and Massachusetts.

The difficulty of discharging dyeing waste streams lies in the color imparted to receiving streams by dyestuff residues. Although most dyes have low BOD values, they add organic carbon and color to the water which is objectionable to the public for aesthetic reasons. Early research studies of color removal from textile effluent streams showed that the trickling filter and activated sludge processes could remove effectively 84 and 93 percent of the color in textile effluents [1], and chemical coagulation of cotton dye waste averaged 80 to 91 percent removal of color [2]. However the color of today's new, improved and virtually non-destructible dyes can only be reduced by 50 to 80 percent [3]. A review of the literature by Porter [4] shows that although carbon adsorption and chemical coagulation processes have been used in treating dyeing and finishing wastes, carbon does not adsorb some vat, disperse, and pigment dyes while coagulation does not remove some acid and reactive dyes of very low molecular weights. Furthermore, both processes require considerable space for the installation of the treatment facility. Extensive control equipments and well-trained personnel are needed. Both hyperfiltration (5) and reverse osmosis are efficient in color removal but very expensive. Even with the water reuse potential fully utilized, the total cost (operating plus amortization) is much higher than other established processes.

Chlorination and ozonation (6) (7) are useful for the removal of certain dye groups but at a slow rate. Similarly radiation processes without the presence of strong oxidizing agents can remove color but at a slower rate (4). Radiation-oxidation process using gamma radiation, Cs137, and chlorine as the oxidizing agent are much more effective than the effect of the two components applied separately. In the presence of 75 mg/l of chlorine, color can be destroyed with a radiation dose of 60 KR within a few minutes. The undesirable features of this process are the high chlorine residue in the effluent which is toxic and the public concern of wide spread use of radioactive material. These undesirable features can be eliminated by using ultraviolet irradiation.

UV-ozone process has been applied for sterilization and reduction of organic matter in wastewater (8), removal of complexed cyanides (9), as well as in the oxidizing of a wide variety of both inorganic and organic species (10). UV irradiation for color removal has advantages over other processes in that neither chemical sludge nor toxic residue is left in the treated effluent. Since most acid dyes and direct dyes commonly used by the textile dyeing and finishing industry are resistant to UV degradation, rate enchancement using sensitizers was investigated in this present study.

BATCH STUDY

In the beginning, all studies were carried out in 8-inch quartz test tubes housed in a Rayonet chamber reactor. Irradiations were carried out at 250 nm using a flux of approximately 10^{16} photon/sec-cm^3, and 12,800 μ watt/cm^2. Distilled water free of iron, organics, nitrogen and turbidity was used to prepare the dye solutions for the study as these contaminants could interfere with the photo-decomposition process. Table 1 shows that all dyes selected for this study, with the exception of Direct green 6, took more than 60 minutes to achieve a visually clear solution. In order to enhance the energy transfer, acetone was added to the dye solutions. An optimal acetone concentration was found to be 5×10^{-2} M for the concentrations of dyes used in the study. Acetone was found to be an excellent sensitizer. It has a high triplet energy (79-82 Kcal/mole). The quantum yield of intersystems crossing is 1.00 (11). It is

Table 1. Dye Photodecomposition

Dye and Concentration	Exposure time required to obtain a clear solution (minutes)			
	Dye alone	With Acetone but no mixing	With Acetone and air mixing	With Acetone and N_2 bubbling
Acid Black 52				
1×10^{-4} M	>60	2	15	1
5×10^{-5} M	>60	1	3	0.5
Acid Blue 40				
1×10^{-4} M	>60	1.5	6	1
5×10^{-5} M	>60	0.75	2	0.5
Acid Red 1				
1×10^{-4} M	>60	5	10	3
5×10^{-5} M	>60	2	6	1
Acid Blue 27				
1×10^{-4} M	>60	0.75	4	0.5
5×10^{-5} M	>60	0.5	2	0.5
Direct Green 6				
1×10^{-4} M	22	13	32	8
5×10^{-5} M	10	6	20	5
Direct Red 80				
1×10^{-4} M	>60	5	10	4
5×10^{-5} M	>60	3	5	1.5
Direct Black 80				
1×10^{-4} M	>60	5	13	2
5×10^{-5} M	>60	2	6	0.5
Direct Red 83				
1×10^{-4} M	>60	4	9	6
5×10^{-5} M	>60	2	5	1.5

non-carcinogenic, readily available and relatively inexpensive. Upon UV irradiation, the triplet acetone transfers its energy to the dye molecule whereupon the triplet dye is decomposed to its colorless end products. Table 1 shows that the exposure time was significantly reduced from >60 minutes to 2~5 minutes (only 50% reduction of exposure time for Direct green 6).

Although photochemical reaction is extremely fast, a collision between the triplet acetone and the dye molecule is necessary to accomplish the energy transfer. The process is collision control or mass transfer rate limiting. Contrary to expectations, air bubbling increased instead of decreased the exposure time required as shown in Table 1. Molecular oxygen has a low triplet energy of 22.5 Kcal/mole (11), consequently making it an effective quencher in the process. Nitrogen bubbling was therefore employed to remove oxygen as well as to enhance the mass transfer rate. Table 1 shows that for most dyes, acetone and N_2 bubbling increased the rate of photodecomposition by 10 to 120 times (2 to 3 times for Direct green 6). Eight dyeing wastewater samples collected from different vat discharges from one textile dyeing and finishing mill showed similar results.

CONTINUOUS FLOW STUDY

Encouraged by the results of the batch study, a multiple-column reactor was constructed. Each column consisted of a Voltare UV-Lux lamp G36T6L inside a quartz jacket surrounded by 1.2 cm thick dye solution. The plexiglass column was lined inside with reflective tape to increase the energy utilization. Nitrogen gas was supplied through four diffusers only to the first column at the bottom. Since the system was operated under slight pressure, the nitrogen bubbles from the first column was carried over to the following columns by the continuous flow of dye solution. Table 2 summarizes the experimental results. Color concentration was expressed in American Dye Manufacturer's Institute (ADMI) color units (12). The interim Federal effluent standards are 220 to 400 ADMI color units depending on the type of industries (13). Rapid removal of color down to the acceptable concentration was achieved for all dyes in one to two columns in series. Table 2 indicates that higher color concentrations could be successfully removed by either extending the exposure time or increasing the number of columns.

The rate of color removal was mass transfer limiting for the system tested. Table 2 shows that the system treated a more diluted dye solution successfully with much lower unit power consumption (see Acid blue 27 and Direct red 83). Future design of the system should improve the

Table 2. Dye Photodecomposition in a Continuous-Flow Multiple Column Reactor

Dye Conc. and ADMI unit	Exposure time (min)	Final ADMI unit Col. 1	Col. 2	Col. 3	Unit Power Consumption per Column 39-W input $\frac{kWh}{m^3}$	13.8-W output $\frac{kWh}{m^3}$
Acid Red 1 - - - - 8 x 10^{-5} M						
(6757)	5	1139	349	359	2.33	0.82
(6554)	7.5	241	169	149	3.50	1.24
Acid Black 52 - 9.25 x 10^{-5} M						
(3201)	15	401	342	262	7.00	2.48
	20	294	234	257	9.33	3.30
Acid Blue 40 - - 4.6 x 10^{-5} M						
(1530)	2	306	256	227	0.93	0.33
	3	212	---	135	1.39	0.49
Acid Blue 27						
9.6 x 10^{-5} M						
(2326)	20	406	240	227	9.33	3.30
7.5 x 10^{-5} M						
(1850)	5	307	242	217	2.33	0.82
5.0 x 10^{-5} M						
(1403)	1	198	176	134	0.46	0.16
Direct Black 80, 4.6 x 10^{-5} M						
(4220)	7	271	220	245	3.17	1.12
Direct Green 6, 5.2 x 10^{-5} M						
(4479)	20	317	205	307	9.33	3.30
Direct Red 83						
9.55 x 10^{-5} M						
(5605)	10	1554	237	218	4.67	1.65
	15	564	218	213	7.00	2.48
6.9 x 10^{-5} M						
(2664)	5	903	203	166	2.33	0.82
2.65 x 10^{-5} M						
(1976)	3	416	155	135	1.39	0.49

mass transfer rate by providing either a better mixing or a thinner thickness of solution for irradiation.

CONCLUSION

This study has demonstrated that UV photodecomposition with acetone as a sensitizer could rapidly and effectively remove color from solutions. Nitrogen bubbling for deoxygenation and for mixing also significantly enhanced the process.

REFERENCES

1. Anon., "Activated Carbon Reclaims Textile Industry's Wastewaters," Environ. Sci. and Tech., 3,314 (1969).

2. Souther, R.H., et.al., "Textile Waste-Recovery and Treatment," Sew. and Ind. Waste, 29 (8), 918 (1957).

3. Masselli, J.W., et.al., "For Better Pollution Control," Textile Ind., 133 (6), 65 (1969).

4. Porter, J.J., "Stability and Removal of Commercial Dyes from Process Wastewater," Pollution Engr., (Oct. 1973).

5. Brandon, C.A., "Reuse of Total Composite Wastewater Renovated by Hyperfiltration in Textile Dyeing Operation," Ind. Water Engr., 14, (Jan. 1976).

6. Stuber, L.M., "Tufted Carpet Dye Wastewater Treatment, Part 1,11,111," Ind. Wastes, (Jan.-June 1975).

7. Nebel, C., et.al., "Ozone Disinfection of Industrial-Municipal Secondary Effluents," J. Water Poll. Control Fed., 45 (12), 2493 (1973).

8. Reuter, L.H., "Research on Wastewater-Reuse System Using UV-Ozone for Sterilization and Reduction of Organics," Proc., 1st Inst. Symp. Ozone Water Wastewater Treat., 476 (1975).

9. Garrson, et.al., "Removal of Complexed Cyanides from Wastewater," Proc., 1st Inst. Symp. Ozone Water Wastewater Treat., (1975).

10. Prengle, H.W., et.al., "Ozone/UV Process Effective Wastewater Treatment," Hydrocarbon Processing, 54 (10), 82 (1975).

11. Murov, S.L., *Handbook of Photochemistry*, Marcel Dekker Inc., N.Y., 1973.

12. Wysechi and Stiles, *Color Science*, John Wiley, N.Y., 1967.

13. Federal Register, Oct. 29, 1979, Part 111, EPA Textile Mills Point Source Category Effluent Limitations Guidelines.

REMOVAL OF SELECTED POLLUTANTS WITH POTASSIUM FERRATE

Sergio Deluca. Department of Civil Engineering, North Carolina State University, Raleigh, N. C.

Allen C. Chao. Department of Civil Engineering, North Carolina State University, Raleigh, N. C.

Charles Smallwood, Jr., Department of Civil Engineering, North Carolina State University, Raleigh, N. C.

INTRODUCTION

Chemical oxidants such as potassium permaganate, ozone and chlorine are widely used in water treatment. Undesirable chemical species are oxidized to either nonharmful or non-objectionable forms. In many cases they are oxidized to insoluble precipitates which can be easily removed from the water with coagulation and sedimentation.

The chemical oxidation process may be utilized for removal of priority pollutants. However, it is not clear at this time that the end-products of oxidation are harmless or whether they may cause more long-term health hazards than the original species. Some preliminary evidence indicates that the oxidation process probably produces new species which are more amenable to the subsequent coagulation process than the original one. Hence, a combination oxidation-coagulation process could be more effective than either oxidation or coagulation alone for removal of the toxic or carcinogen causing species from water.

Potassium ferrate is a strong oxidant. Under suitable conditions, it is capable of oxidizing various inorganic and organic substances. The unique characteristics of this chemical oxidant is that after the oxidation reaction, it forms ferric ions which may further hydrolyze to form hydroxo-metallic polymers. Thus, potassium ferrate is a strong oxidant and at the same time a very effective coagulant. Its combining oxidative-coagulative nature makes it a promising alternative chemical for removal of priority pollutants. The results obtained by Waite [1] on secondary wastewater effluent show that ferrate application can achieve a very high removal of SS , BOD, ammonia, orthophosphates and coliforms.

Potassium ferrate can be produced readily on a large scale with products of high purity. It is innocuous and can be easily and safely stored and transported. Hence it is anticipated by some researchers that if demanded in large quantity the cost of ferrates would be competitive with conventional coagulants. Thus, treatment with potassium ferrate to achieve both oxidation and coagulation may easily become cost-effective and may be easily adopted by conventional water treatment plants for removing priority pollutants. Because of its low initial capital investment, this process may become a desirable alternative for treating industrial wastewater containing priority pollutants.

The potential of potassium ferrate was evaluated at North Carolina State University for the removal of priority pollutants of different chemical structures and characteristics. The work reported here discusses the results obtained for only one of the species studied - the removal of naphthalene with potassium ferrate.

REVIEW OF RELEVANT LITERATURE

Characteristics of potassium ferrate

Potassium ferrate decomposes in neutral and acidic water solutions but is reasonably stable at basic pH levels. The pKa of ferrate is not yet known, but based on similar charged, protonated, tetrahedral oxyanions of similar size such as $H_2VO_4^-$ (8.2), $H_2PO_4^-$ (7.2), $HCrO_4^-$ (6.5) a pKa of 6.60 has been assumed by Lee and Benisek [2]

The standard oxidation potential is 2.20 V in basic solution and 0.74 V in acidic solution [3]. For comparison, the oxidation potentials of ozone in basic solution and MnO_4^- in acidic solution are 2.07 and 1.69 volts, respectively.

The stability of ferrate solutions depends on many factors such as concentration, pH, alkalinity, temperature, turbulence, buffers, catalysts, and organics and inorganics substrates. It is advantageous that dilute ferrate solutions are much more stable than concentrated solutions. Light was shown to have little effect on ferrate decomposition in aqueous solutions. Higher temperature however enhances ferrate degradation in water solutions [4]. Inorganic salts such as NaOH, KOH, KCl, KNO_3, KBr, K_2CO_3, $KClO_3$ have been shown to retard the ferrate decomposition but salts of calcium, magnesium, metals and their oxides increase the rate of decomposition of ferrate. Increasing alkalinity also increases its decomposition rate.

Depending on buffers, pH and organic substrates the oxidation rate curve by ferrate may follow a first or second order reaction pattern. In the absence of significant quantity of other reducing agents, Kalecinski [5] showed that the ferrate decomposes in aqueous solution resulting in the formation of H^+, Hydrated electrons, O_2^-, HO_2^-, H_2O_2, and OH^-. The results obtained by Haire [6] indicated that for a ferrate concentration of 10^{-3}M, its decomposition rate was found to be second order with respect to ferrate ion concentration and first order with respect to pH between pHs 7.7 and 9.5. For ferrate concentrations between 10^{-5}-10^{-4}M the rates of reactions were estimated to be of first order with respect to both ferrate concentration and pH levels. The decomposition rate was found to decrease with increasing pH in the above pH range, then increase rapidly up to pH 12. For pHs lower than 8.0, its rate of decomposition was reported to increase with decreasing pH[1].

Oxidation of organic compounds by Ferrate

The by-products of ferrate decomposition in water such as peroxy radicals, OH radicals, O_2^-, O_2, have been shown in literature to be capable of oxidizing many organic substrates.
Mills [7] showed that phenols, alcohols, aromatic and aliphatic amines, hydroperoxides and polynuclear aromatics are all capable of reacting with RO_2^- and OH^- radicals under suitable conditions. OH^- radicals react by H-atom transfer and addition to double bonds and aromatic rings.
Ferrate oxidizes methanol to formic acid and CO_2; ethanol into acetic acid [8] and propane-diols to hydropropanoic acids [9]. It seems that most aromatics compounds containing a strongly nucleophilic substitutent will enhance the attack by ferrate under basic to acidic conditions.
Waite [10] studied the oxidation of several organic compounds with ferrate. Allylbenzene, chlorobenzene, 1-hexene-4ol, nitrobenzene toluene all reached slowly with ferrate. Benzene was oxidized 18-47%, chlorobenzene 23-47%, allylbenzene 85-100%, when chlorobenzene and benzene were joined a 76% oxidation. Phenol was oxidized 100% at pH 8.0 after 6 min by 20 mg/l of ferrate. Other organic compounds shown to be oxidized by ferrate include histidine, diethylamine and piperidine. Ammonia/ammonium ion is oxidized by ferrate cold at room temperatures, while potassium permanganate only oxidize in hot solutions [11]. Amines are readily oxidized in basic media [8].
The oxidation of several organic compounds with ferrate at pH 8.0 with and without buffer was studied [9]. The average oxidation time for α-hydroxytoluene, 1,2-ethane-diol, and glycerol was about 4 min without buffer and 35 minutes with buffer.
Nitriloacetic acid and possible oxidation by-products were oxidized using potassium ferrate. The reactions were

shown to fit a mixed-order kinetic model with the second order rate being the major component but the rates of reactions showed a strong pH dependence [12].

Oxidation of Priority Pollutants by Potassium Ferrate

Several priority pollutants have been subject to laboratory oxidation studies with ferrate. The results cover the qualitative testing of diethylphthalate, anthracene, 2-4, dinitrotoluene, tetrachloroethylene, and quantitative testing of dichloromethane, trichloromethane, tetrachloromethane, bromodichloromethane, nitrobenzene, dichlorobenzenes, naphthalene, and trychloroethylene.

Volatile trihalomethanes are not oxidized by ferrate but are very well removed by gas stripping with either air or CO_2. Trichloroethylene is easily removed by oxidation and by stripping simultaneously. Nitrobenzene and 1,2-dichlorobenzene react very slowly with ferrate. Naphthalene was completely oxidized under simulated conditions of water and wastewater coagulation processes.

Other priority pollutants may be oxidized by either peroxy or OH^- radicals formed during the degradation (reduction) of ferrate. Among the pesticides, aldrin could be oxidized to endrin, and heptachlor epoxidized to heptachlor epoxide. Acrolein could suffer attack on double bond, like isophorene and 2-chloranaphthalene which has been shown to go to 2-chloronaphthallic acid under oxidation in basic conditions. Heavy metals usually are transformed by ferrate reduction to the III form and then co-precipitated with the $Fe(OH)_3$ floc [13].

Halogenated aliphatics and halogenated ethers could suffer attack on the double bond as it happens to trichloroethylene [14]. Phenols, chlorophenols and nitrophenols have strong nucleophilic substitutents that ease the attack by ferrate oxyradicals [10] in spite of the "protection" offered by nitro and chloro groups. Amines and other miscellaneous compounds all have C-N bonds prone to attack by ferrate reduction agents [12]. Phthalate esters and polynuclear aromatic hydrocarbons could be oxidized to aldehydes or respective carboxylic acids [15].

Ferrate like many other strong oxidizers is specific for certain organic compounds but if enough activation energy is provided through catalysts or pH control it sure will be a more general all purpose oxidizer (and coagulant).

METHODOLOGY

Selection of Pollutants

The pollutant compounds were chosen from the list of 129

"Priority Pollutants" of EPA. The selection was made based on their stability and solubility in water, and organic solvents as well as their mutagenicity in the Ames Test and possible carcinogenicity.

The following results were obtained from a study of the removal of naphthalene. It is a widely used chemical in petroleum refining, paint and ink formulation, auto and laundries, non-ferrous metals, iron and steel manufacturing and timber production. Thus, this substance is commonly found in residential, commercial and industrial wastewaters. It is non-mutagenic but it is a suspected teratogenic and potential carcinogenic compound [16].

Preparation of Samples

Three concentration levels of naphthalene were chosen to be oxidized by ferrate based on a log scale to avoid bias in the experimental output.

Organic free water was produced by double distillation in glass, deionization and by passing the water through XAD resin columns. To this water 0.5 meq/l of $NaHCO_3$ was added to simulate the alkalinity of the soft waters found in southeastern United States.

Reagents and Naphthalene of ACS grade(Fisher) were utilized. HPLC grade organic solvents were utilized throughout the experiment.

Gas chromatography (Shimadzu 6 AM) with FID (Flame Ionization Detector) was employed for qualitative and quantitative analysis. Naphthalene was detected using a stainless steel column 10 ft x 1/8", packed with 100/120 Supelcoport coated with 10% SP-1000, at 130°C with gas flow rate of 60 ml/min of N_2. Internal standard utilized was 1, 2-dichlorobenzene.

Spiked synthetic water samples of 300 ml containing the naphthalene at the pre-selected levels were oxidized or oxidized-coagulated and then transferred to a concentrator for liquid-liquid extraction with Carbon Disulfide/DCM using microextraction technique.

Stock solutions of naphthalene were prepared in methanol and discarded every two weeks.

Method of Oxidation Studies

Oxidation studies were conducted in serum vial bottles of 125 ml capped with teflon lined caps so that no headspace would exist. Water, ferrate and stock solution of naphthalene were added in that order and reacted for the selected times. After oxidation an aliquot of the concentration solution was added to the serum bottle (also containing a bar stirrer) so that 100 X concentration would be obtained.

The concentration step consisted of stirring the mixture for 2 minutes at 300 rpm. An aliquot was then transferred to a 2 ml vial and stored in the refrigerator for further GC analysis.

Separate oxidation-coagulation studies were designed to simulate the detention time and conditions of energy inputs of normal water and wastewater treatment processes. The procedure involved the addition of stock solution of naphthalene to the turbid (bentonite) water samples with simultaneous adjustment of pH to 5.5. This pH was selected based on the optimum point for ferrate coagulation of bentonite in that synthetic water. After 1 min mixing at 300 rpm, the solutions were flocculated 19 minutes at 50 rpm and then settled for 20 minutes. The gradient velocity for flocculation was 43 s^{-1} corresponding to an energy input of 2.02 kw-hr/MG. After settling the clarified samples were transferred to 125 ml serum bottles and concentrated as before.

Blanks were interdispersed with the samples for determination of possible cross contamination and interferences from sample processing hardware. Replication was utilized for the determination of the precision and accuracy of analysis. Factorial design was utilized throughout the coagulation studies. Analysis of variance using SAS [17] was employed to detect possible influences and relationships between variables. Multiple regression analysis was utilized for curve best-fitting to the data.

Kinetic data was studied using a computer model [18] to diagnose the number of terms in the rate law and the values of the pseudo-constants.

The wastes generated in the experiment were disposed by the special services contracted by NCSU for collection and disposing of toxic chemicals and radioactive wastes.

RESULTS

When ferrate is added to aqueous solutions the pH increases due to release of OH^-. This is directly contrary to conventional metallic coagulants like ferric chloride and aluminum sulfate which decrease the pH. On a molarity basis ferrate is as efficient as Alum and a better coagulant than ferric chloride. This confirms the findings [1] that in situ formation of $Fe(OH)_3$ floc from ferrate is very efficient for complexation with turbidity particles.

Table I presents the oxidation of 100 ppb of naphthalene versus time in the pH range of 8.3-8.9. This pH range was chosen because it is the best range for ferrate oxidation of organics [1]. Phosphate buffer was utilized but it showed to decrease the efficiency of ferrate oxidation at that pH range. Because of that and also because all coagulation studies were being conducted without buffers other than the natural water buffering capacity, phosphate buffers were dropped. It is recognized that oxidation among other factors is a function of pH but due to the small variation in pH it was assumed that the effects of pH on organic substrate utilization were not significant.

Table I. Oxidation of 100 µg/l of Naphthalene by Ferrate

Time (min)	Substrate Remaining (S/So)				
	Potassium Ferrate (mg/l)				
	216	90	30	20	10
1	.6998	.7079	.9553	.9772	.9787
2	.4898	.5012	.8532	.9496	.9659
5	.1679	.1778	.7074	.8703	.9294
10	.0282	.0316	.5223	.7497	.8747
15	.0005	.0006	.4246	.6431	.8168
25	.0000	.0000	.2678	.4700	.7398

Table II lists the pseudo-first order reaction constants and respective half-lives for the oxidation of naphthalene. At the pH range tested the oxidation reactions were very slow because of the slow degradation of Ferrate.

Table II. Results of the Kinetic Studies

Ferrate (mg/l)	K_i(*) $(10^{-4} s^{-1})$	Half-Life (min)
216	9.87	11.70
90	8.82	13.10
30	7.80	14.80
20	5.00	23.10
10	1.93	59.86

(*) - K_i - Pseudo first order constant.

The relationship between increasing ferrate concentration and increasing naphthalene concentration is presented in Table III. All oxidations were done in 30 minutes. Analysis of variance (ANOVA) shows that the naphthalene concentration has little influence on the oxidation capacity of ferrate. The reaction rates were much more dependent on ferrate oxidation of water than on ferrate oxidation of naphthalene.

Table III. 30 Min Oxidation of Naphthalene by Ferrate

Ferrate (mg/l)	Substrate Remaining (S/So) Naphthalene (µg/l)		
	100	320	1000
10	.635	.715	.740
20	.405	.615	.734
30	.050	.075	.547
60	.000	.038	.424
90	.000	.000	.302

Table IV shows the results obtained by the jar test coagulation of naphthalene. The average percentage removal was 78%, for the ranges of concentrations tested.

Table IV. Oxidation-Coagulation of Naphthalene with Potassium Ferrate(*)

Ferrate (mg/l)	Substrate Remaining (S/So) Naphthalene (µg/l)		
	100	320	1000
5	.251	.488	.745
10	.125	.283	.558
20	.000	.235	.367
30	.000	.000	0.000
60	.000	.000	0.000
90	.000	.000	0.000

(*) - See text for testing conditions.

If the percentage removals shown in Tables III & IV are compared, it can be seen that the fact of lowering the pH increases the efficiency of removal. While 30 minutes oxidation at pH 8.3 to 8.9 removes 68.5% of naphthalene overall, 1 minute oxidation at pH 5.5 plus flocculation is much more efficient. No adsorption on bentonite has been detected so the increase in efficiency might be due to full use of oxyradical products of ferrate decomposition.

The effect of molar ratios ferrate/naphthalene in the oxidation-coagulation of naphthalene is presented in Table V. A region exists were maximum removal can be obtained, regardless of naphthalene concentration, without using high concentrations of ferrate. Between the molar ratios of 20 to 40 removals varying from 75 to 100 percent can be obtained. It must be remembered however that due to needs for turbidity removal, the ratios oxidant/chemical could be higher and a compromise must be reached.

Table V. Molar Ratios (Ferrate/Naphthalene) and % Removal

Molar Ratios (Ferrate/Naphthalene)	Percentage (%) Removal Naphthalene ($\mu g/l$)		
	100	320	1000
10	22.0	40.0	46.0
20	43.5	66.0	90.0
30	63.2	76.0	100.0
40	75.5	83.5	100.0
60	82.2	95.0	100.0

CONCLUSION

Potassium ferrate is shown to be an efficient oxidant and coagulant for a model priority pollutant.

The current technologies for oxidation (disinfection) and coagulation has been shown to be associated with environmental and health effects. Potassium ferrate as shown might become a general purpose alternative technology for future water and wastewater treatment.

REFERENCES

1. Waite, T. D., NSF-Rann Final Report ENV 76-83897, 1979.
2. Lee, Y. M., and Benisek, W. F., "Inactivation of Phosphorylase b by Potassium Ferrate a new Reactive Analogue of the Phosphate Group," J. Biol. Chem., 251, 1553, (1976).
3. Wood, R. H., "The Heat, Free Energy and Entropy of the Ferrate (VI) Ion," J. Am. Chem. Soc., 80, 2036, (1952).
4. Wagner, W. F., "Factors Affecting the Stability of Aqueous Ferrate Solution," Anal. Chem. 24, 1498, (1952).
5. Kalecinski, J., "The Effect of ^{60}C Gamma Radiation on Aqueous Potassium Ferrate Solutions," Roczniki Chemii, 41, 195, (1967).
6. Haire, R. G., "A Study of the Decomposition of Potassium Ferrate (VI) in Aqueous Solution," Ph.D. Thesis, Michigan State University, 1965.
7. Mill, T., "Structure Reactivity Correlations For Environmental Reactions," EPA 560/11-79-012, 1979.
8. Audette, R. G., "Studies of the Ferrate Ion," Ph.D. Thesis, Univ. of Saskatchewan, 1972.
9. Williams, D. H., Research Report No. 20, WRI, Univ. of Kentucky, 1969.
10. Waite, T. D., and Gilbert, M. J., "Oxidative Destruction of Phenol and Other Organic Water Residuals by Iron (VI) Ferrate," J. WPCF, 50, 543, (1979).
11. Cherwin, M. G., "Part II: Kinetics of the Potassium Ferrate Oxidation of Certain Nitrogen Compounds . . .," Ph.D. thesis, Univ. of Nebraska, 1980.
12. Kelter, P. B., "Kinetic Methods and Kinetics of the Oxidation of Water, Nitriloacetic Acid and Related Organic Substances by Potassium Ferrate," Ph.D. thesis, Univ. of Nebraska, 1980.
13. Murmann, R. K., and Robinson, P. R., "Experiments Utilizing FeO_4^{2-} For Purifying Water," Water Res., 8, 543, (1974).
14. Hendry, D. G., and Kenley, R., "Atmospheric Reaction Products of Organic Compounds," EPA-560/12-79-001, 1979.
15. Augustine, R. L., Oxidation, Vol. 1, 1969.
16. Ambient Quality Criteria for Naphthalene," EPA 440/5-80-059, 1980.
17. SAS User's Guide, 1979 Edition, Version 79.3.
18. Bertrand, R., Dubois, J. E., and Toullec, J., "Non-iterative Microcomputer Analysis of Competing First and m-order Reactions by a Constant Time Interval Method," Anal. Chem., 53, 219, (1981).

WILL ACTIVATED CARBON WORK?

William Brian Arbuckle. Department of Environmental Engineering Sciences, University of Florida, Gainesville, Florida.

BACKGROUND

Most industries have made substantial progress removing conventional pollutants (biochemical oxygen demand and suspended solids) from their wastewaters. Regulatory concern has shifted towards specific chemical contaminants (priority pollutants), with removing them at their source of generation appearing more economical than at the wastewater treatment plant where all wastes are combined. Since activated carbon removes many organic compounds from aqueous solution and it can be easily used, one is frequently asked if activated carbon will work on a specific at-source waste. This report provides an approach to answer the question, "Will Activated Carbon Work?". Answering this question could save the time and effort involved in running an equilibrium test (isotherm) on contaminants unlikely to be adsorbed sufficiently to make adsorption economical.

An isotherm is simple to run (Table I). Recent studies indicate considerably more time is required to achieve true equilibrium [1], but two hours should be sufficient to indicate a compound's adsorbability. If volatile chemicals are present, several precautions should be taken (minimum vapor space, air-tight lid, pressure filtration). The isotherm data are plotted on log-log paper to conform to the Freundlich equation (Figure 1):

$$X = KC^{1/n} \qquad 1.$$

where:

X = loading, mmoles adsorbed/gram adsorbent
C = equilibrium solution concentration, mmoles/liter
K & n = empirical constants

Table I. Isotherm Procedure

1. Weigh 5 to 10 different amounts of pulverized activated carbon into an equal number of 500-ml flasks.
2. Add 200-ml of water to be treated into each flask containing the carbon, plus to an additional flask for use as a control.
3. Shake the sample vigorously for two hours -- a gyratory shaker or wrist shaker is frequently used.
4. Filter sample through 0.45μ filters to remove carbon.
5. Analyze each sample.

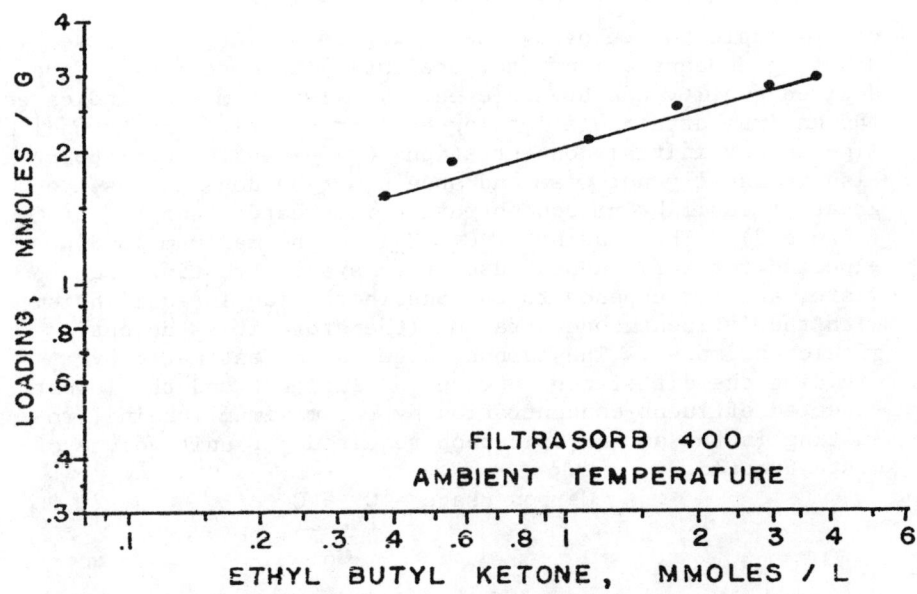

Figure 1. Freundlich Isotherm

The lowest concentration achievable should be assumed to be the lowest one obtained in the isotherm test; even though one obtains a straight line when plotting the data, isotherm shapes frequently change with changing concentration and therefore should not be extrapolated (Figure 2 gives an example). If a lower effluent concentration is desired, the test should be repeated with greater activated carbon doses

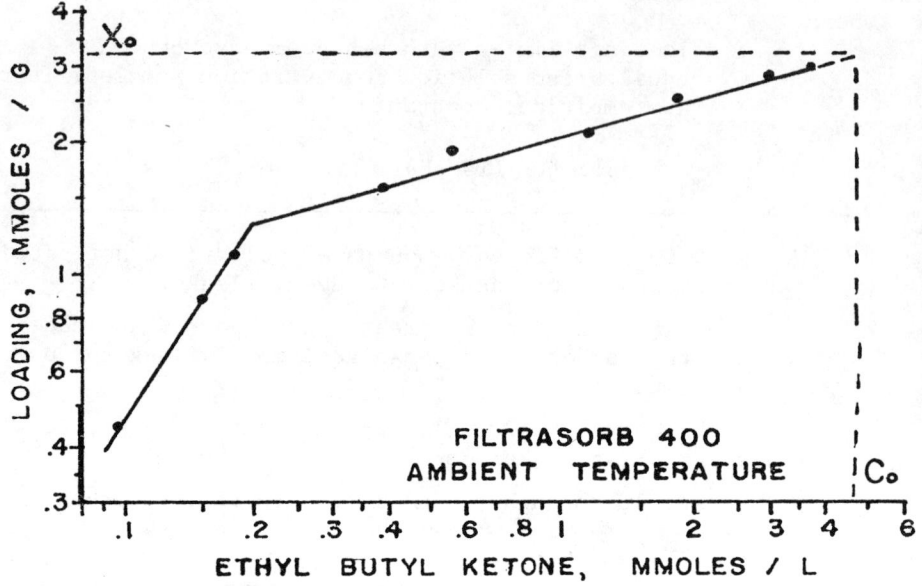

Figure 2. Two-Sloped Freundlich Isotherm

or if single-solute or synthetic wastes are used, diluted starting concentrations (not desirable) to determine if the desired quality can be achieved. The isotherm also indicates the maximum attainable loading by extrapolating the Freundlich line to the initial concentration (C_o) -- again, extrapolation is usually not wise and should not be done unless the greatest equilibrium concentration is greater than 80% of C_o (Figure 2). The loading at C_o (X_o) is the maximum loading expected for a fixed-bed adsorption system treating the waste, and corresponds to the adsorber being in equilibrium with the influent concentration (therefore it is an optimistic estimate). The carbon usage can be estimated by dividing the difference between the influent and the lowest expected effluent concentration by the maximum loading, resulting in the amount of carbon required per unit volume of waste treated:

$$\text{Carbon Usage} = \frac{C_o - C}{X_o} \qquad 2.$$

This carbon usage can then be used to roughly estimate the cost of activated carbon treatment and to determine if continuous column studies are worthwhile.

While the isotherm procedure is simple and the minimum attainable concentration and a rough estimate of carbon usage can be obtained, considerable effort may now be involved in analyzing each sample for the specific contaminant. So before laboratory tests are started, alternative approaches could eliminate activated carbon from further consideration.

LITERATURE ADSORBABILITY

The easiest way to answer the question, "Will Activated Carbon Work?", is to search the literature for the compound of interest; of course, a literature search takes considerable time. But, if five reports are kept readily available, we have access to the adsorbability of about 240 different compounds (References 2 - 6). Dobbs and Cohen [2] list equilibrium data for 139 "toxic" compounds; the actual data are presented so pH, concentration range and isotherm shape can be observed -- their equilibrium concentration may be much lower than many at-source wastes. Giusti, Conway and Lawson [3] studied 93 widely-used industrial compounds; their data are for a single equilibrium concentration/adsorbate loading, but comparisons could be made with a compound known to adsorb well (e.g., phenol). Al-Bahrani and Martin [4] studied 26 aromatic compounds and present their results as Langmuir constants, their data are also presented in figures. Arbuckle [5] presents Freundlich isotherm constants and ranges of applicability for 22 compounds, all studied by Giusti et al. Reimers et al. [6] present Freundlich constants for 15 compounds, range of applicability was not given nor was the actual data; many compounds were also studied by the others. These five references provide a good source of adsorption data to obtain a loading for use in equation 2 to calculate the potential of activated carbon for treating the waste.

CALCULATED ADSORBABILITY

If the compound of interest has not been tested, it would be ideal if adsorbability could be calculated from theory. Equilibrium adsorption models used in the past (Langmuir, Brunauer-Emmett-Teller and Freundlich), just represented the equilibrium data and could not predict loadings. Recently, additional models have been proposed: Solvophobic theory [7], Polanyi Adsorption Potential [8], and Net Adsorption Energy [9], each is potentially useful for estimating equilibrium loadings. Arbuckle used these models to predict the loadings for 13 aliphatic compounds based on equilibrium data for 9 alcohols [5]; both the Polanyi and Net Adsorption Energy methods were easy to apply and resulted in reasonable estimates, at least good enough to determine if laboratory study of adsorption is warranted as an economical treatment method for a waste containing that chemical.

The Polanyi Adsorption Potential Theory views the adsorbent as having a fixed volume of "adsorption space" where adsorption can take place. The adsorption potential is the work required to bring a molecule from infinity to some location within the adsorption space, with the energy being a maximum closest to the adsorbent within the micropores and a minimum (zero) at the edge of the adsorption space. Rather than mathematically describing the relationship, a chacteristic curve is said to exist for each adsorbent which can describe the adsorption of all compounds. In practice, adjustments have to be made to the data to get complete agreement, but it can be accomplished by correcting for polarizability and for volume changes on adsorption. The method plots the logarithm of volume adsorbed (ordinate) versus adsorption potential per unit volume:

$$\frac{RT}{V} \log \frac{C_s}{C} \qquad 3.$$

where:

T is the absolute temperature, °K
V is the solute molar volume, $\frac{cc}{mole}$
R is the gas law constant which is excluded by convention when the data are plotted
C is the equilibrium solute concentration, mmole/liter
C_s is the saturation solute concentration (solubility), mmole/liter

The polarizability correction is obtained by dividing

$$\frac{T}{V} \log \frac{C_s}{C}$$

by:

$$\frac{x}{0.236} - 0.28 \qquad 4.$$

where:

$$x = \frac{n_D^2 - 1}{n_D^2 + 2} \qquad 5.$$

and n_D = refractive index

To use the Polanyi theory to estimate loading, solute solubility, refractive index and molar volume are required. Molar volume (from specific gravity) and refractive index are usually available in the <u>Handbook of Chemistry and Physics</u>, solubilities may be more difficult to obtain. This technique does predict adsorbabilities for 13 compounds (Table II) reasonably well (Figure 3) -- line is the prediction line constructed through adsorbability points for 9 alcohols, whose points are not shown. At the higher loadings, the log scale masks the large absolute differences between the predicted value and the actual values; but, one could still determine if adsorption is a viable option.

The Net Adsorption Energy approach views the adsorption process as one of forming a solute-adsorbent bond and breaking solute-solvent and solvent-adsorbent bonds. Loading at one specific equilibrium concentration is plotted versus Net Adsorption Energy:

$$E_T = E_{js} - E_{is} - E_{ji} \qquad 6.$$

where: E_T = net adsorption energy, kcal/mole

E = energy of affinity, kcal/mole with subscripts, i, j and s relating to solvent, solute, and adsorbent respectively

Table II. Compounds Whose Adsorbability is Predicted

Compound	Actual Loading, (mmole/gm)
acetaldehyde	.01
propionaldehyde	.16
butyraldehyde	.49
ethyl acetate	.51
i-prophyl acetate	.86
propyl acetate	.96
butyl acetate	1.46
acetone	.11
methyl ethyl ketone	.35
diethyl ketone	.77
methyl i-butyl ketone	1.12
ethyl butyl ketone	2.14
isopropyl ether	1.33

The terms on the right-hand-side can be calculated using:

$$E_{js} = V_j \, (\delta_d^{\,j} \, \delta_d^{\,s}) \qquad 7.$$

$$E_{is} = V_i \, (\delta_d^{\,i} \, \delta_d^{\,s}) \qquad 8.$$

$$E_{ji} = V_j \, (\delta_a^{\,j} \, \delta_b^{\,i} + \delta_a^{\,i} \, \delta_b^{\,j}) \qquad 9.$$

where: δ_d = dispersion component, solubility parameter, $(kcal/cc)^{1/2}$

δ_a = acid component of hydrogen bonding, $(kcal/cc)^{1/2}$

δ_b = base component of hydrogen bonding, $(kcal/cc)^{1/2}$

Superscripts refer to components as above.

V = molar volume, cc/mole

Although this appears complicated, δ_a and δ_b values can be obtained from a table in References 5 and 9 and only requires

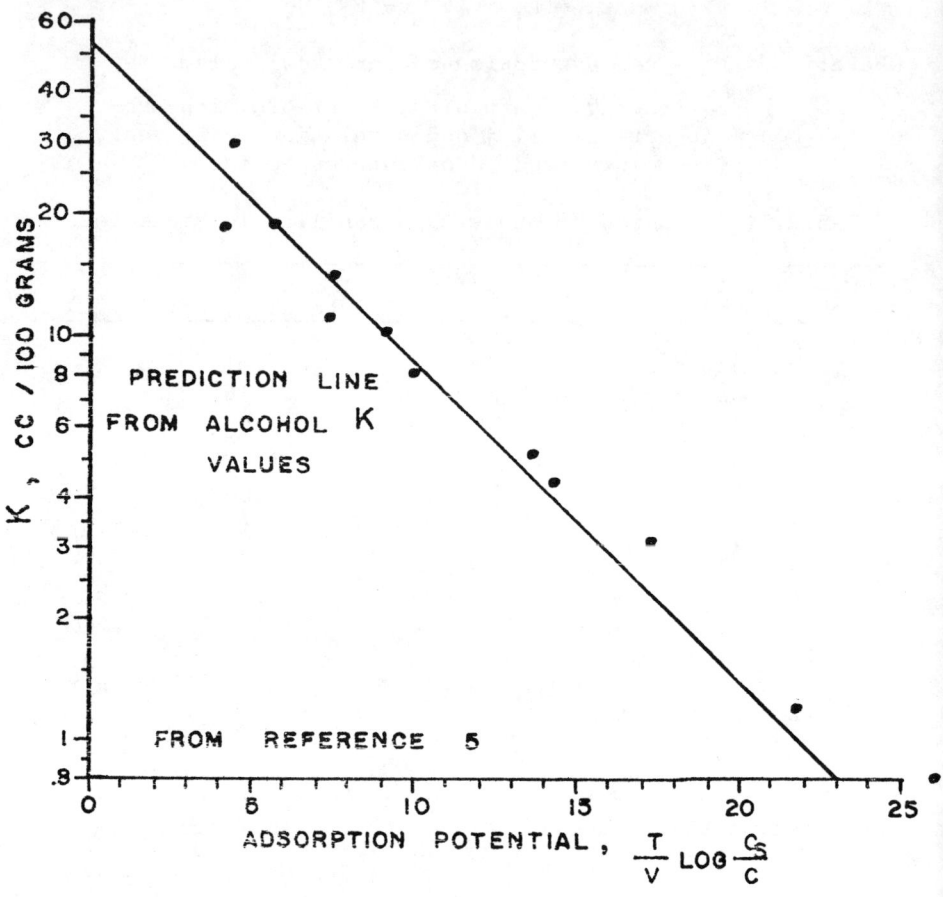

Figure 3. Polanyi Plot for Adsorbability Prediction

an approximation of solubility (slightly soluble, soluble, insoluble, etc.) and δ_d can be obtained from either of the following where x is the same function in the Polanyi approach (based on the fractive index):

$$\delta_d = 30.7x \qquad x < 0.28$$
$$\delta_d = 2.24 + 53x - 58x^2 + 22x^3 \qquad x > 0.28$$

10.

The use of Net Adsorption Energy technique to estimate loading requires refractive index, molar volume (from specific gravity), and a rough estimate of solubility; all are usually available in the Handbook of Chemistry and Physics. This technique also predicts the loading for the 13 values using the lines drawn through the nine alcohol points (not shown in Figure 4). Again, at the higher loadings it doesn't perform as well, but one could still determine if adsorption is a viable option.

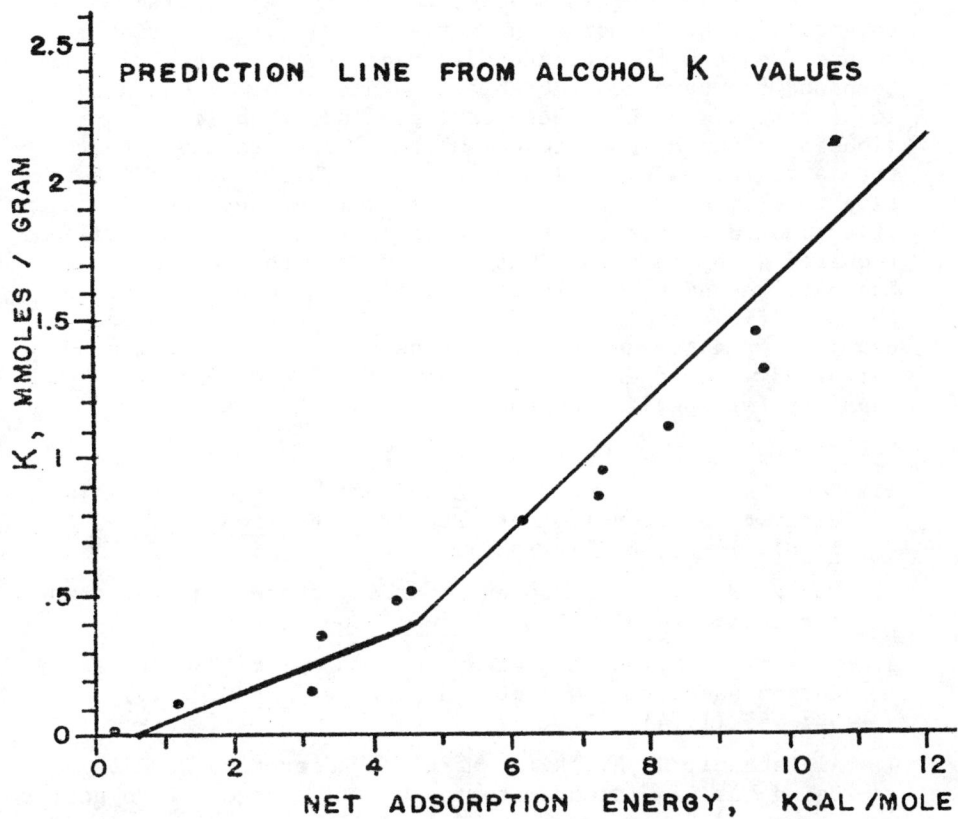

Figure 4. Net Adsorption Energy Plot for Adsorbability Prediction (from Ref. 5)

Single-Solute Isotherm

If isotherm information is unavailable in the literature and if one lacks confidence in the calculation method (after all, no aromatic, halogenated or one of several other families of compounds were tested), one could still simplify his efforts by running an isotherm of the compound in water rather than running the actual waste sample -- this indicates if the loading would be sufficient while greatly simplifying the analytical effort since TOC or COD could be used. The same isotherm procedure outlined previously (Table I) should be used. If distilled water is used, a pH buffer should be considered since large activated carbon dosages can drastically change the solution pH-sensitive compounds.

SUMMARY

Use of 1) literature data, 2) Polanyi Adsorption Potential or Net Adsorption Energy theories, or 3) single-solute isotherm to estimate adsorbability of a specific compound can save considerable time and effort when analytical costs are high. These estimated adsorbabilities (loadings) can be used to calculate carbon usage and therefore the costs for using activated carbon. This provides information to eliminate activated carbon from further consideration. If activated carbon treatment appears feasible, tests will be required with the actual wastes. Tests with real wastes will indicate the adverse effect other organic compounds have on the specific contaminant adsorption. If estimates of carbon use justify activated carbon regeneration, tests should be performed with regenerated carbon since its adsorptive capacity for specific compounds is generally less.

REFERENCES

1. Peel, R. G., A. Benedek, "Attainment of Equilibrium in Activated Carbon Isotherm Studies," Environ. Sci. & Tech., 14(1), 66-71 (1980).

2. Dobbs, R. A., J. J. Cohen, "Carbon Adsorption Isotherms for Toxic Organics," EPA - 600/8-80-023, April 1980.

3. Giusti, D. M., R. A. Conway, C. T. Lawson, "Activated Carbon Adsorption of Petrochemicals," JWPCF, 46(5), 947-965 (1974).

4. Al-Bahrani, K. S., R. J. Martin, "Adsorption Studies Using Gas-Liquid Chromatography - I. Effect of Molecular Structure," Water Research, 10, 731-736 (1976).

5. Arbuckle, W. B., "Estimating Equilibrium Adsorption of Organic Compounds on Activated Carbon from Aqueous Solution," Environ. Sci. & Tech., accepted for publication March 1981.

6. Reimers, R. S., et al, "A Quick Method for Evaluating the Suitability of Activated Carbon Adsorption for Wastewaters," Proceedings 31st Industrial Waste Conf., Purdue University, May 1976, pp 123-142, 1977.

7. Belfort, G., "Selective Adsorption of Organic Homologues onto Activated Carbon from Dilute Aqueous Solutions. Solvophobic Interaction Approach and Correlations of

Molar Adsorptivity with Physiochemical Parameters," *Environ. Sci. & Tech.*, 13(8), 939-946 (1979).

8. Manes, M., "Chapter 2: The Polanyi Adsorption Potential and Its Applications to Adsorption from Water Solution onto Activated Carbon," *Activated Carbon Adsorption of Organics from the Aqueous Phase, Vol. 1* (Eds. I. H. Suffet and M. J. McGuire), pp 43-64, 1980.

9. McGuire, J. J. and I. H. Suffet, "Chapter 4: The Calculated Net Adsorption Energy Concept," *Ibid*, pp 91-116.

SESSION V

HAZARDOUS WASTES
(MANAGEMENT AND CONTROL)

INDUSTRY ROLE IN HAZARDOUS WASTE FACILITY DEVELOPMENT

<u>Janis Stelluto</u>. Executive Office of Environmental
Affairs, Commonwealth of Massachusetts; Boston,
Massachusetts.

INTRODUCTION

    Treatment and disposal of the waste by-products are
industry responsibilities, part of the total operations of
any commerce. Assurance that those wastes are properly
managed to protect the public health and safety, is a government responsibility, a role most clearly implied when the
nature of the waste requires that particular treatment modes
must be employed to prevent threats to the environment and
public health and safety.
    To ensure safe management of this category of wastes,
generically called Hazardous Wastes, government has developed
standards for proper management. Much of industry appears
to be serious in intent to comply with regulations that
protect public health and safety. However, neither sector
can move much farther forward without additional capacity
for hazardous waste treatment and disposal. This can be
done by reducing the amount of waste and thus allow existing
capacity to last longer, and by increasing the capacity,
which requires expansion of present facilities or construction of new ones.
    In order to either reduce waste or increase capacity,
industry must assume a major role. Only industry management
can make decisions that will lead to reduction of waste
products. And only the industries who produce these wastes
can commit the waste to any particular management route, old
or new.
    The prospective consumers of the service should play a
role in the decision making. How involved industry becomes
and in what manner is a choice that industry must make. The

message is clear, however, that industry must play a role in attracting processing firms to expand or construct new capacity for waste treatment and disposal.

ECONOMIC IMPACT OF HAZARDOUS WASTE MANAGEMENT

Hazardous waste has been discussed in the past as an environmental issue, but it clearly has economic implications. Anyone with waste to dispose of knows how costs have escalated in recent years and how that affects individual company policy and profits. Businesses are finding themselves at a competitive disadvantage in many respects, resulting in sluggish expansion in states with strict regulations and inadequate treatment or disposal capacity.

A spokesperson from a large generating firm in Massachusetts testified at a Special Legislative Commission on Hazardous Waste Management, that the parent company corporate headquarters favors expansion or development in locations that afford nearby waste disposal capability, putting the Massachusetts plant at a disadvantage within its own corporate growth plan.

A few processing companies that offer limited options currently control the waste treatment market, allowing them to maintain inflated price demands. Locating new treatment facilities near the sources of generation, operated by new entrants into the processing market, should affect the price of delivery of waste products by decreasing the cost of transportation and increasing competition.

Unfortunately, some generators respond to the high cost of disposal by employing the "midnight hauler" to dispose of the waste product. Continued practice of illegal disposal by a few bad actors gives all industry a tarnished image.

Delayed expansion or construction of facilities imposes further costs of detection and clean-up. When results of improper practices are discovered and the responsible party cannot be identified or is not financially responsible -- as is most often the case -- the government pays the cost of clean up, passing the cost along to taxpayers. Industry picks up a large part of that cost through corporate and employee taxes, again affecting company policy and profit.

With such large stakes in the outcome of facility siting decisions, generating industries have good reason to be part of the planning process. Yet in Massachusetts, experience shows that most generating firms are reluctant to participate in preparing for facility development, avoiding public statement of the fact that they are generators of the feared hazardous waste.

Generators who become a part of that process and admit to producing particular waste products are then committed to paying the higher price of legal and proper disposal. Thus by avoiding a public statement of being among the generators, they retain the opportunity to dump illegally at a lower price.

Each generator seems to fear individual exposure and seeks assurance that enforcement capability will be equally applied to all firms, allowing them to disclose their role as generators collectively. With assurance that all generators are following the same rules, none has a competitive edge.

Government has a responsibility to enforce regulations and to assure that they will be enforced equitably. Generating industry may meet that charge of equitable enforcement by organizing their efforts into a consortium to go forward in a collective mode to perform some tasks necessary to site facilities. And until facilities are built, regulation enforcement and generating industry compliance is costly and unsure.

PROCESSING INDUSTRY INPUT

The Massachusetts siting process assumes privately owned and operated processing firms will propose plans for treatment/disposal developments, reflecting the preference in this country for private enterprise in business. Few government operations are run efficiently and cost effectively; existing state personnel and resources could not manage a waste treatment/disposal operation. Furthermore, companies appear to be ready to enter the waste treatment field, with capital available to invest in the operation. The Massachusetts siting process depends on corporate initiative to locate facilities in the state. That initiative depends on a number of factors, among them a statement of intent by generating firms to do business with the new firm.

Successful siting of facilities implies responsibility by industry at both ends of the waste stream.

GENERATING INDUSTRY INPUT

Only industry can report trends that will affect the characteristic of the total waste stream. Past estimates of waste inventory are in volumes and are vastly imprecise. These have been done by government agencies, and each report is suspect because of generators incentives to reveal as little as possible about their activity. Industry seeking information from industry is likely to yield more accurate data.

As the cost of waste disposal has increased in the past few years, and as regulations require standards of management, firms are investing in equipment and management routines to reduce the amount of waste. Two major generators in Massachusetts have recently installed equipment that reduces waste output by eighty and ninety percent. Similar action by several firms will have a collective effect on the waste stream, and will influence decisions by processors.

In short, the consumers of the service under discussion, manufacturing and service industries with waste to dispose of, should be involved in the planning of waste treatment and disposal facilities. Information critical to making choices for appropriate treatment facilities is available only through the industry. With no waste of their own to negotiate, governmental officials hope to attract processors to the states by assuming that generating industries will patronize the facilities. In fact, only the generating firms may commit waste to a particular treatment mode, and that commitment is sought by prospective developers. If generating industries invest in on-site treatment capacity, further production decisions are constrained by the treatment option that has already absorbed a great deal of capital investment. Having the broader spectrum of treatment/disposal choices that is likely to be available in a centralized facility allows the generator greater flexibility in production planning. However, to ensure that proposed facilities meet those needs, generating industries should be involved in the early stages of facility planning.

Many service and manufacturing firms have more technical staff than state agencies who are officially charged with planning for facility development. Collectively, industry has a wealth of staff resource that should contribute to planning for services to be contracted by industry.

In Bavaria, where generating industries invested real capital in processing facility development, there is an opinion among generators that they should be more involved in current facility management to better ensure that facilities meet the needs of the primary consumers.

SUMMARY

Treatment and disposal of the waste by-products of manufacturing and service processes are industry responsibilities, part of the production process. Assurance that those wastes are properly managed to protect the public health and safety, is a government responsibility, a role most clearly implied when the nature of the waste requires that particular treatment modes must be employed to prevent threats to the environment and public health and safety.

To ensure safe management of this category of wastes, called generically Hazardous Wastes, government has developed a framework for proper management. Industry appears to be serious in its intent to comply with regulations that maximize protection of public health and safety. However, neither sector can move much further forward without additonal capacity for hazardous waste treatment and disposal. This can be done by reducing the amount of waste, allowing existing capacity to last longer, or by increasing the capacity, requiring expansion of present facilities or construction of new ones.

In order to do either, reduce waste or increase capacity, industry must assume a major role. Only industry management can make decisions that will lead to reduction of waste products. And only the industries who produce the waste products can commit the waste to any particular management route. Potential developers of treatment facilities seek information about these generating industry decisions, information that only the industries themselves can supply.

THE IMPACT OF SUBTITLE C OF RCRA UPON INDUSTRIAL WASTEWATER
TREATMENT FACILITIES

Felon R. Wilson, P.E.   MCI-Environmental Engineers, Knoxville,
Tennessee

INTRODUCTION

Subtitle C of RCRA provided for an increased awareness of the management of hazardous wastes from industrial processes and provided a new nationwide definition for hazardous wastes. Included were liquids which previously had been regulated only by existing provisions of other statutes such as the Clean Water Act. Industrial wastewater treatment facilities which handle, treat or store wastes classified as hazardous must comply with certain portions of RCRA and can expect added regulation in the Final EPA programs. Requirements for state programs to obtain primacy are such that they must be at least as stringent as the federal programs; so the EPA standards can be viewed as typical for most facilities.

Industrial officials generally recognize that the first step toward compliance with hazardous waste statutes is an identification of waste steams, either liquid or solid, and an assessment of the characteristics of these wastes relative to the definitions and listings of 40CFR 261. This section of RCRA has been amended or sections finalized several times since May 19, 1980. These changes in hazardous waste listings have often been the result of petitioning by a particular industrial segment. Table I is a summary of the actions regarding the listing of hazardous wastes in 1980. Of special note are the actions related to wastewater treatment operations which comprise many of the amendments. An example of the dynamics of the process lies in hazardous waste No. K074, which results from the treatment of wastewaters from the production of TiO_2.

Table I - 40CFR 261 Actions - 1980

F.R. Date	Location	Action
5/19/80	40CFR 261	Final Rule - Established 4 characteristics of hazardous wastes.
	40CFR 261.31	Interim Final Rule - Listing of hazardous wastes from non-specific sources.
	40CFR 261.32	Interim Final Rule - Listing of hazardous wastes from specific sources.
	40CFR 261.33	Interim Final Rule - Listing of discarded chemical products subject to regulation and toxic wastes.
7/16/80	40CFR 261.32	Proposed Rule - Addition of 11 wastes; 4 wastewater treatment related.
	40CFR 261.31, 32	Interim Final Rule - Addition of 18 wastes.
	40CFR 261.32	Proposed Rule - Proposed addition of 7 wastes; 3 wastewater treatment related.
10/30/80	40CFR 261	Proposed Rule - Removal of trivalent chromium from basis of toxicity listing except in incineration.
	40CFR 261.32	Final action regarding interim final regulation - Deletion of 8 specific wastes which were listed based on toxicity of Cr^{+3}. Note: K074 included; 5 wastewater treatment related.
11/12/80	40CFR 261.31, 32	Final Rule - Finalization of listing of wastes; deletion of 4 wastes; 2 wastewater treatment related.
11/25/80	40CFR 261	Note: Effective date - 11/19/80 Interim Final Rule - Deferment of discarded arsenical treated wood products.
	40CFR 261	Final Rule - Deletion of several discarded chemicals.
	40CFR 261	Temporary Exclusions - Industry site-specific exclusions based on petitioning effort.
12/31/80	40CFR 261	Note: Effective date - 12/24/80 Temporary Exclusions - 8 specific industry exemptions - almost all wastewater treatment related.

This was added to the list on July 13 and then deleted on October 30. The basis of the original listing was the toxicity of trivalent chromium, as was the basis for most of the tannery industry wastewaters. Petitioning efforts by the tannery industry were conducted through trade associations, proving that Cr^{+6} was the chromium species of concern and thus, these wastes were deleted and the EP toxicity testing requirements were structured for the hexavalent chromium form. This action is one of the fundamental options available to industry. National efforts for delisting, primarily through trade associations, is a viable avenue for dealing with regulatory requirements where a genuine action is warranted. Level of effort for delisting varies, but it is generally thought such action requires a mini-environmental impact statement and thus can best be pursued through the combined efforts of many industrial installations such as through trade groups.

SEQUENCE OF APPLICABLE REGULATIONS

The December 18, 1978 Federal Register contained the first set of proposed regulations under Subtitle C. Many comments were presented to EPA, and it was February, 1980, when interim registration standards were promulgated. The major thrust of the hazardous waste program occurred on May 19, 1980, however, when 40CFR 265 (Interim Status) standards, portions of 40CFR 264 (Final Facility) standards, consolidated permit standards, and transportation and hazardous waste listing standards were promulgated. The hazardous waste regulatory process continues to be very dynamic, with several modifications to the May 19 promulgation to date. Portions of these regulations have been directly applicable to industrial wastewater treatment facilities. An explanation follows:

May 19, 1980 - The core of the initial hazardous waste program was established by defining hazardous wastes and creating interim status standards. Mixtures of hazardous wastes and domestic sewage were excluded, after the mixing, as were industrial discharges, after discharge, when covered by an NPDES permit.

November 17, 1980 - A large portion of industrial wastewater treatment systems was excluded, if the system is a "wastewater treatment Unit", which, for definition, must be:
(1) Part of a wastewater treatment facility subject to regulation under either Section 402 (NPDES) or Section 307(b) (pretreatment) of the Clean Water Act.
(2) Receiver, treater, or storer of an influent wastewater which is a hazardous waste as defined by 40CFR 261, or generator or accumulator of a wastewater treatment sludge which is a hazardous waste as defined by 40CFR 261, or

treater or storer of a wastewater treatment sludge which is a hazardous waste as defined by 40CFR 261.
(3) A tank, where tank is defined as: A stationary device, designed to contain an accumulation of hazardous waste which is constructed primarily of non-earthen materials (e.g., wood, concrete, steel, plastic) which provide structural support. Similarly, elementary neutralization units were covered by this action. The tank requirements are once again applicable.

January 12, 1981 - Regulations were promulgated in interim final form which are applicable to surface impoundments built and designed for containment. This applies primarily to those designed and built prior to the effective date of Final Facility Surface Impoundment regulations.

February 5, 1981 - Reproposal of final regulations pertaining to land disposal facilities including surface impoundments along with definitions of seepage facilities were established, based primarily on the function of surface impoundments. Informational requirements for Part B applications were included for these facilities and design standards for new facilities.

February 13, 1981 - Temporary permitting standards were promulgated as an interim final rule, introducing regulatory section 40CFR 267. These standards are primarily to allow the establishment of new land disposal facilities to meet national industrial disposal needs. They serve as a temporary permitting authority for new facilities while final standards are being developed and implemented. No state approvals are allowed under 40CFR 267. If a new surface impoundment is to be used as a storage facility, the regulations proposed on January 12 apply; if it is to be used as a disposal facility, 40CFR 267 applies.

Particular note should be given to the rule concerning wastewater treatment units (11/17/80). While this action excludes many requirements of interim status, this permit-by-rule nonetheless requires several operational assurances on the part of the operator. These requirements are contained in 40CFR 266, a section devoted to standards for specific hazardous waste facilities. The requirements for these facilities follow:
(1) Acquire an EPA identification number.
(2) Establish a security system to prevent contact with waste or disturbance of waste.
(3) Inspect units to prevent unauthorized discharge or danger to human health.
(4) Develop and retain a written inspection schedule and log.
(5) Operate facility to prevent adverse operating conditions.

(6) Prevent abnormal corrosion.
(7) Comply with manifest system and annual reporting.
(8) Remove all wastes at closure.
(9) Report all spills and leakage in writing.

As can be seen, permits-by-rule do not totally exempt facilities from regulation under RCRA.

SUMMARY OF INTERIM STATUS REQUIREMENTS

Table 2 gives a summary of some of the major requirements under the EPA interim program for hazardous waste treatment, storage, and disposal facilities. Requirements, of course, vary depending on the nature of the facility. Post closure plans are required only at disposal facilities, whereas closure plans must be formulated at all facilities. Industrial wastewater treatment facilities must assess the nature of planned final disposition of such units as surface impoundments and the contents, including precipitated solids. A written operational plan is suggested to document the good faith intents of the industry. While most of these plans or requirements have been formulated by most facilities by mid-1981, revisions, modifications, and improvements to draft efforts further the good faith intent on the part of the permittee.

Table 2. Major Requirements for Hazardous Waste Treatment, Storage, and Disposal Facilities Under Subtitle C of RCRA Interim Program

Groundwater Monitoring System	11/19/81 (on-site only)
Groundwater Monitoring Plan	11/19/81
Groundwater Assessment Outline	11/19/81 (on-site, as required)
Groundwater Assessment Plan	As required
Part A Permit Application (Interim Status)	11/19/80
Part B Permit Application	As specified by Administrator
Manifest System (final form)	11/19/80
Annual Generator Report	March 1 of each year, beginning 3/1/82
Surface water run off controls	11/19/81
Waste analysis plan (facility owners)	11/19/80 - can be required anytime after (Part B)
Closure & Post Closure Plans	5/19/81 - Amended 10/30/80 FR72040
Contingency plan	11/19/80 - can be required anytime after (Part B)
Training Programs	5/19/81

IDEALIZED EXAMPLE

Consider the case of XYZ industries, a small manufacturing facility located in the northeast United States. The wastewater treatment scheme is depicted in Figure 1. Three process water flows exit the plant, numbered here as 1, 2, and 3. Stream No. 1 enters an equalization tank, then an aeration basin followed by mechanical clarification with sludge recycle, effluent polishing, and subsequent discharge with an NPDES permit. Stream No. 2 is an acidic organic wastewater which is neutralized in a tank and then combines with Stream No. 1. A portion of the sludge generated in biological treatment is wasted, dewatered, and transported on-site to a land treatment facility. Stream No. 3 undergoes hypochlorite-caustic cyanide destruction with discharge to municipal collection lines. The only RCRA-regulated portions of the entire system are the biological aeration basin and the sludge treatment facilities. Since transportation of the sludge is totally on-site, transporter requirements (manifest, packaging) are not applicable, but the dewatering and land treatment steps are regulated, assuming the sludge is either a listed waste, or retains hazardous characteristics. All other portions of this wastewater scheme are tanks, neutralization units and/or discharges to a POTW. If the clarification occurred in a basin, it would be RCRA-regulated as would the equalization and neutralization units. The CN^- destruction process is possibly further exempted since it is within the confines of the plant, and the definition of the point of waste generation would be questionable. Actually, the wastewater treatment and neutralization facilities are RCRA-permitted by rule.

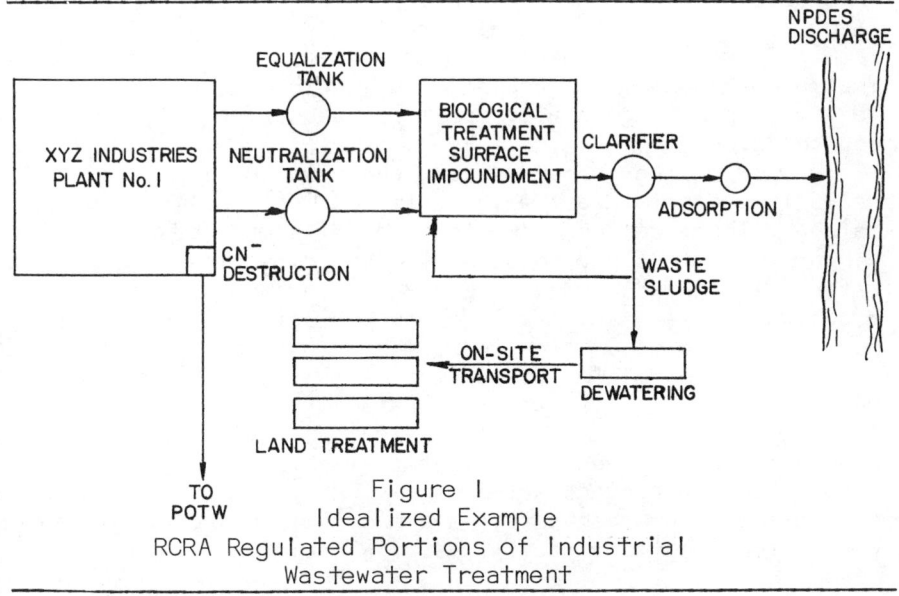

Figure 1
Idealized Example
RCRA Regulated Portions of Industrial Wastewater Treatment

SUGGESTED METHODS FOR INITIAL REQUIREMENTS

Unsaturated Zone Groundwater Monitoring

Interim Status Standards require the monitoring of leachate only at land treatment facilities, a requirement that was to be initiated by November 19, 1980. Temporary and Final Permitting Standards, however, contain various requirements of unsaturated zone monitoring.

The object of leachate monitoring is to intercept the infiltrating liquid as it percolates to the water table. Several methods are suggested:
(1) Simple perforated pipes engineered for monitoring.
(2) Lysimeters.

Each has disadvantages and advantages. Briefly, Method I is generally the most economical. However, some regulatory agencies and hydrogeologists suggest that lysimeters are the most effective method for leachate monitoring. Nonetheless, site specific conditions coupled with sufficient preliminary engineering judgement dictate the proper choice of methods.

Saturated Zone Groundwater Monitoring

The portion of the saturated zone to be monitored is not as straightforward as that for the unsaturated zone and has a regulatory definition for the strata of concern. In the case of hazardous waste treatment, storage, and disposal facilities the "First Significant Aquifer" is to be monitored, although underground injection systems have the added definition of "Surficial Aquifer". This is a broad term and is defined as "The uppermost aquifer with an upper boundary defined by a water table which is naturally recharged from the ground surface and/or from the unsaturated zone - includes formations that are seasonally saturated or may develop a perched water table." This is generally the first subsurface water encountered.

Proper casing, construction, and placing of monitoring wells is essential to the integrity of the results to be obtained. The location of monitoring wells must be preceded by a hydrogeologic study to determine the direction of groundwater flow in order to define up-gradient and down-gradient groundwater. The end result of such a study is usually a potentiometric map, which indicates groundwater flow direction.

Prior to obtaining consulting expertise for the establishment of properly engineered saturated or unsaturated zone monitoring systems, a review of in-plant data is valuable to adequately understand the potential magnitude of the effort. Project engineering departments will usually have foundation boring logs and other data, as a minimum, in the vicinity of wastewater treatment facilities. In addition, inspection reports, or discussions with personnel involved in excavations for surface impoundments will reveal valuable information on soil types and location of water tables.

Specific expertise should be considered when a surficial review of data indicates that groundwater contamination is unlikely. 40CFR 265.90(c) states that all or part of the groundwater (saturated zone) monitoring requirements may be waived if it can be demonstrated that the potential for groundwater contamination from hazardous waste management facilities is low. While utilization of this option will likely be difficult, it should be considered where geologic characteristics and cost-effectiveness (cost of option investigation versus long-term monitoring, etc.) indicate feasibility.

Closure and Post Closure Plans

Depending on the magnitude of the anticipated closure effort, which is generally a function of the type and size of the facility, the preparation of the closure and post-closure plans may or may not be performed in-house. Once again, subsurface data and information on local soil types can be obtained from plant engineering groups. Soil compaction and permeability data are essential, as are post operational characteristics of tanks or impoundments, and can be generally obtained from maintenance or utility engineers. Long-term monitoring for inactive permitted sites containing in-place wastes will necessitate management of technical personnel to conduct the extended groundwater monitoring activities.

Waste Analysis Plan (WAP)

The WAP should be based on process and analytical data obtained for comparison with the listing of 40CFR 261, coupled with process operating characteristics. The characteristics of the wastes should be identified based on existing plant production processes. As plant processes change or major process upsets occur, waste by-products should be analyzed for changes in composition. Discussions with production and maintenance personnel will provide the basis for the WAP. Laboratory personnel should also be consulted to identify changes or abnormalities in the characteristics of the waste.

Contingency Plan

Emergency planning is the objective of the contingency plan. Fire, security, and safety personnel who normally deal with emergency activities should be consulted. These departments should have information concerning the handling of flammable and toxic wastes in addition to having experience in dealing with local emergency response agencies. The Spill Prevention Control and Countermeasure Plan previously prepared by the plant should be relied upon for information on spill controls. The emergency coordinator should be a person having day-to-day knowledge of facility operations, such as a utility engineer. He should also have sufficient plant experience to know lines of responsibility within the plant.

Training Programs

Briefly, simple facilities can best integrate training into regularly scheduled safety meetings for all levels of personnel. More complex hazardous waste management facilities will probably require a specified, written training manual for the hazardous waste facilities and require a more formal implementation.

DEVELOPMENT OF THE HAZARDOUS WASTE MANAGEMENT PROGRAM
FOR PENNSYLVANIA ELECTRIC (PENELEC/GPU) COMPANY

Daniel F. Buss, Program Manager
Ronald Long-Sing Wen, Program Scientist
Camp Dresser & McKee Inc., Milwaukee, Wisconsin

James L. Greco, Ph.D., Environmental Scientist
Pennsylvania Electric Company, Johnstown, Pennsylvania

INTRODUCTION

The goal of Subtitle C of the Resource Conservation and Recovery Act (RCRA) is to insure the proper management of hazardous waste. To meet the goals of RCRA, Pennsylvania Electric Company (Penelec) a member of the General Public Utilities system, undertook an extensive investigation of its electric power generating stations and support facilities to insure compliance with these hazardous waste management regulations. Penelec retained Camp Dresser & McKee Inc. to provide assistance in conducting this investigation. All power generating stations in the Penelec system were studied including a coal cleaning plant for the Homer City Station, and two division repair shops and garages. These stations were located in Western Pennsylvania near Seward, Homer City, Keystone, Conemaugh, Shawville, Wayne, Warren, Seneca, Clairmont, Erie, Williamsburg and Deep Creek, Maryland.
Each study consisted of inventorying all materials at the facility and determining from their chemical constituents whether or not wastes generated following their use, would be considered a hazardous waste. In addition, all process wastes such as industrial waste treatment, fly ash, bottom ash, pyrites or boney, coal pile and ash pile runoff and leachate, metal cleaning wastes, solvents, degreasers, and waste sludges were analyzed to determine characteristics of reactivity, ignitability, corrosivity, and toxicity. Metal cleaning wastes, paint stripper tank sludge, various solvents and degreasers met criteria as being a RCRA hazardous waste. Since metal cleaning

wastes were at times incinerated in the boiler furnaces, the fossil fueled power generating stations obtained interim status for treating this waste through incineration.

A guidance manual was prepared for each station by Camp Dresser & McKee Inc. This manual specifically included instructions for managing wastes considered hazardous; Department of Transportation requirements for manifesting, labeling and placarding the waste; and instructions for compliance with requirements for treating a hazardous waste. To implement the hazardous waste management program and insure compliance with the associated regulations, Penelec designated a technical representative for each station responsible for managing the stations hazardous waste. The hazardous waste management programs at the various generating stations are guided and coordinated through the Penelec Environmental Department located in Johnstown, Pennsylvania.

MATERIALS AND METHODS

The outline of tasks presented in Figure 1 were followed in developing the hazardous waste management programs at the power generating stations in the Penelec/GPU system.

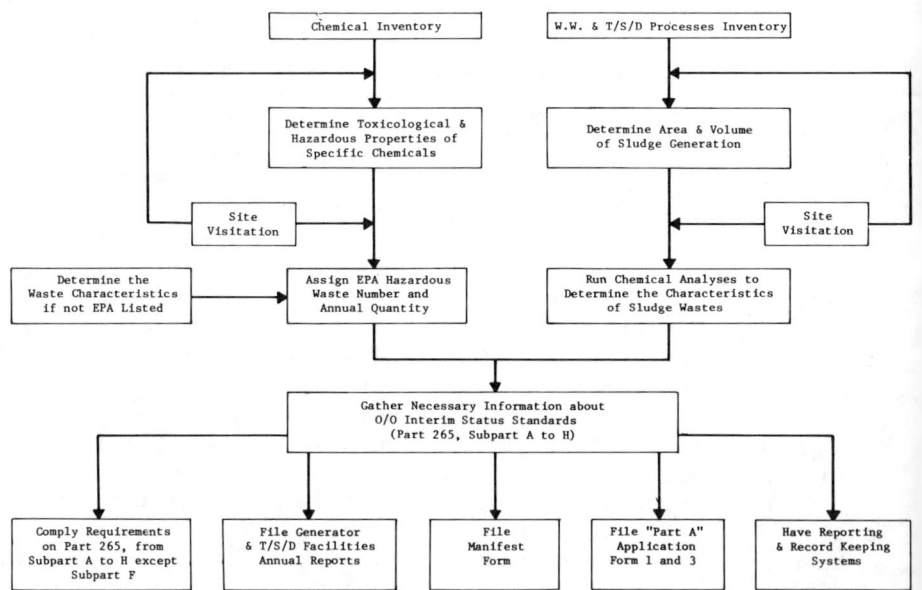

T/S/D stands for Treatment, Storage & Disposal
O/O stands for Owner & Operator
W.W. stands for Waste Water

Figure 1
FLOW DIAGRAM DEPICTING TASKS FOLLOWED IN CONSTRUCTING HAZARDOUS WASTE MANAGEMENT PROGRAMS FOR PENELEC.

During the chemical inventory and process inventory phase, a questionaire was initially prepared to assist in the evaluation of chemicals used at the stations for potential hazardous waste characteristics. The questionaire was primarily directed towards obtaining the necessary information for filling Part A of the hazardous waste permit application. The following information was requested from each station: (1) estimated volume per month for each chemical used; (2) ultimate destination of the waste materials; (3) treatment and disposal scenarios of the wastes; (4) availability of safety data on materials handling information. Since a large number of different chemicals are utilized at the stations, the chemical inventory was prepared according to the category of chemicals. Chemicals were grouped for evaluation into the following use categories:

Boiler feed	Turbine cleaning and lubricating
Boiler laboratory	
Chemical/Metal cleaning	Waste treatment
Coal laboratory	Clarification
Cooling water treatment	Miscellaneous

In addition the processed inventory questionaire requested a wastewater flow diagram, sludge flow diagram, and general layout of the portion of the facility engaged in treatment, storage or disposal. The waste process flow diagram depicting locations for waste characterization is provided for the Keystone station since these locations are most representative of sampling locations at the other generating stations (Figure 2).

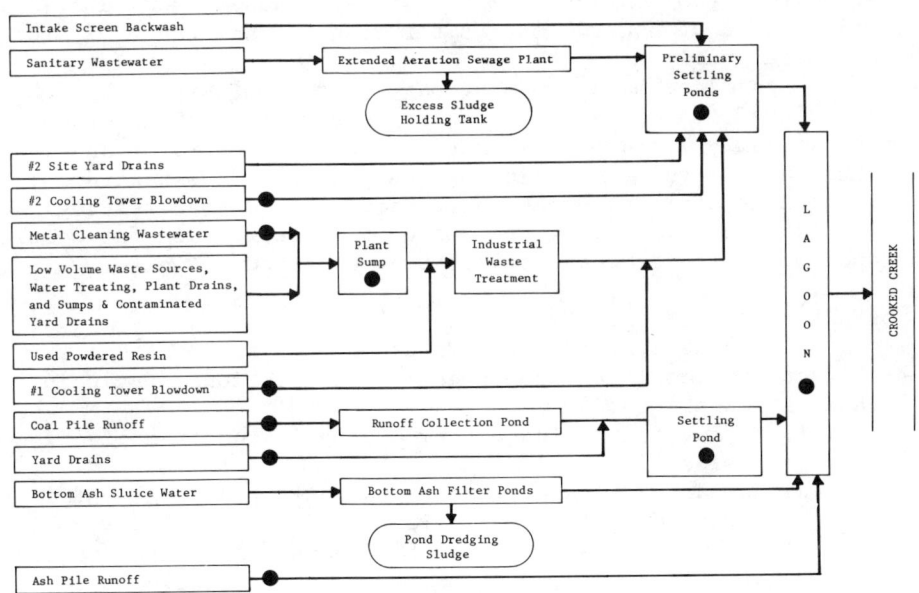

● Sample Collection/Waste Evaluation Locations

Figure 2
SAMPLING LOCATIONS FOR HAZARDOUS WASTE DETERMINATIONS IN PROCESS WASTEWATER STREAMS AT THE KEYSTONE STATION.

Waste streams subject to regulation under Section 402 of the Clean Water Act were not collected because they are excluded from RCRA regulation under Section 261.4(a)(2).

After receiving the information requested in the preliminary inventory phase, all chemicals were screened for constituents that could contribute hazardous compounds to the process waste streams. This initial screening was performed through review of chemical abstracts, chemical specification documents, and contacting suppliers of the chemicals.

In addition, samples of process wastes were collected during the site visitation phase and analyzed for hazardous waste characteristics. The method and equipment used for sampling varied with the form and consistency of the material being sampled, however, it was based on U.S. Environmental Protection Agency (EPA) sampling methodology [1].

All samples were iced (4°C) immediately following collection and returned to the laboratory for analysis. In situ pH measurements were made and ignitability, extraction procedure toxicity, and NACE corrosivity were determined in the CDM laboratory in accordance with EPA methodologies [2]. Trace metals analyses of the toxicity extract were performed by Perkin-Elmer 5000 atomic absorption spectroscopy with a HGA-500 graphite furnace and AS-40 auto sampler.

Quality assurance procedures involved approximately 15 percent of the samples collected which included checking standards, replication, split samples, standard addition, and maintenance of control charts for mean ($\bar{x}$) and range (R) for all metal analyses.

A guidance manual was prepared for each station by CDM after the chemical analyses of process wastes. This manual specifically included instructions for the generator and T/S/D owner and operator managing wastes considered as hazardous; Department of Transportation requirements for manifesting, labeling and placarding the waste; and instruction for compliance with requirements for treating a hazardous waste.

RESULTS AND DISCUSSION

All waste streams were evaluated for hazardous waste characteristics at the power generating stations. Principle waste streams were:
1) Effluent and sludge from coal pile runoff collecting basins;
2) Ash disposal area runoff and collection basin sludge;
3) Ash pond effluent and sludge;
4) Influent to and sludge from industrial waste treatment;
5) Spent solvents, waste oils and degreasing compounds.

Table I presents the concentrations of metals in the aforementioned waste streams following extraction procedure toxicity analysis. Metals concentrations in all waste streams were below the maximum concentration of contaminants for characteristic EP Toxicity. Although there was a large variability among the metals concentrations in each waste stream, barium and chromium and in certain instances arsenic, were present in greatest concentrations for the various wastes with selenium, silver and mercury generally present in trace concentrations. This trend has also been observed in metals concentrations of coal from southern Illinois and western Kentucky [3].

Table I. Metals Concentrations in Process Wastes Following EPA Extraction Procedure Toxicity Analysis

Waste Type	Metals	$\bar{X}$		σ	X_{min}	X_{max}	Waste Type	Metals	$\bar{X}$		σ	X_{min}	X_{max}
Coal Pile Runoff	As	0.68	(7)	0.86	<0.001	1.8	Ash Disposal Area Runoff Collection Pond Sludge	As	0.038	(2)	-	<0.001	0.076
	Ba	0.069	(7)	0.083	0.010	0.23		Ba	0.10	(2)	-	0.027	0.18
	Cd	0.060	(7)	0.11	0.0038	0.30		Cd	0.0053	(2)	-	0.0010	0.0095
	Cr	0.095	(7)	0.090	0.0022	0.26		Cr	0.036	(2)	-	0.017	0.054
	Pb	0.034	(7)	0.042	0.004	0.12		Pb	0.018	(2)	-	0.014	0.023
	Hg	0.0059	(7)	0.0095	<0.0002	0.020		Hg	0.0002	(2)	-	<0.0002	0.0002
	Se	0.001	(7)	-	<0.001	0.001		Se	0.002	(2)	-	<0.001	0.002
	Ag	0.009	(7)	0.005	<0.001	0.015		Ag	0.001	(2)	-	<0.001	0.001
Coal Pile Runoff Collection Pond Sludge	As	0.001	(3)	-	<0.001	0.001	Ash Disposal Area Composite	As	0.14	(7)	0.24	0.002	0.66
	Ba	0.039	(3)	0.026	0.062	0.043		Ba	0.21	(7)	0.18	0.018	0.42
	Cd	0.0010	(3)	0.0006	<0.0005	0.0014		Cd	0.0066	(7)	0.0011	0.0055	0.0090
	Cr	0.0044	(3)	0.0062	0.0006	0.0011		Cr	0.017	(7)	0.022	0.0019	0.060
	Pb	0.0033	(3)	0.0032	0.001	0.007		Pb	0.0038	(7)	0.0042	0.0016	0.013
	Hg	0.0003	(3)	0.001	<0.002	0.0003		Hg	0.0007	(7)	-	<0.0002	0.0007
	Se	0.001	(3)	-	<0.001	0.001		Se	0.011	(7)	0.0092	<0.001	0.025
	Ag	0.001	(3)	-	<0.001	0.001		Ag	0.004	(7)	-	<0.001	0.004
Ash Pond Effluent	As	0.008	(2)	-	0.007	0.009	Coal Pyrite	As	0.0099	(8)	0.011	0.001	0.028
	Ba	0.032	(2)	-	0.031	0.033		Ba	0.12	(8)	0.18	0.028	0.55
	Cd	<0.0005	(2)	-	<0.0005	0.0005		Cd	0.0022	(8)	0.0015	0.0006	0.0047
	Cr	0.0006	(2)	-	<0.0005	0.0006		Cr	0.0091	(8)	0.019	0.0007	0.056
	Pb	<0.001	(2)	-	<0.001	0.001		Pb	0.014	(8)	0.016	0.001	0.049
	Hg	<0.0002	(2)	-	<0.0002	0.0002		Hg	0.0009	(8)	-	<0.0002	0.0009
	Se	0.001	(2)	-	<0.001	0.001		Se	0.0038	(8)	0.0028	<0.001	0.007
	Ag	0.001	(2)	-	<0.001	0.001		Ag	0.001	(8)	-	<0.001	0.001
Ash Pond Sludge	As	0.14	(6)	0.24	0.005	0.62	Industrial Waste Treatment Influent	As	0.015	(9)	0.019	0.001	0.048
	Ba	0.88	(6)	0.87	0.068	1.9		Ba	0.053	(9)	0.031	0.007	0.099
	Cd	0.0047	(6)	0.0038	0.0007	0.023		Cd	0.0078	(9)	0.013	<0.0005	0.0043
	Cr	0.016	(6)	0.010	<0.0005	0.028		Cr	0.0034	(9)	0.0052	<0.0005	0.015
	Pb	0.0049	(6)	0.0037	0.001	0.011		Pb	0.0021	(9)	0.0013	0.001	0.005
	Hg	0.0005	(6)	-	<0.0002	0.0005		Hg	0.0008	(9)	0.0006	<0.0002	0.0014
	Se	0.006	(6)	-	<0.001	0.006		Se	0.003	(9)	-	<0.001	0.003
	Ag	0.001	(6)	-	<0.001	0.001		Ag	0.0038	(9)	0.0043	<0.001	0.010
Ash Disposal Area Runoff	As	0.013	(4)	0.019	<0.001	0.035	Sludge From Industrial Waste Treatment	As	0.022	(4)	-	<0.001	0.022
	Ba	0.028	(4)	0.015	0.020	0.050		Ba	0.073	(4)	0.0076	0.063	0.081
	Cd	0.0014	(4)	0.0005	0.0008	0.0020		Cd	0.0032	(4)	0.0015	<0.0005	0.0045
	Cr	0.0077	(4)	0.010	<0.0005	0.020		Cr	0.0039	(4)	0.0022	<0.0005	0.0060
	Pb	0.0065	(4)	0.0097	0.001	0.021		Pb	0.0018	(4)	0.0013	0.001	0.004
	Hg	0.0003	(4)	0.0001	<0.0002	0.0004		Hg	0.0017	(4)	-	<0.0002	0.0017
	Se	0.001	(4)	-	<0.001	.0.001		Se	0.001	(4)	-	<0.001	0.001
	Ag	0.0013	(4)	0.0006	<0.001	0.002		Ag	0.001	(4)	-	<0.001	0.001

- Insufficient data to calculate standard deviation.
( ) Number of power generating stations sampled.

Arsenic was present in high concentrations in liquid waste streams, i.e. coal pile runoff, ash pond effluent, ash disposal area runoff, and industrial waste treatment influent, but not in most sludge samples. It appears that most of arsenic in coal pile and ash disposal area waste leached out to solution by rainfall and not followed with the particulates which had settled in collection basins. It has been reported that arsenic exists as arsenate (H_3AsO_4, $H_2AsO_4^-$, and $HAsO_4^{-2}$) under oxidizing conditions [4]. Therefore, the arsenic in coal wastes has probably undergone oxidation in the presence of oxygen and water forming arsenate which is soluble in water.

Wastes that met criteria that characterized them as RCRA hazardous waste were primarily spent solvents, degreasers and metal cleaning wastes. Certain solvents had a flash point below 140°F (60°C) and thus met the ignitability characteristic. Metal cleaning wastes, primarily the first cleaning were considered hazardous wastes because their corrosivity and EP Toxicity test for chromium. Recommendations were made to substitute solvents with a higher flash point for those solvents ignitible below 140°F. Since metal cleaning wastes were incinerated in the furnaces of the boilers, interim status was obtained for incineration as a treatment process of managing metal cleaning wastes.

SUMMARY AND CONCLUSIONS

These studies have shown that wastewater and sludge generated in the various processes at the Penelec power generating stations are not RCRA hazardous wastes. Metals concentrations following EP Toxicity testing were well below their respective maximum concentration limit, corrosivity tests for the wastes were much less than 6.35 mm/yr, and pH was between 2 and 12.5. No process wastewater streams met the ignitability characteristic, however, the following wastes were considered hazardous wastes and handled according to RCRA regulations. High volume wastes were:
1) Metal cleaning wastes generating from the cleaning of boilers and condensers;
2) Waste solvents, e.g. Darsol, 1.1.1-Trichloroethane;
3) Highly contaminated waste oils;
4) Hydrazine if discarded.

The low volume wastes found to possess hazardous characteristics were used in the laboratory for chemical tests. These componets were: vanadium pentoxide, ammonium metavanadate, xylene, acetone, benzene, toluene, freon and toluidine.

The characterization of the metal cleaning waste as hazardous was based on high chromium concentrations as determined from previous historical data. Solvents to be discarded and residues of degreasing operations were segregated from other waste streams and managed by proper hazardous waste management procedures. Rags used in degreasing operations were treated as a hazardous waste since the flash point was below 140°F (60°C).

The following alternatives were addressed to handle Penelec's hazardous wastes:
1) Waste oils and solvents could be:
   a) reclaimed and reused on site;
   b) used for energy recovery;
   c) handled through off-site oil reclamation service.
2) Metal cleaning wastes could be managed by:
   a) incineration in the boiler;
   b) handling through a contractor for proper treatment or disposal;
   c) chemical and physical treatments on site to reduce metal concentrations.
3) Low volumes laboratory wastes should be disposed of

properly and should be accompanied by a hazardous waste manifest when transported, treated, stored, or disposed off-site.

REFERENCES
1. deVera, R.R. et al., "Samplers and Sampling Procedures for Hazardous Waste Streams," EPA-600/2-80-018, January (1980).
2. U.S. Environmental Protection Agency, "Test Methods for Evaluating Solid Waste," SW-846 (1980).
3. Costle, D.M., Schaffer, R.B. and Lumm, J., "Development Document for Effluent Limitations Guidelines and Standards for the Steam Electric," EPA 440/1-80-029-6, September (1980).
4. Thurston, R.V., et al., A Review of the EPA RED Book; Quality Criteria for Water, Water Quality Section, American Fisheries Society, April 1979.

REMEDIATION AT AN INACTIVE MUNICIPAL DISPOSAL SITE CONTAINING
PCBs

Warren V. Blasland, Jr., P.E., Vice President
G. David Knowles, P.E., Sr. Project Engineer
Edward R. Lynch, P.E., Sr. Project Engineer
Robert K. Goldman, Project Engineer

O'BRIEN & GERE ENGINEERS, INC.
SYRACUSE, NY

INTRODUCTION

The development of a remedial action plan, based on established criteria, for an inactive waste disposal site containing hazardous and/or toxic wastes requires a carefully planned field investigation program and an evaluation of feasibility alternatives. This paper presents the approach taken to develop a remedial action plan for an existing inactive site known to contain Polychlorinated Biphenyls (PCB's).

BACKGROUND

The municipal open dump site was operated from the mid 1950's until the mid 1960's. Local residents, several municipalities and area waste haulers used the site to dispose of mixed municipal refuse, (i.e., trees, shrubs, bottles, cans, etc.). During the same period of time, independent waste haulers also used the site to dispose of spent and reject electrical components which contained Polychlorinated Biphenyls (PCB). The dump site is 2.5 acres in size and has been inactive since the mid 1960's.

EXISTING INFORMATION

Prior to developing a field investigation work plan, all existing regional and site specific historical information was reviewed. This information included aerial photographs of the site prior to and during the years in which the site was in operation. The aerial photographs aided in identifying the manner and mode in which the site was operated. Countywide soil and groundwater surveys, local groundwater well logs and USGS topographic mapping were also reviewed.

SAFETY PROTOCOL

A safety protocol was developed prior to initiating the field investigations for each work activity to be conducted at the site.

The safety protocol was implemented as follows:

1. During the drilling of test borings, the installation of groundwater monitoring wells, the digging of surface trenches, and the sampling from groundwater wells, a dual carbon respirator, disposable rubber gloves and boots and a disposable acid resistant suit were worn.

2. During other field investigations, the safety equipment listed above was optional, unless a distinct chemical odor was noted, in which case the dual carbon respirator was worn.

With the exception of the dual carbon respirator, new safety equipment was worn each day (when required) and buried at the site at the end of each day's work. Field personnel were instructed to replace filters when odors became apparent. A hard hat was also required to be worn at all times, and a Scott Air Pack was available on site for use if needed.

FIELD INVESTIGATIONS

The goal of the field investigations was to determine the limits of the site and to define the ways and extent by which

PCB's were migrating from the site. The field investigations conducted at the site included a topographic survey, test borings, test pit trenching, a magnetometer survey and ground and surface water monitoring.

A <u>topographic survey</u> was conducted at the site and immediate surrounding area to determine the location and elevation of significant features at the site.

<u>Test borings</u> were drilled at the site to determine the vertical extent of wastes deposited at the site and to determine the nature of the underlying soils and the elevation, type and integrity of the bedrock.

<u>Test pit trenching</u> was conducted at the site to determine the location of PCB waste materials in the site and to determine the vertical and horizontal limits of the site.

A <u>magnetometer survey</u> was conducted at the site to confirm the horizontal limits of the site as previously determined by the test pit trenching operations.

<u>Groundwater monitoring</u> was conducted to determine the direction and rate of groundwater flow. Groundwater samples were collected and analyzed to quantify the mass of PCB leaving the site via the groundwater transport mode.

<u>Surface water monitoring</u> was conducted to determine runoff rates and to detect waste losses through erosion and/or leachate discharge.

RESULTS OF FIELD INVESTIGATIONS

The site topography showed that the northern and eastern slopes were very steep (45^0 slopes), while the western side was a more gradual slope between the top and toe of the site. Runoff from the site and adjacent areas flows into intermittent drainage streams which ultimately discharge into a major navigable water course.

The subsurface materials encountered at the site were consistent with those reported in the county soil survey.

The subsoils encountered at the site included a 0 to 15 foot layer of medium to very stiff brown to gray-brown varved clay. The upper one to two feet of this material contained varying amounts of roots and organic matter. The second strata encountered was a medium stiff gray varved clay with a trace of silt. The top of this material is generally encountered at 10 to 15 feet below grade. This material is softer and moister (often wet) than the brown to gray brown clay. The borings drilled into bedrock encountered a gray, very dense, silty weathered shale. Depth to bedrock under the disposal area and along the periphery ranged from approximately 20 to 35 feet.

Test pit trenches dug at the site showed that the waste materials were located throughout the dump site and therefore established that a remedial program would need to encompass the entire limits of the site. The horizontal extent of the site was defined from visual observations during the test pit trenching and was verified by the magnetometer survey. The vertical limits of the site were defined by data obtained from the trenching operations and the test borings drilled through the site.

The field investigations established two distinct zones of groundwater beneath the site. A shallow zone of perched water occurred at the interface of the refuse and the underlying clay. This groundwater is discharged as leachate through the face of the site and into the intermittent stream. The permanent groundwater zone occurs at a depth of 15 to 20 feet below grade immediately above the shale within the dumping area and flows generally northerly. The discharge of groundwater from the vicinity of the site therefore may occur as leachate flow out of the refuse or as direct flow through the deep saturated clay.

Based on investigations of groundwater flow rates and chemical analyses, the peak quantity of PCB discharged from the site in the leachate flow would range from 0.002 to 0.7 lbs/year and 0.00015 to 0.056 lbs/year from the permanent groundwater system.

Data obtained from analyzing surface runoff samples for PCB from the drainage channels during a rainfall event was insufficient to accurately quantify PCB losses. They indicated, however, that the actual transport of wastes via the erosion transport was very small.

ALTERNATIVES CONSIDERED

Three alternatives were considered for remedial action as follows:

1. In-place containment.
2. Removal to new on-site secure landfill.
3. Removal to off-site secure landfill.

Under each of these alternatives there are a number of options pertaining to materials, surface drainage, groundwater control, leachate collection, and control of volatilization. The criteria used to evaluate each alternative included compatibility with program goals, technical feasibility, ease of implementation, potential for regulatory agency acceptance, and estimated costs.

Within certain limitations, each alternative was considered technically feasible. However, there were certain advantages and disadvantages with each alternative. The primary advantage of in-place containment is that the existing waste deposits would not be disturbed. Under the other two alternatives, the excavation and transfer of the refuse greatly increases the exposure, and therefore the potential for release, of PCB material to the environment. The period of major exposure would be during the excavation phase, and the potential for volatilization losses is greatly increased. An intense rainstorm during excavation could result in an extremely large loss of PCB to the environment through erosion. For removal to an off-site facility, the period of potential exposure is extended to include the transportation to the secure facility. Another advantage of in-place containment is that no additional resources (i.e. land) are irretrievably committed beyond the area of the existing site.

RECOMMENDED REMEDIAL PROGRAM

The recommended remedial program includes a clay cap with vegetative cover, a relocated surface drainage system, a leachate collection system, a groundwater control system and a gas venting system. The plan also included provisions for maintaining and monitoring the facilities to measure the effectiveness of the remedial program.

New surface water channels will be formed at the base of the new slopes. The material excavated for creation of the new channels will be used as borrow for regrading and for the final cover. The new channels will be installed to a depth sufficient to prevent the permanent groundwater table from rising and contacting the refuse.

After placement of the final cover, the infiltration of surface water into the refuse layer and the subsequent leaching of this water through the face of the landfill will be eliminated. A pan lysimeter will be installed to measure the effectiveness of the cap. Leachate could continue to be discharged from the site, however, until the local zone of saturation at the interface of the refuse and the underlying clay is completely drained to eliminate this potential uncontrolled discharge. A leachate collection system will be installed around the perimeter of the site.

The leachate collection system will discharge to an underground holding tank, the contents of which will be periodically pumped to a portable tanker for transportation to an existing approved off-site treatment facility.

In a typical municipal landfill, methane gas is produced by the biological degradation of organic materials under anaerabic conditions. Following the installation of the cap, the generation of gases from the refuse could result in pressure induced stresses which could damage the cap. The gases could also migrate from the site, and pose a potential explosive hazard.

To positively ensure that the cap is not damaged by gas pressures, and to prevent a potentially hazardous situation, the recommended remedial program includes a passive venting system. The atmospheric vents will contain an adsorbent-type treatment unit.

The site maintenance program will include periodic inspections of the site, and implementation of a mowing and herbicidal weed control schedule.

A thirty (30) year monitoring program was developed to determine failure of the remedial program. Failure of the remedial program is deemed to have occurred if any of the following conditions are observed:

1. Any portion of the cap is eroded, allowing wastes to be carried away by surface runoff; or

2. Surface water percolates through the cap and into the refuse, and creates a source of leachate; or

3. Waste materials are volatilized through the cap in significant quantities; or

4. The groundwater table rises above the bottom of the refuse, resulting in water saturation of the refuse and the formation of leachate.

Criteria in the monitoring program to determine site failure is as follows:

1. Periodic inspections to determine significant cap erosion;

2. Flow in the leachate collection system (after the shallow groundwater has drained) to determine cap failure;

3. PCB concentrations significantly above baseline concentrations which will be established following the installation of the clay cap;

4. Groundwater levels rising to contact the refuse by continuously monitoring groundwater elevations in the vicinity of the toe of the site.

By undertaking the monitoring activities outlined above, it will be possible to measure the effectiveness of the remedial program and to detect any failure of the system.

It was not proposed to define failure of the remedial plan by analyzing groundwater or surface water samples for PCB. The very low concentration levels in the groundwater, presently below the refuse layer may well differ at different sampling locations.

A variation in PCB concentration at any downgradient groundwater or surface water monitoring location could simply be the result of a higher or lower zone of PCB concentration moving past the sampling point. Since these PCB concentrations may vary considerably and since such variation would not necessarily be the result of a failure of the remedial program, it was not proposed to attempt to define failure of the remedial program by attempting to interpret such analytical results.

SUMMARY

The work accomplished is summarized below:

1. Field investigations were undertaken to determine the condition of the site and the extent of the wastes.

2. Sampling and analyses were undertaken to define groundwater flow patterns and rates, and to estimate the magnitude of migration of wastes (PCB) from the immediate dumping area via groundwater flow. It has been shown that the loss of PCB by this mechanism is insignificant.

3. Three alternative remedial programs for the site were evaluated: in-place containment, removal to a new on-site facility, and removal to an off-site facility. The criteria used to evaluate these alternatives include compatibility with program goals, technical feasibility, ease of implementation, and compliance with regulatory requirements. In-place containment has been designated as the recommended remedial program.

4. A preliminary plan for the recommended remedial program has been presented. The program is based on in-place containment of the wastes, including a clay cap with vegetative cover, relocated surface drainage system, leachate collection system, and gas venting system. The plan also includes provisions for maintenance of the facilities and monitoring to measure the program's effectiveness for a thirty (30) year period.

POLICY TOWARD SITING OF HAZARDOUS WASTE MANAGEMENT
FACILITY: EFFECTIVENESS OF STATE PREEMPTIVE SITING
AUTHORITY

*Edward Yang. Resources Program, Environmental Law
Institute, Washington

Amy Horne. Resources Program, Environmental Law Institute, Washington, D.C.

INTRODUCTION

The siting of hazardous waste management facilities (HW MF) has recently become an important issue to the federal and state government officials charged with implementing the Resource Conservation and Recovery Act (RCRA). The Act, which calls for "cradle-to-grave" tracking of hazardous waste by generators, transporters, and disposers, began to be implemented November 19, 1980, the effective date of regulations promulgated May, 1980. A critical ingredient for the success of the program is that the quantity of hazardous waste treatment, storage, and disposal capacity available to sufficient to meet demand. Recent attempts to site hazardous waste treatment, storage and disposal facilities, however, have been rebuffed by an alarmed public [1]. The government is unable to decide where facilities should be sited, having been given no authority under RCRA to enforce these decisions. If a lack of new facilities occurs, generators may be forced to dispose of their waste at older, inadequately engineered facilities or dump their waste illegally. If this happens, then the goal of RCRA, stated by Congress as the proper management of hazardous waste and protection of human health and environment, will be jeoparidized.

*We are grateful to Oscar O. Albrech, U. S. EPA for his comments and suggestions.

This paper addresses the policies for resolving the fundamental difficulty in siting hazardous waste management facilities: public opposition. In particular, we will focus on the recent development of state siting authorities to preempting local government's right to reject the siting. In answering the question of whether such siting authorities will be effective in arriving at the socially desirable number and location of hazardous waste management facilities (HWMF), we must first develop a theoretical framework with which to determine the appropriate roles for the private sector, state and federal governments in siting HWMF. Based on this framework we will argue that the locality where the siting is proposed should decide for itself whether to accept a siting and with how much compensation. When it appears that the transaction costs of negotiating a settlement between the locality and HWMF developers will be high and result in no negotiation, the state has the duty to reduce such transaction costs through information provision and mediation. Given this theoretical framework, the assumptions underlying the effectiveness of preemptive siting authorities seems questionable from both economic and legal perspectives.

THE THEORY OF GOVERNMENT PARTICIPATION IN SITING HAZARDOUS WASTE FACILITIES

As the magnitude of the siting difficulty grows, facility developers and government officials have embarked in two directions toward resolving these problems. One direction is the use of local traditional tactics, such as planning, risk mitigation, consultation, mediation and government ownership, in siting undesirable land use facilities. The other is the development of institutional mechanisms, such as preemptive siting authorities [2], that enable quick response to siting problems.

This paper will address only the institutional aspect of siting policy. This is for several reasons. First, many of the siting tactics have been in existence for siting other types of locally undesirable facilities and discussion on their uses can be found in other places [3]. Second, the choice of the institutional framework precedes the choice of siting tactics. That is, if the inappropriate institutional framework is selected, no siting tactic is likely to be effective. Thirdly, state officials, frustrated by their inability to successfully site hazardous waste facilities, are moving quickly to give greater authority to the state and less authority to local governments. It is imperative that this institutional change be examined. This trend suggests a belief that siting difficulties, such as public opposition, can be overcome given the right political maneuvers and suitable institutional structure. However, such an emphasis overlooks the critical need to examine the underlying role that government should assume. Without an answer to the question: What is the legitimate role of the government in siting hazardous waste facility, one can not determine where the present tactics are leading us to. The much needed examination of this question, can be performed if we have a conceptual framework to reduce the siting problem to its basic elements and the relationships among them.

The conceptual framework used here is an economic one. This is not to imply that political and institutional factors are less significant. These latter factors can be incorporated into the framework and will affect its functions. But, they are likely to be misused if the basic framework is not understood.

There are several points we wish to show here. First, no government should consider the development and operation of a hazardous waste facility as one of its inherent functions. A hazardous waste facility in itself is not a public good [4]. Rather, the service has the two characteristics of a private good; that is, one person's consumption of the hazardous waste treatment or disposal service decreases the quantity of service available to other consumers and those not paying for the service can be excluded from consuming it. As a consequence, the private sector has incentives to, in the long-run, provide that quantity of hazardous waste treatment or disposal capacity for which it can charge the long-run average cost. The fact that the private market is having difficulty in siting the facility, despite the social need, does not mean that the only alternative is for government to be entirely responsible for its supply. However, we will argue that the difficulty of the private market can be resolved with the help of government since it is merely a case of an externality. Before we prove this point we must consider the second point.

One needs to identify the consumers of the site as the developer, and the supplier of the site as the local community. The demand for the site is derived from the generators of the wastes. The supply of the site is derived from the local community's willingness to sell the right for siting. The concept of willingness to sell in siting HWMF needs some explanation; it is the minimum amount of compensation that will make the community as well-off as without the siting.

The compensation consists of two components 1) a sum equal to the expected value of damages resulting from the siting, and 2) a sum equal to the willingness to take the risk. The second component is necessary due to the fact that most of us are risk averse. It is an especially important component because risk from the operation of a hazardous waste facility often eludes accurate assessment. Being risk averse, the community needs to be compensated for taking the risk, as well as for the expected value of damage. Yet this component is often ignored, leading to the notion that communities seek exorbitant compensation to cover for possible damages.

Third, if the community's willingness to sell the site equals the developer's willingness to pay, that is the maximum they are willing to pay for the siting, the siting is truly successful, and a trade beneficial to both sides occurs. Alternatively, if both sides do not agree on the willingness to pay and to sell, that particular siting attempt fails; the developer can look elsewhere, and the community simply waits for another offer from another developer. But this is still a socially efficient result. The private market fails as a system only when trade does not occur <u>anywhere</u> despite a need for trade from some members of the society.

Fourth, the community and developer can not reach agreement for siting when certain transaction costs that are necessary to bring the two parties into trading position are so large that neither side thinks that the gain from the trade is worth the cost. Many of the transaction costs are prohibitive in the siting of HWMF. One is the cost of organizing the residents of the community to reveal their collective willingness to sell. Another transaction cost comes from the lack of essential information without which the residents simply refuse to consider the trade. Since the willingness to sell depends on expected value of the negative externality, the community needs to assess the risk of release and exposure, the health effect of the exposure, and the possible environmental damage before it can discuss an offer from the developers. Mistrust of developers and government officials [5], long term scientific uncertainty, and worries of imported wastes all aggrevate the situation through higher transaction costs. When the transaction costs are always higher than the gain from trade, no siting can occur voluntarily.

The final point is that state government is one of the beneficieries of siting HWMF, and it is not clear that the state bears an adequate amount of the cost of siting. The benefits to the state government are the tax revenue, employment, and other social and economic contributions from the industrial activities of the waste generator. Lacking a nearby HWMF, the generators must transport their wastes long distances for treatment and disposal. This practice imposes high transport costs and a greater risk of a hazardous waste spill during rransport. An industry deciding where disposal at each potential location. It will prefer, all other things being equal, areas offering low-cost hazardous waste disposal. States without hazardous waste facilities may, over time, lose industries which generate the waste thereby reducing the state's jobs and tax base. A powerful examply of the crucial role HWMF can play in attracting industry location can be found on the front page of the Washington Post (Dec. 16, 1980) in a story about how the Allied Chemical plant in Baltimore, Md. threatened to move out if no landfill site were established for its chemical wastes.

Given the above points the appropriate role for government can now be derived. First, it is obvious that the state government should contribute toward resolving siting problems, so that benefits and costs of siting HWMF are better aligned. But should the state government act as an arbitrator or even the decision maker? We think not. By taking such a role the state is assuming the position of the supplier of the site and, in essence claiming to know the willingness to sell of all potential host communities. This position is counter-productive because it damages the incentive to reach voluntary and mutually beneficial agreements. In addition, it is socially offensive because of its infringement on the democratic process where one has the right to determine his or her own willingness to sell.

However, there is an appropriate and effective role for the state government as a provider of various non-intervention services; such as paying for mediators, assessing environmental and health risks conducting geological surveys, organizing local referendum and helping industries select sites. These services mainly reduce the transaction costs that impede developers and communities from reaching the stage where negotiation can take place. In answer to the question of how much the state should provide these services, it should compare the expenditures to the benefits it expects from the siting of HWMF. This is albeit, a difficult task. However, one can find an element of cost-effectiveness in this arrangement since there is an economies of scale for the state to provide these services rather than communities and developers.

In summary, the optimal division of responsibilities for hazardous waste siting HWMF is as follows: local governments should determine its willingness to sell a hazardous waste facility based on the community's preferences, and state governments should absorb the costs of providing local governments the information and support services necessary for the communities and developers to enter negotiation. This division of responsibilities best provides the optimal number of hazardous waste facilities and furthers the objectives of proper hazardous waste management.

ASSESSING POTENTIAL EFFECTIVENESS OF STATE PREEMPTIVE AUTHORITY

The use of state preemptive authority in siting HWMF is based on the perceptions that 1) government intervention in siting HWMF is justifiable and, 2) by circumventing the communities right to reject a siting, government can both reduce transaction costs and ignore public opposition.

Government Intervention is Justifiable

State control of siting authority seems to be based on the assumption that, since the private market often fails to site the much needed facilities, the supply can be guaranteed by removing veto authority from the locality. There are many occasions when market failures constitute an economically justified case for government intervention, including when the market is a natural monopoly, the good is a public good, market has barriers to entry, or the property right needs to be assigned. At first glance, a HWMF seems to have some characteristic of a public good, since the disposal or treatment of hazardous waste is becoming a legitimate need on state and national levels. But as was shown earlier, a hazardous waste facility does not qualify as a public good; its capacity is allocated by the price attached to its service. Nor does the facility seem to face decreasing average cost curves in which case it would be classified as a natural monopoly. Siting does not occur because the operation of a HWMF imposes an external cost on the surrounding residents and the externality is usually not incorporated due to prohibitively high transaction costs. Governmental action to reduce this transaction cost is justifiable, particularly since some of the benefits of the facility accrue to the state as a whole in the form of increase in employment and tax revenue. However, it is important to distinguish the proper role of state government reducing transaction costs as a third party to facilitate negotiations between the generator and receiver of the externality, from the role of becoming one of the negotiating parties.

Reduction In Transaction Cost

Another key assumption underlying the shift of siting authority toward the state is that the substantial transaction cost of negotiating or persuading the locality to accept the siting can be avoided by simply circumventing the locality. However, a hidden cost may counter-balance this gain and a non-preemptive role can reduce the transaction cost as effectively. Through the use of preemptive authority, the state, in effect, assumes the role of deciding whether the risk level of operating the facility is acceptable to the residents. The divergence between those who make the decision and those who suffer the consequences of the decision creates a new externality that can result in significant social costs. The form of this social cost is not hard to imagine. When the community's preference is mis-estimated, the cost can be found in all forms of local opposition after the facility is sited, such as protests, court actions, strikes, shutting down of the facility, and even violent disruptions throughout the community. Such a cost, although occurring after the siting, cannot be ignored.

Overriding Local Opposition

The ultimate reason for the perceived effectiveness of preemptive siting authority lies in its assumed power to override local opposition. However, this needs to be further scrutinized from a legal perspective.

In general, the state possesses authority over local governments in the absence of a general home rule grant or explicit delegation of zoning control to local governments by the state's constitution or statutes. Even where home rule or zoning authority can be preempted for particular purposes by appropriate legislative action at the state level. Constitutional grants of authority to local governments are generally broad and vaguely worded, and are also relatively easy for state legislatures to override [6]. The specificity with which the state legislature addresses the preemption issue often determines the outcome of court challenges to state control.

But the approach is misleadingly optimistic. As a practical matter, local opponents often engage in procedural tactics which prolong or delay siting decisions. Organized protests, civil disobedience and public relations campaigns may also be undertaken to persuade proponents to reconsider. Site opponents may file suits alleging that there has been inadequate public participation in the siting decision, amounting to a denial of due process. Site opponents may also argue that industry practice, the economy, and siting needs have changed since the initial siting decision. Earlier findings of need and safety sometimes can be challenged and reargued years later, further prolonging the final siting decision.

California's attempts to site a liquified natural gas (LNG) facility 63, while not involving a hazardous waste site, is illustrative of the pressures that local communities can exert in waste siting decisions [7]. California's preemptive LNG siting statute directed the state to select and rank potential LNG sites, precluding direct local participation in this selection process. One of the effects of the statute was to undercut several years of site evaluation begun by several local communities under their zoning authority.

Thus, while hazardous waste site opponents may lose local siting control via state preemption, an effective challenge may be undertaken by raising procedural or technical legal points, invoking laws unrelated to hazardous waste siting, or introducing new legislative restrictions on siting. Faced with determined local opposition, hazardous waste site developers may have to abandon their development plans.

In summary, the three necessary assumptions underlying the effectiveness of preempting the locality's right to reject siting of hazardous waste facility are challengeable. The breakdown of these assumptions implies that there may be indirect costs which are not being recognized, and the seeming effectiveness of preemptive siting boards may disappear in an actual application of the preemptive authority.

REFERENCES

1. See Centaru Associates, Siting of Hazardous Waste Management Facilities and Public Opposition, EPA Publication (SW-809), Nov. 1979.

2. MI., Md., Conn., Minn., and Penn. are the states with preemptive siting boards.

3. Literature exists on many of the mechanisms. For a review of such literature see Draft Report on "mini-assessment of Problems Related to Siting of Hazardous Waste Management.

4. For definition of public good see William Banmol, Economic Theory and Operations Analysis, 4th Ed. Prentice-Hall Inc. p. 521, 1977.

5. Randall Smith, Problems of Waste Depository Siting: A Review. Seattle, Washington: Battelle Human Affairs Research Centers, September, 1979.

6. The power of the state is illustrated in Rollins Environmental Services of LA. Inc. vs. Iberville Parish Policy Jury, 371 So. 2d 1127 (1a. 1979).

7. See L. Susskind and S. Cassella, "The Dangers of Preemptive Legislation: The Case of LNG Facility Siting in California," Environmental Impact Assessment Review of (1980) p. 19.

INCINERATION OF HAZARDOUS WASTES AT SEA

Donald A. Oberacker. U.S. Environmental Protection Agency, Industrial Environmental Research Laboratory, Cincinnati, Ohio.

Lissa A. Martinez. U.S. Department of Commerce, Martime Administration, Washington, D.C.

Robert J. Johnson. TRW, Inc., Redondo Beach, California.

SUMMARY

This paper summarizes recent EPA studies of design recommendations for a shipboard at-sea hazardous waste incineration system and the design requirements for a waterfront facility to support chemical waste incinerator ships.

These engineering studies assessed the key aspects of at-sea incineration of hazardous chemical wastes. Included are evaluations of alternative shipboard incineration systems, requirements for a waterfront support facility, and all phases of waste disposal from waste selection to final disposition of any effluent or residues produced. A preliminary evaluation was made of generic incinerators potentially applicable to shipboard operation. As a result, liquid injection incinerators were selected for destruction of pumpable wastes, and a rotary kiln was recommend for experimental evaluations of shipboard incineration of solid hazardous wastes. Fluidized-bed, molten salt, multiple hearth, multiple chamber, and starved air incinerators are all limited in operating temperatures and/or waste type capability compared with the rotary kiln.

INTRODUCTION

An estimated 57 million metric tons of industrial hazardous wastes were produced in the United States during 1980.

Many of these wastes, particularly the organic chemicals, are incinerable; but only a limited number of commercially available, land-based hazardous waste incinerators exist in the United States, and public opposition to additional sites is increasing.

Thermal destruction of chemical wastes at sea by a European vessel has been studied by the U.S. Environmental Protection Agency and was found to be an environmentally acceptable and cost-effective means of disposal for many types of liquid combustible chemical wastes. EPA is now encouraging private industry to build a U.S. incinerator ship that is capable of destroying liquid wastes at sea and that can also explore extending the capability of at-sea incineration to solid and semi-solid materials.

EPA's recent studies assessed the key aspects of at-sea incineration and issue a study report (1). An engineering evaluation was performed on several alternative forms of shipboard incineration systems, including waste selection and the disposition of any effluent or residues produced. Design requirements were also developed for a waterfront support facility for incinerator ships.

SHIPBOARD INCINERATION SYSTEM

Types of Incinerators

Major characteristics of candidate incinerators for shipboard, at-sea application are compared in Table 1. The capability of each incinerator to destroy different types of waste material is noted, along with maximum operating temperatures, relative maintenance, and commercial applications.

The liquid injection incinerator can be used only for pumpable liquids; however, it is the most effective means of incinerating liquid wastes at high feed rates, and it is capable of attaining the temperature required (up to $1600^{\circ}C$) for highly efficient destruction of toxic materials. This incinerator can also serve as the afterburner for a solid waste incinerator. Maintenance requirements are low because there are no moving parts within the high temperature zone. Liquid injection incinerators are widely used in commercial applications and represent the only technology well proven at sea for destruction of hazardous wastes.

Rotary kilns are the most versatile type, capable of handling any combination of liquids, slurries, tars, or solids, including containerized wastes. Temperatures as high as $1600^{\circ}C$ can be attained in the kiln. Rotary kilns represent

Table 1. Comparison of Candidate Incinerators for Shipboard At-Sea Application

Item	Liquid Injection	Rotary Kiln	Fluidized-Bed	Molten Salt	Multiple Hearth	Multiple Chamber	Starved Air
Waste Capability:							
Pumpable Liquids	X	X	X	X	X	X	X
Slurries, Sludges		X	X	X	X		X
Tars		X		X	X		
Solids:							
Granular		X	X	X	X		
Irregular		X				X	
Containerized		X				X	X
Maximum operating temperature, °C	1600	1600	980	980	1100	1000	820
Maintenance requirements	low	medium	medium	medium	high	medium	high
Commercial applications	Widely used, liquid wastes	Widely used, all wastes	Limited use, sludges organic wastes	Demonstration tests only	Widely used sewage sludge	Widely used, refuse	Resource recovery

well proven technology on land, but their use at sea would be a new application. Specially designed equipment will be required to withstand pitch, roll, vibration, and environmental conditions at sea. Maintenance requirements may be higher than for a liquid injection system.

Fluidized-bed incinerators are more limited than rotary kilns in their range of feed materials and are not suited for irregular solids or tarry substances. Maximum operating temperatures are limited to 980°C to avoid fusion of the silica and sand bed material. Higher temperatures of 1200°C are possible using alumina refractory particles as the bed materials.

Molten salt reactor pilot and demonstration units have been used to destroy liquid, slurry, and granular solid waste materials, but no large commercial units are presently in operation. Operating temperature of the salt bed is limited to 980°C. Pitch and roll of the ship could cause sloshing of the molten salt within the reactor. A potential advantage of this system, however, is that it can serve as a combined incinerator/scrubber by retaining particulates and contaminants in the bed.

Multiple hearth incinerators are widely used for sewage sludge disposal, but they can also accept granular solid and liquids (injected through the auxillary fuel nozzles). Operating temperatures are limited to 1100°C because of the internal mechanical components (rotating shaft, rabble arms, etc). Maintenance of internal moving parts would be frequent because of ship motion as well as thermal stress.

Multiple chamber incinerators are used extensively for industrial disposal of bulk solid wastes. Liquid wastes can be injected with the auxiliary fuels. Slurries and sludges may fall through the incinerator grates and are not suitable for this incinerator. Solids/air mixing is not as thorough as in rotary kilns, and high excess air rates are required, resulting in operating temperatures of approximately 1000°C.

Emission Control Devices

Emission control devices commonly used with land-based incinerators have many limitations for shipboard operation, including size, weight, and fresh water requirements. A high-energy venturi scrubber represents the lowest weight, volume, and installed cost emission control system evaluated (1), if sea water can be used for scrubbing and discharged into the sea. A closed-loop system requiring a settling tank would be impractical.

Waste Feed Systems

A shipboard waste feed system is required to retrieve the waste from storage and transport it to the incinerator without spillage under operating conditions of pitch, roll, and vibration. Liquid wastes and some slurries can be transported to the incinerator by conventional pumps, piping, and valves. Solids can be loaded into sealed containers on land: either smaller fiber containers can be fed directly into the incinerator, or larger standard bulk material containers can be discharged directly into a sealed hopper. Handling of 55-gallon drums and shredding operations probably involve too much risk aboard ship.

Waste Selection

Wastes can be selected for at-sea incineration only after a complete analysis of their physical, chemical, and thermal properties. Physical properties must be known to determine if the waste is compatible with the incinerator type and waste handling system. Chemical and thermal properties affect the combustion characteristics of the waste. Normally, a minimum heating value of 4400 to 5540 kcal/kg (8,000 to 10,000 Btu/lb) is necessary to sustain combustion, but this is only an approximate limit. The M/T Vulcanus has burned organochlorine wastes with chlorine content as high as 63 percent and heat content as low as 3,860 kcal/kg (6,950 Btu/lb) without firing auxillary fuel.

Monitoring

A dedicated shipboard laboratory is required to assure operational safety by analyzing the shipboard environment for waste constituents, and by verifying waste destruction efficiency. Environmental monitoring during at-sea incineration is essential to ensure personnel safety and to protect the environment.

WATERFRONT SUPPORT FACILITY

The waterfront facility is a critical part of the entire system for chemical waste disposal using incinerator ships. A dedicated, full-service facility must accommodate wastes in almost any physical form and in several types of containers, some of which may be old, corroded, and possibly leaking. Ideally, the facility would serve waste delivery by truck, rail, and/or barge.

The waterfront facility should be designed to prevent emissions of hazardous materials, to contain spills, leaks, and other accidents, and to minimize harm to personnel and the environment in the event of accidents. Ideally, the facility should be located where transportation time and distance are minimal. Structural standards must be carefully followed; these would normally be defined by the Uniform Building Code and additional location-specific building regulations. The design must also meet safety, health, and environmental criteria, which include provisions for facility monitoring, personnel safety, and contingency planning in the event of both major and minor releases of chemical wastes that have the potential to pollute the land, water, or air. These criteria are generally specified in Federal regulations. It is estimated that 75,000 m^2 (18 acres) of land could be required and a staff of approximately 40 needed to operate on a two-shift schedule.

Liquid waste, solid waste, and ash residue from incineration should be processed and stored separately. Liquid waste in drums and other containers can be sent through a shredder in a dedrumming facility. Liquid from both the containers and the decontamination of the containers can be blended to optimize transfer and combustion processes and pumped to storage tanks. Liquid waste arriving in tank trucks or tank cars, or by barge can also be blended and pumped to the storage tanks along with the tanker decontamination rinse. Ash residue from the at-sea burn should be returned to the waterfront facility and kept in the residue storage area until removed for ultimate disposal, probably in a landfill approved for hazardous waste disposal.

Review of Existing Terminal Facilities

Existing terminal facilities in the United States were reviewed as part of EPA's study. Facilities that regularly accommodate vessels similar in length and draft to the conceptual incineration vessel were identified, without regard to the type of commodity or material being handled. Only those terminals having a minimum water depth at loading berths of 7.6 m (25 ft) were included. This list of terminals was then limited to those that are handling either liquid and/or dry cargoes that are hazardous or commodities that possess physical and chemical characteristics that are similar to those anticipated for hazardous wastes (ignitability, corrosivity, reactivity, toxicity). The latter group of terminals was evaluated from the standpoint of their potential for conversion to a liquid and/or solid hazardous waste marine terminal. Marine terminal and materials handling experts were consulted regarding reasonable and practical alternatives in the development of a hazardous waste marine terminal.

On the east, gulf, and west coasts of the continental United States, there are 139 ports and 1,221 terminal docks, piers, or wharves which have sufficient water depth and space to receive the chemical waste incineration vessel as designed. Of the 1,221 terminal docks, piers, and wharves, 381 handle refined petroleum products or liquefied chemicals (and allied products). These terminals are concentrated primarily in the states of Texas, New Jersey, Louisiana, California, and New York. Except for military terminals, ownership of these facilities is predominantly private.

Most of the major and many of the smaller bulk liquid terminal companies that offer services to the public are members of the Independent Liquid Terminal Association, established in 1974. Through that association, they are relatively well informed of environmental rules and regulations that are being developed in compliance with congressional legislation. Accordingly, these companies are familiar with long-existing local, state, and Federal regulations for handling commercial bulk liquid commodities (which are mainly concerned with fire, explosion, and safety matters), and they are also acquainted with recent EPA regulations for implementing the Resource Conservation and Recovery Act.

RECOMMENDATIONS

The recommended incineration system for destruction of both solid and liquid hazardous wastes at sea is a rotary kiln coupled to a liquid injection incinerator. Two or more identical liquid injection incinerators (depending on the size of the ship selected) should be used for the destruction of liquid wastes. A single rotary kiln should be installed in combination with one of the liquid injection incinerators for evaluation before additional kilns are added.

A high-energy venturi scrubber with a pre-quench and a mist eliminator tower utilizing sea water should be considered for shipboard evaluation. Marine environmental effects of a single-pass sea water scrubbing system must be evaluated, however.

Liquid wastes should be stored in inert, gas-blanketed, lined tanks. Flowmeters are recommended for monitoring the liquid waste feed rate to each incinerator burners. Solid material should be processed on land and loaded into sealed bulk material carriers or incinerable containers compatible with shipboard safety requirements to minimize hazards of waste handling on board ship. Use of 55-gallon drums and shredding operations on board ship are not recommended.

Environmental monitoring during at-sea incineration must be conducted to ensure personnel safety and protection of the environment. A shipboard laboratory should be provided for analysis and identification of effluent waste samples and verification of destruction efficiency.

A waterfront facility is essential to support the operation of incinerator ships. The facility must be designed to receive, store, analyze, process, and load wastes aboard the ship in a safe and efficient manner, and to receive the residues from the incineration process for ultimate dispsoal. Some existing private and military terminals may be used for this purpose.

A U.S. incinerator ship can serve two broad functions: first, it can be used for the destruction of hazardous wastes in a location minimizing the risk to public health and the environment; second, it can provide a safe site to continue EPA research and development efforts in hazardous waste incineration. Furthermore, an incineration vessel would provide needed experience in the large-scale processing of hazardous waste materials. The effects of process variations in a commercial-scale incinerator on hazardous waste destruction efficiencies need to be further investigated, along with many types of wastes not yet tested. In addition to destroying hazardous wastes, the proposed incineration vessel could effectively test incinerator designs, emission control concepts, and improved sampling/monitoring equipment and methods.

For additional details of these studies, the reader is referred to Reference 1.

REFERENCE

1. "Report of the Interagency Ad Hoc Work Group for the Chemical Waste Incinerator Ship Program," NTIS PB/81-112849.

WET OXIDATION OF TOXIC AND HAZARDOUS COMPOUNDS

Tipton L. Randall, Senior Research Chemist
Zimpro Inc., Rothschild, Wisconsin

ABSTRACT

Wet air oxidation studies have been conducted in which numerous hazardous or toxic compounds have been readily destroyed under conventional wet oxidation conditions. Destruction approached or exceeded 99.9% at temperatures of 200°-$320^\circ C$ for these materials, which ranged from hydrocarbon to cyanides to chlorinated compounds including many of the priority pollutants. In general, the toxicity of the oxidized materials was greatly reduced compared with that of the starting material, as shown by acute toxicity tests using Daphnia magna. Several of the more environmentally persistent chlorinated compounds, such as PCB's, DDT or pentachlorophenol were found rather resistant to conventional wet oxidation conditions. It was discovered, however, that by employing a wet oxidation reaction medium with proprietary cocatalysts, destruction of PCB's, DDT or pentachlorophenol in excess of 99% can be achieved at $275^\circ C$. Work is in progress to implement this method in a continuous process suitable for plant operation.

WET OXIDATION

Wet oxidation refers to the aqueous phase oxidation of dissolved or suspended organic substances at elevated temperatures and pressures. Water, which makes up the bulk of the aqueous phase, serves to modify the oxidation reactions so that they proceed at relatively low temperatures, 175° to $325^\circ C$, and at the same time serves to moderate the oxidation rates removing excess heat by evaporation. Water also provides an excellent heat transfer medium which enables the wet oxidation process to be thermally self-sustaining with relatively low organic feed concentrations.

The oxygen required by the wet oxidation reactions is provided by an oxygen-containing gas, usually air, bubbled through the liquid phase in a reactor used to contain the process, thus the commonly used term "wet air oxidation". The process pressure is maintained at a level high enough to prevent excessive evaporation of the liquid phase, generally between 300 and 3000 psi.

Wet oxidation is used primarily as a method for waste water treatment but may be useful in a number of other ways. Destruction of high concentrations of organic substances makes the recovery and reuse of many inorganic chemicals both practical and economical. In addition, the highly exothermic nature of the wet oxidation reaction makes the generation of by product process steam or electrical power possible.

CATALYZED WET OXIDATION

A wet oxidation catalyst functions by lowering the temperature required for oxidation of a given fraction of the organic substances in any waste. Thus a reaction which might proceed to an eighty percent conversion at 300 degrees Centigrade without catalyst might be operated with catalyst at either, 1) the same temperature and a much higher conversion, or 2) the same conversion and a lower temperature.

The literature on wet oxidation contains reports of investigations of a large number of substances for catalytic activity. The substances tested include homogeneous catalyst (metal ions and peroxides) and heterogeneous catalyst (metals and metal oxides, both supported and unsupported). While a number of the substances tested were found to have some catalytic activity, only a few could be used in a practical wet oxidation system. Heterogeneous catalysts are of limited utility for wet oxidation since scaling or plugging of the catalyst bed would cause severe problems.

Of the homogeneous catalysts tested, cupric ion has shown a high level of catalytic activity for a wide range of organic substances over a wide range of reactor conditions. Low cost and ease of handling also make the use of copper as a wet oxidation catalyst attractive.

SPECIFIC COMPOUND STUDIES

There are few studies of the wet oxidation of specific compounds in the literature. Rietema and Ploos Van Amstel[1] studies the wet oxidation of glucose and lysine solutions. The kinetics and the products of wet oxidation of butyric and propionic acid were investigated by Williams[2] and Day[3].

Weygandt[4] studies the wet oxidation kinetics of primary, secondary and tertiary alcohols while phenol wet oxidation was investigated by both Shibaeva[5] and Sadana[6]. All these studies centered on the kinetics and mechanism of the process. The compounds studied were found to be relatively easy to oxidize.

In the consent decree of 1976, the EPA published a list of 65 compounds or classes of compounds having top priority for elimination from effluents. Based on chemical structure we chose a variety of these compounds for study. Our objective was maximum destruction of the chosen compound by wet oxidation[7].

Wet oxidation data was obtained for 10 compounds over the temperature range 150°C - 320°C for 1 hour[7]. All ten compounds were destroyed to a high degree at 320°C and 275°C, although a copper catalyst was required for 2-chlorophenol and pentachlorophenol at 275°C (Table I).

Table I One Hour Wet Oxidations

Compound	Starting Conc.	% Starting Material Destroyed		
		320°C	275°C	275°C/Cu^{++}
Acenaphthene	7.0 g/l	99.96%	99.99%	
Acrolein	8.41g/l	>99.96%*	>99.05%	
Acrylonitrile	8.06g/l	99.91%	99.00%	99.50%
2-Chloro-phenol	12.41g/l	99.86%	94.96%	99.88%
2,4-Dimethyl-phenol	8.22g/l	99.99%	99.99%	
2,4-Dinitro-toluene	10.0g/l	99.88%	99.74%	
1,2-Diphenyl-hydrazine	5.0g/l	99.98%	99.98%	
4-Nitrophenol	10.0g/l	99.96%	99.60%	99.97%
Pentachloro-phenol	5.0g/l	99.88%	81.96%	97.30%
Phenol	10.0g/l	99.97%	99.77%	99.93%

* The concentration remaining was less than the detection limit of 3 mg/l.

The solutions resulting from the wet oxidation of these compounds at 320°C and 275°C/1 hr. were tested for toxicity using Daphnia magna. These results were compared with the toxicity of the starting compound[7]. Toxicity reduction was quite good for all compounds studied. The data also indicates that compound destruction or toxicity reduction are nearly as good at 275°C as that obtained at 320°C[7].

The destruction of the compounds in Table II have been studied at 275°C also. The inorganic cyanide and thiocyanate are easily destroyed as is formic acid. N-nitrosodimethylamine showed slightly better destruction without a copper catalyst, although over 99% destruction was observed in either case.

The hydrocarbons toluene and pyrene were easily destroyed with or without a copper catalyst at 275°C, while the insecticide malathion showed excellent destruction even at 200°C without a catalyst.

The chlorinated aliphatic hydrocarbons carbon tetrachloride, chloroform and 1,2-dichloroethane were destroyed to a high degree with or without a catalyst. Chlorinated anilines were also oxidized using a copper catalyst, while hexachlorocyclopentadiene (C_5Cl_6) required 300°C to obtain 99.8% destruction.

Several of the environmentally more persistent compounds showed resistance to wet oxidation even with a copper catalyst. Kepone and 1,2-dichlorobenzene (1,2-DCB) only gave 30-35% destruction while the PCB Aroclor 1254 was quite resistant to wet oxidation even at 320°C.

These studies also provided some emperical observations concerning a particular compounds susceptibility to conventional (with or without a copper catalyst) wet oxidation based on its molecular structure.

a) Aliphatic compounds, even with multiple halogen atoms, can be destroyed within conventional wet oxidation conditions. Formation of oxygenated compounds, such as low molecular weight alcohols, aldehydes, ketones and carboxylic acids result but these are readily biotreatable.

b) Aromatic hydrocarbon are easily oxidized, such as toluene, acenaphthene or pyrene.

c) Halogenated aromatic compounds can be oxidized provided there is at least one nonhalogen functional group present on the ring, ie, pentachlorophenol (-OH) or 2,4,6-trichloroaniline ($-NH_2$).

Table II Wet Oxidation Studies

Compound	Temp.	Catalyst	% Destroyed
Sodium (NaCN)	275°C	0	99.96%
Potassium Thiocyanate (KSCN)	275°C	0	>99.98%
Formic Acid (CO_2H_2)	275°C	0	99.3%
N-Nitrosodimethylamine ($C_2H_6N_2O$)	275°C	0	99.56%
	275°C	+	99.38%
Toluene (C_7H_8)	275°C	0	99.73%
	275°C	+	>99.96%
Pyrene ($C_{16}H_{10}$)	275°C	0	99.995%
	275°C	+	99.997%
Malathion ($C_{10}H_{19}O_6PS_2$)	200°C	0	99.87%
	250°C	0	99.85%
	300°C	0	99.97%
1,2-Dichloroethane ($C_2H_4Cl_2$)	275°C	0	99.8%
	275°C	+	99.9%
Chloroform ($CHCl_3$)	275°C	0	99.92%
	275°C	+	99.94%
Carbon Tetrachloride (CCl_4)	275°C	0	99.99%
	275°C	+	99.99%
2,4-Dichloroaniline ($C_6H_5NCl_2$)	275°C	+	>99.8%
2,4,6-Trichloroaniline ($C_6H_4NCl_3$)	280°C	+	99.97%
Hexachlorocyclopentadiene (C_5Cl_6)	250°C	0	90.0%
	275°C	0	98.2%
	300°C	0	<99.85%
Kepone ($C_{10}Cl_{10}O$)	280°C	+	31.0%
Aroclor 1254 ($C_{12}H_5Cl_5$)	320°C	0	63.0%
	320°C	+	2.0%
1,2-Dichlorobenzene ($C_6H_4Cl_2$)	275°C	0	2.98%
	275°C	+	32.2%
	320°C	+	69.11%

d) Halogenated aromatic compounds such as 1,2-DCB or Aroclor 1254, a PCB, are resistant under conventional conditions.

e) Halogenated condensed ring compounds, such as the pesticides aldrin, dieldrin or endrin are <u>expected</u> to be resistant to conventional wet oxidation.

f) DDT, although destroyed, results in intractable oil formation on conventional wet oxidation.

g) Heterocylic compound containing O, N or S atons provide a point of attack for oxidation and are <u>expected</u> to be destroyed by wet oxidation.

Recently wet oxidation methods for handling the chlorinated aromatic compounds were investigated using 1,2-dichlorobenzene (1,2-DCB) as a model compound. Even oxidation at $320^{\circ}C$ with a copper catalyst gave only 69% destruction of 1,2-dichlorobenzene (Table II). Further experiments using a proprietary cocatalyst system gave nearly complete destruction of 1,2-dichlorobenzene as shown in Table III. Although catalyst A gave good results alone and catalyst B alone was not effective, later studies revealed that the use of these materials together gave the better results.

Table III Cocatalytic Wet Oxidation of 1,2-Dichlorobenzene

Temperature	$275^{\circ}C$	$275^{\circ}C$	$275^{\circ}C$
Catalyst A	+	0	+
Catalyst B	+	+	0
Oxygen Uptake % of Theory	87.50%	25.37%	79.16%
% Chlorine to Chloride	97.60%	32.14%	99.94%
% Input Carbon to CO_2	88.90%	29.87%	80.87%
% 1,2-DCB Destroyed	98.32%	27.38%	99.83%

Application of this cocatalyst system to three of the more environmentally persistant compounds, a PCB, DDT and pentachlorophenol, was investigated. Results, as seen in Table IV, showed that at $275^{\circ}C$, 99.5+% of these materials was destroyed using catalyst A alone. However, when both catalyst A and B were employed an even higher percentage destruction was achieved. In fact, DDT and pentachlorophenol were destroyed to below the detection limit using high pressure liquid chromatography for the compound analysis.

Table IV Cocatalytic Wet Oxidation

Compound	Aroclor 1254		DDT		Pentachlorophenol	
Temperature	275°C	275°C	275°C	275°C	275°C	275°C
Catalyst A	+	+	+	+	+	+
Catalyst B	0	+	0	+	0	+
Oxygen Uptake, %	84.27%	82.68%	77.50%	73.12%	58.06%	61.63%
% Chlorine to Chloride	112.3%	104.5%	82.13%	86.51%	73.10%	71.06%
% Input C To CO_2	81.73%	86.44%	86.02%	87.62%	86.12%	91.48%
% Starting Material Destroyed	99.55%*	99.62%*	99.93%	>99.95%	99.995%	>99.997%

* Calculated on the basis of the largest peak of the mixture remaining after wet oxidation as determined by capillary column g.c. with electron capture detector.

In conclusion, conventional wet oxidation has been shown to destroy a wide variety of organic pollutants at temperatures of 200° - 320°C. Additionally, a wet oxidation process has been developed whereby the most toxic and environmentally persistant chemical compounds can be destroyed. In the developed process, persistant compounds such as 1,2-dichlorobenzene, pentachlorophenol, DDT or PCB's were oxidized at a temperature of 275°C, resulting in destruction levels in excess of 99% for these compounds. This represents a major step forward in the treatment of toxic and hazardous wastes by wet oxidation.

REFERENCES

(1) Ploos Van Amstel, J.J.A. and Rietema, K., "Wet Oxidation of Waste Water Sludge", Chemie Ingenieur Technik, 42, 981 (1970).

(2) Williams, P.E.L., Day, D.C., Hudgins, R.R., Silveston, P.L., "Wet Air Oxidation of Low Molecular Weight Organic Acids", Water Pollut. Res. Can., 8, 224-36, (1973).

(3) Day, D.C., Hudgins, R.R., and Silveston, P.L., "Oxidation of Propionic Acid Solutions" The Canadian Journal of Chemical Engineering, 51, 733-40, (1973).

(4) Weygandt, J.C., "High Pressure Oxidation of Organic Compounds in Aqueous Solution". Ph.D. Thesis, Kent State University, 1969.

(5) Shibaeva, L.V. Metelista, D.I. and Denison, E.T., "Kinetics of the Oxidation of Phenol with Oxygen", Kinetics and Catalysis (Translation from Russian), 10, No. 5, 1, (1969).

(6) Sadana, A.J., Involvement of Free Radicals in the Catalytic Oxidations of Phenol in Aqueous Solution", Ph.D. Thesis, University of Delaware, 1975.

(7) Randall, T.L. and Knopp, P.V., "Detoxification of Specific Organic Substances by Wet Oxidation", J. Water Pollution Control Fed., 52, (8), 2117-2130, (1980).

IN-HOUSE TREATMENT OF HAZARDOUS WASTES AT SOUTHERN
ILLINOIS UNIVERSITY-CARBONDALE

John F. Meister. Pollution Control, Southern Illinois
University, Carbondale Illinois.

INTRODUCTION

The use of chemicals both operationally and academically within a university is commonplace and accepted without question. University, to a certain extent, is conceptually synonymous with chemicals and chemical wastes. Like any other large institution and/or industrial plant, chemicals are used extensively. However, many of these chemicals are the same as those documented to have caused environmental contamination elsewhere. PCB, dioxin, benzene, mercury, arsenic, phenols and cyanides are just a few of the specific chemicals found in university waste streams. These wastes, if released into the environment, can cause environmental damage as well as health problems. Many of the university's chemical wastes are proven carcinogenic, teratogenic or mutagenic wastes. Others are strongly corrosive, powerfully oxidizing agents or highly reactive or contain leachable metals. The definition of being a hazardous waste generator as set by the U.S. Environmental Protection Agency is clearly met. Another paper (1) details the fact that universities are in fact hazardous waste generators.

Pollution Control (PC), an operational division within the Campus Services division at Southern Illinois University (SIU-C), is charged with the overall compliance of the university regarding environmental matters. In response to recognition of the fact that SIU-C was a generator, Pollution Control has established a hazardous waste contral program.

OPERATIONAL PROGRAM

The hazardous waste program consists of four basic components: identification, storage, treatment and disposal of the wastes. In addition, several support functions are conducted. These include: 1) providing scientific and operational support to various departments on special waste problems. The entire program is staffed and operated by graduate and undergraduate students interested in obtaining "hands-on" operational experience.

Since the scope of this paper is on treatment, details on the other aspects of this program can be obtained in another paper (2).

TREATMENT

The objective of the in-house treatment of hazardous waste is to reduce to an absolute minimum the volume of wastes that must be shipped off-campus to approved disposal sites. This reduces the cost of disposal. Many chemicals can be reclaimed and consequently reused, further reducing costs not only for the disposal but also for the academic departments that reuse the chemicals. Lastly, through treatment, the waste is no longer hazardous and thus does not present a threat to the disposal environment.

Pollution Control has summitted an application for authorization as a Storage-Treatment Facility under the USEPA Hazardous Waste Regulations. No objections have been received. The Illinois EPA, as the local enforcement agency, has been kept briefed on this program and have no objections.

Numerous treatment options are employed. They include evaporation, concentration, distillation, neutralization, carbon absorption, biological stabilization and many others (Figure 1).

The treatment option is chosen based upon the nature and type of chemicals that the waste contains. Generally, the decision is made at the time when the waste is inventoried into Pollution Control for disposal. References such as Sax (3) among others are employed to verify the capability of treatment and to determine proper procedures. All treatment is done in batch reactions and is done in safe areas by trained staff with appropriate safety equpiment. Records are kept of all treatment activities. These treatment records are incorporated in a comprehensive "cradle to grave" manifest on all wastes that SIU-C generates.

Evaporation

A large number of wastes are water soluble, i.e., salts of heavy metals, etc. The water can be evaporated off with the volume and weight of waste requiring final disposal drastically reduced. Likewise, many dilute organic solvents are received which can be evaporated into the atmosphere at a slow controlled rate with ho harm.

The evaporations are performed at a remote site. A lean-to was constructed with lumber and covered with clear heavy guage plastic. Under the lean-to, several plastic evaporating pans were placed. The pans are approximately four feet in diameter and liquids are placed in them to a depth of no more than six inches. The purpose of the lean-to is to prevent rain from reaching the evaporatingliquids. The sludges and residues which remain after evaporation are removed and packed for ultimate disposal.

Evaporation provides a drastic reduction of weight and volume along with the fact that the waste is now a solid as opposed to a liquid thus simplifying storage, packing, and shipping problems. Many of the sludges can be packed together to further conserve space. During cold weather months evaporations are carried out under the fume hood within the PC facility. Likewise, some non-toxic wastes are evaporated in various mechanical rooms.

Precipitation

Many wastes solutions can be chemically treated to precipitate out the actual hazardous element. The precipitate can the be separated and disposed and the supernatant disposed of conventionally. The advantages again are the reduction in volume and conversion of the waste into a solid.

Chemical precipitation is also a process where one waste product can be used to treat another thus eliminating two wastes simultaneously. For example, waste chromic acid can be treated with waste sodium hydroxide. Both wastes are eliminated, a precipitate is formed and the remainder of the solution can be disposed of. Additionally, the precipitate, chromium, is a valuable byproduct which can be sold.

The opposite of precipitation is carried out on some solid wastes. Some solids are soluble in certain solvents which then can be incinerated or combusted in the campus boilers. Solid benzyl compounds are an example of wastes which can be treated in this way prior to disposal.

Neutralization

This treatment process is similar to that of precipitation. Different wastes are mixed together and their chemical properties are such that they will neutralize one another. The classic chemical neutralization, which is extensively used is the mixing together of acids and bases. The end product is generally such that it can be disposed of down the the drain with no harm. All neutralizations are carried out in the laboratory under controlled conditions. The end products are checked before disposal to ensure that no hazardous waste properties remain.

Distillation

Many departments and researchers on campus use considerable amounts of solvents. Many of these solvents become contaminated with various other solvents or wastes. Through distillations the solvents can be separated and recovered with a high degree of purity. These solvents can be recycled back to the generator for reuse. This process not only eliminates a waste from having to be disposed of but saves the institution money. Hundreds of gallons of such solvents as xylene, acetone, alcohol, and benzene have been recycled to various departments on campus.

Cleaning

Many wastes received are chemicals which through use have become dirty. A simple cleaning-filtering, washing or physical separation will return the chemical to usable form. The chemical can either be returned to the generator or another user as needed. Mercury is a waste which is commonly received which, with simple cleaning, can be reused in manometers, barometers, etc.

Exchange

Many chemicals received by the Hazardous Waste program are still pure enough to be directly reused. They are chemicals which a generator no longer has a need for and wishes to dispose of just to conserve shelf space. They are generally in their original containers and thus easily identifiable.

Pollution Control Hazardous Waste division has instituted an exchange program for the above type of chemical wastes. Wastes when received, are evaluated as to their exchange possibilities. Any waste which is in its original container, homogeneous, and is disposed of because it is no longer needed, is set aside for possible exchange purposes.

Periodically, as the amount justifies, a listing of all accumulated exchangeable wastes is prepared and distributed to interested departments on campus. Pollution Control provides no assurance as to the quality of the chemicals but does provide them free of charge. They are delivered on the basis of first come-first served. Not all chemicals are claimed but the response is such that the program is a success. There are several advantages to all parties. The wastes are recycled and thus do not need treatment or disposal. This reduces the cost of the Hazardous Waste program, and ultimately, disposal. The receiving department obtains chemicals at no cost. The concept of recycling is promoted and lastly, the participants are reminded of the benefit of the Hazardous Waste program.

The quantity of wastes recycled is a record of the success of the program. During 1978, over 300 gallons of solvents and 3,000 different chemicals were returned to academic departments. The replacement cost of these chemicals was calculated at over $20,000. Due to the success of past exchanges, many departments now have standing requests for certain chemicals as they are received.

DISPOSAL

Ultimately, there remains residues which must be disposed of. Disposal of wastes and residues is accomplished both on and off campus. On campus disposal consists of sewer dumping, thermal destruction in the campus boilers, encapsulation and special biological treatment. Off campus disposal consists of shipment to an EPA approved hazardous waste landfill and is mandatory for all wastes and residues which are considered extremely toxic or harmful to humans or the environment. All chlorinated compounds are sent off campus. All wastes disposed off campus are packed in 55 gallon drums in accordance with EPA and DOT specifications. Transportation and disposal is done by certified hazardous waste transporters and disposal facilities.

On Campus Disposal

A number of both hazardous and non-hazardous wastes can be safely disposed of via dilution in the sanitary sewer. If the material can be adequately diluted to concentrations low enough and will not reconcentrate or bio-magnify then it is disposed of in this manner. Likewise if the material can be biologically stabilized in the treatment process then it is dumped. Research is done on all such wastes prior to dumping to insure that the waste will not harm the collection system or interfere with the sewage treatment process.

To insure safety and proper dilution, wastes are disposed of directly into a manhole. The sewage flow rate is at least 300 gallons per minute and a 4 inch diameter PVC plastic pipe is used as a funnel to insure that the wastes are injected into the flow not on the sides of the manhole. Monitoring is conducted at a downstream manhole to insure that no adverse impacts are occurring.

Other wastes such as alcohols and waste organi solvents contain heating values. These are taken to the SIU-C physical plant's central steam plant and are thermally destructed. They are sprayed (hand-sprayer) onto the coal as it enters the fire-box. No more than 50 gallons per day or 10 gallons per hour is destructed at once. The dilution (70-100 tons of coal per day) and the high temperature (1700°-2000°F) in the fire-box insures safe and thorough destruction. An animal incinerator is also available for use. Since it reaches higher temperatures, it is used on more refractory wastes. Prior to incineration, the liquids are solidified via absorption onto sawdust.

Many collected wastes, after identification and inventory, are determined to be non-hazardous. Others exhibit a very low degree of hazardousness. Both of these are disposed of via conventional solid waste disposal. It is believed that the dilution with the conventional refuse, primarily paper, and the fact that the disposal rate is controlled makes this a cost effective safe disposal method.

Many other wastes have unknown composition as there are no records or labels providing identification. These wastes are encapsulated in concrete and disposed of in a special section of the local landfill. A used 5 gallon solvent can is used. The top is removed and the can is filled with wet concrete. Reinforcing wire is put in the concrete to strengthen it. The bottles of wastes are inserted directly into the wet concrete. Experiments have shown that the encapsulation is not damaged by the landfill equipment.

A small scale activity sludge biological stabilization plant is being constructed to provide on-campus biological treatment of selected wastes. The capacity will be approximately 100 gallons. Bacteria will be acclimated to feed on such compounds as phenols. Such wastes will then be slowly released into the aeration tank for stabilization. Monitoring will be done to insure complete stabilization prior to discharge to the sanitary sewer.

REFERENCES

1. Meister, J.F. The Need for University Hazardous Waste Disposal Programs, Waste Management in Universities and Colleges Workshop, USEPA July 1980.

2. Meister, J.F. A Hazardous Waste Management Program for University Generated Waste, 3rd Annual Madison Conference of Applied Research & Practice on Municipal and Industrial Waste, Madison WI, 1980.

3. Sax, N.I. Dangerous Properties of Industrial Materials, Van Nostrand Reinhold, P1258, 1975.

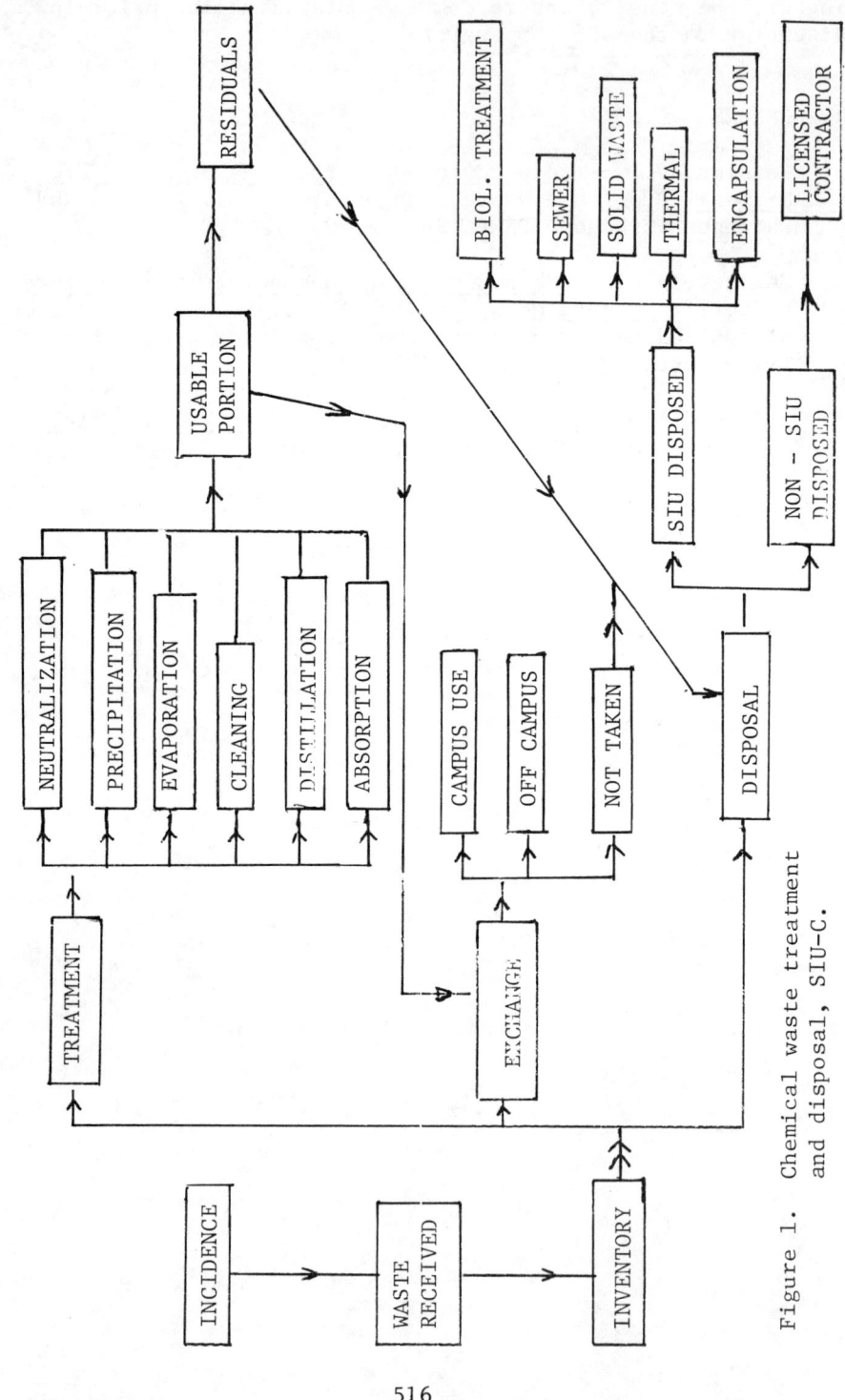

Figure 1. Chemical waste treatment and disposal, SIU-C.

PUBLIC PARTICIPATION: THE MISSING INGREDIENT FOR SUCCESS IN HAZARDOUS WASTE SITING

<u>Eleanor W. Winsor</u>. Executive Director, Pennsylvania Environmental Council, Philadelphia, Pennsylvania

In the past three or four years, industries that have sought to site hazardous waste treatment or disposal facilities have found their projects thwarted or delayed because of strong public opposition. The public's perception of hazardous waste treatment facilities is extremely negative. According to a national public opinion survey released by the U.S. Council of Environmental Quality in October, 1980, 65% of the people interviewed indicated that they were unwilling to accept a disposal site for hazardous waste chemicals within 100 miles of their home. This was almost the same percentage as were unwilling to accept a nuclear power plant. Only 9½% of those surveyed were unconcerned about living near a disposal facility [1]. This public attitude makes hazardous waste siting an extremely vociferous, drawn out and frequently unsuccessful process.

Many sites are never built because of public opposition. Yet the technology to site facilities is available, as is the mechanism to protect the public. What is missing is a way to encourage the public to participate in the siting process so that it is satisfied that the facility will be a good neighbor. In order to do this, a strong public participation program is necessary.

The typical approach to siting hazardous wastes disposal facilities is extremely simple. The sponsor conducts marketing studies to determine the economic feasibility of the facility. Environmental and engineering studies are conducted to identify potential sites within the service area. Once a company locates a site, it obtains an option on the land. In some instances, the option is obtained without the land owner knowing the purpose for which his land will be used. After the option is acquired, the sponsor conducts preliminary discussions with different state, possibly regional and local agencies. After these discussions are held, he is ready to go. Additional engineering work occurs and the application to the state for a permit is made. The public is informed at this point, or shortly thereafter. In some instances, however, the permits have been granted and the facility's construction begun without any public knowledge unless it was necessary to obtain a zoning variance. If such a change is necessary, then the members of the local government will be contacted. The public and local officials are told what is going to happen, but they are not involved in the decision making process.

More and more often the state agencies are requiring some sort of public meeting to notify the public of what is proposed. The development sponsor then takes his project, with neat descriptions of what he is proposing to do, and meets the public in an open forum. These meetings frequently result in shouting matches, because by this time the public has a general idea of what is proposed, but lacks any specific facts. As the developer starts to describe his project, the public concentrates on the worst possible scenario, does not fully listen to what he is saying, and spokesmen for the community groups proceed to present every possible reason why the plant will fail. Local media, which is looking for a good story, concentrates on the opposition to the facility rather than on a full description of what is proposed. As a result, a very volatile situation develops which increases the difficulty of obtaining the site.

Enacting this scenario has proven to be an extremely costly and ineffective process for many hazardous waste treatment companies. Yet, with a shortage of hazardous waste treatment disposal facilities nationwide, and the strong penalty procedures which are outlined in the Resource Conservation Recovery Act of 1976 (RCRA) and many of the state laws which were passed subsequent to RCRA, it is crucial that sites be developed which generating industries can use to dispose of their wastes safely. The only way that this is going to be done, is to overcome public opposition to treatment and disposal facilities. The best way to do this is to ascertain the public's concerns at an early enough point in the project so that they can be involved in the siting procedure and the facility modified to provide the protection that the public is demanding. This necessitates government and industry utilizing new techniques to reach the public and to involve them in the siting procedure.

In October, 1980, the Pennsylvania Environmental Council, (a statewide environmental action organization that seeks to get environmentalists and industries to work together for responsible environmental regulations that are economically sensitive), the Leauge of Women Voters of Pennsylvania, and the Pennsylvania Chamber of Commerce in cooperation with the Pennsylvania Department of Environmental Resources (DER), asked representatives of some forty anti-hazardous wastes disposal organizations to meet together. Our objective was first to have them explain to DER the problems of hazardous waste management facilities (HWMFs) which the groups perceived so that these could be answered in statewide siting criteria, secondly, to help the organizations understand the technical states of the art for treating and disposing of hazardous wastes, and third, to develop a dialogue and hopefully trust between the state and the citizen organizations so that siting in the future could occur in an atmosphere that assured the protection of environmental quality and responded to social concerns while permitting the facilities that are necessary for industry to continue to operate. At the first meeting, the groups identified their major areas of concern. These are listed in Table 1. While this list is not all inclusive, it does reflect the major public concerns citizens voice about hazardous waste treatment or disposal facilities. Public participation programs must address these issues if they are to be effective.

"Why our community?" is the first question raised in everybody's mind when they hear that a treatment or disposal facilitiy is to be located nearby. While people are quite willing to use the products which are produced elsewhere, they are extremely unwilling to be the recipient of the residues of the production of those products. In fact, most people do not perceive the connection between the medicine they take, the stove they cook on, the electricity they use, or the clothes they wear and the creation and/or disposal of hazardous wastes. This lack of perception, coupled with the belief that "our community is different from others because it has unique resources and characteristics" makes the people feel that it should not be used for any purpose which could detract from amenities that they found when they first moved there. Whether these amenities are real or perceived is immaterial. The important thing is that the people who live there perceive them to exist and feel that they merit protection and consideration. The fact that certain soil types or conditions may exist in a particular community that do not exist elsewhere and are suitable for disposal of hazardous wastes also is ignored.

When considering a HWMF in their community, people want to know what is in it for them. As far as they are concerned their community is being used as a "dump" and has to take all the environmental and health dangers that are associated with "dumps" while receiving nothing in return. They express concern over changes in zoning, increased noise and traffic and possible pollution. Property values are another major concern. Will values decline because they are near a HWMF? "Why our community?" becomes a broad rallying cry of those who are opposed to the facility.

Table I. General Areas of Public Concern
Regarding Location of Hazardous Waste Management Facilities

A. Why Our Community:

1. Our community has unique resources/character
2. Solid waste from out of state or region
3. Equity - what is in it for us?
4. Zoning; noise, traffic

B. Doubt of the "State of the Art":

1. Potential Pollution of water supply
2. Fear of industrial waste
3. Why not recycling/waste exchange/process modification
4. Odors, rats and other vermin
5. How long will landfill liner last?
6. Want guarantees pollution will not occur

C. Character of Facility Operators:

1. History of past violations
2. Financial responsibility/capability of operator
3. Moral character of operator/financial backers
4. Accuracy of Self-monitoring of waste samples
5. Willingness to permit citizen inspection

D. Credibility of Government Agency:

1. Public notices of proposed sites are inadequate
2. Agency fails to check veracity of statements made on application
3. Agency's job is to protect the citizens, not industry
4. Honesty of inspectors
5. Agency fails to take legal action when notified of violations
6. Former Agency employees work for disposal industry
7. Agency ignores public concerns and issues permit
8. Agency inspection inadequate

Based on Public Concerns expressed at Hazardous Waste Roundtable sponsored by the Pennsylvania Environmental Council, League of Women Voters of Pennsylvania, Pennsylvania Chamber of Commerce and Pennsylvania Department of Environmental Resources, October, 1980.

The second major area of concern is doubt over the state of the art. Experiences such as Love Canal, Valley of the Drums, and other well publicized failures of hazardous waste management disposal are common knowledge. The failure of hazardous waste operators to handle waste responsibly in the past has led to public questioning of the state of the art of treatment and disposal. Laymen lack on understanding of industrial processes and fail to realize that everything cannot be recycled, exchanged or eliminated. In this instance, a broad public education program describing hazardous wastes and the ways to treat them is necessary, not to spotlight a particular community, but to give the broader public a greater understanding of the state of the art. Effective public participation will increase when the public gains a better understanding of the state of the art for treatment.

The character of facility operators bothers many people. It is a valid concern, based on the record of some treatment operators. Unscrupulous operators have failed to provide environmental and health safeguards. They have disposed of wastes in the cheapest possible way, irrespective of the damages that may result. Other operators lack the financial capability to operate a facility properly or to rectify damages should they occur. Consequently, the character of all facility operators looms large in public perception as to whether or not a facility should be sited.

In the long run, the greatest concern of the public is the credibility of the government agency which is to regulate the HWMF. Throughout the United States the public feels that government agencies, which are supposed to enforce regulations, are not doing so, have not done so, and will not do so in the future. Many citizens express a willingness to have facilities within their community. They will accept the state of the art and the operator, but they have absolutely no confidence in the ability of the state agency or U.S. Environmental Protection Agency to police all sites with the degree of consistency which is necessary to protect the public. They have had too much experience with state agencies which fail to shut down an operator because they say there is no place else for the waste to go. This is, in part, because of the agency's perception of its role in hazardous waste treatment and disposal facility siting. The agency considers itself responsible for making sure that industry has a place where it can dispose of its wastes, not protecting the public. Private industry can play this role better because it has the capital, the technology, and the incentives to site and operate HWMFs. To do so, they need to know the criteria they must meet to safeguard the public and that they can obtain permits based on their technical and financial ability to handle the wastes. Therefore, the primary responsibility of the agency should be the consistent enforcement of the hazardous waste management regulations. If the agency responds to the applications for permits for facilities promptly and carefully, HWMFs will soon exist. Then if the enforcement procedure makes sure that the sites are operating properly and closes them down for infractions of the rules, citizens will be more willing to accept HWMFs.

If we agree that these four general areas of concern are valid, then in order to site HWMFs we must develop a HWMF public participation program that addresses the questions that they raise. Such a program is the missing ingredient in the siting process. Public participation programs have emphasized providing the public with information. In order to respond to these concerns, it cannot be just an information program. It must also make sure that the developer modifies his plan to reflect the public's concerns.

Often planners, government officials and project sponsors feel that the best place to start a public participation program is with the engineering team. This team determines the number and types of reports that would be made public for each task on the critical path from start to completion of the project. For each report, the sponsor determines "what can we do for the public with this project information?" The reports are summarized, simplified, and distributed at meetings or sent to people on a mailing list. Where appropriate a newsletter is started which highlights the overall process of the study. If the technical reports contain some newer significant information about the site, press releases are prepared. As the project nears completion of the planning and engineering stages, the appropriate personnel are used for radio talk shows or small workshops and seminars. These forms concentrate on the directly impacted public, those living closest to the project. They are informational in nature and while each report is designed to reach a particular audience, there is little effort to provide for project modification based on the comments received.

This sort of public participation program is fine for non-controversial projects. However, siting of HWMFs is far from non-controversial. Because of their real or perceived impact on the community, people do not want just to be informed; they want to have a say in what occurs. A public participation program which involves the public in the actual development of the criteria used to locate and develop the site or facility is the missing ingredient for success in hazardous wastes siting. The public must participate in the beginning phases otherwise the project has very little chance of success.

Obviously it is not possible to involve everybody in the community in the siting process. What is needed is a public advisory committee made up of leaders of local environmental and civic organizations. This group, of no more than 15 or

20 people, provides a liaison between the community and the HWMFs' sponsor. The committee reflects the local sense of uniqueness.

A hospital consultant, to whom I talked, stated that fund raising efforts were far more effective when the hospital board or an advisory committee participated in the development of the program, even though 97% of the recommendations were what the consultant would have suggested in the first place. The time consumed in having the committee develop the criteria resulted in greater support for the campaign. The same thing is true in siting HWMFs. If local citizen representatives are involved in the development of the criteria for site selection, they will identify factors that are of particular concern to local people. They will have assurances that their interests and concerns will be addressed adequately at a time when it is not difficult to make modifications in the project.

And, it is important to remember that modifications will have to be made. The Pennsylvania Power & Light Company, which used a Public Advisory Committee composed of representatives of civic groups to help determine the site for an 800 megawatt coal fired generating station, found that there were a number of valid considerations the Advisory Committee raised which the company had previously ignored. These issues, when considered, resulted in modifications to the project. The company incurred additional costs in planning and engineering but the result was a far better sited facility. This early involvement of the public and modification of the project to reflect their interests is a new concept for industry but it is an important one which will result in far greater public acceptance of company projects.

If the public is involved in the development of the site when it becomes necessary to present the program for approval to state and local agencies, there will be a far greater probability of facility acceptance by the government agencies. There will, in all probability, still be local opposition, but, hopefully, the leadership of the local organizations, who have participated in the study, will be able to transmit their increased knowledge of HWMFs and the guarantees which the committee has worked out to protect the public to their organizations. Then the HWMFs' sponsor has a body of local people who understand and will speak up on the issues. These people are not representatives of the corporation. They are independent individuals, who represent their community and are interested in what is best for the community. They will, therefore, bring back to the HWMFs' sponsor the concerns of the local community and will press for their consideration in the planning, development, and even the operation of the facility.

This sort of public involvement is a unique experience for the industry but because of the controversial nature of

HWMFs, it is the only way that this sort of project can succeed.

1. Council on Environmental Quality, et al, <u>Public Opinion on Environmental Issues: Results of a National Public Opinion Survey</u>, p.30-31, 1980.

THE STORAGE AND HANDLING OF TOXIC AND HAZARDOUS MATERIALS
A REGULATORY APPROACH

J. H. Pim, W. C. Roberts and H. W. Davids
Suffolk County Department of Health Services

    The problem of controlling groundwater contamination resulting from the storage and handling of toxic and hazardous materials was wrestled with for many years by the Suffolk County Department of Health Services before a definite course of preventative action was chosen. Dozens of municipalities around the country were consulted and information from many foreign governments was gathered in order to learn what concepts were used by others in the prevention of accidental spillage or leakage from storage facilities containing fuels and other hazardous liquids.
    To our dismay, we discovered that there were no comprehensive regulations in existence in this country to use as a model. Only in Europe had the subject been adequately addressed through regulation and we were faced with great difficulties in translation. It became quickly evident that if we were to develop the much needed control of toxic and hazardous materials storage in Suffolk County, we would have to create our own unique law.
    Suffolk County has a population of 1.3 million people and comprises an area of over 900 sq. mi. located on the eastern end of Long Island. The only source of drinking water supply is the groundwater which is supplied from the precipitation which falls on the island and percolates through the porous, sandy soil. The Nassau/Suffolk region was one of the first areas in the country to be designated as a sole source aquifer by the U.S. Environmental Protection Agency.
    For many years a vigorous effort has been expended in Suffolk County in attempting to control the conventional

point sources of discharge to our groundwaters. We have earned a reputation as having a very strong water pollution control program. The New York State Pollutant Discharge Elimination System Permit program has been actively pursued. In spite of these efforts, evidence has developed that our groundwaters are becoming increasingly contaminated with organic chemicals. In the past 5 years over 9,000 private and public wells have been analyzed in the county for organic contaminants. This has resulted in the discovery of 22 public and 324 private water supply wells, which exceed the N.Y.S. drinking water guidelines for organic contaminants.

Starting in 1977, an active organic sampling program was initiated. Chemicals found have been both in the volatile halogenated and non-halogenated categories. They are mostly of the type found in petroleum fuels and common industrial solvents such as benzene, toluene, xylene, trichloroethylene, trichloroethane, tetrachloroethylene and methylene chloride. In most cases it has not been possible to locate the specific source of an individual well contamination problem.

Clearly, the effort to control point source discharges alone was not sufficient to prevent the continual degradation of our precious groundwater resources.

Coincident with the increase of well contamination problems, we have experienced a rapid increase in the discovery of substantial leaks and spills from containers of hazardous materials, particularly fuels. Our investigation of such incidences went from 4 in 1976 to 155 in 1980. The thousands of underground steel gasoline and fuel oil storage tanks commonly used throughout the county were beginning to fail at an alarming rate.

Additionally, we found many incidents of tanks being overfilled, pipes leaking, spillage from transfer operations and discharges resulting from fires. Serious contamination was found from drums stored on bare earth due to rusting from exposure and bursting from the pressure developed by the summer sun.

Under the hydrogeologic conditions that exist on Long Island, when a petroleum product leaks from an underground tank, the fuel drops straight down until it reaches the groundwater table where it collects and floats and begins to move slowly downgradient on the groundwater surface. This underground rate of movement is quite slow, from 1 to 2 feet per day. Eventually the plume of floating and dissolved product that develops may reach a well, storm drain or a basement, creating water contamination or fire safety problems. We found that these plumes could be defined by the installation of a large number of observation wells to the groundwater surface and that a substantial quantity of the floating product could be recovered by the installation of sophisticated draw-down wells and automated skimmers located within the plume area. However, the cost of each recovery operation is astronomical and even after skimming is complete, a huge area of permanently contaminated groundwater containing the soluble fractions of the spilled product such as benzene, toluene and xylene is often left behind. To attempt to clean up the dissolved portion too, requires an entirely separate operation and more than doubles the cost.

The situation is even worse with a soluble chemical material such as a metal plating solution which will, upon reaching the groundwater table, diffuse directly into the aquifer rather than floating on the water table. The only possible remedy at that point involves the pumping and treatment of endless millions of gallons until the entire mass of affected water is removed. An example of a successful effort of this type in Suffolk involved the removal of some low-level radioactive waste inadvertently pumped into a well.

The decision was made that such spills and leaks simply could not be tolerated and that the existing methods of toxic and hazardous materials storage had to be radically altered to require construction standards, which would minimize to the greatest degree possible, the inadvertent discharge to the ground of these products.

The Suffolk County Department of Health Services was charged with the task of creating the necessary legislation as a new section to the Suffolk County Sanitary Code. The Sanitary Code is the local health law created by the County Board of Health and enforced by the Department of Health.

Independently and after consulting foreign regulations, we came to the conclusion that the concept of a double-walled containment for all hazardous materials storage was essential to provide maximum protection. This was the basic tenant around which our regulations were developed.

This concept was to be applied as nearly as possible to all situations where hazardous materials were found including above and below ground tanks, portable containers and transfer facilities.

For political and economic reasons, however, and for the lack of existing technology in the field, it became necessary to retreat from this ideal in the case of petroleum products stored underground.

For the storage of gasoline and fuel oil in underground facilities, the law allows single-walled tanks but they must be non-corrodible, have leak detection devices installed and be equipped with over-fill protection. All other facilities, above or below ground, for petroleum or any other toxic materials, must embody the double-walled concept. In the case of underground tanks, this usually means a tank in a product-tight concrete vault.

We have had various configurations installed so far including steel, fiberglass and concrete tanks in concrete vaults. The concrete has been sealed with chemical resistant coatings and flexible membranes like swimming pool liners. All tanks must be leak tested by filling with water.

For the underground single-walled fuel tanks we have accepted fiberglass reinforced plastic, fiberglass coated steel and properly factory coated and cathodically protected steel.

Various leak detection devices have been employed to meet the requirements of the law. The most common system used on the single-walled underground fuel tank is what has come to be known as a "U" tube. It consists of a 4" PVC pipe which is installed vertically down one end of the tank, then horizontally under the invert of the tank and finally vertically to the surface at the other end of the tank. The top half of the horizontal portion is slotted and this portion is pitched slightly to one end where a tee fitting provides a small collection sump. Any leakage from or spillage around the tank will find its way, in part at least, into the "U" tube and be detectable by opening the tube at the surface and inspecting visually, or by odor, or with an explosive vapor detector. The tube also lends itself to the installation of an automatic detection mechanism, which can be wired to lights or an alarm. Accurate inventory control is required for all storage facilities, and usually requires sticking of underground tanks to check the product level each day. When this is performed, the "U" tubes are supposed to be checked as well for evidence of leakage.

An alternate plan is in use where the tank lies in or near the groundwater, since the "U" tube will not serve its function if submerged. In this situation, an 8 foot high skirt of flexible, reinforced plastic sheeting is installed vertically in the excavation completely surrounding the tank and placed to be always partly below and partly above the groundwater table. It thus acts as an underground dam or wall, capturing within its circle, any leakage or spillage from the tank. Four observation wells are then installed permanently within the skirt to detect floating product.

Leak detection for the double-walled underground tanks involves monitoring the annular space between the two containers. Many possibilities exist and new ideas keep coming in. Various probes can be used, such as those sensitive to pH change or conductivity, for example, depending on the material being stored. Simple level alarm devices can be used such as float or mercury switches. No one system has proven superior as yet.

Overfill protection is provided by high level alarms on tanks that are filled by pump. However, on gravity filled tanks such as normal service station gasoline tanks, a simple floating ball check valve device is used inside the tank at the vent line outlet. The ball closes as the fluid approaches 4" from the top of the tank thus shutting off the air exhaust from the tank and also nearly stopping the gas flow into the tank giving the operator time to realize the problem and shut the valves. The small space in the tank still leaves enough room to drain the fill hoses without spillage.

A 15 year period was provided for the upgrading of existing underground facilities to the new construction standards but during that period they must be leak tested in accordance with the Final Test of the National Fire Protection Association's Recommended Practices commonly known as the Kent-Moore Test. The frequency of testing during this period depends on the age of the tank and varies from every 2 to every 5 years. If a tank is found to be leaking, it must be removed and replaced with a tank meeting the new construction standards.

For above ground storage facilities, the same double-containment concept applies though it is more easily implemented. Above ground tanks must be built on an impervious pad surrounded by an impervious dike sufficient to contain 110% of the volume of the largest tank enclosed. The pad may be made of concrete or a thick layer of clay. Membranes and other materials are possible but have not yet been investigated. The impervious layer under the tank makes it possible for leaks through the bottom of the tank to seep out to the side where they can be observed rather than straight down into the ground under the tank, undetected. Above ground tanks must also be provided with high level alarms to prevent overfilling.

Large existing above ground storage tanks, which were built on bare earth can not easily be raised to install an impervious barrier beneath them to meet the new construction standards. For these, another exception was provided. It was determined that the best that could be accomplished was to make the existing single layer tank bottom as leak-proof as possible. Therefore, we require that tanks of this type be entered, cleaned, inspected, tested, sand blasted and glass reinforced plastic coated on the bottom. Before coating, all plates must be ultrasonically tested and seams vacuum tested. This must be accomplished within 5 years. Thereafter, at least every 7 years, each tank must be re-entered, inspected and the bottom repaired as necessary.

Leaks from bulk storage tanks have been found. A leak from the bottom of one tank resulted in a 4 foot layer of fuel oil floating on the groundwater. Recovery operations are still progressing after nearly 2 years of operation.

Registration of all tanks is required and the initial phase is presently under way. All information is being computerized and will provide for automatic retesting notification.

The law also calls for the removal or filling-in of all abandoned tanks. Our information so far indicates that this will be a monumental task with apparently more abandoned tanks in existence than active tanks. We have found that a very large percentage of abandoned tanks still have product in them. Abandoned tanks are generally the oldest and, therefore, the discovery and removal of them is an urgent matter.

The law applies the double containment concept to drum storage as well. (Indoor storage is mandatory unless prevented by local fire regulations). All portable containers of hazardous materials must be stored on a paved, diked, and roofed area protected from weather, vandalism, freezing temperature and damage. The number of containers that can be stored is limited to the approved storage area design number.

Transfer facilities and pipelines are also included. Truck fill stands must be roofed, paved and diked with a holding tank for drainage. Pipelines must be protected from corrosion and have leak detection. Additionally, the truck driver making a delivery has been made responsible for ensuring that all conditions are safe for a transfer before beginning the operation.

The county itself is a major owner of hazardous materials storage facilities. With over 500 county-owned or leased buildings, simply bringing county tanks into compliance is a major and very expensive task. We have provided ourselves with Kent-Moore testing equipment, a van and crew for the testing of all county-owned tanks. In addition, we own and operate 2 drilling rigs with crews and appurtenant equipment, which we use for installation of observation wells for contamination plume definition.

We had to equip and install our own recovery operation when a leak was discovered at a county gasoline facility. An initially declared 500 gallon spill was later more accurately estimated at something in excess of 70,000 gallons. After two years of operation, we have spent about $250,000 not including labor. Over 33,000 gallons of gasoline have been recovered and returned to the cars for use. The recovery wells so far have pumped nearly a half billion gallons of water maintaining the draw-down necessary for the recovery system to work.

The law also covers bulk storage of dry materials. They must be stored in a manner which will prevent the formation of a toxic or hazardous leachate. Road salt, for example, must be stored in buildings or on diked pads with brine collection systems. In some cases we have permitted construction of unprotected salt storage areas near the shore line with the drainage going to salt water bodies.

Throughout the development of this law, we enjoyed the

strong cooperation of the American Petroleum Institute and many other major organizations directly effected. Most recognized the inevitable need for a change and generally supported the approach that was developed. The industry contributed many suggestions, which helped to make the legislation practical and workable. After over a year of practice, we have seen the need for many minor changes and refinements in the law but no problems have developed with the basic concept. The process of writing rules and regulations to define the details of the law will go on for some time.

By taking a local legislative approach, we have filled a gap that still exists elsewhere at all levels of pollution control. A massive effort of capital improvement construction is under way in Suffolk County to permanently eliminate the insidious problems created by leakage and spillage of toxic and hazardous materials.

So far, after only fifteen months of operation, nearly 400 underground tanks have been installed meeting our new requirements. Most are glass fiber reinforced plastic. More than 500 tanks have been tested for leakage. All the major oil companies have embarked on programs to replace their existing tanks. Fuel terminal operators have begun the process of opening, inspecting and coating their above ground tanks. There are 17 active spill recovery operations under way and 70 spill cases in some stage of investigation.

We are just learning the lessons that were recognized in Europe nearly 20 years ago. Hopefully, other areas of the country will take similar action before our nation's precious groundwater reserves are further destroyed.

ACKNOWLEDGMENT

The authors J. H. Pim, W. C. Roberts, and H. W. Davids are Supervisor of the Hazardous Waste Management Section, Chief of the Bureau of Environmental Pollution Control, and Director of the Division of Health, respectively, of the Suffolk County Department of Health Services, Hauppage, NY 11787.

SESSION VI

MANAGEMENT

(Social, Legal and Economic Aspects;
Reuse and Recovery; Fate and Transport)

CITIZEN RESPONSE TO INDUSTRIAL WASTE

<u>Paul D. Bush</u>. Associate Writer for the Valley Advocate newspaper and Contributing Editor for the Dangerous Properties of Industrial Material Report

There has been a major change in the industrial waste field in the last decade. The effects of this change are more far reaching than any technical innovation or scientific advance that may be discussed here or developed in the coming years. While its importance and the need to adapt to it have not been completely recognized, this change has already been felt by both government and industry. If it is not visible in the local press or on television, one need only look out the factory gate to observe it, in many cases. What is this new development?: the involvement of the public in industrial waste issues.

Although it has not been manifested in every area of industrial waste, citizen involvement is clearly on the rise and will have to be considered by everyone in the field. Industry and government can no longer deal with technical matters, regulatory questions or even day-to-day decision making without receiving citizen input. The failure to recognize and to accommodate this change has serious consequences; it has already led to the disruption of waste treatment, transportation and disposal. The power of the public and the many forms that it can take are substantial.

Examples of citizen response to industrial waste abound. Most companies are accustomed to telephone or written complaints, but protests and demands to be heard are taking more active forms. Picketing is one method, as firms such as Rollins Environmental Services have learned. Neighbors of the disposal corporation's Logan Township, New Jersey, facility have picketed several times in recent years. They received broad press coverage on two occasions when they demonstrated after a fire at the site in 1977 and during PCB incineration tests in 1980. Community health questions had been raised in both incidents and local residents responded to what they considered a lack of company concern by protesting.

In November, 1980, an overflow crowd at the Enfield, Connecticut, high school auditorium responded to a Browning Ferris Industries plan for a sludge treatment and storage facility in their community. They loudly denounced the planning board, town officials and BFI representatives. At one point when a BFI spokesman said, "I think this is a good company," someone in the crowd yelled back, "That's what they said about J. R. Ewing."

Area residents formed CASE, the Citizens Alliance for a Safe Environment, almost immediately upon announcement of plans for the facility. By the night of the town meeting CASE had collected 10,000 signatures protesting the proposal. Five days later a march around the perimeter of the site attracted 2000 people, and talk began of raising a million dollars for a court fight. The reaction was not a simple response of "not in my backyard," but was fueled by BFI's failure to properly notify the townspeople, as well as by the common perception that the company was practicing deception to gain the site. Today CASE claims 14,500 members, uses a computer to maintain its files and has plans to organize 24 chapters in New England and the Middle Atlantic states, according to president Michael Scalzo.

The residents of Wilsonville, Illinois, took their protests a step further. Incensed by reports of PCB-laden wastes being transported to a nearby disposal site and enraged by the company's apparent callousness, citizens chose to take direct action. According to one government document, "Some residents had armed themselves and were prepared to use these arms in some way against either the facility or the trucks bringing in the wastes . . . Some in the crowd reportedly had sticks of dynamite under their belts and threatened to blow up the facility." Violence was averted, with the conflict being transferred to the courts, but the site was closed and the final resolution is still being awaited, pending a decision by the Illinois Supreme Court.

Hazardous waste is not the only issue that generates citizen activism. Landfills are being picketed, waste water treatment plants are being targeted for law suits and even on-site storage or disposal facilities are being subjected to neighbors' protests. These examples of citizen response are being repeated all across the nation. Where once people would not question an industry's treatment or disposal techniques, today they are vociferously expressing their opinions.

The press is blamed by some for sensationalizing the industrial waste issue, creating a mass hysteria that is at the base of citizen involvement. This perception is false and only reinforces a continued refusal to recognize the true roots and real dimensions of the citizen response.

Citizen involvement is not the creation of the evening news or the daily papers, it arises instead from community problems with transportation, treatment, storage or disposal of industrial waste. In many cases residents have discovered local incidents of improper treatment or disposal and attempted to correct them, long before the press or government have become involved.

In Belchertown, Massachusetts, neighbors of a tank-truck washing company became alarmed at its practices in 1977. The firm, Belchertown Bulk Carriers, was pouring rinse water from up to 40 trucks a day into two domestic type septic tanks. (Many of the vehicles came from Matlack Trucking and other chemical haulers.) Unable to get local government to act the neighbors formed BRACE, Belchertown Residents Action Coalition for the Environment, and increased their lobbying efforts. When the Massachusetts Department of Environmental Quality Engineering finally tested for ground water pollution, two years after BRACE originated, 150 chemical compounds were discovered. "We predicted contaminated soil, contamination in Bachelor Brook and the possibility that wells beyond the brook would be contaminated, and we were right on all three counts," said BRACE member Bernie Kubiak. If government and industry had reacted to this instance of citizen involvement, the closing of many nearby private wells and the resulting law suits might have been prevented.

The Bayview Civic Association, a community group in Baltimore, Maryland, played a role similar to BRACE, but got more immediate results. Parents concerned about waste dumping at "Crystal Hill," an area where children commonly played, found that lead contamination reached 400,000 parts per million. They held press conferences, pressured health officials and lobbied city government until the city halted lead waste dumping by the Pemco Corporation, an SCM subsidiary.

Citizen response is growing and becoming more coordinated. A meeting of citizen groups in Washington, D.C., in November, 1980, gave ample evidence of that. Almost 200 people, representing over 30 environmental organizations, unions and citizens groups, came from 8 states and the District of Columbia to share information and discuss tactics. Groups such as GASP, Citizens Against Waste, B-cause and Concerned Citizens of Southwestern Pennsylvania were present; all had been organized around community problems, but were interested in making contact with others on a regional and national level. Meetings like it are being duplicated around the country under the auspices of Waste Alert!, a program begun by several environmental and citizens organizations.

The Love Canal Homeowners Association, once responding to a problem in its backyard, is now helping establish a national citizens' clearinghouse on hazardous waste. According to Lois Gibbs, one of the founders of the Association, over 50 different groups have already been addressed and a strong backing for the clearinghouse exists.

New issues are being raised by citizen involvement. New standards for corporate responsibility are being raised and requirements for public participation are being implemented. In Philadelphia a pioneer right to know law was recently passed, requiring notification from companies of toxic substances used, produced, stored or disposed of within the city. A key element of the bill is that information is available to anyone working or living in Philadelphia. (The law was drafted and lobbied for by 42 community, labor and environmental groups who had come together under the name of the Delaware Valley Toxics Coalition.) These issues and their impact on industrial waste disposal are certain to multiply in the coming years.

Trends toward political conservatism are unlikely to affect citizen involvement. Even as President Reagan was being elected national polls were showing an almost universal concern with the handling and disposal of hazardous industrial waste. Respondents to a poll conducted for the Chemical Manufacturing Association said they thought waste disposal was the most important health and safety issue relating to the chemical industry. A Resources for the Future poll found that 64% of Americans were "concerned or worried a great deal" about the disposal of industrial chemical wastes. And a June, 1980, Louis Harris poll showed that 93% of Americans wanted stricter hazardous waste regulations.

The response and involvement of the public cannot be ignored by industry or by regulators. Citizens' opposition can be no more than a hindrance, but disposal and storage sites have been shut down, new facilities prevented or delayed and transportation methods changed by community activism. State and local legislators are responding to the popular imperative with new legislation, changing the way industrial waste has been dealt with in the past. The continued denial of the existence or the validity of the citizen response will only result in more serious consequences.

As a reporter and an environmental writer, I have witnessed this change taking place in New England. There have always been concerned neighbors with fears or complaints, but now they are organizing when industry and local government do not respond to their concerns. The success of their efforts varies, but they can play a positive role, as DEQE has learned from BRACE, and they cannot be disregarded, as CASE has shown BFI. If the example of New England and the Middle Atlantic states holds true, in the coming years citizen response and involvement may become the key to managing industry's waste.

DEVELOPMENT OF EFFLUENT GUIDELINES
FOR THE TEXTILE INDUSTRY

Debra J. Moore. College of Engineering, University of
South Carolina, Columbia, South Carolina.

INTRODUCTION

The textile industry produces a wide range of products
essential to our society including: fibers, dyes, cloth,
clothing, carpet, and many more. Since the industry's products
are population-dependent, industry growth has paralleled population growth. In the 1972 Census of Manufacturers, 7,203
separate manufacturing locations were identified for textiles.
[3] Each process at a location may involve several steps, each
of which generates unique waste. Typical steps might include
sizing, kiering, desizing of woven materials, bleaching, mercerizing, dying and printing. Production of synthetic fibers
and blends also contributes to the complexity of the problem,
since these products produce wastes which is a function of the
chemicals which form the fiber.

Historically, industry has discharged its wastewaters into
adjacent water bodies. Only since the late 1950's has sufficient public pressure been generated to precipitate legislation
mandating wastewater treatment throughout the country. Since
many textile mills have been in existence for 50 or even 100
years, clean-up of their wastewater discharges has been complicated due to previous discharge practices. For example, it
is not uncommon to find a textile mill adjacent to a stream
with 50 or more discharges into that stream.

The high BOD of many textile wastes causes a rapid consumption of the dissolved oxygen in a stream. This may be
further accentuated by benthic decomposition as a result of
the high suspended solids load in the waste. Benthic decomposition may yield the formation of methane, hydrogen sulfide and
other dissolved gases and, by disrupting the benthic layer,
results in additional biochemical oxygen demand on the stream.
The wastewater discharges from free-chlorine bleaching processes
may result in reduction of the waste assimilative capacity of

the stream by reducing the microorganism population and diversity. Heavy metals may also have a similar effect in the reduction of microorganism population and diversity and thereby reduce the decomposition rates for many of the organics that are undergoing oxidation. Dye stuffs that are sometimes associated with discharges from textile manufacturing may prevent light from reaching plants and thus inhibit or substantially reduce their photosynthetic activity which, in many instances, may be enhancing the oxygen supply to the fresh water body. Dyes further have been found to sometimes have toxic effects on fish and other aquatic organisms. The discharge of dye may have a deleterious effect on down-stream users that require a low color water for their industrial operations. [11] Thus, from the foregoing description, it can be perceived that the waste from many textile operations is unsuitable for discharge either to a POTW or a free-flowing water body.

The textile industry generally falls into Standard Industrial Code 22. This includes the industrial processes necessary for the conversion of natural and man-made fibers into fabrics and other textile products. [16] Discharge from this Standard Industrial Code category was addressed by the Federal Water Pollution Control Act and subsequent federal legislation in an attempt to reduce the impact of industrial discharges. The present strategy for the textile industry is one that attempts to implement technology to achieve the requisite criteria on the discharges in light of the waste generated, the technology available to treat the waste, the economic energy and other impacts of implementing the technology. It is the objective of this paper to address the textile industry in terms of the present effluent guidelines and standards mandated for the industry and how these criteria were developed.

AUTHORITY TO SET GUIDELINES AND STANDARDS

In 1972, the Federal Water Pollution Control Act Amendments [PL-92500, Section 10(A)] established a comprehensive program to "restore and maintain the chemical, physical and biological integrity of the nation's water." [8] The goals of PL-92500 and the subsequent regulations promulgated as a function of the legislation required all industries using water in their processes to treat to an acceptable level any wastewater discharged to rivers and streams. More specifically, Section 301(B)(1)(a) required all industries to achieve the best practical technology (BPT) by July 1, 1977. Section 301(B)(2)(a) required existing industries to achieve best available technology (BAT) by July 1, 1983 (later postponed to July 1, 1984). New industries were controlled under Section 306, the New Source Performance Standards (NSPS) to meet BAT. Section 307 (B) and (C) accounted for pollutants which were discharged from an industry into a Publicly Owned Treatment Works (POTW). Finally, Section 402 defines the National Pollutant Discharge Elimination System which accounts for new, direct discharges not regulated by any other section. This legislation was indicative of the movement toward eventual industrial pollutant elimination. Promulgation of the standards by the EPA administrator was made through Section 304(B) for the BPT and the BAT; Section 304(C) and 306 for the NSPS and Section 307(A) for toxics. Sections 304(S), 307(B), and 307(C) promulgated pretreatment standards for new and existing sources (PSNS and PSES) discharging into POTWs. In addition, the clause in Section 501(A) gives the administrator the power to promulgate any other regulations "necessary to carry out his function." [8]

It is apparent that the complexities of PL-92500 are such that many of the regulations that EPA was required to promulgate were not addressed during the period prescribed within the legislation. For that reason, a substantial revision of the strategy was accomplished in 1976 when EPA set standards for 65 priority pollutants with associated BAT, NSPS, PSNS, and PSES to be promulgated according to a schedule for 21 industries. [8] The law was further modified in 1977 with the Clean Drinking Water Act, Section 304(E) which recognized that toxic pollutants from non-point sources, such as plant site run-offs, spillage or leaks, sludge waste disposal, drainage from raw material storage in some way associated with the manufacturing process may, in fact, effect the environment. [17]

THE TEXTILE INDUSTRY AS A WATER POLLUTION SOURCE

To consider the Textile Industry as a water pollution source, it is first necessary to evaluate the basic industry profile including: age, production, geographic location, type of discharge, raw materials utilized, production processes, final products, in-plant controls, end-of-pipe controls, wastewater data, and all associated costs for textile wastes. Historical wastewater data was collected from EPA regional offices, state water pollution control agencies, municipalities, reviews of the literature and current research. [3] A 208 Data Request (Industry Survey) was conducted to update the existing data base. [5] A master list was formed from the *Davison's Textile Blue Book* [4] which classified mills as "wet" or "dry." Wet mills were surveyed during February, March and April of 1977 and followed up by a telephone survey. [5] The results of the investigation defined 2,000 mills as "wet processing," thus producing wastewater. The majority of these mills were located in the Southeastern United States and knitting mills are the single largest product class. [5] A screening and verification sampling program of "wet" plants was conducted between March 1977 and October 1978 on a total of 50 mills to determine the presence of 129 toxic pollutants. "Sampling and Analysis Procedures for Screening of Industrial Effluents for Toxic Pollutants" and "Analytical Methods for the Verification Phase of the BAT Review" were the procedures followed in screening and verification studies. [5]

Control and treatment technology information was collected from sources such as "EPA research information, published literature, information furnished by individual textile firms and government agencies, and on-site visits including sampling programs and interviews at representative textile plants throughout the United States." [5] Treatment technology was evaluated on the basis of amounts and kinds (physical, chemical and, or biological) of pollutants present. Costs associated with various recommended technologies were assessed for a range of textile mill sizes. [5]

CATEGORIZING THE TEXTILE INDUSTRY

Due to the diversity of materials and processes utilized in the textile industry, it was deemed necessary to divide the industry into similar groups which consumed comparable amounts of water and produced similar amounts and quality of wastewater. A first evaluation for such categories is made by segre of processes for differing fibers which are eventually converted into fabric. The three most significant fibers were determined to be wool, cotton and synthetics.

The method chosen for the industry categorization was the Wilcoxon Two Sample Test. The statistical analysis was conducted on parameters including: water usage rate, BOD, COD, and TSS and considered around 50 combinations of manufacturing processes, types of discharge, production quantity, geographical location, mill age and amount of automation. The results of these analyses determined the subcategories of the textile industry listed in Table I [5] which are to be addressed by appropriate treatment strategies after further investigation.

TABLE I

Selected Subcategories of the Textile Industry

1. Wool Scouring
2. Wool Finishing
3. Low Water Use Processing
4. Woven Fabric Finishing
   a. Simple Processing
   b. Complex Processing
   c. Complex Processing Plus Desizing
5. Knit Fabric Finishing
   a. Simple Processing
   b. Complex Processing
   c. Hosiery Processing
6. Carpet Finishing
7. Stock and Yarn Finishing
8. Non-Woven Manufacturing
9. Felted Fabric Processing

TEXTILE WASTE CHARACTERISTICS

The pollutants resulting from textile manufacture are classified into three groups: conventional, nonconventional, and toxic as listed below:
- Conventional: Biochemical Oxygen Demand (BOD), Total Suspended Solids (TSS), Oil and Grease.
- Nonconventional: Chemical Oxygen Demand (COD).
- Toxics: Sixty-five classes of toxic pollutants were addressed as a result of the Consent Decree in <u>NRDC vs. Train</u>, 8ERC-2120. [5] Toxic pollutants were defined as:
  - substances for which there is substantial evidence of toxicity, carcinogenicity, mutagenicity, and/or teratogenicity. [5]

CONTROL AND TREATMENT TECHNOLOGY

The two criteria upon which treatment technology is based are: (1) the characteristics of the wastewater that is to be treated, and (2) the effluent requirements of the treated wastewater. Two basic approaches to treatment strategy exist: (1) in-plant controls, and (2) end-of-pipe treatment.

<u>In-Plant</u>

Several in-plant options have been successfully implemented in the textile industry. These include water re-use, water reduction, process chemical substitution, and reclamation of useful substances. Water re-use in a textile mill usually involves recycling of cooling water or reuse of a water from a process, such as cooling water as rinse water or perhaps the use of final washwater as scrubber water makeup. Water reduction is achieved by the use of conservation, basic good-housekeeping practices, and automation.

The substitution of "less polluting" chemicals in some textile processes may improve effluent treatability and/or strength. Such a substitution might include replacement of soap with a synthetic detergent for cotton or wool scouring. It should be emphasized that advantages and disadvantages of substitutes must be considered. Specifically, when detergent replacement is practiced, a concentrated biodegradable soap is replaced by a less degradable synthetic. Thus subsequent effects on stream quality must be considered to determine the better alternative.

Recovery of process by-products or chemicals is an important aspect of in-plant control. Slashing starch, which is used to give cotton tensile strength during weaving, but removed after that process, and lanolin, which is a basic component of wool grease, are recovered. [3]

<u>End-of-Pipe</u>

Most textile plants use end-of-pipe treatment for their combined waste streams. Newer facilities may choose to segregate individual waters to affect smaller, less complex treatment operations or material recovery. This approach usually results in lower treatment costs and higher treatment efficiency. [5] Table II indicates the end-of-pipe treatments which are typically employed in the textile industry. [5]

TABLE II

End-of-Pipe Treatment Technology

1. Preliminary Measures
   a. screening
   b. neutralization
   c. equalization
2. Biological Processes
   a. aerated lagoons
   b. activated sludges
   c. stabilization lagoons
3. Chemical Processes
   a. coagulation
   b. precipitation
   c. oxidation
4. Physical Separation
   a. filtration
   b. hyperfiltration/ultrafiltration
   c. dissolved air flotation
   d. shipping
   e. electrodialysis
5. Sorption Systems
   a. activated carbon
   b. powdered activated carbon

COSTS, ENERGY, AND OTHER NON-WATER QUALITY ASPECTS OF TEXTILE WASTEWATER POLLUTION CONTROL

During the process of regulatory development, costs and energy benefits were evaluated for each process control strategy, both for new and existing plants discharging into POTWs or natural water bodies. Curves were developed for each treatment strategy which, in concert with cost-effective summaries, permitted a comparative evaluation of strategies.

The environmental impact of textile processing included consideration of sludge management, due to the increased amount of sludge which will be produced by the removals of excess solids from process wastewater and air pollution associated with some modes of treatment was also recognized, though its incidence was found to be limited. The data for such evaluation of the other impacts of processes was assimilated from surveys of existing textile facilities. [5]

A separate economic impact analysis was performed entitled *Economic Impact Analysis of Proposed Effluent Limitations Guide-*

lines, New Source Performance Standards and Pretreatment Standards for the Textile Mills Point Source Category. [7] Data was gathered from EPA document(s) and other representative financial documents and compiled to relate specific economic impacts effecting the textile industry including such factors as:

"1. prices profitability & growth
2. extent & determinants of capitalization
3. number, type and size of plants,
4. production and employment, and
5. community & balance of trade affects."

SUGGESTED EFFLUENT REDUCTION APPROACHES

The various control strategies considered by EPA for each textile process, its associated wastewater and implementation impact are discussed here. [5]

Best Practicable Technology

The effluent limitations implemented July 1, 1977 specified the degree of effluent reduction attainable through the application of BPT. This was generally based upon average best existing performance of exemplary plants of various ages, sizes, and unit process within the textile industry.

Consideration was also given to the total cost of technology application as related to benefits achieved; size and age of equipment and facilities involved; processes employed; engineering aspects related to application of various control techniques; process changes; and non-water quality environmental impact. [3]

Best Available Technology

BAT are effluent limitations which are scheduled to be achieved by July 1, 1984. These limitations were determined by identification of the best control and/or treatment technology utilized by a specific point source within an industrial category, subcategory, or similar technology industry. Therefore, BAT was defined as the "top of the line" current technology with some consideration for cost. Implementation has allowable associated risks which are, in most cases, outweighed by efficiency or other innovative qualities. In-plant controls are suggested, but not required. End-of-pipe technology is addressed as to levels of attainment:

1. Biological treatment (extended aeration and activated sludge, or BPT technology).
2. Biological treatment and filtration.
3. Biological treatment and chemical coagulation.
4. Biological treatment, chemical coagulation, and filtration.

Ozone and activated carbon were considered, but rejected due to cost. [5]

Specific recommendations have been made under BAT for some textile processes such as wool scouring which must meet level 4, biological treatment criteria plus chemical coagulation, and filtration to comply with BAT. Limits are also defined by subcategory and associated control requirements. Effluent limitations are based upon minimum daily and 30-day average loadings.

EPA-sponsored plant surveys revealed that many plants in compliance with BPT will meet BAT regulations with only minor modifications. These modifications might include in-plant controls, filtration, and/or coagulation processes. Determination of BAT guidelines, for a particular process, is accomplished using NPDES data, full-scale operating data, and pilot plant data.

Data for the aforementioned surveys also indicated regulations for nonconventional pollutants and toxics including color, COD, total phenol, chromium, copper, and zinc (toxics) which are typically found in textile effluents. Sampling for other toxic pollutants is also required for specific types of textile operations; these include nickel, selenium, silver, antimony, arsenic, lead, cadmium, and mercury. Total suspended solids is used as a precursor indicator for justification of further evaluation of toxic components.

The environmental impact of BAT also includes greater treatment cost, energy utilized and sludge production. Cost will increase due to both capital and operating and maintenance cost of the more advanced facilities. Energy costs are projected to increase to .02 and .03 percent of current cost for filtration and .2 to .5 percent of current cost of coagulation. Greater quantities of sludge will require more handling and disposal which utilizes larger facilities and more personnel at increased cost. [5]

Best Conventional Technology (BCT)

Best Conventional Technology replaced BAT after a cost reasonableness test was conducted and recorded in the *Federal Register* #50732 on August 29, 1979. [17] It was found that BAT cost was prohibitive for controlling the conventional pollutants BOD, TSS, fecal coliform, pH, and any others to be classified in the future as conventional. Therefore, "Congress directed EPA to consider the best measures, end-of-pipe treatment technologies and other alternatives that reduce pollution to the maximum extent feasible, including where practical, a standard permitting no discharge of pollutants." [5] Thus, BCT criteria were developed in terms of the cost of treatment per increment of benefit attained. BAT level 2, biological treatment and filtration, was defined as the requisite treatment for BCT. [5] BCT was a positive modification to BAT environmental inputs since it cost less, required less energy, and produced less sludge.

New Source Performance Standards (NSPS)

NSPS apply to industrial facilities upon which construction begins after proposed regulations are determined. Typically NSPS are comprised of the BAT as a base with modifications that consider improved production processes and/or treatment techniques. Thus, new industries must consider best in-plant and end-of-plant process control technology. [3]

NSPS utilize secondary biological treatment as a basis for its control strategy. The levels of treatment identified under NSPS are: [5]

Level 1 - Biological treatment (extended aeration and activated sludge)
Level 2 - Biological treatment, chemical coagulation, and filtration
Level 3 - Segregation of toxics from waste stream for separate treatment and disposal. Provide chemical coagulation, filtration, and activated carbon for remaining waste stream.

The NSPS regulations were developed using median water usages for industry (not averages). It is projected that 3 to 11 percent of new plant capital investment will be required for pollution control equipment and facilities. [5]

Pretreatment Standards for Existing Sources (PSES)

PSES is the treatment strategy which applies to wastewater discharged from existing textile plants into a POTW. Plant characteristics, engineering aspects, and non-water quality impacts (cost, energy, etc.) were considered during the development of PSES. The development rationale for pretreatment standards was to achieve BAT after subsequent treatment at the POTW. Thus, pretreatment was to provide processes through filtration while the POTW was to provide biological treatment. Levels of PSES pretreatment include:

Level 1 - Current level of pretreatment (preliminary treatment, including screening, equalization, neutralization).
Level 2 - Preliminary and chemical coagulation.
Level 3 - Preliminary, chemical coagulation, and filtration.

Utilization of activated carbon and ozone was once again considered but rejected due to cost. Total cost for implementation of PSES is lower, since many plants already have sufficient technology to comply with regulations.

Pretreatment Standards for New Sources (PSNS)

PSNS are similar to PSES, since special in-plant controls are not required. However, if in-plant measures are not utilized, end-of-pipe treatment for new sources must meet higher requisite quality. End-of-pipe technologies implemented include: [5]

Level 1 - Current level of pretreatment
Level 2 - Pretreatment with the segregation of toxics for further chemical coagulation and filtration.
Level 3 - Pretreatment with the segregation of toxics for further chemical coagulation, filtration, and activated carbon absorption.

Ozone, as with other conditions, was found too costly for inclusion.

PSNS are similar to PSES, since they attain BAT through pretreatment and the POTW. Cost for pollution control is projected at 1 to 8 percent of the plant capital cost. [5]

PUBLIC AND INDUSTRIAL COMMENT

Solicitation of public and industrial comment upon proposed guidelines for pollution control regulations and incorporation of these comments into the development of guidelines and standards is vital to maximize overall acceptance and minimize errors and misallocation of resources. It would be short-sighted not to consider the opinion of industry when they are responsible for the pollution. In June of 1978 a draft report, "Second Interim Report: Textile Industry BATEA, NSPS, PSES, PSNS Study," was distributed to interested parties. The American Textile Manufacturers Institute, Northern Textile Association, Carpet and Rug Institute, Individual Firms, State Water Pollution Control Federation Agencies, and some municipal authorities were included. The same document was subsequently released, in a revised form in November of 1978. A public hearing was held on December 12, 1978 in Washington, D.C. At the public hearing, presentation of the technical merit of proposed regulations was made. The minutes of the public meeting are published in the *Federal Register*. [17] Topics including methodology of determining the significance of toxic pollutions, concern that variability in wastewater characteristics, within subcategories, would invalidate cost and performance projections, low priority placed upon health effects as related to toxics, and concern that cost was underestimated.

Subsequent public comment was solicited through the *Federal Register*. Through this mode, the textile industry was asked to identify deficiencies in proposed regulations, evaluate toxic components in their wastewater, evaluate sludge characteristics and predict economic impacts.

Bibliography

1. Callely, A. G., Forester, C. F. and Stafford, D. A., eds., *Treatment of Industrial Effluents*, Hodder and Stoughton, 1977.

2. "The Clean Water Act: Showing Changes Made by the 1977 Amendments," Water Pollution Control Federation.

3. Cooper, Sydney G. Cooper, *The Textile Industry Environmental Control and Energy Conservation*, Noyes Data Corporation, 1978.

4. Davison's Textile Blue Book, 111th Edition, Davison Publishing Company, 1977.

5. "Development Document for Effluent Limitations Guidelines and Standards for the Textile Mills Point Source Category," EPA 440/1-79/022b.

6. "Development Document for Proposed Effluent Limitations Guidelines and New Source Performance Standards for the Textile Mills Point Source Category," EPA 440/1-74-022.

7. "Economic Impact Analysis of Proposed Effluent Limitations Guidelines, New Source Performance Standards and Pretreatment Standards for the Textile Mills Point Source Category," EPA 440/2-79-020.

8. "Federal Water Pollution Control Act Amendments," *Federal Register* (33 USC 466 et. seq.) U.S. Environmental Protection Agency.

9. *Industrial Waste Water*, International Congress on Industrial Wastewater held in Stockholm, Sweden, 2-6 November 1970, Butterworths, 1972.

10. Kemmer, Frank N., ed., *The Nalco Water Handbook*, McGraw Hill Book Company, 1979.

11. Koziorowski, B. and Kucharski, J., *Industrial Waste Disposal*, Pergamon Press, 1972.

12. Miller, I. and Freund, J. E., *Probability and Statistics for Engineers*, 1965.

13. Nemerow, Nelson L., *Liquid Wastes of Industry: Theories, Practices, and Treatment*, Addison-Wesley Publishing Company, 1971.

14. Rudolfs, Willem, ed., *Industrial Wastes, Their Disposal and Treatment*, Rheinhold Publishing Corporation, 1953.

15. "Sampling and Analytical Procedures for Screening of Industrial Effluents for Toxic Pollutants," U.S. EPA, Cincinnati, March 1977, revised April 1977.

16. "Standard Industrial Classification," Office of Management and Budget, Statistical Policy Division, 1972.

17. "Textile Mills Point Source Category Effluent Limitations Guidelines, Pretreatment Standards, and New Source Performance Standards; Proposed Regulations," *Federal Register* Vol. 44, No. 210, U.S. Environmental Protection Agency.

ENVIRONMENTAL IMPAIRMENT LIABILITY INSURANCE

C. Fred Gurnham, Peter F. Loftus Corporation (Illinois), Chicago, Illinois

A person or a company that disposes of a dangerous waste is normally and legally responsible for any harm that may result from the material. The harm could be injury to persons, damage to property, or some other form of liability. Conscientious persons and companies have always disposed of their hazardous by-products and unwanted materials in what was conceived to be a safe and proper manner. Some disposers have been less conscientious and, as a result, we have turned to laws and regulations instead of morals to assure public safety.

In the past, it has been believed that "Nature" would take care of harmful substances if she were given sufficient time. The natural processes involved included chemical reactions such as neutralization and biodegradation, supplemented ultimately by dispersion and dilution. Thus, if corrosive or flammable or toxic wastes were dumped in far out-of-the-way places, or buried beneath the soil, the rest of the deactivation could safely be left to natural processes. There is a certain rationale and logic to this reasoning; it overlooks, however, the matter of quantity or degree. At any particular site and time, the healing effects of nature can be overloaded, and the hazardous material may remain as a hazard over a wide area and a prolonged time, even to miles and centuries. There are no longer any out-of-the-way places, and buried wastes may be brought back into mankind's environment by leaching and other processes of nature.

CURRENT LEGISLATIVE SITUATION

In the United States, in recent years, numerous laws have been enacted and regulations have been promulgated for the purpose of protecting the public from possible harmful effects of hazardous waste disposal. Although not generally stated, these rules seek to achieve their objective in two manners: to introduce an awareness that disposal techniques formerly considered satisfactory are no longer acceptable and perhaps never truly were; and to control by punitive measures, if necessary, those who dispose of wastes in an unapproved manner.

The Resource Conservation and Recovery Act (RCRA) is a lengthy and complex piece of legislation; the present paper will go into only one significant portion of the Act. RCRA is directed to the control of generators and transporters of hazardous waste materials, as well as of those who store, treat, destroy, or otherwise provide for their disposal. The particular subject of this paper is financial responsibility. The generator of a dangerous material is not released from responsibility when he turns the waste over to a scavenger in order to remove it from his property. The transporter, in turn, retains some degree of responsibility after he dumps the material at a treatment or disposal site.

The disposal site operator, it would appear, bears the major responsibility, regardless of whether he incinerates or otherwise treats the material to eliminate or reduce its hazard, or whether he disposes of the substance by burial or other form of disposal onto or into the soil. But the earlier handlers, both the waste generators and the transporters, share this responsibility and must, under RCRA and similar legislation, provide assurance of financial reliability if damage to the environment should occur.

FINANCIAL RESPONSIBILITY

In the initial version of RCRA, it was required that financial responsiblity be provided by a trust fund. This fund should be large enough to cover any damages to persons or property that could be anticipated in the event of a spill or leak, including long-term seepage. The revised RCRA legislation permits alternative evidences of financial responsibility; of these, insurance is often the most realistic.

Almost every company, regardless of the nature of the materials it processes and handles, carries some form of Comprehensive General Liability (CGL) insurance. This provides financial response to most injuries or damages to persons or property, other than the owner's employees (who are covered by Workmen's Compensation insurance) or the owner's property. The latter may be included in the CGL policy by an extension, or is taken care of by fire insurance and related coverage. However, it is significant that the conventional form of insurance policy, including CGL, protects only against sudden or accidental events.

In the environmental field, a Comprehensive General Liability policy would provide financial reimbursement, within the limits of the policy, for such events as an accidental spill of hazardous material from a highway truck, the rupture or bursting of a tank containing a hazardous substance, or the breaking or overtopping of a lagoon dike impounding a pollutional liquid. However, CGL insurance commonly would not provide costs for nonsudden or gradual impairment of the environment, such as the slow seepage of lagooned liquids into the groundwater or alleged injury to a neighbor or to his property by prolonged exposure to corrosive gases from a continuous emission.

NONSUDDEN OCCURRENCES

Insurance against nonsudden environmental risks is a comparatively recent development, not yet well understood by many companies that could benefit by it. Environmental Impairment Liability (EIL) policies, covering nonsudden occurrences, were first offered in 1966, but were not widely accepted. They were withdrawn a few years later, because insurance underwriters were disturbed by several major disasters, including a series of oil spills from large tankers and oil production facilities.

HISTORY

In 1976, EIL insurance was again introduced into the United States. It is being accepted rather slowly, but it does serve a real need, which has been emphasized by the financial responsibility requirements of RCRA and similar legislation.

EIL policies are currently offered in the United States by two underwriting firms. Other underwriters will enter the field soon, and may already have done so; the first two are Howden Agencies, Ltd., of Cranford, New Jersey, and Shand, Morahan and Company, Inc., of Evanston, Illinois. Each of these has a favored nationwide brokerage firm that is experienced with EIL policies. These are, respectively, Wohlreich and Anderson, and Alexander and Alexander. Both of these brokers have offices in most major cities. The ethics of the insurance industry requires that these broker/underwriter relationships are not exclusive. Any insurance broker, including local representatives and independent brokers, should be able to offer and to compare EIL policies from both (or all) underwriters.

COVERAGES AND EXCLUSIONS

The coverage provided by EIL insurance is, as noted earlier, limited to the nonsudden or gradual escape or effect of a hazardous waste. It might be possible, however, to combine the EIL and CGL policies into a single contract; the author does not know if this has been done. The EIL policy, like most CGL policies, covers injury to persons other than employees, damage to off-site property, other impairment of legal rights, and litigation expenses.

Tyically, Environmental Impairment Liability insurance has certain "exclusions," or specified items that are not covered by the policy. These vary from one underwriter to another, and may be subject to negotiation at the time that the policy is written. Some of these exclusions are easily understood, including the nonduplication of other insurance policies such as CGL, workmen's compensation, and product liability. All EIL policies include a lower limit or "deductible" to discourage nuisance claims, and an upper limit. The latter may be as high as $50 million. Punitive costs, such as fines and penalties, are commonly not included in policy payments, as this would be in conflict with the public interest. Intentional acts, of course, are not included unless such acts have been legalized under the terms of a variance or special permit. Finally, as has been observed, on-site damages are commonly excluded, whether injury to persons, damage to property, or on-site clean-up costs.

PREMIUM COSTS

The premium, or price, of an EIL insurance policy is always established on an individual or case-by-case basis. There is not a sufficient information bank on the history and magnitude of claims to establish a sound mathematical support for the amount of premiums. Further, the premium can be greatly affected by various negotiable points, including, in particular, the amount of the deductible and the underwriter's estimation of the degree of risk.

If there is such a thing as a "typical" risk, with average liabilities, and with a deductible from $5000 to $100,000, the annual premium could be expected to fall between $10,000 and $40,000. Each prospective purchaser should evaluate this cost in comparison with his own estimate of the consequences of possible uninsured occurrence and its financial liability. The probabilities and the costs should be explored with an insurance reprosentative, to obtain more specific information than can be supplied in this general paper.

A critical part of any evaluation, whether conducted by the underwriter or by the owner of the risk, is a thorough inspection of the site by an independent engineer. Such an inspection, with a written report, is always required by the underwriter prior to his quoting on a premium. Ths report is made available to the client, whether or not the EIL policy is accepted. It can be a valuable information source for the owner of the facility studied.

The author is aware of four claims that have been reported on EIL policies. Competition among underwriters prevents any estimate of the total number of policies that have been written, or of additional claims, or of the current status of existing claims. It is known that at least one of the four claims reported has been settled; the others are presumably in the necessary investigation stage.

SUMMARY

Environmental Impairment Liability insurance is a development of the past few years. If it had been available earlier, we might have eased the financial problems that exist because of the Love Canal episode and the contamination of rivers by Kepone and polychlorinated biphenyls. Certainly EIL is worthy of investigation by any firm that generates, transports, stores, treats, or disposes of hazardous materials, or even of conventional pollutional substances.

DAIRY INDUSTRY WASTE-OPTIONS AND ECONOMICS

ManMohan Varma. Department of Civil and Environmental Engineering, Howard University, Washington, DC.

Ramesh C. Chawla. Department of Chemical Engineering, Howard University, Washington, DC.

INTRODUCTION

   Many industrial discharges with nutritional value and/or potential for biological conversion are currently being treated as wastestreams. These may be recycled for resource recovery, i.e. production of food or energy. Food industry group is a major example of this class of wastewaters. Dairy industry waste forms a considerable portion of the total food industry waste. Furthermore, a large portion of oxygen demand of dairy wastewaters is due to whey. In this paper major emphasis will be given to whey treatment and utilization because the other portion of dairy waste is weaker. Thus an economic strategy for handling whey would, in large part, dictate the ways of handling dairy waste.
   Whey is the greenish-yellow fluid left after separating curd-like material in the cheese manufacturing process. Whey contains approximately 6 percent solids, mainly composed of lactose, water soluble proteins of milk, inorganic salts, nonproteinacious nitrogen and some vitamins. One pound of manufactured cheese is accompanied by approximately six to nine pounds of whey. It is estimated that the annual world wide production of whey is over 160 billion pounds and increasing [1]. In 1979, 37.4 billion pounds of whey were produced in the United States. It is estimated [2] that in 1990 the annual production of whey in this country will be approximately 59 billion pounds.
   Sanitary engineers are concerned about the large amounts of oxygen demand that whey imposes on the receiving water. If untreated whey is discharged into the municipal sewers it

produces a heavy burden on the publically owned treatment works (POTWs). It is estimated that 1,000 gallons of raw whey, if discharged into a stream, would require the dissolved oxygen in over 4,500,000 gallons of unpolluted water for its bio-oxidation. The BOD_5 of whey varies from 30,000 to 60,000 mg/ℓ. The high organic constituents of whey lead to treatment and disposal problems. On the other hand, whey can potentially be converted to useful end products such as single cell protein (SCP), food and beverages, animal feed, alcohol and methane etc.

Because of high BOD_5 of whey, it will require better than 99 percent treatment. Even at 99 percent removal efficiency, the residual BOD_5 in whey will vary from 300 to 600 mg/ℓ. The former is equivalent to raw domestic sewage.

The pH of sweet whey ranges from 5 to 7, and that of acid whey is between 4 to 5. The pH of the acid whey will tend to lower the pH of the receiving waters. The acquatic organisms require an optimum range of pH from 6.5 to 8. On either side of this value there are detrimental effects to the aquatic life.

The choices of whey disposal are:

- discharge to surface waters after treatment at site, or
- treatment of whey in conjunction with domestic sewage at POTWs and/or
- land disposal via irrigation.

Alternatively, whey may be processed by one of the various methods for subsequent utilization. Utilization of whey may be accomplished by several means such as:

- directly as animal feed,
- human and animal food after drying
- conversion to single cell protein (SCP)
- manufacture of wine and alcohol via fermentation
- as fertilizer

Before discussing the economics of treatment vs. utilization, a brief review of literature in these areas is presented below.

WHEY TREATMENT

Whey fluid produced in cheese manufacturing processes is generally non-pathogenic and non-toxic but has high BOD. The characteristics of whey are similar to that of milk waste with the exception that the organic constituents in whey are highly concentrated. It can be treated biologically either exclusively or in conjunction with domestic wastewater. However difficulties are experienced because the solids are in a finely divided state which do not settle readily, and the BOD is high. In the past not much work has been done on the treatment of whey per se, because major attention was given to milk waste.

Fulton, et al. [3] used a 7-stage rotating biological contactor (RBC) for 156,000 gpd dairy waste.

Micro-flora generated in the biological unit appears to be incapable of degrading the whey proteins. Incorporation of whey in aeration units causes excessive foaming, which is due to the combination of microbial cells and β-lactoglobulin-whey protein. In treating whey, at times, there is nitrogen deficiency and exogenous nitrogen is required. In this plant, ammonia was added to satisfy the nitrogen requirement. The removal efficiencies for BOD_5 and suspended solids were 99 percent. In 1977 dollars, the cost was 8.5¢ per pound of BOD_5 removed. The city of Frederick, Maryland in 1972 found that a cheese manufacturing plant was generating 2.5 million gallons of acid whey plus another 2.5 million gallons of wash and rinse water [4]. The city's sewage disposal plant was treating a high concentration of whey along with other dairy waste. The combined BOD from the milk plant was 15,000 mg/ℓ.

In order to control the pollution problem several treatment alternatives including building a treatment plant large enough to take care of all the plant's liquid wastes were considered. However, due to lack of adequate space it was decided to pass the whey through a treatment system which would lower the BOD_5 to such an extent that pretreated effluent could be safely discharged into the municipal sewers. The treatment system employed included: (a) an equalizer tank, (b) a primary tank, (c) two aeration tanks, and (d) a sedimentation tank.

The whey effluent from the cheese manufacturing operations pumped into a collection tank via an equalization tank where defoamer and caustic soda were added to control foam formation and pH. Thereafter whey was pumped to two aeration tanks (in series) and then to a clarifier. A recycle loop was provided from the sedimentation tank. The BOD_5 was reduced by 98 percent from 15,000 mg/ℓ to 300 mg/ℓ.

Guo, et al. [5] compared the performance of extended aeration system to the conventional activated sludge process. It comprised of two oxidation ditches followed by clarifiers. In one system the sludge recycle ratio was 1.2 to 1, and in the other the settled sludge was recirculated at 40 to 250 percent of the plant influent. In the activated sludge the aeration was achieved by diffusers, and the recirculation was 200 percent of the plant influent. The results are summarized below.

    1. The dairy waste can be treated by extended aeration oxidation ditch. BOD_5 and suspended solids removal

rates of as high as 99% and 95% respectively can be expected. In 1977 dollars, the cost for oxidation ditch was 16 cents per pound of BOD_5 removed.
2. Hydraulic surges can be taken care of, by providing equalization tank in the treatment system; this improves the performance of the activated sludge process.
3. A supplemental nitrogen source should be provided for the treatment of wastes which are deficient in this nutrient.

Quirk and Hellman [6] investigated the treatability of large quantities of whey in conjunction with a relatively small quantity of domestic sewage. They used two stage packed tower filter in series followed by coagulation and settling. The effluent was chlorinated before discharging to surface water. The sludge was vacuum filtered. Some of the authors' conclusions are summarized below:
1. Trickling filtration is effective in treating whey or a mixture of whey and sewage, both, on absolute terms, and on a relative basis, when compared with activated sludge treatment processes.
2. When packed tower trickling filters were employed, the treatability levels achieved for whey compared well with that of other industrial effluents.
3. Nutrient additives other than ammonia nitrogen did not result in increased rates of biodegradability. Similarly, addition of ferrous iron also did not increase the treatability level. However, addition of nitrogen resulted in a 40% increase in the BOD removal rate.
4. The high acidity of whey effluent had no adverse effect on the performance of the trickling filters packed with high porosity media. Rather, a higher BOD removal rate was achieved with a decrease in pH from 7.0 to 4.5.
5. The performance of trickling filter was found more sensitive to flow variations than BOD variations.

In England [7] dairy waste was treated using alternating double filtration (ADF) process. In this process the wastewater was passed through two filters in series with settlement after each filtration. Periodically the order of filters but not of the sedimentation tanks was reversed. The effluent was recirculated. The various loading rates and recirculation ratios produced an effluent with BOD ranging from 4 to 8 mg/ℓ. The corresponding influent BOD ranged from 250 to 700, thus accounting for greater than 99 percent BOD reduction.

Scheltinga [8] has also described alternate double filtration and plastic medium biofiltration. According to the author, ADF provided better effluent than that obtained by filtration.

In dairy waste whey is more difficult to treat because of its high BOD, and finely divided suspended matter which does no readily settle. The various treatment methodologies, along with the BOD range, and percent BOD reduction for dairy plants are summarized in Table I [9]. The average BOD reduction by aerated lagoon was 76 percent, and activated sludge provided the highest reduction of 89 percent.

Table I. Comparison of Treatment Methods for Dairy Industry Waste [9]

Method of Treatment	BOD_5 in mg/ℓ Untreated		Treated		% Reduction in BOD_5	
	Range	Aver.	Range	Aver.	Range	Aver.
Activated Sludge	84–2,920	404	7–632	85	17–99	79.08
Activated Sludge	180–15,000	1903	1–1700	163	57–99.8	88.7
Trickling Filter	130–32,000	2846	10–1400	81	35–99.5	77.4
Aerated Lagoons	75–10,000	286	25–800	191	52–89.5	76

WHEY UTILIZATION

Because of high nutritional content whey can be fed directly to animals. The protein content of whey (0.9 percent of liquid or 21 percent of dry matter) is similar to that of barley but in terms of quality it is much better. Two of the major limiting amino acids methionine and tryptophan are similar in whey protein and barley, but the third, lysine is twice as much in whey as in barley. Therefore, lower quality proteins or less of high quality proteins may be needed to supplement whey than are needed to supplement cereals. Toullec, et al. [10] found that it is feasible to replace all skim milk protein by whey proteins and the milk substitute for feeding preruminant veal calves.

If whey is fed to animals, it eliminates the disposal problems, and cost of drying. However the animals need adaptation to whey. The excessive whey feeding may cause diarrhea, bloat, or scours, which may even result in death.

Whey can be dried and concentrated by drying or evaporation for its use as human food. The largest use of whey in a dairy product is probably in ice cream. Federal regulations for ice cream standard allow replacement of up to 25 percent of the non-fat dried milk with whey. The other uses include bakery products, beverages and soups.

Whey can also be concentrated by membrane processes

(ultrafiltration/reverse osmosis). They produce two streams, a lactose-rich permeate and a protein-rich concentrate. Permeate is further used for the production of lick-blocks for animal feed, and protein concentrate can be processed for human consumption.

Whey can be biologically converted into gaseous or liquid energy via anaerobic pathway. Alternatively the aerobic organisms can convert whey into single cell protein.

On a theoretical basis, if 37.4 billion pounds of whey, produced in the U.S. in 1979, were converted to ethanol, it would yield 152 million gallons, this could cut down the import of crude oil by 275 million gallons.

On a stoichiometric basis, one pound of liquid whey will give 0.014 lb. of CH_4 or 0.314 ft.^{3} (STP) each of CH_4 and CO_2 in the product gas mixture, with corresponding heating value of 314 BTU/(lb. of whey) [6280 BTU/(lb. of lactose) assuming 5% lactose in whey]. The potential heating value of U.S. whey in 1979 would have been 12×10^{12} BTU/year. Assuming fuel gas @ \$3.00/million BTU, the potential monetary value is 36 million dollars.

The total U.S. whey production in 1979 represents a potential nutritional content in terms of lactose and protein of 1.75 billion pounds and 280 million pounds respectively. In the U.S. the daily recommended protein intake is approximately 60 grams for an average adult weighing 150 pounds. Assuming the conversion efficiency to edible protein from whey is 50 percent, the total U.S. whey can provide the equivalent protein requirement for 4.7 million people yearly.

The protein intake in the developing countries is far less than that recommended in the U.S. For instance, in Central Africa the average daily protein intake is about 40 g/day. Therefore the extracted whey protein in the U.S. can supplement about 14 million people of Central Africa.

Whey which is high in nutrient value can be utilized as a fertilizer especially for corn and grassland. Whereas repeated use of artifical fertilizer can deplete important physical components in the soil; the organic constituents of whey can improve the structure and moisture holding qualities of topsoil, thereby reducing wind and water erosion. A recently conducted research study [11] has shown that an application of one acre-inch (27,000 gallons/acre) provides almost 320 lbs. of nitrogen, 100 lbs of phosphorus, and 400 lbs. of potassium which are enough to produce a yield of 150 bushels of corn. Christie [12] and Magnusson, et al. [13] have found from their respective studies that whey and dairy effluents can be utilized for spray irrigation on pasture without any kind of significant environmental risks.

ECONOMICS

If whey is treated as wastewater, it will require contribution of heavy expenses by the industry to meet the effluent guidelines. The useable portion of whey can be converted to marketable end products. The final decision is dependent to a large extent on the economics of the process selected while complying with the environmental regulations.

Assuming the estimated whey generated is 37.4×10^9 lbs/yr, and that BOD_5 varies between 30,000 to 60,000 mg/ℓ, in 1980, the capital cost to build treatment facilities, would be approximately 1.5 to 3.0 billion dollars and the operation and maintenance cost would be 370 to 740 million dollars annually.

One pound of dried solids is derived from 16.6 pounds of liquid whey. Assuming a treatment plant for 10 mgd of domestic sewage (the BOD equivalent of 20,000 lbs.) the operating and maintenance cost of this size plant is 70 cents per 1000 gallons. Extrapolating this to treat 16.6 pounds of liquid whey the O&M cost will vary between 17 to 34 cents, depending on the BOD_5 of the whey, (30,000-60,000 mg/ℓ). On the average the treatment cost is 1.5 ¢/lb. of liquid whey. This is equivalent to 50¢/lb of BOD_5 removed.

In 1973, from experimental data, Boxer[14] projected the cost of drying to be 7 cents per pound of dried whey. This is equivalent to 12.25 cents on 1980 basis.

Carver-Greenfield[15] process reduces this cost by about 33 percent. Other recent innovations in two-stage drying and energy conservation promise further savings.

Whey fermentation studies in the Milbrew plant[16] compared the gross margins for SCP and alcohol production with spray dried whey on CWT basis. Total Anerobic fermentation was almost 50 percent more profitable than total aerobic fermentation and compares evenly with the spray dried whey.

The 1980 estimated[17] cost of wine production is $2.95/gallon which is far less than the cost of commercial wines. However, the taste acceptance may be a problem and need further exploratory studies.

NEOCYTE™[18] has projected pay back period of less than a year for fuel alcohol production plant from whey.

Based on a recently conducted study[2], spray drying, land application and fermentation provide very attractive utilization alternatives. Ultra-Fermentation (a combination of ultrafiltration and fermentation) is a promising area which will completely utilize both the protein and the lactose fractions of whey. With the stronger environmental regulations expected in the future and an economy geared towards greater resource recovery, whey utilization is likely to increase from its 60% value in the United States. The icing on the cake is that it also makes good economic sense.

ACKNOWLEDGMENT

This work was performed under EPA Subcontract No. 68-01-5772 from E. C. Jordan Company to Environmental Consultants, Inc.

REFERENCES

1. Kosikowski, F.V., in Cheese and Fermented Milk Foods, 2nd Ed., Edward Brothers, Inc., Ann Arbor, MI, 1978.
2. "Whey Utilization and Disposal Study," A report submitted to E. C. Jordan Co. by Environmental Consultants, Inc., EPA Contract No. 68-01-5772, September 1980.
3. Fulton, C.W., Mulyk, P.A., and Haskill, J., "A Summary of Cold Climate Operating Experiences with Biological Waste Treatment Systems In the Food Processing Industry", in Proceedings Ninth National Symposium on Food Processing Wastes, Denver, CO, pp 1-60, March 1978.
4. Sall, H., "How Capital Milk Converts a 15,000ppm BOD tiger Into a 300ppm BOD Kitten", American Dairy Review, p. 14, 1976.
5. Guo, P.H.M., Fowlic, P.J. and Jank, B.E., "Activated Sludge Treatment of Wastewaters From the Dairy Products Industry", Proceedings Ninth National Symposium on Food Processing Wastes, EPA Cincinnati, Research and Development, Denver, Colorado, pp 178-202, March 1978.
6. Quirk, T.P. and Hellman, J., "Activated Sludge and Trickling Filtration Treatment of Whey Effluents", Proceedings Second National Symposium on Food Processing Wastes, EPA, Research and Monitoring, Water Pollution Control Research Series, 12060-03/71, Denver, Colorado, pp 447-499, March 1971.
7. Wheatland, A.B., "Treatment of Wastewaters from Dairies and Dairy-Products Factories - Methods and Systems", J. Soc. Dairy Technol., 27, 71 (1974).
8. Scheltinga, H.M.J., "Measures Taken Against Water Dairy and Milk Processing Industries", Pure Applied Chemistry (Eng), Vol. 29, No. 1-3, pp 101-111, 1972.
9. Harper, W.J. and Blaisdell, J.L., "State of the Art of Dairy Food Plant Wastes and Waste Treatment", Proceedings Second National Symposium on Food Processing Wastes, EPA Research and Monitoring Water Pollution Control Research Series, 12060-03/71, Denver, Colorado, pp 509-545, 1971.
10. Toullec, R., et al., "Utilization of Whey Proteins by Preruminant Veal Calves. II - Digestibility and Utilization for Growth", Ann. Zootech, 23(1), 75, 1974.

11. Candela, et al., "Land Application Year 'Round', Whey 'Ploughed In' for Better Utilization", *Food Processing*, 58-59, February 1980.

12. Christie, C.J., "The Disposal of Dairy Factory Waste on Pasture", *New Zealand Soil News*, Vol. 18, No. 4, p. 129, 1970.

13. Magnusson, F. et al., "Spray Irrigation of Dairy Effluent" *Ann. Bull.*, Intl. Dairy Federation, No. 77: p. 122, 1976.

14. Boxer, S., "Elimination of Pollution from Cottage Cheese Whey by Drying and Utilization", *Environmental Protection Technology Series*, EPA-600/2-76-254, September 1976.

15. Silverman, T., "Whey Dehydration Process", Whey Products Conference, Atlantic City, 76, 1976.

16. Everson, T.C., "Whey Derived Gasohol--A Reality", Whey Products Conference, Minnesota, p. 62, 1978.

17. Palmer, G.M. and Marquardt, R.F., "Modern Technology Transforms Whey into Wine", Food Product Development, Feb., 1978.

18. Armstrong, G.M., "Fuel Alcohol Production for the Cheese Industry", Paper presented to the Wisconsin Cheesemaker Assoc., Lake Geneva, Wisc., November 1979.

PRIORITIZATION OF PRODUCTION RELATED WASTEWATER TREATMENT PROBLEMS IN THE TEXTILE INDUSTRY AND COST-EFFECTIVE RECOMMENDATIONS FOR THEIR SOLUTION

Michael S. Bahorsky. Chemical & Environmental Research, Institute of Textile Technology, Charlottesville, Virginia

INTRODUCTION

The task of secondary wastewater treatment in the textile industry is in the optimization stage at this time. Historically, it became clear by the early 1970's that extended aeration activated sludge was going to be the preferred choice for conventional biotreatment of all textile sub-categories. In July 1972 the "EPA" provided the textile industry with preliminary guidelines (treatment requirements) which were based on a small sampling population. It was mutually agreed between the EPA and industry representatives that more suitable figures would be desirable in order to correlate with practically available treatment technology. The American Textile Manufacturers Institute and the Carpet and Rug Institute enlisted the services of the Institute of Textile Technology and Hydroscience, Inc. to do an intensive and comprehensive study of the treatment capabilities of in place and "well operating" facilities in all sub-categories of the textile industry. The study results produced the effluent figures by sub-category shown in Table I.

It should be emphasized that these results reflected, for the most part, systems that were at the time operating reasonably well. Furthermore, there are relatively few wool finishing, stock and yarn dyeing operations, and especially few wool scouring mills, so those sampling populations were very small. Most wet finishing mills in the industry are woven or knitting mills. It was also generally clear that knit finishing mills discharged an effluent which was more amenable to biological treatment than woven mills. This was eventually to be reflected in tighter permit requirements.

Firstly, the detention periods of aeration narrowed from the early 1970 examples of Table II to a tighter range of from 1.3 to 7.5 days. The mills in Table III with the designation UG represent the upgraded versions of those coded in Table II. Power levels ranged from 44 to 150 HP/MG in the l 1970's, as compared to 9 to 353 HP/MG in the early 1970's. Aeration basin solids (MLSS) ranged from 1400 to 5800 in Table III, while early 1970's levels ranged from 200 to 65,000 mg/l. (It might be explained that due to a lack of sludge wasting capability, fairly common in early system designs, the 65,000 mg/l system aeration basin resembled an aerobic digester with mounds of drying solids building around the aerators). Also, with the exception of Mill E, the narrowing of design criteria ranges resulted in better than 90% BOD removals. Mill E remained one of the systems still without sludge wasting. It was instructive to see how many years some of the extended aeration systems could inventory solids. While the trends of design criteria optimization may yet be only suggestive from Table II and III, another approach was used to support the developing picture of optimization. In this approach we sought systems which performed consistently well in effluent quality produced over about a 10 year period. These systems were termed "exemplary" systems. The goal was to isolate key unit operations in these systems, and with the background of the more "average" group of facilities, generate a model system (or models, if indicated) by which designers could derive guidelines, and poorer designed and operating systems could peg up to. These facilities are illustrated in Table IV.

The most obvious example in Table IV is Mill AL with a final effluent TSS of 275 mg/l. Typically, the final TSS represents biological solids carry over. However, in this case the final TSS is primarily wool fiber as no prior fiber screening steps were taken at this facility. Interestingly, the 275 mg/l TSS was within permit requirements. The main point of Mill AL is that even with high BOD strength wool scouring wastes (grease especially), they can be well treated within the detention ranges of other systems-given higher MLSS levels, higher power levels, and simple pretreatment steps, such as screening.

One of the primary intentions or goals of this study has been to obtain the maximum benefit out of the biological system with the least expenditure of capital. This is to say that beyond the maximum oxidative assimilation capability, there is much to be determined about biofloc absorptive capacities (such as for dyes), physical entrapment capacities (such as for latexes), and sludge blanket aspects in the clarifier (again pin-floc entrapment, additional organic removal, and so on).

IN-PLANT MANAGEMENT AND WASTEWATER TREATMENT

   a. General Concerns

   After the industry/EPA study was completed, the Institute of Textile Technology continued its water pollution control efforts for its membership. These efforts of the Institute focused mainly on closely following up selected facilities from the standpoint of biological treatment optimization, because considerable potential was seen in this area. At this time EPA studies focused primarily on research concepts involving advanced treatment concepts.

   Concerning the woven-knit area, Table II illustrates some of the key design and operational factors of mills under the Institute's study purview in the early 1970's.

   Ranking in Table II is by aeration basin detention time. At this stage in the application of the relatively new activated sludge concept, it was still vague as to how extended aeration systems were defined. Extended aeration in municipal systems was often spoken of in terms of 10 to 12 hours. These factors were to have inevitable influences in design practices. From Table II, a spread from 14 hours to 17 days aeration basin detention times can be observed. Although all of these systems possessed clarifiers with sludge recycle, their power levels (HP/MG) and aeration basin mixed liquor suspended solids (MLSS) levels would more aptly define them as aerated lagoons. Interestingly, very high power levels (353 HP/MG, Mill Code BW) and "short" detention times (14 hours) did not guarantee high quality effluents. On the other hand, long detention times and moderate power levels did not appear to guarantee quality effluents either. Extremely variable MLSS levels for another contributed to the difficulty of making optimization considerations in the early 1970's. The potential for improvement was obvious however. Our goals were to attempt to at least more closely define extended aeration system design criteria, i.e. detention times, power levels, MLSS, and so on, for all textile sub-categories. Eventually, we wished to specifically define for our membership where biological treatment should leave off and pretreatment or advanced treatment begin. However, with the most cost/effective approach primary in mind, the Institute always initially examined in-plant approaches, i.e., water conservation, waste strength reduction, reclamation and reuse, etc. since this was also most consistent with the unique insights the Institute possessed into textile production processes which generated wastes. By the late 1970's definite trends were beginning to develop as can be seen in Table III.

It is curious that the EPA has historically focused on much more expensive alternatives, such as advanced treatment add-on concepts, before comprehensively determining the biological concepts maximum capabilities.

Some of the primary findings and observations of these study findings included that with properly applied bar and fine screens, equalization (occasionally with MLSS and/or mixing), about 50-60 HP/MG (high strength wastes more, low strength wastes less), 3 to 6 days aeration detention, suitable MLSS for the task, "quiet" uniform flows to the clarifier, clarifier overflow rates of 250-300 GPD/ft^2, and rational sludge recycle and wasting regimens, one could reasonably expect a close to optimal basic biological treatment system design, i.e., one that would perform consistently well over the years with minimal testing and operational controls.

b. Specific Problem Areas

Because the potential for improved extended aeration system design and operation was and is so extensive, it was indispensible to have the preceding perspective in order to isolate true production-related problems and provide relevant cost-effective solutions. The aforementioned fundamental guidelines provide one with a frame-of-reference design/operation concept to which production problems may be related, and from which their solutions may be derived. Each o these three aspects (problem-design/operation-solution) are so intimately related that one cannot be legitimately discussed without reference to the other two.

In reference now to the optimally designed and operated system, our survey has indicated the following production-related problems in order of most concern to be:

      high-alkaline strength discharges
      foam
      color
      high organic loadings
      lint
      grease
      latex

It should be understood that while many people experience problems from high-alkaline strength discharges, and a few experience grease or latex problems, the grease or latex problem may be as profound a problem for the individual concerned as the high pH problem is to other individuals. It should also be noted that, in general, the problems of mill-owned facilities are similar to problems experienced by mills discharging to city-owned facilities.

High-Alkaline Strength Discharges

The following series of figures gives some insight into the high pH problem. In early June 1974, we received inquiry from a member concerning drastically reduced BOD removal capability, loss of aeration basin solids (MLSS), and extremely high final solids (TSS). On visiting the site, it was determined that the underlying factor was that the mill had inadequate data utilization. In other words, good samples were taken and good tests performed, but the data were filed away without suitable attempts at interpreting developing trends. In Figure 1 it can be seen that the MLSS, which was being increased to obtain better color removal, dropped precipitously around May 29th. Other sets of data were plotted and clearly (and relatively simply) indicated the problem was caused by a caustic slug on May 28th, Figure 2. Also, most instructively, it could be seen that within 5 to 7 days, the aeration basin recovered from the caustic overload, and that BOD removal capacity returned to previous levels in about 12 days. It was also gratifying that much earlier we had recommended aeration basin pH monitoring to this mill because without this information the intrepretations would be much more difficult.

During pilot-scale studies of the Rotating Biodisc System on textile discharges, we examined the effects of hydroxyl alkalinity on BOD removal. These results are shown in Figure 3.

In Figure 3 it can be seen that when raw waste hydroxyl alkalinity presented to these particular organisms exceeds 100 mg/l, the BOD removal deteriorates.

Later studies of "acclimated" organisms, both in the field and in respirometer studies, showed that consistent raw waste pH levels (11.0-11.5) produced alkaline-resistant strains that gave very high quality effluents, i.e., 98-99% BOD removal, 80 to 90% COD removal, and final TSS below 25mg/l. Wide pH swings were, on the other hand, invariably a problem.

For some time, as a result of the prevalence and apparent logic of the approach, we recommended neutralization for those with pH problems. The graph in Figure 4 illustrates typically why this approach is no longer strongly preferred. This mill called in early 1976 for help to rectify consistent permit pH violations (and others that followed from it). After installing manually operated acid neutralization, the peak around January 1977 illustrates best that, in the face of highly variable discharge pH, manual control can be tedious and relatively unsuccessful. Around mid-1977 automatic control installation showed steady improvement as experience was gained and tighter in-plant controls were applied. However, as the peak around January 1978 illustrated, even automatic pH controls were not as successful as could be desired.

This led to a scrutiny of equalization basin effectiveness in pH control. Of the data generated, Figure 5 is most representative of one important feature of equalization.

Out study eventually indicated that one day's equalization was a good starting rule-of-thumb, with stronger discharges requiring more heroic efforts, such as longer detention times, and/or mixing, or low level MLSS (produce CO_2 for "natural" neutralization), and, finally, acid addition as a last resort. In one mill visit it was observed that acid was being added manually while the raw waste pH had dropped to 5.6.

Equalization demonstrated its technical and economic value as the most preferred treatment method for providing pH uniformity, leveling of organic loading, and for dampening hydraulic surges through a treatment system. It is again important to emphasize, that from our accumulated experience, in-plant processes could often be "tightened" before end-of-pipe changes were recommended, and in this way better treatment would result, in addition to manufacturing cost savings. In essence it was found that high-strength alkaline discharges were found to originate mainly from cotton mercerizing processes in which 20 to 25 percent sodium hydroxide (caustic) is used. The key finding was that caustic recovery systems in most cases were not operated optimally. Moreover, closer investigation showed wide caustic recovery design and operating variations for what was essentially the same procedure. Recommendations, from survey data and discussions with vendors and mill engineers, were made so that mills might be assisted in optimizing their recovery operations and save manufacturing costs. Finally, recommendations on the most cost-effective treatment approaches observed for handling unavoidable caustic residuals were defined. Similar approaches were taken for each of the major problem areas to be reported on.

Foam

Foam "clauses" appear almost invariably in permits to the effect that:

> "There shall be no discharge of floating solids or visible foam in other than trace amounts."

Wet-textile operations almost all utilize a surface-active material, and some may involve as many as four different types in the same process. It would be reasonable to assume that almost all the surfactant put into a wet process is generally rinsed out. Moreover, misapplication of practices, such as the "foam finishing" process seen in Figure 6, add to the often overwhelming impact of foam on the treatment system.

Typically, foam becomes a problem immediately in waste treatment. Figure 7 shows sprays in a bar screen chamber wetting foam down, while a foam-coated treatment basin is seen in the background.

In a detailed paper [1] addressing the impact of foam on wastewater treatment, a series of slides, similar to Figure 8, showing extremely foamy biotreatment system discharges were used to open the presentation. Each of the facilities shown had exceptionally good effluent quality when looking at BOD, COD, and TSS removals.

Prolonged cloudy days and egg beater-like aerators intensify the foam problem. EPA investigators Garrison and Hill [2] in 1972 demonstrated that partial breakdown on phenol-containing surfactants resulted in an 80 fold (0.05 mg/l to 4.0 mg/l) increase in p-nonylphenol after biotreatment at a carpet mill. The inference is that partially metabolized by-products may be more "toxic" than parent compounds. Finally, Tables V and VI show how foam-hampered treatment can be corrected by substitution and reduced usage to result in better treatment and reduced manufacturing costs.

Tables V and VI illustrate the preferred cost-effective approach best. Sprays and anti-foams treat the symptoms, but the problem often recurs downstream. An example of a cotton/polyester package dyeing operation (packages of wound yarn are dyed in a large kier) is used to guide interested parties into an exploration of foam problem solving at the mill level. First, the package has been scoured and bleached. It probably retains some small amount of surfactant. It is then generally "wet out" with 0.25-0.5% surfactant. Then a disperse dye (for polyester) containing surfactant in the dye powder is added along with as much as 2% surfactant (leveling agent), and perhaps as much as 1-2% of a anti-agglomerating agent (surfactant). Then the disperse dyeing is rinsed off with some (0.15-0.25%) surfactant. Next, the vat dye (cotton) may be applied. Finally, (0.25-0.5%) a soap-off is given. The main point is that mills often take a chemical vendors recommendation as to the amounts of surfactants needed without evaluating actual needs. All of the surfactants used eventually end up in the drain. There are numerous opportunities to reduce surfactant usage in the mill, if time is taken to explore possibilities.

Color

Color resulting from dyeing rinses and exhausted baths causes treatment problems for most biological treatment systems. Attempts to handle the problem have included granular carbon columns, powdered carbon (P.A.C.T. systems), chemical coagulation and settling, in-plant hyperfiltration (reverse osmosis), ozone, and chlorine. Our mill studies have shown chlorine decolorization after biotreatment to be effective, simple to operate, inexpensive in capital and operating costs, and to result in no secondary sludges for disposal or incineration. Problems with residual chlorine and chlorine by-product toxicity are yet to be resolved however. The approach most preferred is to increase solids (MLSS), as indicated by the trend shown in Figure 1. Our studies have shown much increased color removal when dyes are treated at higher MLSS concentrations and longer detention times. Figure 9 illustrates filtered liquors after being aerated 6 to 24 hours at 10,000 mg/l MLSS. Samples aerated with 1500 mg/l MLSS showed no differences after 24 hours and are not pictured.

Experience has indicated some designs have a relatively low capacity for holding beyond a certain level of MLSS. Some problems, such as poor recycle pump capability, can easily be rectified. Certainly, systems which indicate capability for holding high MLSS solids, such as the Carrousel system shown in Figure 10 may be studied with advantage.

High aeration basin solids levels are not only effective in color removal, but also produce better reduction of foam and slug organic loads than lower MLSS operating levels. Furthermore, levels of lint, grease, and latex beyond those normally tolerated at MLSS levels of 1500-4000 mg/l are other benefits of higher MLSS. Grease is metabolized, while lint and latexes are partially degraded and removed (physically) in the interstices of the bioflocs. In certain cases, lint and small amounts of latex have been said to assist in sludge dewatering.

Lint

The removal of lint, when warranted, is best accomplished by fine screens. Certainly, excessive lint clogs pumps, pipes, and tangles aerator shafts. Manmade fibers especially can pass through a low MLSS system to deposit in quiet stream zones. Cotton fibers and lint can catch on guy wires and other objects, to make an unsightly mess of the treatment system. Experience indicates that certain steps should be taken before simply buying what the facility "down the road has," or what the vendors representative suggests. louvered screens are commonly seen, with varying performance levels, in the textile industry. Pilot-scale studies are indispensable to selecting proper screen sizes. Once screen sizes have been confidently established, flows to the screens should be uniform. Several installations have been seen that capture lint into troughs that are, in turn, washed back into the system when a periodic flow surge overflows the screens. By-passes are also recommended. Vibrating screens, of course, are energy consumers, and some equipment in the early 1970's suffered from vibration stress and failed frequently. A model which involves a sloping 5 ft. x 8 ft., 60 mesh screen, cleaned by an eccentric bar vibration, has been seen to work very effectively over the years with little maintenance requirements.

Grease and Latex

The method of choice for removal of excessive grease or latex is preliminary dissolved air flotation (DAF).

SUMMARY

In summary, a textile production-related problem can be very different if one is concerned with a poorly designed and operated system versus the impact on a well-designed and operated system. This paper focuses on the definition of an optimized textile biotreatment system, then addresses the effect of production-related treatment problems, and preferred solution approaches gained after extensive industry-wide study and ten years experience.

REFERENCES

1. Bahorsky, M.S., "Impact of Foam on Textile Wastewater Treatment," Amer. Dyestuff Reporter, 68:26-29, 51 (Oct. 1979).

2. Garrison, A.W. and Hill, D.W., "Organic Pollutants from Mill Persist in Downstream Waters," Amer. Dyestuff Reporter 61:21,24-25 (Feb. 1972).

TABLE I. RANGE OF EFFLUENT CONCENTRATIONS
GENERALLY ATTAINABLE BY BPCTCA
(Best Practical Control Technology Currently Available)

Sub-Category	Effluent, mg/l		
	BOD	COD	TSS
Wool Scouring	80-120	1000-1500	120-160
Wool Finishing	25- 40	250- 400	40- 60
Greige Mill	40- 50	100- 150	40- 60
Woven Finishing	30- 45	250- 300	40- 60
Knit Finishing	20- 30	200- 250	25- 40
Carpet Operations	20- 80	200- 800	40-160
Stock and Yarn	15- 30	150- 200	25- 40

TABLE II. EARLY 1970's BASIC DESIGN DATA AND
PERFORMANCES OF TEXTILE BIOTREATMENT FACILITIES

Mill Code Processing Type	Flow, MGD	Raw BOD Load, lbs/day	Aeration Detention Time	HP/MGD	MP/MG	MLSS mg/L	% Removal BOD	% Removal COD	Effluent TSS mg/L (Monthly Averages Used)
BW, knit	0.6	2,000	14 hours	200	353	3,500	95	74	43
Sp, woven	14.0	70,000	18 hours	86	120	5,000	79	50	200
B-L, woven	5.0	6,500	38 hours	210	88	400	83	48	48
A, knit/woven	0.5	550	2.8 days	200	77	850	90	27	28
M, knit/woven	2.5	5,600	2.9 days	180	63	4,000	89	65	103
C-W, woven	1.5	2,400	4.0 days	167	41	1,500	98	87	20
UD, knit	0.7	1,000	5.0 days	200	21	<500	92	67	30
C-D, woven	2.5	2,800	6.4 days	60	9	<500	73	67	not measured
B, woven	0.9	2,000	10.0 days	356	37	65,000	75	negative	269
P, knit/woven	1.3	3,000	16.0 days	460	29	<200	<10	negative	131
G/E, knit/woven	0.7	1,060	17.0 days	570	33	1,500	88	58	69

TABLE III. LATE 1970's BASIC DESIGN DATA AND PERFORMANCES OF TEXTILE BIOTREATMENT FACILITIES

Mill Code, Processing Type	Flow, MGD	Raw BOD Load lbs/day	Aeration Detention Time-Days (MG)	HP/ MGD	HP/ MG	MLSS mg/L	% Removal BOD	% Removal COD (Monthly Averages Used)	Effluent TSS, mg/L
CW, knit	2.5	4,800	1.3 (3.2)	200	150	5,800	$\frac{7}{230} = 97$	$\frac{210}{939} = 78$	10
Sp-UG	7.2	-	2.1 (15)	250	120	5,000	(9)*	(513)*	16
P-UG	1.7	3,000	2.2 (3.8)	150	66	1,400	$\frac{21}{210} = 90$	$\frac{438}{969} = 55$	85
E, woven	0.38	350	3.2 (1.2)	400	125	3,700	$\frac{91}{112} = 19$	$\frac{2175}{879}$ = Neg.	904
G/E-UG	1.0	1,400	3.6 (3.6)	160	44	2,000	$\frac{9}{172} = 95$	$\frac{135}{500} = 73$	24
B-L	3.1	-	3.9 (12)	350	92	2,000	(11)	(126)	31
Bv, woven	2.1	6,100	5 (10.5)	300	57	-	$\frac{3}{350} = 99$	(135)	52
AL, woven	1.0	2,300	7.5 (7.5)	600	80	4,500	$\frac{4}{275} = 98$	$\frac{150}{670} = 78$	42

TABLE IV. BASIC DESIGN DATA AND PERFORMANCE OF EXEMPLARY BIOTREATMENT FACILITIES

Mill Code Processing Type	Flow, MGD	Raw BOD Load, lbs/day	Aeration Detention Time-Days (MG)	Connected HP/ MGD	Applied HP/ MG	MLSS, mg/L	Removal BOD	Removal COD	Effluent TSS, mg/L
B, woven	0.085	480	5.3	235	44	12,000	99	95	15
T, woven	2.5	7,400	5.6	320	57	3,800	99	91	6
K, woven	1.0	3,320	3	320	53	2,500	96	85	9
Al, wool scour	0.25	2,400	6	960	106	25,000	97	90	275
AmT, stock and yarn	1.2	930	3.3	150	46	<750	95	75	8

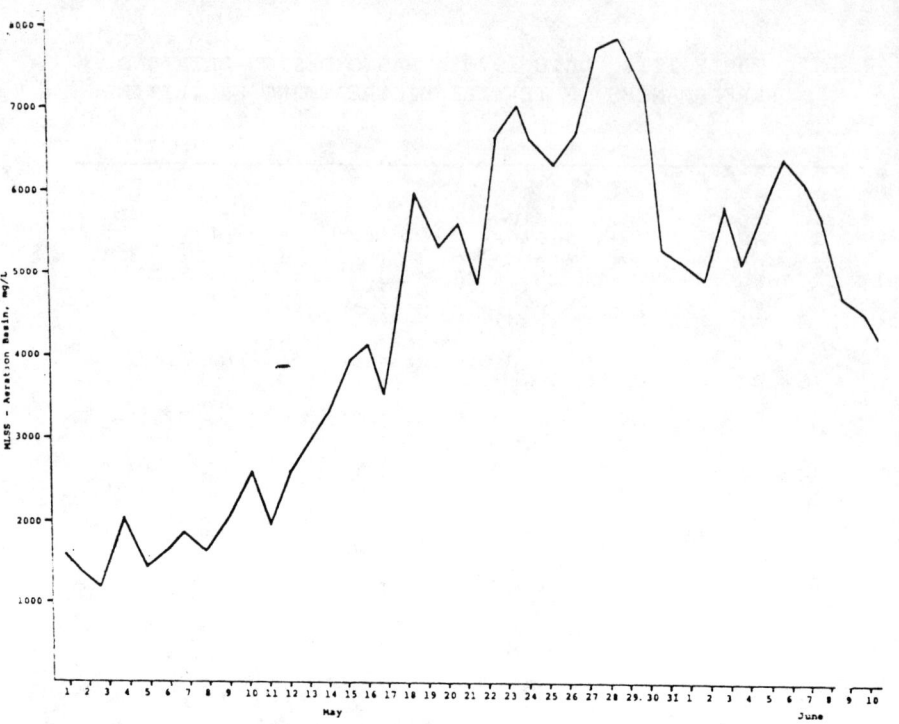

Figure 1. MLSS Development and Precipitous Loss On May 28

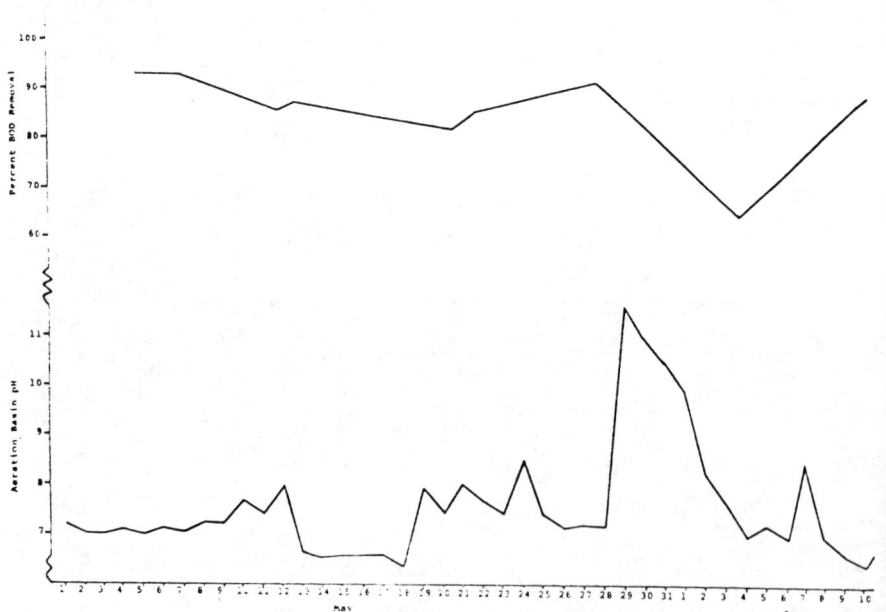

Figure 2. Effect of pH Shock on BOD Removal.

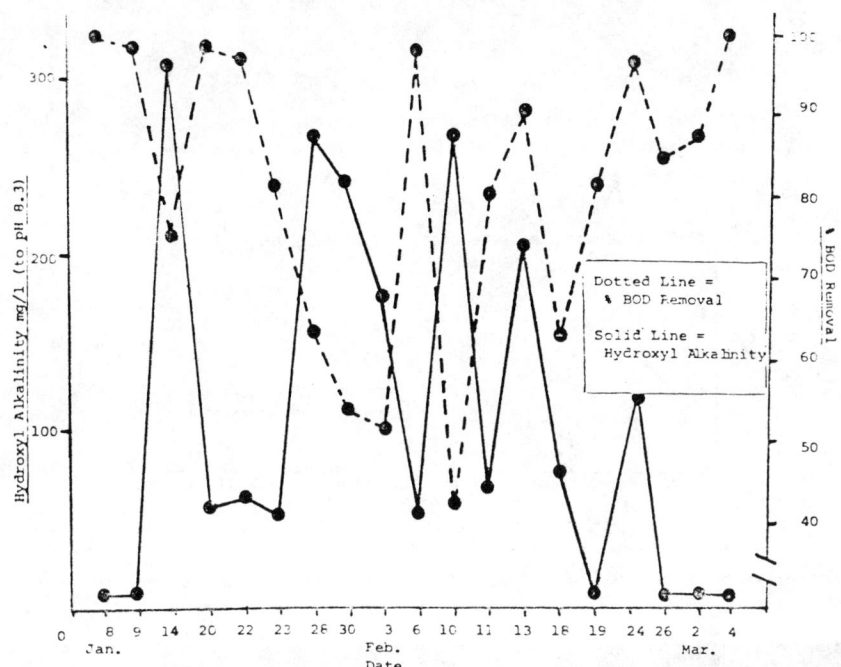

Figure 3. Effect of Hydroxyl Alkalinity on BOD Removal.

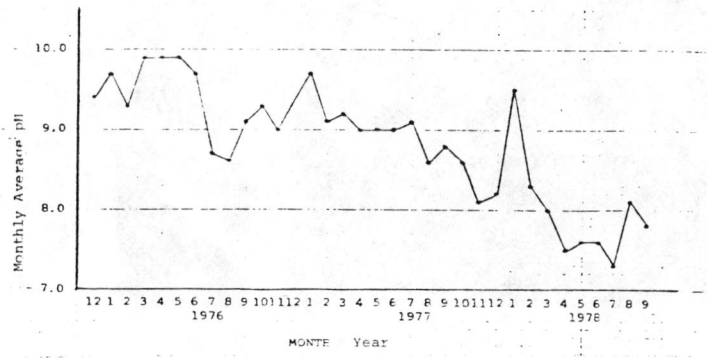

Figure 4.

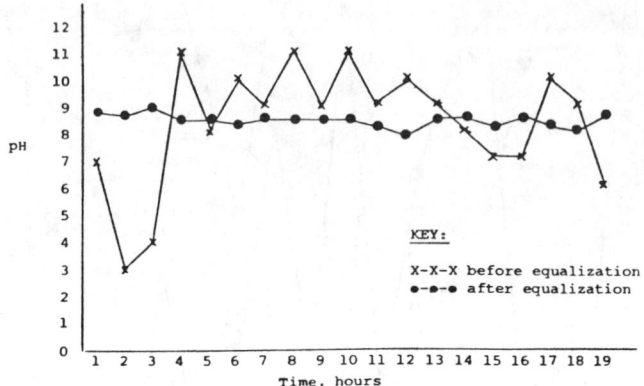

Figure 5. pH Before and After Equalization.

Figure 6. Foam finishing operation overflows copious amounts of foam to the drain.

Figure 7. Sprays keep foam down in the bar screen chamber (foreground), foam covered treatment basin in background.

Figure 8. Illustrates a commonly found foamy textile discharge after biological treatment.

TABLE V. FOAM-HAMPERED PERFORMANCE DATA

Date	Influent BOD's, mg/l	Effluent BOD's, mg/l	Removal Efficiency, percent	MLSS, mg/l
11/2/72	400	120	70	620
11/9/72	390	130	67	420
11/16/72	340	120	65	520
11/22/72	390	120	69	540
11/29/72	400	130	68	440
12/4/72	369	141	62	420
12/14/72	360	130	64	420
12/22/72	390	130	67	420
12/29/72	380	120	68	400
1/4/73	400	130	68	400
1/11/73	370	110	70	410
1/18/73	380	120	67	410
Average	381	125	67	

TABLE VI. CHANGE IN PERFORMANCE WITH FOAM CONTROL

Date	Influent BOD's, mg/l	Effluent BOD's, mg/l	Removal Efficiency, percent	MLSS, mg/l
2/22/73	370	120	68	440
3/1/73	370	130	65	370
3/8/73	360	130	64	890
3/9/73				1,200
3/14/73	357	70	80	1,280
3/22/73	300	50	83	1,400
3/29/73	329	70	79	1,500
4/5/73	330	75	77	1,670
5/10/73	360	75	79	1,760
6/7/73	340	60	82	2,180
6/13/73	317	52	84	2,299
6/18/73	359	65	82	2,340
6/26/73	362	65	82	2,370
Average	339	65	81	

Source: In-Plant Control of Pollution, Manual 1, EPA 625/3-74-004, Oct. 74.

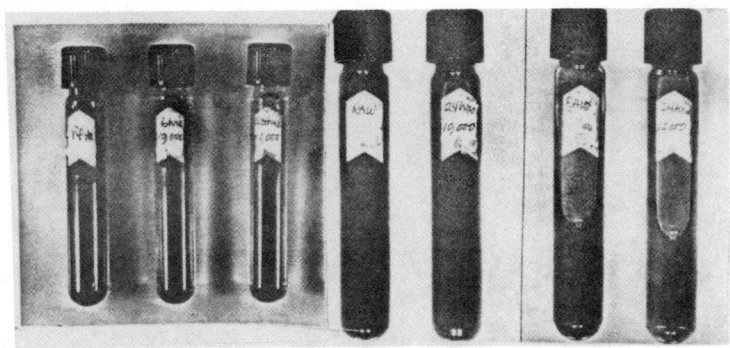

	6HRS	24HRS				
RAW	Aeration	Aeration	Raw	24HRS	RAW	24HRS

Figure 9. The effect of aeration basin solids at 10,000 mg/l MLSS on color removal.

Figure 10. Carrousel system is compact, holds high solids levels, and has variable energy aerators.

RECOVERY OF ORGANIC MATTER FROM SEAFOOD PROCESSING
WASTEWATERS BY ULTRAFILTRATION AS AN ALTERNATIVE METHOD

Allen C. Chao. Department of Civil Engineering, North
Carolina State University, Raleigh, North Carolina

Gary Davis. Department of Civil Engineering, North
Carolina State University, Raleigh, North Carolina

INTRODUCTION

This paper presents some preliminary results of a research project funded by the Sea Grant College Program, University of North Carolina, on the use of an ultrafiltration technique as an alternative method for treating seafood processing wastewaters. Landing and processing seafood is an important source of income for residents of the nation's coastal states. Annually more than 5 billion pounds of seafood are landed in the United States while 2 million pounds are landed in North Carolina. Seafood processing industries have an estimated annual water consumption of 2.5 billion gallons. Nearly all of the water consumed in processing the seafood harvest is discharged as seafood processing wastewater.

Current treatment practices are limited to dissolved air flotation and screening with fine mesh wire which provide partial removal of suspended and colloidal particles. The BAT technology recommended by the EPA include extended aeration and lagoons. Seafood processing plants are often small scale operations scattered in coastal areas where land is expensive and expansion space is limited. Additionally, ground water table considerations are an inhibitive factor in applying biological treatment methods. Since seafood processing operations are dependent on seafood catch, processing plants are often not operated on a continous 12 month basis. Therefore, a biological treatment system may not fit the economies of scale, the physical location, or simply may not operate well for many of the seafood industries.

The ultrafiltration process employs a semi-permeable membrane to separate macromolecular substances from water and substances of small molecular weight. It has been successfully used in food and textile industries for separating macroorganic matter from water and other small molecules (2; 6; 7; 9).

Unlike the many municipal and industrial wastewaters, the seafood processing wastewater does not contain toxic or known carcinogenic substances. The organic constituents are valuable nutrients which can be separated from the water by use of an ultrafiltration process. This also achieves pollution abatement.

EQUIPMENT

Ultrafiltration Membrane

The membrane filters used in this experiment are a combination of the hollow fine fiber and the single tube configuration manufactured by Romicon. Four types of membrane cartridges are used in these studies. The PM 50 cartridge has a nominal cut-off molecular weight of 50,000 daltons with an apparent pore diameter of 50 angstroms. Three other cartridges used in this experiment have a molecular weight cut-off of 30,000, 10,000, and 5,000 daltons and are designated as PM 30, PM 10 and PM 5, respectively.

The membrane filter cartridges were mounted vertically on a unistrut structure and connected to the feed pump with 1 1/2" I.D. pipe. A 2 HP pump (Triclover C114D501-S) is capable of delivering 60 gpm at 25 psig. A backflushing pump was connected to the lower permeate port with 1" I.D. PVC pipe. The backflushing water can be delivered at a rate of 5 gpm at 5 psig pressure.

METHODS

The raw wastewaters used in the studies were collected directly from the discharges of several seafood processing plants. These include a hand-picked conventional blue crab processing plant, a minced fish processing plant, a hand picked calico scallop processing plant, and a mechanical picked calico scallop processing plant. The wastewater samples were transported to the laboratory in 5 gallon plastic containers and temporarily stored in a refrigerator at $4^\circ C$ until used.

Samples were placed in a 200 liter reservoir. The temperature was adjusted to the original level with hot water passing through a copper tubing heat exchanger submerged in the tank. After heating the wastewater was continuously pumped through the filters. The concentrate stream was returned to the feed tank. The operation was continued until the raw wastewater was concentrated ten times. Samples of the permeate and the concentrate streams were collected at periodic intervals. Permeate flux and temperature was also checked periodically.

The samples were analyzed for BOD-5, COD, TOC, TSS, TS, and TKN in accordance with the 14th edition of STANDARD METHODS (1).

RESULTS AND DISCUSSION

Permeate Flux

The permeate flux across the membrane is the most influential factor on the capital and operating costs of an ultrafiltration process. (3). A higher permeate flux will certainly result in a smaller capital investment and operating cost. Generally, a permeate flux of 8-15 GPDSF is considered minimum for the ultrafiltration process to be economically feasible for treatment of wastewaters.

During the operation, the inlet pressure was maintained at the maximum recommended level of 25 psig and the average pressure exerted on the membrane cartridge is determined by taking the average of the inlet and outlet pressures. But, a higher outler pressure had the undesirable effect of reducing the flow rate through the cartridge. For outlet pressures maintained above 17 psig a serious reduction of the average flow velocity in the filter fibers was observed (5). Therefore, the inlet and outlet pressures were maintained at 25 psig and 17 psig respectively. Under these conditions, the wastewater was circulated through the various cartridge membrane fiber bores at about 27 gpm and an actual flow velocity of 8.8 fps.

The flux rate was also influenced by the liquid temperature. Higher liquid temperatures resulted in a higher flux rate. A much larger flux rate can be expected if the ultrafiltration system is operated at its maximum allowable temperature of $75^{o}C$. No attempt was made to maintain a higher operating temperature than would be normally expected at each sampled seafood processing plant.

Under conditions of constant temperature and pressure, the permeate flux decreases as the wastewater is concentrated. The concentration factor (CF) is calculated by dividing the initial sample volume by the volume of remaining wastewater in the feed tank (3). The variation of permeate flux rate has been shown to be a linear function of the log of CF (2). This is confirmed by the results of this study.

Quality of Permeate

The quality of the permeate obtained is indicated by the measured TOC, COD, BOD_5, TS, and TSS levels. Tables I-V contain a summary of the test results from the five seafood processing wastewaters.

The permeate quality is not affected by the operation time but rather is a function of the membrane selected and the nature of the organic matter. Since most of the seafood processing wastewaters contain soluble proteins of smaller molecular weight which pass through the membrane of the smallest pore, the permeate quality is not affected by the type of membrane tested.

Ammonia nitrogen because of its small molecular weight cannot be removed from the permeate. On the other hand, suspended solids were removed very effectively in most cases.

TABLE I. Summary of Tests Results Employing Membrane Cartridges for Treating Conventional Blue Crab Processing Wastewater.

	Raw Waste	Concentrate	Permeate
COD	26,100	62,600	13,100
BOD_5	10,000–14,000	50,000–60,000	4,000
TS	28,000	61,000	19,500
TSS	2,600	14,400	400
pH	7.0–7.5	(All except pH are in mg/l)	

TABLE II. Summary of Test Results Employing Membrane Cartridges for Treating minced Fish Floor Drain Processing Wastewater.

	Raw Waste	Concentrate	Permeate
COD	570		110
BOD_5	550		60
TOC	90		20
TKN	59		85
TS	570	680	410
TSS	70	90	13
pH	7.0–7.5	(All except pH are in mg/l)	

The percentage removal efficiency is shown in table VI. The efficiencies are close to the efficiencies reported for treatment of laundry waste with microfiltration techniques by Santo, et al., (8). Noteworthy is the fact that soluble organics found in seafood processing wastewaters have not been successfully removed by screening or dissolved flotation methods.

TABLE III. Summary of Test Results Employing Menbrane Cartridges for Treating Fish Bleach Tank Drain Processing Wastewater

	Raw Waste	Concentrate	Permeate
COD	1880		300
BOD_5	1810		330
TOC	390		90
TKN	241	539	81
TS	1330	2150	1120
TSS	170	580	15
pH	7.0-7.5	(All except pH are in mg/l)	

TABLE IV. Summary of Test Results Employing Menbrane Cartridges for Treating Hand Picked Calico Scallop Processing Wastewater.

	Raw Waste	Concentrate	Permeate
COD	600		240
BOD_5	420		195
TOC	220		110
TKN	64	116	35
TS	1670	1920	1415
pH	7.0-7.5	(All except pH are in mg/l)	

TABLE V. Summary of Test Results Employing Membrane Cartridges for Treating Mechanically Picked Calico Scallop Processing Wastewater.

	Raw Waste	Concentrate	Permeate
COD	3300	11570	1020
BOD_5	2340	6270	1340
TOC	1280	3710	630
TKN	409	1164	
TS	6780	13300	5730
TSS	680	6620	70
pH	7.0-7.5	(All except pH are in mg/l)	

TABLE VI. Membrane Filter Efficiencies in Percent Reduction.

	COD	BOD_5	TOC	TKN	TS	TSS
Conventional Blue Crab	50	50			30	85
Minced Fish Floor Drain	80	90	80	70	30	80
Minced Fish Bleach Tank	85	80	75	65	15	90
Hand Picked Calico Scallop	60	55	50	45	15	
Mechanically Picked Calico Scallop	70	45	50		15	90

Concentrate Quality

The ultrafiltration system has been operated in these tests runs to achieve a concentration factor of 10. The concentrate stream from the minced fish floor drain and the hand picked calico scallop processing wastewaters do not show specially large concentrating effects of total solids or organic matter. However, the conventional blue crab processing wastewater concentrate stream and the mechanically picked calico scallop processing wastewater concentrate stream showed excellent concentrating effects. In the case of the crab processing wastewater concentrate stream the total solids were concentrated from about 3 percent to about 6 percent. The mechanically picked calico scallop processing wastewater concentrate stream showed a two fold increase in total solids concentration to about 1.4 percent.

Perhaps of more interest is the nutrient content of the concentrate streams of the ultrafiltration processes. Experience with poultry processing wastewater indicates that the ultrafiltration processes can treat the wastewater to obtain a concentrate strean containing 5,000-6,000 mg/l total solids. The concentrate was further centrifuged at 3,000 X g for 30 minutes and then vacuumned at 50 C for harvesting of nutrients. Approximately 108 grams dry solids can be obtained out of 20 liters of concentrate. The solids contain 40% protein and 23-45% fat(9).

The concentrate obtained from treating blue crab wastewater contains 61,000 mg/l total solids or about 6 percent solids. The potential for recovering large quantities of valuable nutrients is obvious. A method used to estimate the protein content of any solution is to multiply the value of nitrogen as determined by the Kjeldahl nitrogen by the factor 6.25. This is acceptable since the nitrogen content of any protein varied from 15-18 percent and averages about 16 percent. When this method of estimating protein content is applied to mechanically picked calico scallop processing wastewater concentrate protein content is calculated at about 7300 mg/l or about 55 percent of the total solids concentration.

The ultrafiltration system can be operated longer to obtain a greater concentration of solids and nutrients in the concentrate stream to facilitate the subsequent dewatering step for recovering the nutrients. Further studies are underway to evaluate the feasibility of recovering nutrients for compensating the operating costs while eliminating the sludge disposal problems.

CONCLUSIONS

The preliminary results of the study indicate that ultrafiltration techniques are feasible for treating some types of seafood processing wastewaters. Under the conditions of moderate temperature and a concentrated wastewater stream the permeate flux rate is sufficiently high for the ultrafiltration process to be considered economically feasible.

The concentrate obtained by use of the ultrafiltration process on seafood processing wastewaters can be high in nutrients. The potential exists for dewatering the concentrate streams that are rich in nutrients. Based on experience with poultry wastewater the yield could be significant. Thus, additional benefits could be obtained from the recovered nutrients to compensate for operating costs while eliminating sludge disposal.

If properly managed the ultrafiltration process can be an alternate method for treating some types of seafood processing wastewaters. The results demonstrate the potential for pollution abatement and nutrient recovery. Further studies are necessary to make the system applicable for field use.

ACKNOWLEDGMENT

The work upon which this publication is based was supported partially by the Office of Sea Grant, NOAA, U. S. Department of Commerce under Grant No. 04-8-Mo1-16, the State of North Carolina, Department of Administration and the Department of Civil Engineering, N. C. State University. The U. S. Government is authorized to produce and distribute reprints for governmental purposes notwithstanding any copyright that may appear hereon.

REFERENCES

1. APHA, AWWA & WPCF, "Standard Methods for the Examination of Water and Wastewater", American Public Health Association, New York, 1975.
2. Breslau, R. R., and Kilcullen, B. M., "Hollow Fiber Ultrafiltration of Cottage Cheese Whey; Performance Study", Jour. Dairy Sci., 60, 1379 (1977)
3. Breslau, R. R., Agranat, E. A., Testa, A. J., Messinger, S., and Cross, R. A., " Hollow Fiber Ultrafiltration a System Approach for Process Water and By-Product Recovery", Paper Presented at the 79th American Institute of Chemical Engineers National Meeting, Houston, TX, 1976.
4. Chao, A. C., Unpublished data, 1979.
5. Chao, A. C., Machemehl, J. L., and Galarraga, E., "Ultrafiltration Treatment of Seafood Processing Wastewater", Paper Presented at the 35th Purdue Industrial Waste Conference, 1980.
6. Goldsmith, R. L., Roberts, D. A., and Burre, D. L., "Ultrafiltration of Soluble Oil Wastes", Jour. Water Poll. Cont. Fed., 46, 2183 (1974)
7. Riddle, M. J., and Shikaze, K., "Characterization and Treatment of Fish Processing Plant Effluents in Canada", Proc. 4th National Symposium on Food Processing Wastes, EPA 660/2-73-031, 1973.
8. Santo, J. E., et al., "Regeneration of Wastewaters for Reuse through Hydroperm Microfiltration", Paper Presented at the Water Reuse Symposium, Washington, D.C., March 25-30, 1979.
9. Shih, J.C. H., and Kozink, M. B., "Ultrafiltration Treatment of Poultry Processing Wastewater and Recovery of Nutrient By-Product", Jour Poultry Science (in press).

RECYCLING WASTEWATER IN COLORADO - THE ISSUES, THE PROBLEMS AND THE SOLUTIONS

Thomas G. Sanders. Department of Civil Engineering, Colorado State University, Fort Collins, Colorado.

INTRODUCTION

Wastewater reuse in Colorado is not a new issue. Wastewater has been recycled for agricultural purposes for many years ever since effluents from wastewater treatment plants have been discharged into the many small streams and irrigation ditches. At one time when the population was small and dispersed, this de facto reuse of wastewater apparently posed few problems. But now as the population is increasing at a rapid rate (the front range is one of the fastest growing areas in the country), the dangers of not addressing uncontrolled wastewater reuse policies of the state is becoming critical. Colorado is on the threshold of institutionalizing a method of unrestricted wastewater reuse that could mean the use of unsafe secondary wastewater effluent for the irrigation of human consumable food.

ISSUES

Fortunately Colorado has had abundant resources of pristine, unpolluted water as the majority of its water originates from melted snow. Historically, a large portion of this water has been and is still being used for agriculture which is one of the major industries and revenue producers in the state. Therefore, it has been public policy to maintain a viable agricultural economy, that is, not drying up farmland (removing the water) to meet the water needs of a rapidly expanding population.

Not only is there a policy to maintain an agricultural economy, an urban policy has been adopted in several front range cities to maintain green areas and lawns. These two apparent necessities of the people of Colorado have forced the continued need for developing more water supplies to meet the increasing demand of the urban areas.

Developing new water supplies by trapping and containing more water previously lost during spring runoff or finding and developing new nontributary groundwater sources is becoming very expensive. Therefore if new supplies are limited, and there is a recalcitrance to remove water from the agriculture sector for domestic purposes; it is obvious that water reuse and specifically domestic wastewater recycling appears to be an attractive alternative for meeting the water demands of the future.

There are many categories of water reuse which can be effectively used to meet the many water uses in a state [2], however, only the reuse of domestic wastewater for agricultural purposes is being addressed in this paper.

In order to illustrate some of the concerns of wastewater reuse for agricultural purposes, a project which is presently being developed in an area north of Denver will be discussed. The Northglenn Colorado Water Reuse Project presented as innovative and a means of retaining an agricultural economy in harmony with a rapidly expanding population, borrows farmer's irrigation water and uses it for domestic purposes. The sewage which is approximately 70 percent of the original volume of the borrowers water is returned to the farmers after secondary treatment along with 30 percent makeup water and a 10 percent bonus. The Northglenn treatment system, while almost identical to one in Muskegon, Michican which is a controlled land application system to treat and dispose of domestic wastewater, has its primary objective as the reuse of the domestic wastewater and not disposal. This difference may make a system that is applicable in the well-watered east not so attractive for the arid and semiarid west.

The Northglenn plan appears to be a valid way to retain an agricultural economy nearby and still meet the cities needs. However, what are the problems?

PROBLEMS

Firstly and probably the most important problem with wastewater reuse for agricultural purposes in general and the Northglenn, Colorado plan in particular is the immense public health implications. A treatment process suitable in a plan for disposing of wastewater may not be suitable in a plan for reusing wastewater. Whenever a wastewater having only secondary treatment or its equivalent is used and comes into intimate contact with human food, mans wastes and particularly his pathogenic bacteria, viruses and protozoa are introduced into his food chain possibly precipitating the resurgence of many diseases, such as the typhoid fever epidemics in Denver in the early 1900's. Countries which have sewage in their irrigation water such as Egypt have chronic health problems with such diseases as schistosomiasis and typhoid fever, both of which most probably will never be eradicated until the sewage is removed from the irrigation water.

Secondly, wastewater reuse in this fashion will certainly slow down but will ultimately not stop the transfer of water from farmland. The original Northglenn plan envisioned that most of the makeup water would be developed from new sources, however the majority of the 40 percent makeup water will come from existing farmland [5]. A small portion of the makeup water will also come from street runoff which may contain lead, zinc, asbestos, hydrocarbons and other heavy metals.

Unfortunately there has been an evolving attitude in this country that public health is a relatively minor aspect in water resources development, most noticeably manifested in removing the responsibility of water pollution control from the states public health institutions to water resource institutions. And, although not a reality in Colorado, it is an issue of continuing concern.

In Colorado the lack of concern for public health is manifested in the regulatory vacuum that exists regarding the licensing and control of sewage treatment plant effluents to prevent undue risk to public health. Existing regulations in Colorado for the quality of an effluent of a sewage treatment plant are listed in Table 1.

Table 1. Effluent Limitations [1]

	Parameter Limitations	
Parameter	7 Day Average	30 Day Average
BOD	45 mg/ℓ	30 mg/ℓ
Suspended Solids	45 mg/ℓ	30 mg/ℓ
Fecal Coliform	As determined by the Division of Administration of the State Health Department to protect public health in the stream classification to which the discharge is made.	
Residual Chlorine	Less than 0.5 mg/ℓ	Less than 0.5 mg/ℓ
pH	6.0 to 9.0	6.0 to 9.0
Oil & Grease	10 mg/ℓ and there shall be no visable sheen	

Only fecal coliform and residual chlorine are parameters which may directly assess the public health implications of the water. For the Northglenn Project, the state initially recommended a fecal coliform count of 1000 counts/100 mℓ which is their usual limit of instream fecal coliform densities for agricultural purposes [4]. A 1000 fecal coliforms/100 mℓ as a public health criteria or standard is not sufficient to protect public health if the water were to be used for spray irrigation on human consumable food and/or a raw water source for a potable water supply. Historically in the United States, water that is known to have fecal contamination (effluent from a sewage treatment plant and/or a high fecal coliform bacteria count) is just not used as a public water supply source. There are many who think that this extra precaution is too conservative. However, this is the only country in the western world with the possible exception of England and Canada to have consistently safe water.

The EPA, after reviewing the Northglenn plan, also considered "secondary treatment with disinfection at 1000 fecal coliforms/100 mℓ is insufficient to protect the public." As a result a fecal coliform count of 200/100 mℓ was made a requirement by EPA before funding was available [5]. In the same report it was noted that both the EPA and the State of Colorado do not have a policy stating criteria or establishing standards for wastewater application on human consumable and raw edible food crops.

Although certainly an improvement over 1000 counts per 100 mℓ, many experts believe 200 counts/100 mℓ to be inadequate. The FDA recommends that sewage not be used on human edible food crops. The World Health Organization (WHO) which has developed criteria to be applied to under developed countries which can barely grow enough food to survive, recommends sand filtration following secondary treatment and a coliform bacteria count of 100/100 mℓ. The State of California where water reuse is practically a way of life recommends adequate disinfection, preceded by extensive treatment including coagulation and filtration having an effluent with a median coliform organism count of 2.2/100 mℓ for agriculture irrigation water to be used on unprocessed human consumable foods [3].

Obviously there is a basis to look again at the bacteria quality criteria of a wastewater to be used on human consumable foods but there is also an equally important problem and that is the type of wastewater treatment that precedes wastewater reuse. Unfortunately it appears to be the conventional wisdom (by those making decisions in water reuse) that secondary treatment is adequate and that even the timeless aerated lagoons will generate an effluent of acceptable quality to protect public health. (This statement assumes that all decision makers responsible for wastewater reuse are not intentionally ignoring public health ramifications.) The Northglenn system as it is presently designed is not a disposal system but instead a recycling system. It is completely different than past applications of most land treatment systems, where the levels of pretreatment process design and levels of disinfection were adequate for the controlled application and disposal of wastewater on the land but not adequate for the uncontrolled discharge of wastewater which can come into human contact and can be used for irrigation of human consumable food prior to filtrating through the soil mantel.

This problem of using inappropriate treatment trains stems from two sources: 1) the very high real costs of treatment beyond secondary treatment, and 2) the growing feeling that recycling, going back to basics (the land) and disputing the efficacy of our "water factories" is the way to go.

The former is valid and needs no further explanation other than it will be very costly to properly treat domestic wastewater that will be used to irrigate human consumable food crops without increasing the risk of disease. The latter, although appearing ludicrous and needing no discussion, is in fact manifested in the state and federal funding procedures. The EPA following requirements mandated by much of the recent water quality legislation actively funds "innovative" wastewater treatment alternatives one of which is the aforementioned aerated lagoon with land application proposed by Northglenn. Lagoons are a technology applied in the early 1900's which generally cannot meet existing secondary effluent criteria summarized in Table 1 and are applicable to small discharges with little concern for effluent quality. The addition of induced aeration increases and prolongs the useful life of lagoons at a very small cost. Although a tremendous improvement over existing overloaded lagoons in Colorado, aerated lagoons certainly are not the "innovative" wastewater treatment alternative of the future.

The incentive is also great to look at a system similar to the aerated lagoon because by law land application alternatives must be looked at in the preliminary design phase; the cost can be 115 percent higher than the best alternative and still be funded by EPA and 85 percent of cost of the project instead of 75 percent can be funded by EPA as well. And finally, if the "innovative" process will not work, the law implies that the EPA will pay for 100 percent of costs of modifying or replacement of any innovative or alternative treatment process to meet the new standards. (This capability that EPA will fund 100 percent incompetent engineering is an abomination to professional engineering.)

Colorado also encourages the development of the alternative treatment processes by giving 45 points for projects promoting water reuse in their priority point system which decides which projects get funding. A project that alleviates a designated health hazard receives only 40 points.

Another problem can result if plans similar to Northglenn are developed in Colorado. Instead of the regionalization and coordination of basin wide wastewater management as mandated by the 1972 Water Pollution Control Act Ammendments, Section 208, there will be a proliferation of an inefficient process that is basically the same as those used at the turn of the century.

A possible scenario if the state of Colorado does not pursue an active policy of defining, controlling and regulating agricultural reuse of domestic wastewater and sludges is the categorical rejection by large food processors of any food crop which utilizes wastewater. A similar case has already happened in the Yakima Washington area [6].

SOLUTIONS

There are several solutions that are being promulgated today to the problems raised concerning the Northglenn project. To protect the public health the effluent concentration of fecal coliforms is 200 counts/100 mℓ, a more strict requirement than is currently enforced for sewage treatment plant effluents in the state. There is no requirement for other treatment except that it is equivalent to secondary treatment: BOD = 30 mg/ℓ and SS = 30 mg/ℓ. Because it appears highly possible that aerated lagoons will not even meet the SS concentration for secondary treatment the Water Quality Commission of Colorado relaxed this requirement by not applying the limit to aerated lagoons [1]. Although the new standards are less strict than those promulgated by EPA, it is allowable as it is assumed that the irrigation ditch water is not considered navigable water and hence not under the jurisdiction of the federal standards.

If the domestic wastewater coming from the proposed aerated lagoon system were to be used to irrigate human raw edible food, the city of Northglenn at one time during the evolution of the project suggested that it would buy all the crops. In addition, in order to get the necessary funding from the EPA, the city of Northglenn will issue annually a notice to farmers and others of the possible health effects of watering gardens having human consumable foods with the reused wastewater [5]. And finally if all else fails, as stated previously, the EPA will provide 100 percent funding to make the system meet the design standards.

The solutions that are needed in Colorado and quite possibly other states considering or using wastewater reuse systems (with the exception of California) are as follows.

First and foremost is to remove the technical decision making capability from laymen and political activists in the area of wastewater reuse. It is obvious when a commission responsible for wastewater reuse decisions does not have the technical competence and experience in the field of water, and more specifically wastewater reuse and if it does not actively solicit the opinions of professionals, the public will not be protected from the undue risks of reusing wastewater for agricultural purposes.

Next, develop the control through laws and regulations and administered by the State Department of Public Health over all wastewater effluents to be reused. This would include not only specifying the minimum quality levels for each reuse category (for agriculture reuse, different criteria can be applied for each category: animal food, and human consumables including processed, unprocessed and raw) but also specify the exact treatment processes or their equivalent to produce the required effluent. For example, California required not only a 2.2 coliform bacteria count/ 100 mℓ for water used for irrigating processed human consumable foods, but also requires that the wastewater is disinfected, oxidized, coagulated, clarified and filtered prior to use. (It should be noted that even with this treatment, which is an order of magnitude better than that treatment required in Colorado, this water still cannot be used as a raw water source for a potable water supply.) Specification of treatment process trains or their equivalent and 2.2 coliforms/100 mℓ as the criteria for wastewater to be reused will generally eliminate the necessity for BOD and SS criteria. To meet this coliform requirement both must be close to zero anyway, otherwise 2.2 counts/100 mℓ cannot be met.

Finally, the state and federal funding procedures should be altered to not so heavily advocate returning to the land and alternative and innovative waste treatment processes without properly addressing the public health implications. It should be remembered that very few counties in the world have such safe water supplies as the United States; you can drink the water and eat raw edible food anywhere and not get sick. It would not take too long to reverse this tradition of safe water and uncontaminated food if wastewater reuse for agriculture is not recognized as a definite area of public health concern and not controlled properly.

REFERENCES

1. Colorado Department of Health, "Regulations for Effluent Limitations," Section 10.1.0-10.10.1.16 Ammended March 5, 1979.

2. State of California, "Wastewater Reclamation Criteria," California Administrative Code Title 22, Division 4, Environmental Health, 1979.

3. Turner, Charles D., "Planning Water Reuse," Dissertation in partial fulfillment of the requirements for the degree of Doctor of Philosophy, Colorado State University, Fort Collins, Colorado, Spring, 1981.

4. U.S. Environmental Protection Agency, "Northglenn Water Management Plan," For distribution at a public meeting in Frederick, Colorado, August 30, 1979.

5. U.S. Environmental Protection Agency, "Environmental Impact Statement Final, Northglenn Water Management Program," City of Northglenn, Colorado, EPA 908/5-79-002B, June, 1980.

6. U.S. Environmental Protection Agency Region 10 - Seattle, Washington, "Land Application of Sewage Sludges," Memo from Robert S. Burd to Chris Beck, August 25, 1980.

THE PRODUCTION OF INSULATING FIBERS FROM POWER PLANT WASTES

William H. Buttermore, David J. Akers and Edward J. Simcoe.
Coal Research Bureau, West Virginia University, Morgantown,
West Virginia.

INTRODUCTION

Increased interest in energy conservation in the United States has led to an increased demand for insulating material. One common insulating material in use today is wool fiber produced from blast-furnace slag or siliceous limestone. The Coal Research Bureau has been involved in research in past years designed to evaluate the use of coal ash from conventional power generation stations as the source material for insulating wool production. These studies have indicated that many coal ashes can be successfully converted into insulating wool fiber similar in appearance to commercial 'Rockwool'.[1] Based on these studies, limestone fluidized bed flyash (LFBFA) was examined to determine its suitability as a source material for insulating wool fiber production.

Physical property testing was performed on the LFBFA wool, and results compared with physical properties of wool fiber produced from pulverized flyash, commercial glass wool ('Fiberglas')[2] and commercial mineral fiber wool ('Rockwool'). The results of these preliminary tests indicated LFBFA to be comparable to commercial products.

---

[1] Rockwool is a product of Rockwool Industries of Denver, Colorado.

[2] Fiberglas is a trade name of Owens-Corning of Toledo, Ohio.

Production of Fiber

Mineral wool is made commercially by heating the source material until it becomes molten and sufficiently fluid for further processing. The choice of melting equipment used is determined in part by the size and nature of the raw materials and by the type of energy source or fuel to be used. The most common furnace used in commercial mineral wool production is the water-jacketed cupola type. Some manufacturers have used oil or gas fired reverberatory furnaces or submerged arc resistance furnaces. In the current testing program, a Detroit electric 10 lb. capacity, 17 KVA, carbon arc rocking furnace (Figure 1) capable of being rotated on its horizontal axis was used. This type of furnace was utilized because the molten flyash can be poured directly from the hot furnace chamber as is done in industrial practice. During the heating of the flyash, the furnace was gently rocked to produce a homogeneous melt by mixing the various components. Rocking also increased the efficiency of the furnace as the heat energy was more evenly distributed among the particles in the chamber. Temperature measurements were made using an optical pyrometer.

Figure 1

Laboratory Testing Furnace

When the molten flyash became sufficiently fluid, it was poured in a narrow stream (1/4 to 1/2-inch in diameter) into the path of a "blowing-gas" stream. The gas was directed through a nozzle system located at approximately a right angle to the stream. In commercial practice super-heated steam is often used as the blowing gas in order to extend the time period during which the molten particles cool after pouring. This practice can result in greater fiber production as compared to the amount formed when bottled compressed air is used. In the testing program, however, bottled air was used as there are fewer safety hazards for the operator who must work in close proximity to the blowing apparatus. In industrial practice, no workers are near the blowing apparatus.

The nozzle used in these tests was a flat-spray blow-off nozzle with a slot width of 0.023 inches. At 95 psig it delivered approximately 10 standard cubic feet per minute of compressed air. The nozzle was hand held to maintain proper spatial alignment (air-jet perpendicular or at a slight upward acute angle) to the slag stream with the nozzle orifice approximately 1/4-inch from the stream as shown in Figure 2.

Figure 2

Blowing Wool

The compressed air or blowing-gas has a shearing action upon the slag stream, breaking it into individual drops and propelling the drops into a collection area. As the drops of molten slag are propelled through the air, friction causes them to develop long "tails". These "tails" instantaneously solidify into insulating fibers and any remnants of the original drops are called "shot". If the pouring and blowing conditions have been properly controlled, almost all of each individual drop will be consumed in the formation of the fibers. If conditions are not optimum, i.e., air pressure too low or melt viscosity too high, shorter, coarser and more brittle fibers will be formed. In addition to these less desirable fibers, large shot particles will occur. When the melt becomes thin, fiber elongation does not occur and short interlocking fibers are produced resembling snow. Figure 3 is a photograph taken during a blowing operation using molten limestone fluidized bed flyash. In the left portion of the picture wool fibers may be seen floating to the floor.

Figure 3

Fiber Elongation

Figure 4 is a photograph of a sample of the wool made from limestone fluidized bed flyash (LFBFA). This wool was produced at a temperature of 1375°C without the addition of any fluxing agents. In appearance, the wool is light gray in color, fine textured, with a moderate shot content.

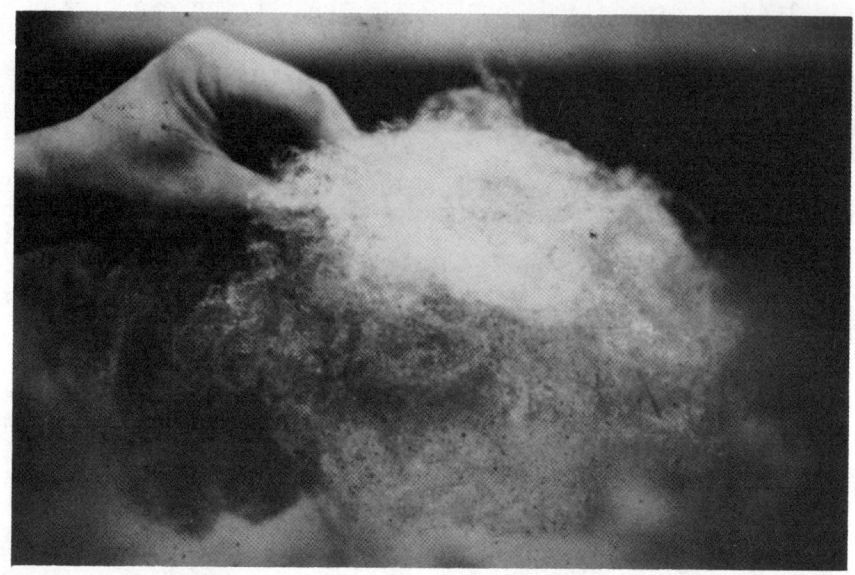

Figure 4

LFBFA Wool

Photomicrographic Analysis

Figure 5 contains scanning electron photomicrographs of five samples of insulating wool identified as follows: Fiberglas (drawn glass, commercial); Rockwool (blown mineral, commercial); LFBFA (limestone fluidized bed flyash, blown, laboratory); E.PFA (eastern pulverized flyash, blown, laboratory); and W.PFA (western pulverized flyash, blown, laboratory). Magnification is indicated by the micron scale on each. As can be seen from the photographs, there is very little difference in structure apart from a moderate percentage of attached shot apparent in the laboratory produced samples. It is also apparent that the blown wool samples are all smaller in average particle diameter than the glass fiber sample which has an estimated 20 micron average particle diameter. Furthermore, it can be seen that the laboratory samples of blown wool, i.e., E. PFA, W. PFA, and LFBFA, are slightly finer than commercial 'Rockwool' (an estimated mean particle

 FIBERGLAS

 LFBFA

 ROCKWOOL

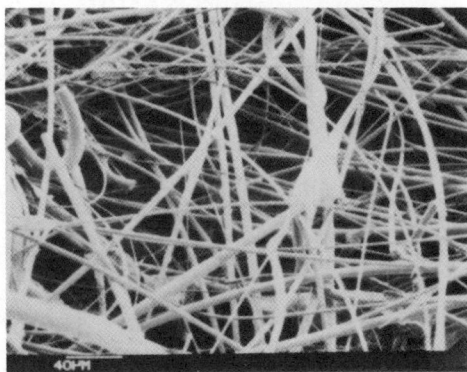 E. PFA

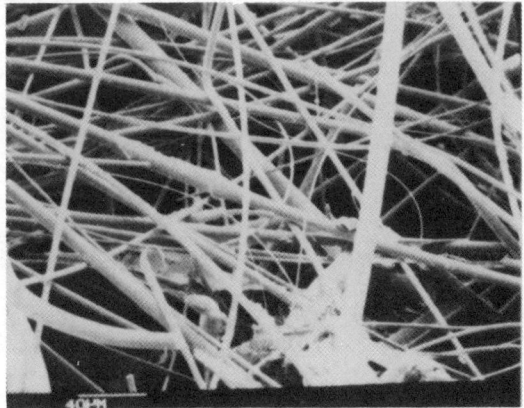 W. PFA

Figure 5

Photomicrographs

diameter of 10 versus 15 microns). Most significant is the similarity in structure in all samples pictured indicating that a similarity in physical insulating properties can be expected.

Thermal Conductivity Test

The most important property of insulating fiber is its ability to prevent heat transfer. To effectively insulate, a material must prevent heat transfer by conduction and convection. In the case of insulating wool this is accomplished by trapping air (which has an extremely low thermal conductivity) in such a way that convection losses are minimized. If air is free to move, heat will be efficiently transferred through convection. In an effort to compare the relative insulating efficiencies of commercial 'Fiberglas', 'Rockwool' and laboratory blown insulating wool, a special test was devised. A thermal conductivity apparatus was modified according to the diagram, Figure 6.

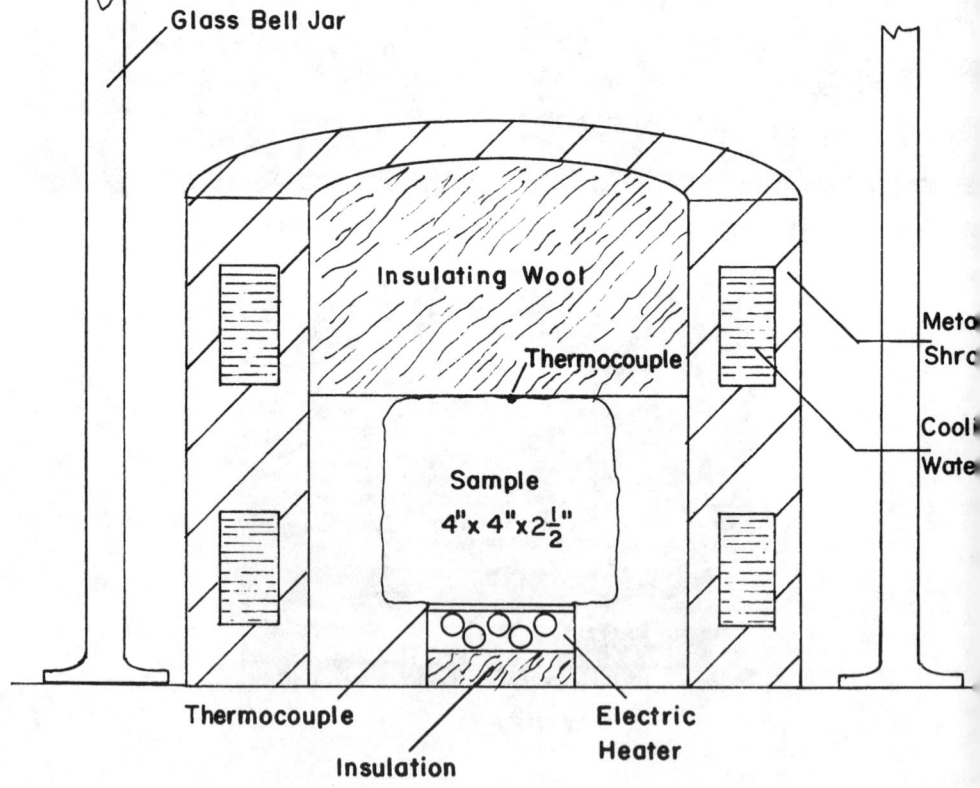

Figure 6

Thermal Conductivity Apparatus

For the test, a 4" x 4" x 2-1/2" piece was cut from a standard 'Fiberglas' batt and 'Rockwool' batt. For in-house produced mineral wools, i.e., those produced from limestone fluidized bed flyash (LFBFA) and pulverized flyashes from eastern and western United States sources, a 4" x 4" x 2-1/2" piece was cut with scissors from a bulk gathered sample. A 2-1/2" height was used because it best represented the actual thickness of the commercial batts of 'Rockwool' and 'Fiberglas'. The test samples were placed within the conductivity apparatus. A temperature of 500°F was maintained in the lower heater and the temperature of a metal plate above the heater in contact with the bottom of the sample was read and compared with the temperature of the probe on top of the sample. Each test was run for a time sufficient to allow stabilization of temperature to occur, usually about 5 hours. As a control, the top probe was suspended in the chamber 2-1/2" above the heater plate to indicate the comparative insulating value of air. The results given in Table 1 show a similarity in insulating ability among samples tested. Convection losses are indicated by the smaller temperature difference for the air gap compared to the wool samples.

The results indicate no significant difference in insulating properties between commercial 'Fiberglas', 'Rockwool' and LFBFA wool. A lower insulation value (reported as a 29°F improvement over air gap) was recorded for wool produced from western pulverized flyash. Duplication of the test with different gathered samples of WPFA wool might be expected to produce results more consistent with the other wool samples. Since convection functions as the major heat transfer mechanism in this test, it is possible that a gap, hole, or space between discrete pieces of "gathered" wool in the test sample provided a convective path for air resulting in reduced insulating efficiency.

Corrosion Test

Preliminary testing was performed to determine the effects of close contact of mineral wool insulating fibers with metal surfaces and to indicate if any special corrosion problems exist with LFBFA. Certain applications may require contact with steel, zinc (as in galvanized pipe or electrical conduit), copper, lead, iron, aluminum and other metals. These metals are normally found exposed in plumbing and wiring systems within walls where insulation is likely to be used. Steel was chosen to represent the effects of wool contact with metal surfaces. For the test program, six one-inch wide strips were cut from a 22 gauge sheet of mild steel. The surface of each strip was treated by removing all grease and oil with trichlorethylene, pickling in dilute hydrochloric acid and grinding one side with 60 grit paper in a

Table 1

Insulation Properties

2-1/2" Thickness Insulating Wool Type	Top Temperature °F	Bottom Plate Temperature °F	Temperature Difference °F	°F Improvement Over Air Gap	% Improvement Over Air Gap
'Fiberglas'	198	512	314	37	13
'Rockwool'	196	512	316	39	14
LFBFA	196	512	316	39	14
W. PFA	206	512	306	29	10
Air Gap Control	235	512	277	0	0

belt sander. These steps were performed to insure uniformity of initial corrosion (by pickling), and to present an identical surface (by grinding) for subsequent corrosion. Each strip was numbered as shown in Figure 7 and a piece of wool 4" x 3" x 2" thick was prepared for five of the strips. The sixth strip was left uncovered as a control. Table 2 indicates the source of each wool material and the weight of the test strips before and after 49 days exposure to 90-100 percent relative humidity at 70°F.

Figure 7

Corrosion Samples

The data presented in Table 2 indicates that aside from the control strip #6, strip #1 covered with LFBFA wool lost the least weight, 1.29 percent due to corrosion, while strip #5, covered with wool produced from Eastern pulverized flyash lost the most weight, 3.93 percent. The percent weight loss is seen to differ only slightly among the remaining strips, 2, 3 and 4. It is important to note that such high humidity is not likely to be encountered in normal use except in cases where the material is improperly placed or misapplied.

Table 2

Corrosion Test Weight Loss

Source Material	#	Initial Weight	Final Weight	% Weight Loss
LFBFA	1	23.3 gr.	23.0 gr.	1.29
'Rockwool'	2	22.9	22.1	3.49
'Fiberglas'	3	24.0	23.1	3.75
Western PFA	4	22.0	21.2	3.64
Eastern PFA	5	22.9	22.0	3.93
Control	6	22.2	22.0	0.90

Acid-Base Leaching

It was desired to determine if contacting insulating wool with water would produce undesirable acidic or basic solutions. A test was designed in which 10 gram samples of mineral wool and commercial wool were immersed in beakers containing 500 mls of distilled water for one week and pH measurements made at increasing time intervals as shown in Table 3.

As can be seen from the table, wool produced from limestone fluidized bed fyash (LFBFA) and commercial 'Rockwool' both caused the water to become slightly acidic initially while the water containing 'Fiberglas' wool was mildly basic. After a week in the water the pH had stabilized to slightly alkaline for all three samples. These results suggest no significant difference in aqueous acid-base reaction among the wools tested.

Table 3

pH Leaching Study

Time Interval	LFBFA Wool	'Fiberglas' Wool	'Rockwool'
1 Minute	6.5	8.8	6.4
2 Minutes	6.7	8.8	6.5
3 Minutes	7.0	8.9	6.7
4 Minutes	7.1	8.9	6.7
5 Minutes	7.2	8.9	6.7
2 Hours	9.0	8.8	7.3
7 Days	7.7	7.6	7.8

Flame Test

A flame test was performed in which three different samples of insulating wool fibers were held 6" above a Fisher laboratory burner for approximately one minute at a temperature of 1025°C. Exposed edges of the 'Fiberglas' sample melted more quickly than either 'Rockwool' or LFBFA wool. All three wools melted to some extent, however, none supported combustion.

Synopsis of Results

Blown insulating wool produced from limestone fluidized bed flyash and pulverized flyash from the Eastern and Western United States coals are nearly identical in physical properties to commercial 'Rockwool'. Shot content of laboratory blown wool samples was higher than the commercially blown product, but insulating values were found to be nearly identical.

Drawn glass fiber insulation ('Fiberglas'), differs most obviously from blown wool fiber in the absence of shot due to the different method of manufacture. Also, 'Fiberglas' produced for home insulation is lower in bulk density than blown wool fiber although the insulating properties of each are comparable.

THE RELATIVE CONTRIBUTION OF
POINT SOURCES OF AMMONIA TO SURFACE WATERS

William J. Rue, Jr.
James A. Fava
Environmental Toxicology and Chemistry
Ecological Analysts, Inc.
Sparks, Maryland 21152

INTRODUCTION

In January 1980, EPA proposed adding ammonia to the toxic pollutant list, citing the extent of point source discharges as partial evidence for the need to upgrade ammonia's present non-toxic non-conventional pollutant classification. The regulatory effect of the proposed listing would make economic (301(c)) and water quality based (301(g)) waivers from BAT unavailable to industrial dischargers. In response to the Agency's proposed action, an assessment was made of the point sources of ammonia to determine whether more restrictive controls would result in a significant reduction of ammonia discharged into surface waters.*

The industries considered in this paper include all of those listed in EPA's proposal to add ammonia to the toxic pollutant list [1], as well as those industrial point-source categories with Federal effluent guidelines and/or standards for the direct discharge of ammonia [2]. The calculation methodologies employed consisted of combining industry production data with EPA's best practicable control technology currently available (BPT) 30 consecutive day averages. In the cases of POTWs, meat processing and leather and tanning,

---

*This work was funded in part by The Fertilizer Institute and the American Petroleum Institute. A comprehensive review of the sources, chemistry, fate and toxicity of ammonia in surface waters may be obtained through the authors.

average final effluent concentrations were used in conjunction with final effluent flow volumes.

POINT SOURCE DISCHARGERS

Iron and Steel Industry

Maximum ammonia discharges from the iron and steel industry can best be estimated using EPA's October 1979 draft effluent limitation guidelines. There are two basic sources of ammonia from the steel industry--by-product coke making and blast furnace operations. For each of these categories, the draft development document contains information on both capacity and Best Practicable Control Technology (BPT) effluent limitations.

For by-product coke making:

72.1 million tons per year x 0.0921 pounds per thousand pounds (BPT) = 6,640 tons total ammonia per year.

For blast furnace operation:

117.3 million tons per year x 0.0651 pounds per thousand pounds (BPT) = 7,636 tons total ammonia per year.

Therefore, total ammonia in effluents from the iron and steel industry are estimated to be 14,276 tons per year. Assuming 365 working days per year, this value converts to 78,225 pounds per day ($\equiv$ 64,145 pounds total ammonia nitrogen per day).

Fertilizer Industry

Estimates of ammonia in fertilizer plant effluents were computed based on 1978 production statistics for ammonia, ammonium nitrate, urea in solution, and urea solids combined with Best Practicable Control Technology (BPT) effluent reduction factors [3]. Using these data, the U.S. fertilizer industry discharges an estimated 6,565.3 tons of ammonia nitrogen annually (Table I). Assuming 365 working days per year, this annual value converts to 35,974 pounds per day.

TABLE I - ESTIMATES OF AMMONIA IN FERTILIZER INDUSTRY EFFLUENTS

Best Practicable Control Technolocy (BPT)
Point Source Ammonia Discharges

Fertilizer Product	1978 Production in tons [4]	BPT Factor (#/1,000 #)	Calculated Annual Ammonia Nitrogen Discharge (tons)
Ammonia	16,950,654	0.0625	1,059.4
Ammonia nitrate	7,217,539	0.39	2,814.8
Urea--solution	2,768,960	0.48	1,329.1
Urea--solid	2,308,459	0.59	1,362.0

Total Estimated Annual Ammonia Nitrogen Discharge = 6,565.3

Petroleum Industry

Two hundred eighty-five petroleum refineries are currently registered in the United States and its possessions, with a daily production of approximately 14,739,000 barrels of crude oil [5]. Of these 285 refineries

1. 182 are direct dischargers generating a total effluent of 312 million gallons per day (MGD),

2. 48 are indirect dischargers generating a total effluent of 61 MGD, and

3. 55 are zero dischargers [6].

A comparable flow value was calculated by combining the BPT approach presented in Table II with the daily feedstock utilization rate (14,739,000 bbls) to produce a direct discharge flow rate of 328.9 MGD.

Estimates of ammonia in petroleum industry effluents were computed using BPT 30-day average effluent limitations and median BPT wastewater flow rates. As presented in Table II, these subcategory computations combined to generate an industry weighted total ammonia nitrogen concentration of 6.22 mg/l. Using this concentration with the BPT-derived direct-discharge flow value (328.9 MGD) resulted in an estimated total ammonia nitrogen discharge value of 17,070.8 pounds per day.

TABLE II. ESTIMATION OF AVERAGE EFFLUENT AMMONIA CONCENTRATION FOR THE PETROLEUM REFINING INDUSTRY ACHIEVABLE BY BPT TREATMENT

Subcategory	Average BPT Flow[7] (gal/bbl)	Daily Average NH$_3$-N Load[8] (lbs/1,000 bbl)	Estimated Average NH$_3$-N Concentration(a) (mg/l)	Percent of Plants of Total Industry[9]	Average NH$_3$-N Concentration Contribution From Each(b) Subcategory (mg/l)
A - Topping	20	0.45	1.8	38	0.68
B - Cracking	25	3.0	9.6	40	3.84
C - Petrochemical	30	3.8	10.1	9	0.91
D - Lube	45	3.8	6.8	8	0.54
E - Integrated	48	3.8	6.3	4	0.25
Total					6.22 (mg/l)

Industry Average NH$_3$-N Concentration = 6.22 mg/l

(a)Calculated using Columns 1 and 2 and monthly variability factor of 1.5[10] for NH$_3$-N; i.e., for Subcategory A estimated average NH$_3$-N concentration = $\dfrac{0.45 \times 10^{+3}}{20 \times 8.34 \times 1.5}$ = 1.8.

(b)Calculated by multiplying Columns 4 x 5 and dividing by 100, for percent values in Column 5.

This petroleum industry ammonia contribution was verified by combining the flow-weighted final effluent concentration (6.42 mg/l) calculated from data presented in the 1979 EPA development document with the direct discharge flow rate presented above (312 MGD), to generate an estimate of 16,714.4 pounds of ammonia per day.

Although the two approaches do not produce identical final discharge values, considering the different procedures used to obtain these estimates, the final total ammonia nitrogen discharge values (17,070.8 vs. 16,714.4 pounds per day) are very close.

Meat Processing Industry

Estimates of ammonia in effluents from the meat processing industry were provided by U.S. EPA's Effluent Guidelines Division [11]. Approximately 100 NPDES direct dischargers generate an estimated 200,000 gallons per day total effluent containing 40 mg/l total ammonia. These figures yield an estimated 6,640 pounds of total ammonia per day (5,445 pounds total ammonia nitrogen per day) from the red meat processing industry.

Leather and Tanning Industry

Estimates of ammonia in leather industry effluents are based on EPA's effluent development document [12], which indicates that 18 leather tanning and finishing industry dischargers generated 5 million gallons per day total effluent containing 100 mg/l ammonia nitrogen [13]. These figures yield an estimated 4,150 pounds of ammonia nitrogen per day from the leather tanning and finishing industry.

Inorganic Chemicals Industry

Ammonia is listed as a regulated pollutant in effluents from the ammonium chloride production subcategory. An estimate of the amount of ammonia discharged into surface waters from this industry was obtained by multiplying annual ammonium chloride production data by EPA's BPT effluent guideline value of 4.4 pounds of ammonia nitrogen per thousand pounds of ammonium chloride produced. Using this approach and assuming 365 production days per year, the estimated 25,000 tons of ammonium chloride produced annually [14] would result in $\sim$ 603 pounds of ammonia nitrogen per day being released to surface waters.

Nonferrous Metals Manufacturing

Estimates of ammonia in effluents from the nonferrous metals manufacturing industry were provided by U.S. EPA's Effluents Guidelines Division [15]. Of the eight industrial subcategories regulated, only the secondary aluminum smelting industry lists an effluent limitation for ammonia. Combining the sum of the production capacities from the four existing secondary aluminum smelting plants ($\sim$ 195 million tons per year), with the EPA's published BPT value (0.01 pounds of ammonia nitrogen per 1,000 pounds of product), an estimate of 1,950 pounds of ammonia nitrogen per year is calculated. Assuming 365 production days per year, the ammonia contribution from this industry is estimated to be 5.34 pounds per day.

Ferroalloy Manufacturing Industry

Within the ferroalloy manufacturing industry, ammonia is listed as a regulated pollutant in effluents from both the electrolytic manganese products subcategory and the electrolytic chromium subcategory. BPT limits are published for each. An estimate of the amount of ammonia nitrogen contributed by the ferroalloy manufacturing industry to surface waters was provided by EPA's Effluent Guidelines Division [15]. Based on BPT values and industry production data, it was computed that the electrolytic manganese and electrolytic chromium subcategories annually contribute approximately 741 pounds of ammonia nitrogen. Assuming 365 production days per year, the figure yields 2.03 pounds of ammonia nitrogen per day from this industrial point-source subcategory.

Sewage Treatment Plants

The concentration of ammonia in domestic sewage is quite variable. Reported values range from 11 to 46 mg/l [16], with a medium strength sewage containing 25 mg/l of ammonia nitrogen [17]. Ammonia concentrations in treated effluents had an average total ammonia nitrogen concentration of 15 mg/l [17].

Based on EPA's statistics [18], the nation's 14,592 publicly owned treatment works generated 26 billion gallons of effluent per day. Assuming these facilities provide secondary treatment with a conservatively estimated ammonia-nitrogen concentrations of 15 mg/l, POTWs generate approximately 3,237,000 pounds of ammonia nitrogen per day.

Point-Source Discharges: Industrial Versus POTWs

Table III summarizes the industrial point-source estimates computed in this analysis for sewage treatment plants and major identified industrial dischargers of ammonia. This evaluation indicates that the industrial sources examined contribute < 5 percent of the total point-source fraction, while publicly owned sewage treatment plants comprise >95 percent of the total.

These data indicate that stricter industrial ammonia controls would be of little value in reducing the total ammonia point-source load. This point-source distinction is important because, although industrial sources and publicly owned sewage treatment plants are both identified as dischargers of ammonia, they are not considered identically under the provisions of the Clean Water Act.

Specifically, Section 301(b) requires achievement by July 1, 1977, of effluent limitations for point-sources other than publicly owned treatment works based upon the "best practicable control technology currently available" ("BPT").
By July 1, 1983, effluent limitations based upon "best available technology economically achievable" ("BAT") are required. Publicly owned treatment works ("POTWs") must achieve effluent limitations based upon secondary treatment by July 1, 1977, and based upon "best practicable waste treatment technology" by July 1, 1983 [19].

Although publicly owned treatment works may be required to meet water quality standards established pursuant to any state or federal regulations, they are not subject to limitations based on BAT.

LIMITATIONS OF THE STUDY

It should be noted that although the values presented above are based on the best data available, they are national estimates only, and are not applicable to individual discharges which could be either higher or lower. Secondly, it is acknowledged that other industries (and other subcategories within the industries addressed above) may produce wastewaters containing ammonia. However, because they are not federally regulated, they were not addressed in this study.

TABLE III. ESTIMATES OF POINT SOURCES OF AMMONIA

Point Source	Estimated Contribution by Source (pounds total ammonia nitrogen/day)
Sewage Treatment Plants (POTWs)	3,237,000
Steel Industry	64,145
Fertilizer Industry	35,974
Petroleum Industry	17,071[a]
Meat Processing Industry	5,445
Leather Industry	4,150
Inorganic Chemicals Industry	603
Nonferrous Metals Manufacturing	5
Ferroalloy Manufacturing Industry	2
Total	= 3,364,395

Industry ÷ Total Ammonia Estimate = 3.79 percent

POTW ÷ Total Ammonia Estimate = 96.21 percent

---

[a] Using an alternative procedure, a value of 16,714 pounds was obtained.

NOTE: See text for references to individual calculations.

REFERENCES

1. 45 Federal Register 803.

2. 40 Code of Federal Regulations - Sections 405 through 415.

3. 44 Federal Register 9388.

4. Tennessee Valley Authority, "Fertilizer Trends 1979," National Fertilizer Development Center, Muscle Shoals, Ala., 1980.

5. American Petroleum Institute, Basic Petroleum Data Book, Volume 1 Number 1, 1980.

6. U.S. EPA. Memorandum: "Proposed BAT and BCT effluent limitations for existing sources, new source performance standards, and pretreatment standards for new and existing sources in the petroleum refining point source category," prepared by the Assistant Administrator for Water and Waste Management, 27 August 1979.

7. U.S. EPA. Memorandum: "Draft Development Document including the Data Base for the Review of Effluent Limitations Guidelines New Source Performance Standards, and Pretreatment Standards for the Petroleum Refining Point Source Category," (1978).

8. U.S. EPA, "Final BPT Effluent Limitations Guidelines for Petroleum Refining Industry," 40 Fed. Reg. 21939, (1975).

9. U.S. EPA, 1979. "Development Document for Effluent Limits, Guidelines and Standards for the Petroleum Refinery Point Source Category," EPA 440/1-79/014b, (1979).

10. U.S. EPA, "Development Document for Effluent Limitations Guidelines and New Source Performance Standards for the Petroleum Refining Point Source Category," PB-238 612, (1974).

11. Dysinger, C., Project Officer, Effluent Guidelines Division (Red Meat Processing Industry) U.S. EPA, Personal Communication, 30 April 1980.

12. U.S. EPA, "Development Document for Effluent Limits, Guidelines and Standards for the Leather Tanning and Finishing Point Source Category," EPA 440/1-79/106, (1979).

13. U.S. EPA, Memorandum: "Typical total NH3 levels in treated effluents from the red meat and leather tanning and finishing industries," prepared by M. Turvey, Effluent Guidelines Division. Prepared for A. Rubin, Criteria Branch Office. U.S. EPA, Washington (1979).

14. Martin, E., Project Officer, Effluent Guidelines Division (Inorganic Chemicals Industry, and Glass Manufacturing Industry) U.S. EPA. Personal communication. 13 and 17 November 1980.

15. Williams, P., Project Officer, Effluent Guidelines Division (Nonferrous Metals Manufacturing) U.S. EPA. Personal communication. 13 November 1980.

16. Hansen, A. M. and T. F. Flynn, Jr. "Nitrogen compounds in sewage," in Proceedings of the 19th Industrial Waste Conference, Purdue University, Lafayette, Indiana (1964).

17. Metcalf and Eddy, Inc. Wastewater Engineering, McGraw-Hill, New York (1972).

18. U.S. EPA, "Summary of Technical Data for 1978 Needs Survey," EPA 430/9-79-002, (1979).

19. Hall, R. M., Jr., "The evolution and implementation of EPA's regulatory program to control the discharge of toxic pollutants to the nation's waters," Natural Resources Lawyer 10 (3), 507-529 (1977).

INACTIVATION OF ANAEROBIC RELEASE OF PHOSPHORUS
BY WATER TREATMENT SLUDGES

Yousef A. Yousef, Martin P. Wanielista, Harvey H. Harper, and David B. Jellerson. Department of Civil Engineering and Environmental Sciences, University of Central Florida, Orlando, Florida.

INTRODUCTION

Phosphorus has been shown to be a critical nutrient for nuisance algal and plant growth in many lakes surveyed. It is generally the most limiting nutrient in water bodies because of its scarcity in relation to other major nutrients [1]. Phosphorus in the water column is reused several times by biota before it is removed by sedimentation and becomes part of the sediment. Sediment phosphorus content is generally highest near the sediment-water interface and decreases with depth as illustrated from sediment phosphorus profiles studied by various investigators [2]. Also, high concentrations of phosphorus have been detected in the interstitial water. This may have resulted from anaerobic conditions which usually exist in lake sediments. Bacterial excretion of orthophosphate and addition of organic and inorganic acids that are formed during microbial activity account for the high concentration in the interstitial water [3]. This high concentration of soluble orthophosphate can now be recycled into the aquatic system either directly or indirectly through absorption and release by phytoplankton and zooplankton.

In lakes experiencing significant internal loading of phosphorus, treatment of the sediments will be required. Nutrient inactivation through the use of chemical precipitants such as aluminum, iron or calcium is a relatively new procedure in the treatment of lakes, reservoirs and other standing bodies of water. Aluminum sulfate was used to inactivate lake phosphorus in the Netherlands, Sweden, Wisconsin, Washington and Ohio [4,5]. Also, several other treatments including fly ash for nutrient inactivation were reported in literature [6].

The use of waste products, such as water treatment sludges, in treatment of natural systems such as lakes, retention, detention ponds and percolation systems, if successful, would constitute a significant contribution and would become a very attractive alternative. This paper reports on the effectiveness of potable water treatment sludges to prevent the anaerobic release of phosphorus from lake bottom sediments.

FIELD AND LABORATORY EXPERIMENTATION

Study Sites

Two sites were chosen in the central Florida area to investigate the possibility of sediment nutrient inactivation: Lake Eola and Lake Jessup. Lake Eola is a small land-locked lake located in the heart of downtown Orlando and Lake Jessup is a shallow hypereutrophic lake located in Seminole County, Florida. The bottom of Lake Eola has become covered with an accumulation of loose flocculant partially decomposed organic matter which is easily disturbed. The loose nature of this material makes it difficult for rooted submergent plants to exist, and with the exception of a very small area near the shoreline, no rooted plants of any kind were seen in Lake Eola. In areas near the center of the lake, this organic matter, subjected to long periods of anoxic and reducing conditions, has formed into sapropel, complete with the characteristic hydrogen sulfide smell. Sediment material in Lake Jessup has developed into a loose flocculant "soup" which is very easily disturbed. Yousef, et al. [7] reported that Lake Jessup sediments contained an average of 90.7% moisture content and 33.7 % volatile solids. Similar studies by Marshall [8] showed that the average moisture content varied between 23–43.4% and volatile content between 4–31.4% for shallow and deep areas of Lake Eola, respectively.

Experimental Equipment

In situ experiments were conducted using isolation chambers constructed from heavy duty 200 liter polyethylene containers as shown in Figure 1. All tanks were painted on the outside with a semi-gloss black alky-enamel to prevent light penetration and subsequent photosynthetic activity. Chambers were placed inverted on the lake bottom, isolating 0.25 square meters of surface area. Chemical additions were accomplished through a 3/4 inch diameter tygon tube which extended from the top of each tank to the water surface in the lake. The tubing was connected on the top of the tank to an inverted 28 cm diameter polyethylene funnel, as shown in Figure 1, to insure that chemical inactivants were distributed evenly over the sediment bottom. Visual laboratory observations of chemical additions through these funnels in clear plexiglass tanks indicated that dispersion was relatively uniform. All additions were followed by an additional 10 liters of lake water to wash

all chemicals from the tubing into the isolation chambers and
to also provide for a small amount of mixing by way of agitation inside the tank. The large diameter tubing was then capped with a tight fitting rubber stopper and dropped back into
the water.

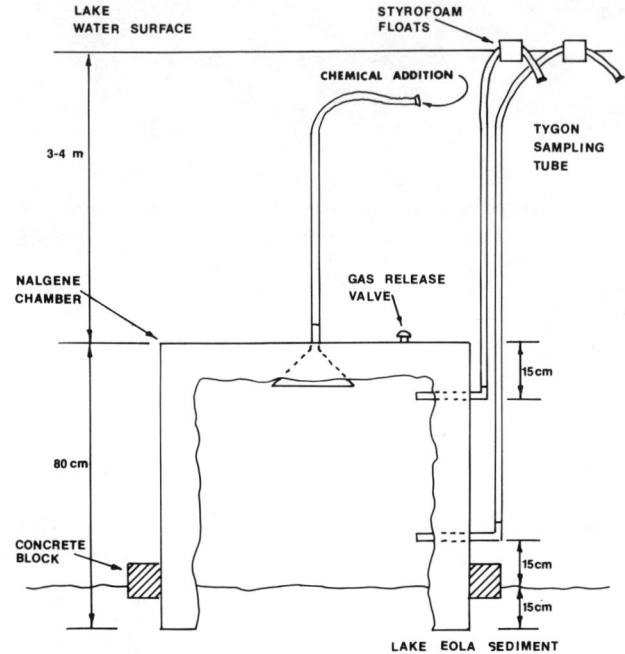

FIGURE 1:   TYPICAL ISOLATION CHAMBER

Samples were collected from the surface tubes using a
peristaltic pump operating at a very low flow rate to minimize
agitation and atmospheric gas exchange within the samples.
Approximately 300-400 ml of sample was initially pumped from
each tube to purge the tubes of any contaminated water and to
make sure that the water which was collected actually came
from inside the chamber. Samples from the top and bottom
ports were combined in equal volumes inside a 500 ml amber
linear polyethylene bottle to form one sample from each chamber on a particular sampling date.

Water Treatment Plant Sludges

A calcium carbonate-based water treatment plant sludge
was obtained from the Clyde Doyle water treatment plant in
Cocoa, and an alum-based water treatment plant sludge was obtained from a treatment process in Tampa, Florida. The moisture content and percent loss on ignition averaged 0.4 and 2.9
percent for the Cocoa sludge. Similarly, Tampa sludge exhibited 89% moisture content and 45.8 percent loss on ignition.
The calcium sludge showed very little moisture content since
it was air-dried in the laboratory with time after collection.

Chemical analysis of the water treatment sludges used as chemical inactivants is presented in Table I. The data shown in this table represent an average of five samples tested.

TABLE I. Chemical Analysis of Water Treatment Plant Sludges

Element	Average Concentration mg/g oven dry weight	
	Cocoa Plant	Tampa Plant
Cd	0.001	.001
Zn	0.006	.056
Cu	0.009	.044
Fe	0.256	8.770
Pb	0.226	0.070
Ni	0.014	0.010
Cr	0.029	0.254
Al	0.379	206.7
Mg	11.40	3.14
Ca	207.2	10.4
P	0.233	0.351
Mn	0.036	0.193
Ba	0.077	0.031

Experimental concentrations of inactivants and corresponding isolation chamber designations are listed in Table II. Dosages were selected to provide a fairly uniform floc. Water samples were collected approximately 24-48 hours after addition of the inactivant after which they were then collected at approximate one-week intervals for one month, with samples collected at 2-3 week intervals thereafter. All samples were returned under refrigeration to the Environmental Engineering laboratory at UCF for analysis.

TABLE II. Experimental Concentrations of Chemical Inactivants

Experiment Number	Location	Chemical Inactivant	Inactivant Dosage	
			(g)	(g/m^2)
1	Lake Eola	CaCO$_3$ sludge	350	1400
	Lake Eola	Alum sludge	2000	8000
	Lake Eola	None	None	----
2	Lake Jessup	CaCO$_3$ sludge	350	1400
	Lake Jessup	Alum sludge	2000	8000
	Lake Jessup	None	None	----
3	Lake Jessup	Alum sludge #1	27.5	110
	Lake Jessup	Alum sludge #2	55	220
	Lake Jessup	Alum sludge #3	110	440
	Lake Jessup	Alum sludge #4	220	880
	Lake Jessup	Alum sludge #5	440	1760
	Lake Jessup	None	None	----

Laboratory Procedures

All samples collected were analyzed for dissolved oxygen, pH, turbidity, specific conductance, alkalinity, ammonia, nitrites, nitrates, dissolved orthophosphorus, total orthophosphorus, and total phosphorus. Heavy metal analysis was also conducted on each sample for the following metals: lead, nickel, copper, zinc, cadmium, chromium, aluminum, iron, and calcium. Analyses were conducted as specified by the U.S. EPA in <u>Methods for Chemical Analysis of Water and Wastes</u> and <u>Standard Methods for the Examination of Water and Wastewater</u> [9,10].

All phosphorus determinations were analyzed using the ascorbic acid procedure. Dissolved orthophosphorus was measured using this technique on samples filtered through Whatman GF/A filters. Total phosphorus was measured on a sample which was first acidified and digested by persulfate digestion.

Heavy metal analyses were performed on concentrated acidified samples using a Spectrometrics Incorporated Spectrospan III Plasma Emission Spectrometer.

RESULTS AND ANALYSIS

Lake Eola

The data collected at Lake Eola indicated that the initial water quality inside all the isolation chambers was essentially the same. The pH values ranged from 8.61 to 8.72, turbidity ranged from 5.8 to 7.1 JTU's, specific conductance ranged from 166 to 181 mho's/cm and the alkalinity ranged from 97.1 to 98.8 mg/l as $CaCO_3$. The initial filterable orthophosphorus, total orthophosphorus, and total phosphorus concentrations ranged from 3 to 5, 17 to 26, and 64 to 73 µg-P/l, respectively. Also, data on water quality characteristics in the open lake adjacent to the isolation chambers' site showed that the ranges during the entire sampling period for dissolved oxygen, pH, specific conductance, turbidity and alkalinity were 9.3-12.6 mg/l, 8.46-8.97 pH units, 182-230 µmho/cm, 4.6-6.7 JTU, and 88.4-103.8 mg/l as $CaCO_3$, respectively. The ammonia, nitrite and nitrate nitrogen varied from 9.7 to 327, below detection limit to 7.3, and below detection limit to 82.1 µg-N/l, respectively. Also, filterable orthophosphorus, total orthophosphorus and total phosphorus in Lake Eola varied from below detection limit to 5, 10 to 18, and 46 to 70 µg-P/l, respectively.

The dissolved oxygen concentration inside the isolation chambers dropped from an initial value averaging 11.2 mg/l to approximately 1.0 mg/l in the first seven days of incubation. Also, the dissolved oxygen (DO) uptake rate during the first seven days averaged 874 mg/sq m/day.

The pH values in isolation chambers dropped from initial values ranging from 8.61-8.72 to 6.90-7.41 after seven days of incubation followed by a slight increase which was generally maintained during the following months of incubation. Specific conductance of water samples collected from isolation chambers before addition of chemicals ranged from 166 to 181 µmho/cm, however, one day later, after adding these chemicals, the values ranged from 221 to 226 µmho/cm. The control station increased from 183 to 214 µmho/cm after the first day with specific conductance increasing gradually with time for the following three months. The maximum values observed were 268 µmho/cm inside the control chamber, 280 inside the chamber dosed with Cocoa plant sludge, and 271 inside the chambers dosed with Tampa wet sludge. The increase in specific conductance may have resulted from the solubilization and release of various ions under anoxic conditions. The maximum increase in specific conductance measured one day after incubation until the end of the study varied between 38 and 57 µmho/cm and averaged 38 µmho/cm during a period of more than four months.

Nitrogen

Ammonia, nitrite and nitrate nitrogen were measured in all samples collected throughout this study. Generally, the ammonia nitrogen concentrations increased gradually from initial values of 22.1 to 32.7 to maximum values of 1558 to 1972 µg-N/l after 140 days of incubation in isolation chambers. The ammonia concentrations in the open Lake Eola outside the isolation chambers were generally lower than 100 µg-N/l. Very small concentrations of nitrite nitrogen ranging from 1.0 to 17.6 µg-N/l were observed during the study period. Also, nitrate concentrations ranged from below detection limits to 165 µg-N/l. The maximum total inorganic nitrogen released varied between 1639 and 2051 µg/l inside the isolation chambers.

Phosphorus

Filterable orthophosphorus, total orthophosphorus and total phosphorus were measured in all samples collected from inside the isolation chambers and from the open lake outside these chambers. The filterable orthophosphorus showed concentrations between below detectable limits and 18 µg-P/l in the open lake and inside isolation chambers treated with alum sludges. The data suggested that alum seemed to effectively inhibit phosphorus release from bottom sediments under anoxic conditions. This was clearly demonstrated by Figure 2 which showed that phosphorus concentrations associated with lake water inside isolation chambers treated with alum sludges were lower than the control chamber. The total phosphorus concentrations were kept below 20 µg-P/l inside alum sludge treated isolation chambers after an incubation period of 140 days, whereas the total phosphorus inside the control chamber and the one treated with calcium carbonate sludge showed a gradual increase to a maximum value of approximately 170 µg-P/1. Similarly, filterable orthophosphorus concentrations were generally below 10 µg-P/l inside alum treated sludges at all times during the study period, whereas the control chamber and the one treated with calcium carbonate showed a gradual increase to a maximum value of approximately 130 µg-P/l.

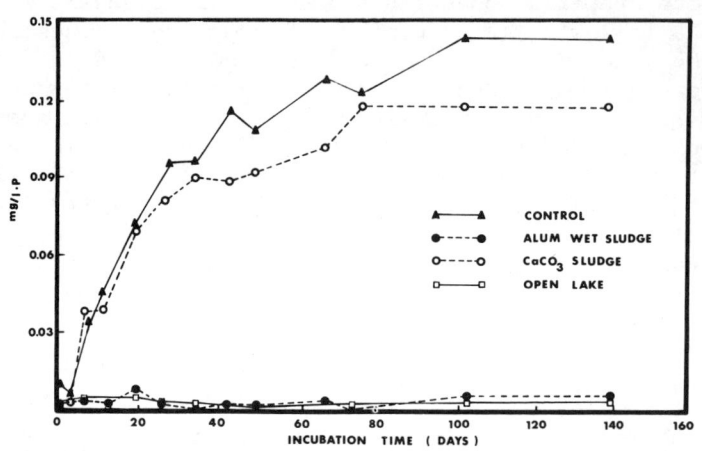

FIGURE 2: EFFECT OF VARIOUS INACTIVANTS ON DISSOLVED ORTHOPHOSPHORUS RELEASED FROM ANAEROBIC BOTTOM SEDIMENTS IN LAKE EOLA.

Lake Jessup

Lake Jessup results generally supported the data obtained from Lake Eola. The alum sludge inhibited the release of phosphorus, while the calcium carbonate sludge did not [11].

Alum sludge was effective in reduction of anaerobic phosphorus release from the bottom sediments of Lake Eola and Lake Jessup to the overlying water. The dosage used in these cases was 2000 grams of wet sludge, from the City of Tampa, Florida, which is equivalent to 880 grams of dry sludge per square meter. This sludge contained 89% moisture. This dosage was excessive but was used to illustrate the usefulness of alum sludge. Another experiment was designed to determine the optimum dosage required to inhibit the release of phosphorus from Lake Jessup bottom sediments. Six isolation chambers were dosed with various concentrations of wet sludge as shown in Table II. This experiment was run for 62 days and changes in water quality parameters inside the anaerobic isolation chambers with various dosages of alum sludge were collected. The data showed the general trends observed in the previous experiments. pH values and turbidity measurements generally decreased, while specific conductance and ammonia nitrogen increased with incubation time under anoxic environment. Phosphorus concentrations increased gradually in the control chamber and declined in isolation chambers treated with alum sludge. It was apparent that alum sludge retained the phosphorus released from the bottom sediments and could also retain the phosphorus associated with the overlying water.

Average concentrations, standard deviations and analysis of variance were calculated for dissolved heavy metals detected in water samples collected from inside the isolation chambers and outside these chambers from the open lake. Most of the results for Lakes Eola and Jessup showed no significant differences. However, Table III indicates that the probability due to chance for calcium and chromium in Lake Eola chambers were generally less than 10%. Calcium concentrations in water samples collected from Lake Eola outside the incubation chambers were lower than those collected from inside the incubation chambers. This trend was reversed in the case of chromium. Chromium inside these chambers was lower than chromium in the open lake. This may be true due to anoxic conditions inside the chambers which result in change of oxidation state for chromium and thus affects its solubility. The release of calcium inside isolation chambers may be associated with the release of phosphorus under anoxic conditions.

TABLE III. Variations in Metal Concentration Detected in Water Samples Collected from Chambers Incubated in Lake Eola

Metal	Station Designation	No. of Observations	Average Concentration (mg/l)	Standard Deviation	AOV Chance Prob.
Calcium	Control	11	44.3	2.5	0.013
	Open lake	8	39.9	3.6	
	Alum sludge	10	46.5	4.4	0.003
	Open lake	8	39.9	3.6	
	$CaCO_3$ sludge	10	44.2	4.0	0.029
	Open lake	8	39.9	3.6	
Chromium	Control	12	0.015	0.012	0.062
	Open lake	8	0.027	0.012	
	Alum sludge	11	0.014	0.015	0.061
	Open lake	8	0.027	0.012	
	$CaCO_3$ sludge	11	0.016	0.018	0.147
	Open lake	8	0.027	0.012	

The results generally conclude that variations in heavy metals in water samples from inside isolation chambers were insignificant and there was no significant release from the sludges to the water environment.

Optimum Dosage

Changes in optimum orthophosphorus released from Lake Jessup bottom sediments treated with various concentrations of water treatment sludges from the Tampa plant under anaerobic environment for two months were presented in Figure 3. The orthophosphorus released was calculated by multiplying the difference between the initial phosphorus concentration in the water column beneath the isolation chamber and the maximum concentration reached during the study period times the water volume of the isolated column and divided by the bottom sediment area. It was interesting to see that the smooth curve shown in Figure 2 intercepted the zero optimum orthophosphorus released in mg-P/sq m at alum sludge dosages between 110 and 220 grams which were equivalent to 2.0-4.0 tons of wet sludge per acre of bottom sediments. Above these dosages, not only was the release of orthophosphorus from bottom sediments inhibited, but the orthophosphorus content in the overlying water column was also reduced.

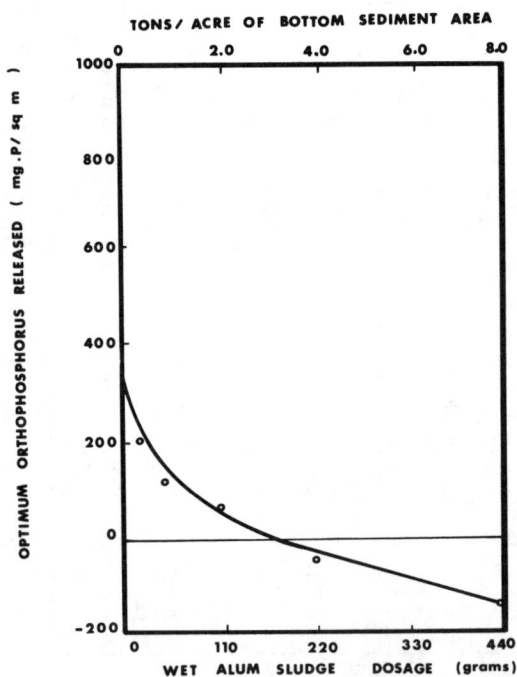

FIGURE 3: CHANGES IN OPTIMUM ORTHO-PHOSPHORUS RELEASED WITH VARIOUS CONCENTRATIONS OF ALUM SLUDGE DOSAGE.

To normalize the sludge dosage, a graph was constructed on a log-log scale between the orthophosphorus retention ratio (ORR) and alum sludge dosage. The ORR was calculated as the maximum orthophosphorus retained in mg-P/sq m during the study period divided by sludge dosage in grams of oven-dry sludge per square meter of bottom sediment treated. The maximum retained orthophosphorus reflected the difference between orthophosphorus concentration in the treated isolation chamber and the control chamber at the end of 62 days of incubation, as shown in Table IV. The relationship between ORR and alum sludge application rate was represented as follows:

$$Y = 434.2 \, X^{-0.86}$$

where: Y = orthophosphorus retention ratio, mg-OP/g sludge
         X = oven-dry alum sludge dosage, gm/sq m

TABLE IV. Changes in Dissolved Orthophosphorus Retention with Applied Various Alum Sludge Dosages

Sludge Dosage (gm/sq m)		Orthophosphorus Concentration (mg-P/l)		Ortho-Phosphorus Retained (mg/sq m)	Phosphorus Retention Ratio
Wet	Dry	Control Chamber	Treated Chamber		
110	12.1	1.57	0.566	602.4	49.8
220	24.2	1.57	0.427	685.8	28.3
440	48.4	1.57	0.234	801.6	16.6
880	96.8	1.57	0.234	801.6	8.3
1760	193.6	1.57	0.039	918.6	4.7

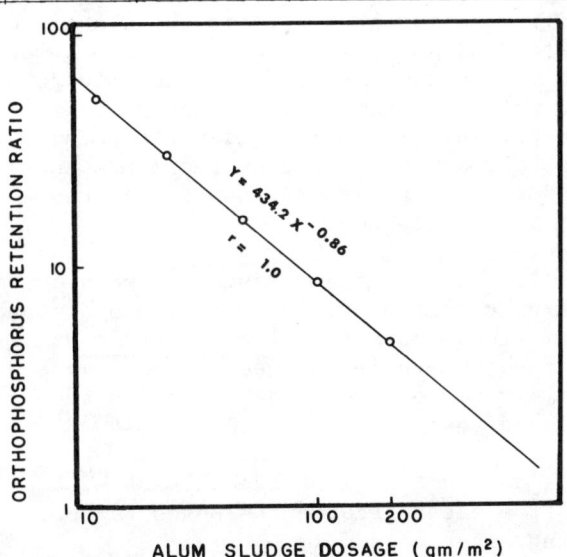

FIGURE 4: PHOSPHORUS RETENTION BY VARIOUS DOSAGE OF ALUM SLUDGE TO TREAT ANAEROBIC SEDIMENTS

SUMMARY AND CONCLUSIONS

The feasibility of using water treatment sludges to inactivate anaerobic phosphorus release from lake bottom sediments has been studied. The sludges used were calcium carbonate based sludge from Cocoa plant and alum based sludge from Tampa plant. The sludges were dosed through plastic funnels inserted inside nalgene chambers isolating approximately 150 liters of the water column overlying a 0.25 square meter of bottom sediments in Lakes Eola and Jessup of the Central Florida area. From these studies, the following conclusions were reached:

1. Alum sludge was effective for inactivation of anaerobic release of phosphorus from bottom sediments. However, the calcium sludge was not effective.

2. The control isolation chambers showed a gradual increase in orthophosphorus concentrations in water and reached maximum values within two to three months of incubation time. The maximum orthophosphorus released was 136 mg-P/sq m for Lake Eola sediments and 805 mg-P/sq m for Lake Jessup sediments. Generally, all phosphorus released appeared to be in the orthophosphorus form.

3. The orthophosphorus and total phosphorus released from bottom sediments to the contained lake water column inside the isolation chambers treated with various dosages of alum sludge decreased with increasing sludge dosage. In some cases, the sludge inhibited the release of phosphorus from the bottom sediments and also reduced the phosphorus content in the overlying water column.

4. An excellent correlation existed between orthophosphorus retention ratio by bottom sediments and alum sludge dosage. Alum retention of orthophosphorus varied between 4.7 and 49.8 mg-P/gm sludge (oven dry) depending on the alum sludge dosage.

5. The data does not suggest the release of heavy metals from water treatment sludges to the overlying water column.

ACKNOWLEDGEMENT

The authors wish to acknowledge the financial support granted by the Engineering and Industrial Experiment Station at the University of Central Florida. Without this support, it would have been impossible to complete this research.

REFERENCES

1. U.S. Environmental Protection Agency, "National Water Quality Inventory/1975 Report to Congress", EPA-440/9-75-014, Washington, D.C., Government Printing Office, (1976).
2. Stewart, E.A., "A Study of Differences in Vertical Phosphorus Profiles within the Sediments of Selected Florida Lakes as Related to Trophic Dynamics", M.S. Thesis, Civil Engineering and Environmental Sciences Department, Florida Technological University, Orlando, (1976).
3. Fillos, J., and W. Swanson, "The Release Rate of Nutrients from River and Lake Sediment", J. Water Pol. Con. Fed., 47 (5), pp. 1032-1042, (1975).
4. Dunst, R.C., S.M. Born, P.D. Uttormark, S.A. Smith, S.A. Nichols, J.O. Petersen, D.R. Knauer, S.L. Serns, D.R. Winter and T.L. Winth, "Survey of Lake Rehabilitation Techniques and Experiences", Department of Natural Resources Bulletin, No. 75, Madison, WI, (1974).
5. Peterson, S.A., W.D. Sanville, F.S. Stay and C.F. Powers, "Nutrient Inactivation as a Lake Restoration Procedure", EPA-660/3-74-032, Environmental Resources Center, Corvallis, OR, (1974).
6. Funk, William H., and Harry L. Gibbons, "Lake Restoration by Nutrient Inactivation", published in Proceedings of a National Conference on Lake Restoration, August 22-24, 1978, Minneapolis, MN, EPA 440/5-79-001, pp. 141-151, (1979).
7. Yousef, Yousef A., Waldron M. McLellon and Herbert H. Zebuth, "Changes in Phosphorus Concentration Due to Mixing by Motorboats in Shallow Lakes", Water Research, Vol. 14, pp. 841-852, July, (1978).
8. Marshall, Frank E., III., "Phosphorus Dynamics of Lake Eola Sediments", Thesis, University of Central Florida, Orlando, (Winter, 1980).
9. U.S. Environmental Protection Agency, "Methods for Chemical Analysis of Water and Wastes", EPA-625/6-74-003, Office of Technology Transfer, Washington, D.C., (1976).
10. American Public Health Association. Standard Methods for the Examination of Water and Wastewater, 14th ed., (1975).
11. Yousef, Y.A., H.H. Harper, and D.B. Jellerson, "Inactivation of Lake Sediment Release of Phosphorus", a Final Report submitted to the UCF Engineering and Industrial Experiment Station, Research Grant #111699034, Dec. 1980.

Volatilization Problem: Land Disposal of Industrial Wastes

Thomas T. Shen
New York State Department of Environmental Conservation
Albany, New York 12233

Introduction

Volatilization is a process by which a substance is transferred from a liquid or solid phase to a vapor phase. It occurs readily from most organic compounds at a relatively low temperature. The rate of volatilization of an organic compound depends mainly on its vapor pressure, molecular weight, diffusion coefficient, mass transfer coefficient, concentration, exposed surface area of the specific compound, and the surrounding environment such as temperature, pressure, and wind speed. Volatilization also involves desorption of vapors from absorbents such as soils or other materials. Generally, low molecular weight organic compounds are more volatile than high molecular weight organic compounds. Recent studies have revealed that even very stable chlorinated hydrocarbons with very low vapor pressure, such as PCB's and pesticides, do volatilize (1). The results of field monitoring data have shown also that PCB concentrations were fairly high in the ambient air and in vegetation near the PCB dump sites and certain contaminated dredge spoil sites along the Upper Hudson River of New York State (2). Thus, volatilization of hazardous wastes from lagoons and landfill sites can be an important source of air pollution.

In New York State, there are approximately 4,000 industrial sources producing hazardous wastes at a rate of about 1.2 million tons annually (3). Large quantities of such wastes are acids, pickling liquor, organic solvents and sludges as shown in Table 1. The predominate practice of disposal has been lagoons and landfills. There is an increasing concern regarding continuous emissions of organic compounds containing toxic substance from industrial waste lagoons and landfill sites. Atmospheric transport is the major pathway by which toxic emissions are carried to locations far from those sites into urban areas. They are transported either as a vapor or absorbed/adsorbed onto fine particles and eventually deposited on land or into bodies of water by particle fallout and precipitation.

The identification of air quality degradation by waste volatilization from such lagoons and industrial waste landfills requires air monitoring. However, routine monitoring of toxic air contaminants associated with land disposal of hazardous wastes is costly and it is generally not done unless there is an immediate threat to public health.

This paper introduces empirical equations to determine volatilization problems by calculating emission rates of toxic contaminants and their ambient concentration. Calculation of essential variables such as vapor pressure, equivalent vapor concentration, saturation vapor concentration, diffusion coefficient are also illustrated. The methods presented in this paper would be useful for an air quality screening process in the absence of monitoring data. A case study of the Caputo PCB waste dump site near South Glens Falls, New York is presented to illustrate the application of empirical equations associated with industrial waste volatilization problems.

Emission Rate from Landfills

The volatilization or emission rate of organic compounds from industrial waste buried sites can be predicted, assuming that diffusion in the vapor phase is the only transport process operating. If transport in moving water and degradation of the organic compound in the site are considered insignificant, the volatilization and emission rate of a specific organic compound can be estimated using the following equation:

$$E_i = D_i C_{si} A P_t^{4/3} W_i (1/L) \qquad (8)$$

Where:
- E_i = emission rate of a specific compound in the waste (g/sec);
- D_i = diffusion coefficient of a specific compound (cm^2/sec);
- C_{si} = saturation vapor concentration (g/cm^3);
- A = exposed area (cm^2);
- P_t = total soil porosity (dimensionless);
- L = effective depth of soil cover (cm); and
- W_i = weight percent of a specific compound in the waste (%).

The soil porosity may be calculated by the following equation:

$$P_t = 1 - d_b/d_p \qquad (9)$$

Where:

d_b = soil bulk density (g/cm^3); and
d_p = particle density (g/cm^3).

Air-filled porosity is the significant soil parameter affecting the final steady state mass flux through a soil cover. The bulk density and volumetric water content of the soil determines its air-filled porosity. Consequently, highly compacted wet soil covers are most effective in reducing volatilization. Other soil parameters such as organic matter content and texture, which may affect adsorpotion and absorption by soil, will influence the emission rate. Experiments of hexachlorobenzene (HCB) volatilization showed that increasing the relative air-filled porosity 13.4% increased specific HCB volatilization rate 45%, and that air-filled porosity has an exponential effect on the HCB volatilization rate through soil (7).

In general, the particle density of most soil is about 2.65 g/cm^3 and the soil bulk density varies between 1.0 and 2.0 g/cm^3. If the soil is assumed to be dry and has a bulk dnsity of 1.2 g/cm^3, the soil porosity would be 0.55 which represents a maximum volatilization rate. If the soil is assumed to be compacted or has a high moisture content, its porosity may decrease to 0.35.

Emission Rate from Dump Sites

Dump sites are landfills which have no covering material and cause greater volatilization problems. Based on Arnold's (8) diffusion equation, the volume of vapor generation of a pure organic compound under steady state conditions may be calculated by the following equation:

$$V = 2 \, C_e A (Dt/\pi F_v)^{\frac{1}{2}} \tag{10}$$

Where: V = volume of vapor (cm^3);
t = time (sec);
C_e = equilibrium vapor concentration (%);
A = exposed area (cm^2);
D = diffusion coefficient (cm^2/sec); and
F_v = correction factor (see Fig. 1).
Ref. 4

The rate of toxic emissions at dump sites is strongly affected by wind speed. Zigler (9) modified Arnold's equation to include wind speed for estimation of the rate of volatilization. His modified equation for estimation of emission rate from dump sites is as follows:

$$dV/dt = 2 \, C_e \, W \, (DLv/\pi F_v)^{\frac{1}{2}} W_i \tag{11}$$

Where:

dV/dt = emission rate (cm^3/sec);
W = width of the dump site (cm);
L = longest dimension of the dump site (cm);
v = wind speed (cm/sec); and
W_i = weight percent of a specific compound in the waste (%).

A strong wind can increase the emission rate; however, it also increases the dilution factor. Thus, the net-effect of wind speed on ambient concentration becomes compensative and depends on location of the receptor. It can be computed by the PAL atmospheric diffusion model (10).

Concentration

Once the emission rate of a toxin release from lagoons or landfills into the atmosphere is known, the ambient concentration of which can be calculated by the computer programmed PAL atmospheric diffusion model (10). PAL is a multisource Gaussian-Plume atmospheric dispersion used directly in the computation for point, line, and curved path sources, and in a modified form for area sources. Concentration calculations are based on hourly meteorology, and averages can be computed for averaging times from 1 to 24 hours. The PAL model was not designed for estimating concentrations from waste disposal sites but it is a very useful tool which can be applied as a screening process to predict ambient toxin concentrations from the waste disposal sites under various meteorological conditions at various locations of receptors.

Since toxin release from the site is continuous and the duration of release is much greater than travel time, ground level concentrations along the centerline of the plume, with no effective plume rise, may be calculated by the following simplified equation (9):

$$X = Q/\pi \, \sigma y \, \sigma z \mu \qquad (13)$$

Where: X = the toxin concentration (g/m^3);
Q = the uniform emission rate of the toxin (g/sec);
σy = standard deviation of horizontal plume concentration (m);
σz = standard deviation of vertical plume concentration (m); and
μ = wind speed (m/s).

Values of σy and σz can be found readily from Turner's workbook of atmospheric dispersion estimates under various atmospheric stability classes (11).

Case Study

The Caputo land disposal site was an open dump located near South Glens Falls, New York. The exposed surface area of the site was estimated to be about 35,000 square meters as shown in Figure 2. The irregularly shaped site has its longest cross dimension of 300 meters and short cross-dimension of 180 meters approximately. The site was used until 1968 to receive polychlorinated biphenyls (PCBs) wastes from two General Electric capacitor plants. The in-place wastes were estimated to contain about 5,000 ppm of PCBs as Aroclor 1242 in soil. Ambient air sampling for PCBs was conducted at the site beginning in early November of 1977. Results of the air monitoring are presented in Table 4. The maximum ambient concentrations of PCBs were 5.9 $\mu g/m^3$ in November of 1977, 300 μ/m^3 in August of 1978, and 540 $\mu g/m^3$ in April of 1979. After covering with layers of 6 inches of manure, 12 inches of paper mill sludges, and 2 inches of topsoil beginning in late May of 1979, the ambient concentration of PCBs were greatly reduced to less than 1.0 $\mu g/m^3$ in June of 1979.

To estimate the potential emission of PCBs at the Caputo site, Equation (11) was applied as follows:

$$dV/dt = 2 C_e W(DLv/\pi F_v)^{1/2} W_i$$

Where: W = 18,000 cm_2 and L = 30,000 cm
D = 0.0519 cm^2/sec at 30°C from Equation (4)
C_e = p/P = $4 \times 10^{-3}/760$ = 0.53×10^{-5}
v = 400 cm/sec (assumed)
F_v = 1 (from Reference 4)
W_i = 0.005

By substituting these know values into Equation (11), we get

$dV/dt = 2 \times 0.53 \times 10^{-5} \times 30,000 (0.0519 \times 18,000 \times 400/\pi \times 1)^{1/2} \times 0.005$
      = 0.548 cm^3/sec.

This volumetric emission rate may be converted into mass emission rate Q as below.
$Q = 0.548 \times (258 \times 10^6 \mu g/mole)/(24,860 cm^3/mole) = 5,687$ $\mu g/sec$ @ 30 C.

By applying Equation (13) for ground level concentrations, the ambient concentration of PCBs will be

$$X = 5,687/\pi(7)(4.7)(4) = 13.76 \ \mu g/m^3 \gg \text{Acceptable ambient level}$$

Assuming "D" atmospheric stability class and wind speed 4 m/s, the values of $\sigma y=7$ and $\sigma z=4.7$ can be found from Figure 3-2 and Figure 3-3 of Reference (11).

The calculated ground level concentration is based upon soil temperature of 30°C, windspeed 4 m/s, atmospheric stability class D, and in-place PCB concentration of 5,000 ppm. Since these assumed values may involve appreciable errors, one should consider the results of emission estimation as a first approximation for determining the extent of volatilization problems of land disposal sites.

When volatile emissions exceed the acceptable ambient concentration such as the above case, the best immediate short-term remedial action would be to place a cover above the wastes. Equation (8) may be used to calculate the depth of cover that will produce an acceptable level of landfill emissions.

$$L = D\, C_s\, A\, P_t^{4/3}\, W_i/E \qquad (14)$$

If we use 0.1 µg/m^3 as an acceptable ambient concentration of PCB's, a minimum depth of cover may be computed as follows:

$$Q = X\pi\, \sigma y\, \sigma z\, \mu = 0.1 \times \pi \times 7 \times 4.7 \times 4 = 41.34\ \mu g/sec$$
$$L = 0.0519 \times (5.5 \times 10^{-2}) \times (3.5 \times 10^{8}) \times 0.4^{4/3} \times$$
$$0.005/41.34 = 42.6\ cm$$

Where C_s = pM/RT = $4 \times 10^{-3} \times 258/62.3 \times 303 = 5.5 \times 10^{-5}$ g/l
$= 5.5 \times 10^{-2}\ \mu g/cm^3$.

In June 1979, the site was covered with a total of 50.8 cm (20 inches) of 3 layers consisting of manure, paper mill sludges, and topsoil as previously mentioned. The calculated ambient concentration of PCB after covering the site ws 0.083 µg/m^3 based on the following computation.

$$E = 0.0519 \times 5.5 \times 10^{-2} \times 3.5 \times 10^{8} \times 0.4^{4/3} \times$$
$$0.005/50.8 = 34.6\ \mu g/sec.$$
$$X = 34.6/\pi \times 7 \times 4.7 \times 4 = 0.083\ \mu g/m^3$$

The ambient monitoring data for PCBs showed to have been greatly reduced to less than 0.7 µg/m^3 as a result of applying the 20-inch cover.

Volatilization of PCBs from soils will increase exponentially with increases in soil temperature due to the effect of increasing vapor pressure of the PCBs. For example, each 10 C rise in soil temperature will increase PCB (1C1) vapor pressure more than double as indicated in Table 2. Thus, a disposal site located in an area where soil temperature is 10 C warmer than Glens Falls of New York would have to use a soil cover at least twice thicker to achieve the same degree of emission reduction, assuming everything else is constant.

Summary and Conclusion

Air pollution from hazardous waste landfills and lagoons is largely unknown. Toxins as a vapor or absorbed/adsorbed onto fine particles can be carried to locations far from the disposal sites and finally deposited on land or into bodies of water by particle fallout and precipitation.

Routine monitoring of toxic air contaminants associated with industrial waste lagoons and land disposal sites is costly, the methods and empirical equations presented in the paper would be useful for air quality assessment in the absence of monitoring data. As indicated in the case study, the estimated emission rate and ambient PCB concentrations, before and after the cover, appear to be satisfactory when compared with actual monitoring data which also showed wide variations in the order of 10 to 50 times. It is concluded, therefore, that the empirical equations are useful as a screening process to examine the question of whether or not volatilization is to be significant for a given contaminant. And also, they may be valuable to evaluate permit applications concerning waste volatilization problems.

TABLE 1
MAJOR VOLUMES OF HAZARDOUS WASTES BY TYPE IN NEW YORK STATE

Type	Approximate Hazardous Waste Generation (Million Gallons)	% of Total From Survey
1. Acids	19.0	7.7
2. Alkalies	2.4	1.0
3. Pickle Liquor	14.6	5.9
4. Etching Solution	1.1	0.5
5. Halogenated Solvents	1.0	0.4
6. Other Solvents	50.7	20.6
7. Heavy Metal Sludges	14.5	5.8
8. Wastewater Treatment Sludges	52.4	21.3
9. Other Sludges	16.6	6.8
10. Waste Oils	15.7	6.4
11. Still Bottoms	3.3	1.4
12. Photographic Chemicals	1.9	0.8
13. Rinse Water & Other Wastewaters	22.9	9.3
	216.1	
Use 10 lb/gal conversion to tons	1,080,500 tons subtotal	
14. Leather tanning wastes	59,890 tons subtotal	4.9
Grand Total	1,140,390 tons	92.8

which is @ 92% of Total Hazardous Waste Generation from Survey

Table 4

Ambient Monitoring Data at Caputo Site

Date	No. of Samples	Ground Temp., °F	PCB Concentration, ug/m3
Aug. 1978	3	75-85	300, 246, 260
Sept. 1978	2	75	46, 10
Nov. 1978	2	35	18, 4
Apr. 1979	7	65-72	540, 0.90, 0.80, 390, 60, 350, 80
May 1979	3	60-85	44.5, 0.08, 0.07
June 1979*	3	95	0.68, 0.31, 0.36
July 1979*	1	90	0.07
Sept. 1979	3	80	0.11, 0.12, 0.19

*After covered with layers of manure, paper mill sludges and topsoil.

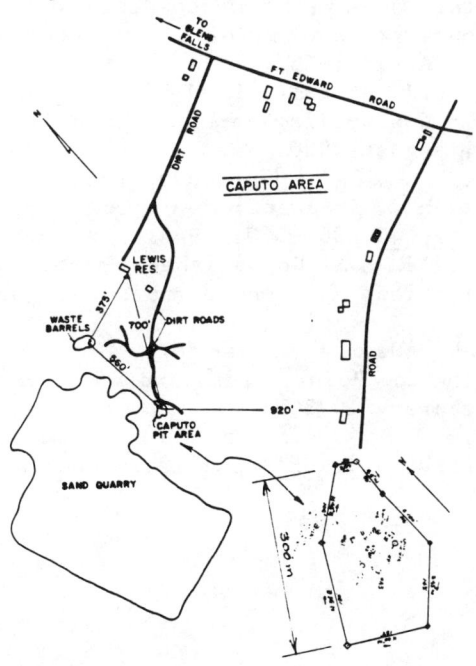

Fig. 2   A Sketch of Caputo PCB Disposal Site

References

1. Shen, T.T. and T.J. Tofflemire, "Air Pollution Aspects of Land Disposal of Toxic Waste" J. of the Environmental Engineering Division of ASCE, Vol. 106, No. EE1, pp. 211-226, February 1980.
2. Tofflemire, T.J. and T.T. Shen, "Volatilization of PCB from Sediment and Water: Experimental and Field Data", Proceedings of the 11the Mid-Atlantic Industrial Waste Conference, University Park, Pennsylvania, July 16-17, 1979.
3. Iannotti, J.E. et. al., An Inventory of Industrial Hazardous Waste Generation in New York State. New York State Department of Environmental Conservation Report, June 1979.
4. Shen, T.T., "Emission Estimation of Hazardous Organic Compounds from Waste Disposal Sites", Pre-print paper No. 80-68.8 presented at the 73rd Annual Meeting of the Air Pollution Control Association in Montreal, Canada, June 22-27, 1980.
5. Mackay, D., "Solubility, Partition Coefficients, Volatility, and Evaporation Rates", The Handbook of Environmental Chemistry, Edited by O. Hutzinger, Vol. 2, PP. 31-45, Springer-Verlag, Berlin Heidelberg New York, 1980.
6. Neely, W.B., "Predicting the Flux of Organics Across the Air/Water Interface" Proceedings of the 1976 National Conference on Control of Hazardous Material Spills, New Orleans, April 1976.
7. Farmer, W.J. et.al., "Land Disposal of Hexachlorobenzene Wastes - Controlling Vapor Movement in Soil", EPA-600/2-80-119, August 1980.
8. Arnold, J.H., "Unsteady-State Vaporization and Absorption", Transaction of American Inst. of Chem. Engineers, 40, 361-379, 1944.
9. Ziegler, R.C., Unpublished paper through personal Communication, Calspan Corporation, Buffalo, New York, 1979.
10. U.S. EPA, User's Guide for PAL - A Gaussian - Plume Algorithm for Point, Area, and Line Source, EPA-600/4-78-013, Feburary 1978.
11. Turner, D.B. Workbook of Atmospheric Dispersion Estimates. U.S. EPA Publication No. AP-26, Revised 1970.

ASSESSMENT OF REMOTELY SENSED DATA FOR THE DETECTION OF
IMPOUNDED   CHEMICAL WASTES

Mohan M. Trivedi. Department of Electrical Engineering,
Louisiana State University, Baton Rouge, Louisiana.

John M. Hill. Department of Civil Engineering, Louisiana
State University, Baton Rouge, Louisiana.

Peter A. Allain. Department of Civil Engineering, Louisiana
State University, Baton Rouge, Louisiana.

INTRODUCTION

   Proper management of chemical wastes is of utmost importance to our highly industrialized nation. According to the Environmental Protection Agency's (EPA) estimates in the year 1980 alone the United States has produced approximately 57 million metric tons of hazardous waste (1). It is also estimated that about 90% of such waste was improperly disposed in the environment (1). Improper handling which includes treatment, storage, transportation and disposal of the waste can result in irreversible damage to the environment and human health. The research reported in this paper deals with the application of remotely acquired data for the purposes of waste management.
   Remote sensing offers two unique capabilities which make methods utilizing remote sensing potentially attractive waste management tools. Firstly, it offers a better perspective of the scene by viewing the complete scene almost at once; and secondly these techniques can be quite cost and time effective compared with the conventional on the ground sampling methods.
   The State of Louisiana has a large number of heavy processing industries mostly related with the production of petrochemicals. Aside from being a major domestic oil and natural gas producer and exporter, Louisiana also has the potential to engage in the massive extractive industry of lignite mining. These processing and extractive industries generate both hazardous and non-hazardous wastes in the forms of solids, liquids, and gases.

Due to the large number of waste producing industries located in the state, Louisiana has been ranked very high in the amount of hazardous materials produced relative to the other States. Louisiana has many sites to handle liquid wastes. Methods of disposing of such wastes range from sophisticated on-site chemical treatment processes to injection wells to simple earthen pits, ponds, and lagoons that are no more than open dumps. A common feature to many of these sites appears to be the storage of waste in open impoundments. These impoundments are generally used for one of the following three functions: (i) to store and collect waste for eventual removal to a final disposal site, (ii) as holding tanks in which waste treatment is performed, or (iii) as the disposal sites for treated or untreated wastes.

The state board of Mineral Resources conducted a letter survey which requested Louisiana industries to supply information related to surface impoundments. The information requested consisted of - location of the site, type of waste and disposal procedures. The Louisiana Geological Survey located these sites on topographic maps and assessed the ground-water contamination potential. The Remote Sensing and Image Processing Laboratory (RSIP) at the Louisiana State University was awarded a contract to verify the letter survey using aerical photographs.

The objective of the research reported in this paper is to assess, both qualitatively and quantitatively, the utility of color IR imagery for the detection of impounded chemical wastes. Details of the location verification study are presented in the next section, followed by a section on the densitometry study of the color infra-red (CIR) imagery. Finally discussion of the results and conclusions are presented.

LOCATION VERIFICATION STUDY

The aerial photographs utilized in the study were acquired by EPA from an aerial camera aboard NASA'S U-2 aircraft flown on the October 9th 1978. The CIR imagery had a resolution of 1:60,000 and is recorded in 9x9 inch transparency format. The film's spectral response range was 0.5 to 0.9 micrometers, thus such imagery is specifically useful for recording reflected energy off ground objects in the green, red, and near infra-red portions of the electromagnetic spectrum.

The verification study was limited to the southern Louisiana, primarily due to the availability of the aerial photographs. This area also happens to have the largest number of industrial waste generators, as well as a large and active oil and gas industry. The southern half of the state was further divided into six study areas determined by two parameters: (i) the type of industry, and (ii) the geological surface upon which these waste sites are located. The first category called industrial consisted of heavy processing and petrochemical industries. The second category was termed "oil and gas" and generally represented a brine generating facility, which is usually related to oil and gas extraction. The geological parameters for the selection of study area identified the Lake Charles, Abbeville and Jennings areas as those located upon the Pleistocene Prairie Terrace, the Carville and Hahnville areas situated upon recent Mississippi river Alluvium, and the Terrebonne area as one composed of recent coastal marshes and swamps.

The verification procedure involved of comparing the coordinates given by the letter survey with the coordinates determined from the aerial photographs. Corrected coordinates were plotted on the original letter survey topographic maps.

The results of the location verification study can be summarized as follows: 63.0% of the examined sites were found within 0.25 miles of the reported location, 10.5% were between 0.25 miles and 0.50 miles, 3.7% were between 0.5 and 1.0 miles, 3.1% were between 1.0 and 2.0 miles, 1.9% were mislocated by at least 2 miles, and 18.5% were not locatable at all. Also, 69.7% of the ponds reported coincided with the number of ponds sited on the photographs, more ponds were reported than viewed 7.2% of the time, less ponds were reported than viewed 6.4% of the time, and 16.6% of the ponds were unlocatable. Further details of the location verification study can be found in reference (2).

DENSITOMETRY STUDY

The location verification study reported earlier is based completely upon the subjective interpretation of the photointerpretor. It is basically a qualitative study where successful inferences drawn about the waste impoundments depend to a great extent on the experience and skills of the interpreter. Also such techniques tend to be slow and tedious. In order to overcome these drawbacks and to ensure generating quantitative and objective results about the remote sensing problem digital data analysis approach seems very attractive. Digital techniques can efficiently produce consistent results utilizing the processing capabilities of a computer. Unfortunately digital image processing and pattern recognition techniques are relatively new and a completely automatic approach to the detection of impoundments seems to be infeasible as of now. A semiautomatic technique where a user can interactively use a computer seems to the optimal approach. The objective of this part of the study is to provide preliminary feasibility evaluation of digital pattern recognition techniques for the impoundment detection problem.

In order to apply quantitative methods for image analysis and pattern recognition it is imperative that image data should either be acquired or transformed into digital form. Densitometry allows one to convert film data into digital form (3). The CIR film basically senses reflected scene energy in the broad 0.5-0.9 micrometer region of the electromagnetic spectrum. Therefore such an image can be considered as a record of reflected energy from ground objects in the following three bands:

    (i)   green,   0.5-0.6 micrometers
    (ii) red,   0.6-0.7 micrometers, and
    (iii) near infra-red, 0.7-0.9 micrometers

we used Macbeth TD-102 transmission densitometer to acquire density measurements in three separate bands which roughly correspond to the green, red, and near infra-red bands. Density measurements relate inversely to the transmittance of the film at a certain point, which is determined by the "darkness" or "lightness" of the film at that point (3). We employed three filters Wratten #94, #93, and #92 to filter the light source in such a way that the blue-forming, green-forming, and red-forming layers of the CIR emulsion are isolated, respectively. Density measurements of a point on the image, thus provides three readings: blue filter readings - corresponding to the green reflectance, green filter readings corresponding to the red reflectance and red filter readings corresponding to the infra-red reflectance of the scene.

In this study we are trying to establish whether the energy reflected from the various objects in the scene - like impounded chemicals and associated background, uniquely characterize those objects or not. The photographically recorded image is a function of the reflected energy off the scene objects, solar illumination conditions, atmospheric effects, and film processing parameters. To ensure that the solar illumination and atmospheric conditions do not affect the inferences drawn in the study, the flight records of the data acquisition phase were consulted. Only those images which were recorded within a period of approximately one and a half hours were used to derive the data set for the analysis. Flight records also indicated that the skies were clear with no cloud cover during the entire period. Also to ensure that the film processing parameters, affecting the density of the image, remain constant for the entire data set, only those images which were derived from a single roll of film were used in the analysis. Finally, to eliminate any common mode variations (those affecting all objects of the scene unifromly) each density reading of blue, green, and red filter were normalized by dividing them by the visible filter (Wratten #106) reading of the corresponding point.

Detailed observation of the aerial imagery indicated that the impounded chemicals are generally situated in two classes of backgrounds. We have identified them as (a) Vegetation class, which corresponds to trees or grass, and (b) Manmade class, which corresponds to roads, concrete structures, or gravel. Also it was observed that significant number of impoundments contained dry, solid material. This type of contents usually showed up as yellow, white or greyish material on the CIR imagery. To account for dry and liquid wastes separately subclasses of chemicals which correspond to these materials were introduced.

A quantitative summary of the data set is presented in Table I. Means and standard deviations of the normalized density measurements for each of the three bands and five classes are presented in the table. Notice the relatively low mean for the normalized red filter reading for the vegetation class. This implies a high transmittance value for the corresponding spectral band, which in turn is associated with the high reflectance of vegetation in the near infra-red region. Also notice that green and red measurements for chemicals and vegetation are significantly different and the liquid chemical class is considerably different from manmade class in the blue and red values.

CLASS	# of Samples	NORMALIZED DENSITY					
		Blue		Green		Red	
		Mean	S.D.	Mean	S.D.	Mean	S.D.
Chemicals	125	.75	.15	1.09	.16	1.11	.21
Vegetation	79	.95	.17	1.45	.28	.77	.15
Manmade background	70	.83	.10	1.00	.08	1.18	.13
Chemicals (liquid)	76	.70	.12	1.11	.16	1.08	.20
Chemicals (dry)	49	.81	.17	1.07	.17	1.15	.21

Table I. Statistical summary of the five class data set.

The final step in the analysis of the densitometry data relates with the design of Bayesian classification scheme (4). This design is based upon the multivariate data analysis model where each class can be uniquely specified in a multidimensional space. For our problem the dimensionality of such space is three, since there are three spectral bands under consideration. In such a model each sample point, for which density measurements with blue, green, and red filters are known, is described as a vector in three dimensional space. The data collected from the known samples of chemicals, vegetation, and manmade objects are used to specify the "clusters" (or regions) in the space, one associated with each class. Such specification when used together with an assumption of the samples being distributed according to Gaussian distribution law requires evaluation of mean vector and covariance matrix for each class. Once this process of specification (usually known as "training") is completed then the classifier assigns an unknown measurement vector to a class which it is most likely to have belonged. This type of classification also assures that the overall probability of error due to misclassification will be minimized. Results of the classification performed on the data set are presented in Table II. Here all the chemicals (dry and liquid) are considered in one class. The percent correct classification is calculated by dividing the number of samples which were classified correctly by the total number of samples in a particular class. Overall classification accuracy is derived by dividing the number of samples correctly classified in all three classes by the total number of samples in the data set. Notice, for the chemical class, from a total of 125 samples, a significant number of samples were misclassified: 11 in the vegetation and 33 in the manmade class, resulting in a 65% correct classification. Closer scrutiny of the chemical samples misclassified in the manmade class indicated that most of these samples were associated with dry, solid material, which tend to look like manmade background objects to the computer. Therefore, a classification study involving only the impounded liquid chemical samples was performed. Results of such study are presented in Table III. Notice a considerable improvement in the correct classification of chemical class. The results also show that from a total of 76 chemical samples tested only 3 were classified in the vegetation and 10 were classified in the manmade class, resulting in 83% accurate classification. Also, the overall classification accuracy has improved from 76.3% to 85.3%. These results are based upon utilizing the same data set for the training as well as testing of the classifier. Such scheme for assessing the classifier performance make optimal use of the available data, but the results tend to be optimistic.

CONCLUSIONS

The research reported in this paper performed a preliminary feasibility study to evaluate remote sensing techniques for the detection of impounded chemical wastes. The data set was derived from color infra-red imagery. The study showed utility of such data to verify the waste site locations. By utilizing densitometry methods we were able to apply digital

CLASS	CLASSIFIED AS			TOTAL	% CORRECT CLASSIFI-CATION
	Chemicals	Vegetation	Manmade background		
Chemicals	81	11	33	125	65%
Vegetation	12	63	4	79	80%
Manmade background	2	3	65	70	93%

Overall classification accuracy = 76.3%

Table II. Classification results.

CLASS	CLASSIFIED AS			TOTAL	% CORRECT CLASSIFI-CATION
	Chemicals (liquid)	Vegetation	Manmade background		
Chemicals (liquid)	63	3	10	76	83%
Vegetation	12	63	4	79	80%
Manmade background	1	3	66	70	94%

Overall classification accuracy = 85.3%

Table III. Classification results with liquid chemical class.

pattern recognition methods to analyze data which roughly related to the reflectance characteristics of the objects in the green, red, and near infra-red regions of the electromagnetic spectrum. The results of such a study can be used to conclude that the liquid chemical, vegetation, and manmade class exhibit unique refelectance characteristics, also a Bayesian classification approach utilizing these characteristics provides favourable results. A study where data can be acquired directly in digital form using multispectral scanner, with accurate "ground-truth" information should be the next step in the research program aiming at the utilization of remote sensing technology for the waste management.

REFERENCES

[1] Everybody's Problem: Hazardous Waste, United States Environmental Protection Agency, Office of Water and Waste Management, Washington, D.C., SW-826, 1980.

[2] Hill, J. M. and Trivedi, M. M., "Location Verification of Waste Impoundments in Louisiana Using Aerial Imagery," Proceedings of National Hazardous Waste Conference, New Jersey Institute of Technology, Newark, N.J., June 1980.

[3] Lillesand, T. M. and Kiefer, R. W., *Remote Sensing and Image Interpretation*, Chapter 6, John Wiley and Sons, New York, 1979.

[4] Duda, R. O. and Hart, P. E., *Pattern Recognition and Scene Analysis*, Chapter 2, John Wiley and Sons, New York, 1973.

INDEX

Abudu, O.T.
  Synfuel Retort Waters 217
Activated Carbon
  Organics Removal 444
Activated Sludge
  Control Parameter 160
  Treatment of Pulp and Paper
    Mill Wastewater 177
Activated Sludge Processes
  Performance, Organic and
    Pesticide Loadings 250
  Treatment of Coal Gasifi-
    cation Wastewaters 228
Activated Sludge Systems
  See Settleability 396
Adsorption
  Cadmium(II) Plating
    Wastewaters 89
Aerobic Treatment Processes
  High Organic Solid Waste-
    waters 239
  By Product Utilization 115
Akers, D. J.
  Power Plant Wastes 596
Albrecht, C.R.
  Chrome Residual Contain-
    ment 316
Allen, J.
  Methane fermentation
    Process 208
Aluminosilicate
  See Adsorption 89
Ammonia
  Contribution to Surface
    Water from Point Sources.
    608
Anaerobic Reactor
  Treatment of Tannery
    Beamhouse Waste 197
Analysis
  See Fermentation Process
    208
Aquifer
  See Leachate 323
Arbuckle, W. B.
  Activated carbon 444
Atypical Industrial Wastes
  Procedures for Processing
    295

Authority
  State Preemptive Siting 484
Bailey, D. G.
  Tannery Beamhouse Waste 197
Bailey, S. W.
  Heavy Metals in Sludge 15
Bahorsky, M.
  Production Related Waste-
    water 563
Baratta, N. D.
  Cooperative Concept 288
Blasland, W. V.
  Remedial, Municipal Disposal
    Site 476
Belt Presses
  Selection, Sludge Dewatering
    341
Bowers, A. R.
  Treatment of Heavy Metal-
    Laden Wastewater 51
Brantner, K. A.
  Heavy Metals Removal 43
Brunker, R. L.
  Heavy Metals Removal by foam
    Flotation 72
Buckley, D. B.
  Treatment of Pulp and Paper
    Mill Wastewater 177
Bush, P. D.
  Citizen Response to Indus-
    trial Waste 533
Buss, D. F.
  Hazardous Waste Management
    Program for a Power Company
    469
Buttermore, W. H.
  Power Plant Wastes 596
By-Product
  See Yang, P. Y. 239
Calcium Scaling Tendencies
  See Evaporation 26
Chao, A. C.
  Oxygenation coefficient 151
  Potassium Ferrate Treatment
    434
  Seafood Processing Waste-
    waters 580
Chavalitnitikul, C.
  Heavy Metals Removal by
    Foam Flotation 72

Chawla, R. C.
 Dairy Industry Waste 554
Chin, G.
 Treatment of Heavy Metal-
 Laden Wastewater 51
Christensen, G. L.
 Sludge Conditioning 404
Chrome and Cyanide Waters
 Treatment, Programmable
 Controller 115
Cichon, E. J.
 Heavy Metals Removal 43
Citizen
 Response to Industrial
 Waste 533
Clarifier
 Performance Affected by
 Wastewater Characteristics
 370
Color
 See Photodecomposition 427
Compositing
 Industrial Residues 264
Controllers
 Treatment of Chrome and
 Cyanide Wastes 115
Cooperative Concept
 Industrial Waste Management
 288
Cost
 See Design 159
Cost-effective
 See Prioritization 563
Crossflow Filtration
 See Hydroperm® 378
Cyanides
 See Ozone 105
Davis, G.
 Seafood Processing Waste-
 waters 580
Deluca, S.
 Potassium Ferrate Treat-
 ment 434
Denn, M. M.
 Cost and Sensitivity of
 Activated Sludge System
 189
Design
 Cost and Sensitivty, Acti-
 vated Sludge System 189

Sulfide Precipitation
 System 63
DiLoreto, J.
 Influence of Pretreatment
 on Sludge 76
Ditaranto, V. M.
 Pretreatment of Heavy Metals
 Wastewater 2
Eckenfelder, W. W.
 Biologically Active Sand
 Filters 128, 139
Economics
 Dairy Industry Waste 554
Edmondson, B. C.
 Heavy Metals and pH Control
 35
Edwards, G. L.
 Activated Sludge Processes
 Control Parameter 160
Elliott, H. A.
 Retention of Heavy Metals
 by Soils 95
Evaporation
 Concentration of Wastewater
 26
Fava, J. A.
 Point Sources 608
Fermentation Process
 Cellulosic Substrate Utili-
 zation 208
Fibers
 Production from Power Plant
 Wastes 596
Flotation
 Dissolved Air, Upgrading 386
Flower, F. B.
 Land Uses 302
Foam Flotation
 Removal of Heavy Metals
Formaldehyde
 Treatment of Industrial
 Wastewaters 416
Fossum, G. O.
 Coal Gasification Wastewaters
 218
Foster, D. H.
 Synfuel Retort Waters 217
Friedman, A. A.
 Tannery Beamhouse Waste 197

Fuentes, H. R.
 Biologically Active Sand
  Filters 128, 139
Gagnon, A. P.
 Chrome and Cyanide Wastes
  115
Galarraga, E.
 Oxygenation Coefficient 151
Gehr, R. L.
 Dissolved Air Flotation 386
Gilman, E. F.
 Land Uses 302
Goel, K. C.
 Belt Presses 341
Goldman, R. K.
 Remedial, Municipal Disposal
  Site 476
Gormley, E. F.
 Influence of Pretreatment
  on Sludge 362
Grandin, W. T.
 Cooperative Concept 288
Graner, W. F.
 Leachate Detection 323
Greco, J. L.
 Hazardous Waste Management
  Program 469
Guideline
 Effluent, Textile Industry
  537
Gurhham, C. F.
 Environmental Impairment
  Liability Insurance 549
Harper, H. H.
 Anaerobic Release of Phos-
  phorus 518
Harvell, G.
 Performance of Activated
  Sludge Processes 250
Hazardous Waste
 Incineration At Sea 693
 Industry Role, Faculty
  Development 455
 Laboratory, In House Treat-
  ment 509
 Management, Specialty
  Organic Chemical Plant
  273
 Siting Policy 484

 Public Participation in
  Siting 517
Heavy Metals
 Antomated Pretreatment
  Facility 2
 Control of in Treated
  Effluents 35
 In Municipal Sludge 15
 Removal, Comparison of
  Processes 43
 Removal, by Foam Flotation 72
 Removal, Sulfide Precipi-
  tation System 63
 Retention, by Northeastern
  U.S. Soils 9
Henry, J. G.
 Dissolved Air Flotation 386
Horne, A.
 Hazardous Waste Siting Policy
  484
Howe, R. H. L.
 Oxygenation Coefficient 151
Hroncich, T.
 Leachate Detection 323
Hsu, S.
 Belt Presses 341
Huang, C. P.
 Cadmium (II) Plating Waste-
  water 87
 Treatment of Heavy Metal-
  Laden Wastewater 51
Hung, Y. T.
 Coal Gasification Waste-
  waters 228
Hydroperm®
 In Removing Suspended
  Solids 378
Impact
 Subtitle C on Industrial
  Wastewater Treatment
  Facilities 459
Incineration
 Hazardous Waste At-Sea 493
Industrial Waste
 Citizen Response 533
Industrial Wastewaters
 Treatment with Formalde-
  hyde 416
Industry
 Textile 537, 563

Dairy 554
Seafood 580
Poultry Processing 350
Insurance
  Environmental Impairment
  Liability 549
Jellerson, D. B.
  Anaerobic Release of Phosphorus 618
Johnson, R. J.
  Hazardous Waste 693
Kennan, P. H.
  Atypical Industrial Waste 295
Kerr, J. E.
  Latex Waste 331
Khararjian, H.
  Performance of Activated Sludge Processes 250
Knowles, G. D.
  Remedial, Municipal Disposal Site 476
Krupka, M. J.
  Settleability of Biological Solids 396
Kulowiec, J. J.
  Management at Specialty Organic Chemical Plant 273
Kunicki, W.
  Ozone Treatment of Cyanide Wastewaters 105
Land Disposal
  Volatilization Problems 630
Land Fills
  Chrome Residual Containment, Hawkins Point, MD. 316
  Developing Parks, Golf Courses, and Recreation Areas on 302
Latex Waste
  Problem, Company Solves 331
Leachate
  Detection 323
Leone, I. A.
  Land Uses 302
Levine, S.
  Concentration by Evaporation 26
Liability
  Environmental Impairment 549

Liberati, M. R.
  Retention of Heavy Metals by Soils 95
Liu, J. K.
  By-Product Utilization 239
Liu, Lt. L.
  Crossflow Filtration System 378
Lockhart, C. H.
  Latex Waste 331
Lutz, W. A.
  Composting 264
Lynch, E. R.
  Remedial, Municipal Disposal Site 476
Mahmud, Z.
  Performance of Final Clarifier 370
Management
  Hazardous Waste, A Power Company 469
Manas, R.
  Concentration by Evaporation 28
Martinez, L. A.
  Hazardous Waste 493
McKeown, J. J.
  Treatment of Pulp and Paper Mill Wastewater 177
Meister, J. H.
  Laboratory Hazardous Wastes 509
Methane
  See Fermentation Process 208
Mital, H. K.
  Belt Presses 341
Mohr, R. T.
  Atypical Industrial Wastes 295
Moore, D. J.
  Textile Industry 537
Noble, R. D.
  Synfuel Retort Waters 217
Oberracker, D. A.
  Hazardous Waste 493
Options
  Dairy Industry Waste 554
Organic
  See Activated Sludge Processes 250

Organic Chemicals
  Integrated Management 273
Oxygenation Coefficient
  Reevaluation Temperature
    Effect 151
Oxygen Uptake
  Activated Sludge Process
    Control Parameter 160
  Rates, Influence of Water
    Component 217
Ozone
  Treatment of Cyanide Wastewater 105
PACT
  Treatment of Synfuel Retort
    Waters 217
Paulson, L. E.
  Coal Gasification Wastewater 228
Pekin, T.
  Chrome and Cyanide Wastes 115
Pesticide
  See Activated Sludge Processes 250
Pim, J. H.
  Toxic and Hazardous
    Materials 252
pH
  Control of, in Treated
    Effluents 35
Phosphorus
  Inactivation of Release by
    Water Treatment Sludge 618
Photodecomposition
  Of Color in Dying Wastewater 427
Plating Wastewater
  Cadmium (II), Treatability 89
  See Ozone 105
Precipitation
  Heavy Metals Removal, 43
  Line-Neutralization, Heavy
    Metal Removal 51
  Sulfide For Heavy Metals
    Removal 63
Prescott, M. K.
  Cooperative Concept 288

Pressman, M.
  Sulfide Precipitation System
    for Heavy Metal Removal 63
Pretreatment
  Heavy Metals Wastewater 2
  Influence on Sludge Quality 362
Point Sources
  Contribution of Ammonia to
    Surface Water 608
Policy
  Hazardous Waste Criteria 480
Pollutants
  Removal, With Potassium
    Ferrate 439
Poon, C. P. C.
  Dying Wastewater 427
Power Company
  Hazardous Waste Management 469
Power Plant Waste
  Production of Insulating
    Fibers 596
Prioritization
  Textile Industry Waste Treatment Problems 565
Randall, T. L.
  Toxic and Hazardous Compounds 501
Recovery
  See Ultrafiltration 580
Recycling
  Wastewater in Colorado 588
Regulatory
  Approach, Toxic (Hazardous)
    Materials 525
Remedial
  Municipal Disposal Site, PCB_S 476
Residues
  See Composting 264
Retention
  Of Heavy Metals by Soils 95
Ritter, W. F.
  Sludge Management, Poultry
    Industry 350
Roman Seda, R. A.
  Biologically Active Sand
    Filters 128, 139
Rue, W. J.
  Point Sources 608

Ruetenik, K.
  Methane Fermentation Process 208
Sanders, T. G.
  Recycling Wastewater 588
Sandfilters
  Biologically Active 128, 139
Sensitivity
  See Design 189
Settleability
  See Solids 396
Sewage Influent
  See Heavy Metals 72
Shen, T. T.
  Land Disposal of Industrial Wastes 630
Sherrard, J. H.
  Activated Sludge Process Control Parameter 160
Simcoe, E. J.
  Power Plant Wastes 596
Siting
  Hazardous Waste, Public Participation 517
  Policy, Hazardous Waste 984
Sludge
  Conditioning 404
  Conditioning Iron and Lime U.S. Polyelectrolytes 404
  Quality, Influenced of Pretreatment 362
  Management, Poultry Processing Industry 350
  Waste Treatment, Inactivation of Phosphorus Release 618
Sludge Dewatering
  Belt Processes 341
Smallwood, C. Jr.
  Potassium Ferrate Treatment 434
Smith, J. W.
  Performance of Activated Sludge Processes 250
Soils
  See Retention 95
  Biological, Dynamic Response in Activated Sludge System 396

Stelluto, J.
  Industry Role, Hazardous Waste 455
Subtitle C
  Impact on Industrial Wastewater Treatment Facilities 459
Textile Industry
  Effluent Guidelines, development 537
Thomas, J. M.
  Settleability of Biological Solids 396
Toxic (Hazardous) Materials
  Storage and Handling, Regulatory Approach 525
Toxic (Hazardous) Compounds
  Wet Oxidation 501
Tunick, M. H.
  Tannery Beamhouse Waste 197
Ultrafiltration
  Treatment of Seafood Processing Wastewater 580
Varma, M.
  Dairy Industry Waste 554
Vittimberga, B. M.
  Dying Wastewater 427
Volatilization
  Land Disposal of Industrial Wastes 630
Wahielista, M. P.
  Anaerobic Release of Phosphorus 618
Walker, M., Jr.
  Heavy Metals and pH Control 35
Warner, M. G.
  Heavy Metals and pH Control 35
Waste
  Tannery Beamhouse 197
Wastewaters
  Coal Gasification 228
  Heavy Metal-Laden 51
  Inorganic Chemical Industries 35
  Pulp and Paper Mill 177
  Recycling in Colorado 588
  Textile Industry 563

Wavro, S. G.
  Sludge Conditioning 404
Wet Oxidation
  Of Toxic and Hazardous
  Compounds 501
Whang, J. S.
  Sulfide Precipitation System
  for Heavy Metal Removal 63
Willson, W. G.
  Coal Gasification Wastewater
  228
Wilson, F. R.
  Impact of Subtitle C 459
Winsor, E. W.
  Hazardous Waste Siting,
  Public Participation 517
Wirth, P. V.
  Cadmium (II) Plating Wastewater 89
Wu, Y. C.
  Performance of Final Clarifier 370
Yang, E.
  Hazardous Waste Siting
  Policy 484
Yang, P. Y.
  By-Product Utilization 239
Young, D.
  Sulfide-Precipitation System
  for Heavy Metal Removal 63
Yousef, Y. A.
  Anaerobic Release of Phosphorus 618
Zimomra, D. C.
  Heavy Metals in Sludge 15

*Trivedi, M. M.
  Remote Sensing,
  Chemical Wastes 639
*Hill, J. M.
  Remote Sensing,
  Chemical Wastes 639
*Allain, P.A.
  Remote Sensing,
  Chemical Wastes 639
*Detection
  Chemical Wastes,
  Remotely Sensed Data 639
*Chemical Wastes
  Detection,
  Remotely Sensed Data 639